Dean Lapsley
Wed - 2:30

party: 2601 Dana

Kinematic Analysis of Mechanisms

McGRAW-HILL SERIES IN MECHANICAL ENGINEERING

KARL H. VESPER, *Stanford University, Consulting Editor*

BARRON · *Cryogenic Systems*
BEGGS · *Mechanism*
CAMBEL AND JENNINGS · *Gas Dynamics*
CSANADY · *Theory of Turbomachines*
DURELLI, PHILLIPS, AND TSAO · *Introduction to the Theoretical and Experimental Analysis of Stress and Strain*
ECKERT · *Introduction to Heat and Mass Transfer*
ECKERT AND DRAKE · *Heat and Mass Transfer*
GRÖBER, ERK, AND GRIGULL · *Fundamentals of Heat Transfer*
HAM, CRANE, AND ROGERS · *Mechanics of Machinery*
HARTENBERG AND DENAVIT · *Kinematic Synthesis of Linkages*
HARTMAN · *Dynamics of Machinery*
HINZE · *Turbulence*
JACOBSEN AND AYRE · *Engineering Vibrations*
JUVINALL · *Engineering Considerations of Stress, Strain, and Strength*
KAYS · *Convective Heat and Mass Transfer*
KAYS AND LONDON · *Compact Heat Exchanges*
LEIGH · *Nonlinear Continuum Mechanics*
LICHTY · *Combustion Engine Processes*
MARTIN · *Kinematics and Dynamics of Machines*
PHELAN · *Dynamics of Machinery*
PHELAN · *Fundamentals of Mechanical Design*
RAVEN · *Automatic Control Engineering*
SCHENCK · *Theories of Engineering Experimentation*
SCHLICHTING · *Boundary-Layer Theory*
SHIGLEY · *Dynamic Analysis of Machines*
SHIGLEY · *Kinematic Analysis of Mechanisms*
SHIGLEY · *Mechanical Engineering Design*
SHIGLEY · *Simulation of Mechanical Systems*
SHIGLEY · *Theory of Machines*
SPALDING AND COLE · *Engineering Thermodynamics*
STOECKER · *Refrigeration and Air Conditioning*
SUTTON AND SHERMAN · *Engineering Magnetohydrodynamics*
WILCOCK AND BOOSER · *Bearing Design and Application*

Kinematic Analysis of Mechanisms

JOSEPH EDWARD SHIGLEY
Professor of Mechanical Engineering
The University of Michigan

Second Edition

McGraw-Hill Book Company

New York St. Louis San Francisco Toronto London Sydney

Kinematic Analysis of Mechanisms

Library of Congress Catalog Card Number 68-9559

56868

1234567890MAMM7543210698

Preface

Since the first edition of this book was published ten years ago there has been a revolution in computer technology, and we now find machine computation used extensively in both the practice and the study of engineering. Furthermore, the basic engineering-mechanics sciences, which are prerequisites for the study of kinematics and dynamics, are now being taught, almost universally, by using vector methods. These two changes alone would have justified a complete revision of the original book. But along with these advances, there have been so many significant contributions to the field of kinematics that our total fund of knowledge has more than doubled. It is for these reasons that I found it necessary to write an entirely new book.

The book is planned especially for the undergraduate engineering student who wants to begin a serious study of modern kinematics. But because of the emphasis and because so many new developments are included, the book should be a convenient reference for practicing engineers too.

The subject matter is developed by using vector mathematics, algebra, and graphics in a combined approach. Each method of solution reinforces the other. Graphic methods are used extensively because they provide more insight into the physics of the problem. But a good grounding in the use of vector mathematics and analytical methods must be provided to lay a foundation for advanced studies in kinematics. Furthermore, the analytical methods are necessary when machine computation is to be employed in the solution of problems in kinematics.

To complement the emphasis on analysis, the fundamentals of curvature theory and basic notions relating to the geometry of mechanisms are introduced in Chapter 6. And in this edition, an entire chapter is devoted to the synthesis of linkages. So many different synthesis techniques are available in the literature that it was difficult to make a selection. It is believed, however, that those which have been included are representative and are adequate for the solution of most problems.

In acknowledging the assistance of others in the preparation of a new book most writers know from experience that reviewing a manuscript is a wearisome and monotonous task with little if any monetary reward. It is done out of the goodness of one's own heart. And so I am very proud to acknowledge the assistance of Professors Ferdinand Freudenstein of Columbia University, Allen S. Hall, Jr., of Purdue University, John D. Kemper of the University of California at Davis, Phillip Lovett of The Pennsylvania State University, and Charles McLarnan of the Ohio State University in planning the book and reviewing the preliminary and final manuscripts. I am indeed grateful to these gentlemen for their careful and critical analyses and helpful comments.

I want also to acknowledge my indebtedness to my friend and associate Dr. Milton A. Chace of The University of Michigan. Dr. Chace's original contributions in kinematics have been used extensively in the book, and I am also grateful to him for his advice and assistance with many parts of the book.

Joseph Edward Shigley

Contents

CHAPTER 6 THE GEOMETRY OF MOTION

CHAPTER 7 CAMS

CHAPTER 8 SPUR GEARS

CHAPTER 9 HELICAL, WORM, AND BEVEL GEARS

CHAPTER 10 MECHANISM TRAINS

CHAPTER 11 SYNTHESIS OF LINKAGES

CHAPTER 12 SPACE MECHANISMS

Kinematic Analysis of Mechanisms

1
Introduction

The theory of machines is an applied science which allows an understanding of the relationship of geometry and motion in machines. Though not strictly necessary, a tremendous simplification results if this science is divided into two parts. The first of these is *kinematics*, in which only the physical units of *length* and *time* are involved. And the second is *dynamics*, in which the units of *force* are added to those of length and time.

1-1 KINEMATICS AND MECHANISMS

A *mechanism* is a set of machine elements arranged so as to produce a specified motion. The piston, connecting rod, and crank of an internal-combustion engine or of an air compressor comprise a mechanism. So do a pair of meshing gears or two pulleys connected by a V-belt.

Kinematics is the study of motion, quite apart from the forces which produce that motion. More particularly, kinematics is the study of position, geometry, displacement, rotation, speed, velocity, and acceleration. The study, say, of planetary or orbital motion is also a problem in kinematics, but in this book we shall concentrate our attention on kinematic problems which arise in studying the behavior of machines.

In designing a new engine, the automotive engineer must find answers to many questions. For example: What is the relation between the motion of the piston and the motion of the crankshaft? What is the sliding velocity at the lubricated surfaces for various positions of the piston? What are the direction and magnitude of the acceleration of the piston, and how do they change through a complete cycle? What is the path of the center of gravity of the connecting rod? What is the difference in the motion characteristics between a short-stroke and a long-stroke engine? These and many similar questions may be answered by a kinematic analysis of the mechanism.

But mechanisms must also be synthesized. That is to say, the geometries must be determined so that they will fit into the required spaces and deliver the motion characteristics which are desired.

Consider, for example, the problem of weighing and bottling a few grams of a drug costing, say, $50 per ounce, at the rate of hundreds of times per minute. If too much of the drug is weighed out the company loses money. If too little is weighed out the Federal government complains. This is a problem which involves extreme accuracy. The loading of the powder charge for small-arms ammunition is a similar problem. Here the kinematician's task consists in devising a mechanism which will weigh and load the material accurately at a very high speed.

In summary, *kinematic analysis* is simply the means used to arrive at answers relating to the motion of mechanisms, and *kinematic synthesis* is the means used to find the geometry of a mechanism which will yield a desired set of motion characteristics.

1-2 MECHANICAL DESIGN

The creation of a plan for the construction or assembly of a machine, a device, or a mechanical system or process is called *mechanical design*. The design of machines consists in conceiving an arrangement of parts which will accomplish the desired purpose, and then defining the geometry of each part and the material, the processing, and the assembly down to the last detail. There is still a good deal of creative activity involved in the design process. The design engineer must first define the actual "thing" to be designed. Then he must consider such factors as wear, heat, friction, processing, utility, cost, safety, noise, appearance, flexibility, control, stress, strength, rigidity, deflection, and lubrication, as well as a host of others. Having assessed the relative values of these factors to his design, the designer employs a decision-making process, which may or may not be a logical one, to arrive at a final design.

A great deal of analysis will inevitably be employed in the decision-making process in order to compare the merits of the various possibilities which may arise. But there may also be a good deal of judgment, experience, and experimentation involved in the procedure.

In the design of machines the arrangement of parts decided upon will usually consist of one or more mechanisms which will have to be analyzed in order that they can be evaluated and properly designed. To select these mechanisms the designer should be acquainted with a large variety of them. He must also be able to analyze them for their kinematic characteristics as well as synthesize them to obtain particular motions.

1-3 CLOSURE

The material in this book has been arranged with the expectation that the reader will proceed continuously from the beginning to the end. At first some material will be found which has been studied previously or is familiar. This has been included to establish the nomenclature, as a foundation for the balance of the book, for review purposes, and as a ready reference source. Unless a review is necessary, it is sufficient to scan this material lightly and pass on to new material.

2
Vectors and Scalars

In this chapter the reader will find the basic material which is required for an understanding of the balance of the book. The sign conventions, notation, and graphical and analytical tools for problem solving will all be explored. The chapter should be studied exhaustively in order to acquire confidence and familiarity with the methods to be employed.

2-1 COORDINATE SYSTEMS

In kinematics the position of a point is of vital importance, and we shall use both stationary and moving reference systems as methods of defining the location of a point.

Each of the two coordinate systems of Fig. 2-1 may be used to define the position of point P. Figure 2-1a is a *rectangular-coordinate system,* and P is located by its rectangular coordinates x and y. Both x and y are shown in their positive directions. In Fig. 2-1b the point P is located by its *polar coordinates,* which are r and θ. When polar coordinates are

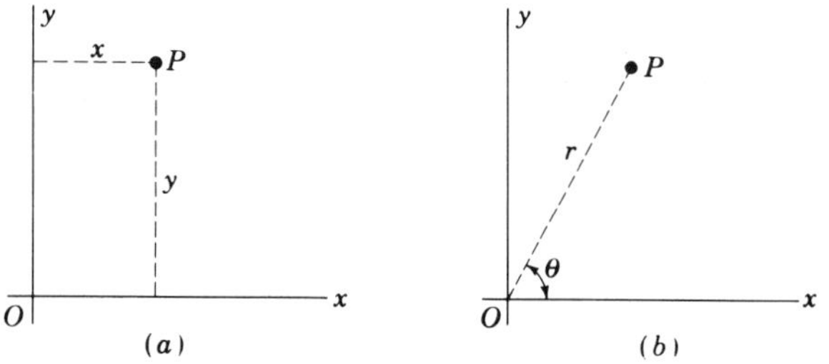

Fig. 2-1 (*a*) Rectangular-coordinate system; (*b*) polar-coordinate system.

used, r will always be taken as positive; θ will be positive if measured counterclockwise from the positive x axis, and negative if measured clockwise.

Both of the systems shown in Fig. 2-1 are two-dimensional systems because they describe points which are restricted to occupy a two-dimensional space. Only two coordinates are necessary to define the position of a point, and the movement of the points is necessarily restricted to a plane. While much of our investigation in kinematics will be restricted to the movement of a point in a plane, it is necessary that we be able to think and to express our ideas in terms of three dimensions.

Figure 2-2 is a three-dimensional coordinate system. Here, three coordinates, x, y, and z, are necessary to locate the point P, and P is now free to move in space. It is convenient to call the system of Fig. 2-2 a *right-handed coordinate system*. The reason for this is as follows: If the x axis is rotated 90° into the y axis, the motion will be counterclockwise when viewed from the positive z axis. A way to remember this is to hold

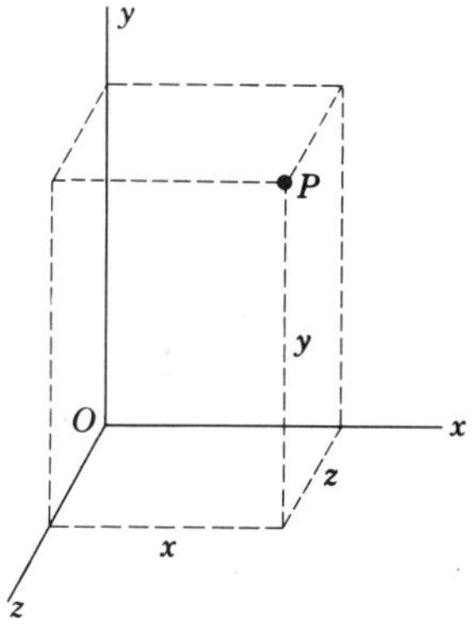

Fig. 2-2 A right-handed, three-dimensional coordinate system.

right-hand system: $(\overline{+x}) \times (\overline{+y}) = \overline{+z}$

the fingers of the right hand pointing from the x axis counterclockwise to the y axis. Then the thumb will point in the positive z direction.

2-2 SCALARS

A scalar is any quantity which can be specified by an ordinary real number. Scalars are the simple numbers (positive, negative, or zero) used to specify such quantities as time, temperature, energy, and volume. In this book lightface italic letters such as A, B, C or a, b, c will be used to represent scalars.

2-3 VECTORS

A vector is a quantity which has a magnitude, a direction, and a sense. In this book a vector will always be shown as a line having an origin and a terminus with an arrowhead, as shown in Fig. 2-3. The length of the line represents the magnitude of the vector. The direction in which the arrow points will be used to represent the direction of the vector quantity.

It is important to amplify the meaning of magnitude, direction, and sense. A 3-in. line segment drawn on a sheet of paper *without* an arrowhead has a magnitude and a direction, but no sense. Its magnitude is 3 in. Its direction is north *and* south or east *and* west. Adding the arrowhead is equivalent to specifying the *sense* of the vector. We can now say, for example, that the vector has a magnitude of 3 in., it has a north *and* south direction, and it has a sense *to* the north.

We shall follow conventional practice in this book by using boldface letters, such as **A**, **B**, **C** or **a**, **b**, **c**, for vectors. Also, since the magnitude of a vector is a scalar it will be convenient to use the same character for the magnitude as that used for the vector itself. Thus, we shall use r to designate the magnitude of **r**.

It is often necessary to assign standard directions, such as the x, y, or t direction, to subscripted vectors. In these cases we shall place the direction as an exponent on the character representing the vector. For example, the vectors

$$\mathbf{R}_A{}^x, \ \mathbf{S}_B{}^y, \ \mathbf{R}_A{}^t$$

have the x, y, and t directions, respectively.

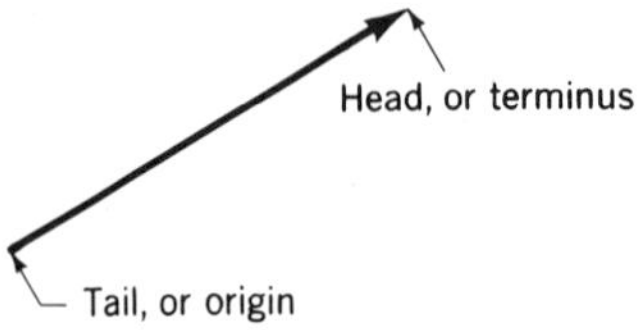

Fig. 2-3

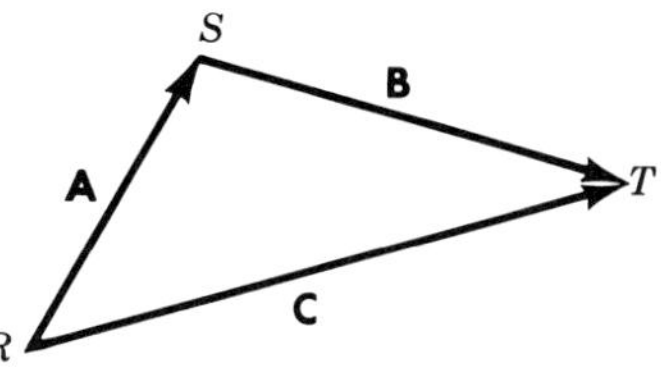

Fig. 2-4

2-4 VECTOR ADDITION

The operation $\mathbf{A} + \mathbf{B}$ is called the *vector addition* of **A** and **B**. As shown in Fig. 2-4, the operation produces a new vector

$$\mathbf{C} = \mathbf{A} + \mathbf{B} \qquad (a)$$

Figure 2-5 shows the *parallelogram rule* for vector addition. We observe that the same result is obtained whether **B** is added to **A** or **A** is added to **B**. In vector algebra this is called the *commutative law,* and it is stated as follows:

$$\mathbf{A} + \mathbf{B} = \mathbf{B} + \mathbf{A} \qquad (b)$$

To investigate the addition of three or more vectors let us consider the sum $(\mathbf{A} + \mathbf{B}) + \mathbf{C}$. This expression means that we are to form the vector $\mathbf{A} + \mathbf{B}$ first and to add the vector **C** to this sum. Figure 2-6*a* shows the three vectors **A**, **B**, and **C**, and the operation $(\mathbf{A} + \mathbf{B}) + \mathbf{C}$ originates at the tail of **A** and terminates at the head of **C**. Figure 2-6*d* shows that the operation $\mathbf{A} + (\mathbf{B} + \mathbf{C})$ gives the same result. This is called the *associative law* and may be stated:

$$(\mathbf{A} + \mathbf{B}) + \mathbf{C} = \mathbf{A} + (\mathbf{B} + \mathbf{C}) \qquad (c)$$

The commutative and associative laws of vector algebra mean that when several vectors are added the result is the same regardless of the order in which they are combined.

The addition of a group of vectors to obtain a single equivalent vec-

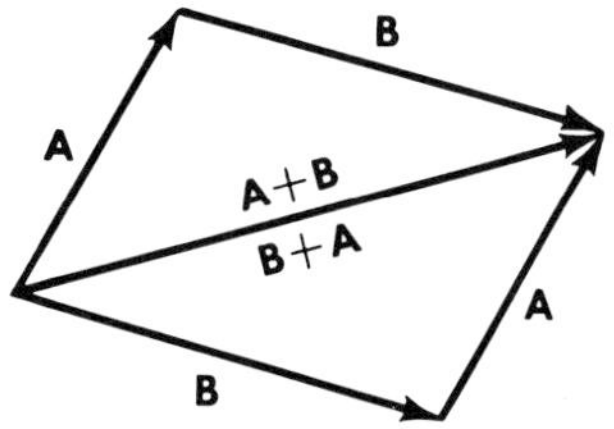

Fig. 2-5 The parallelogram rule.

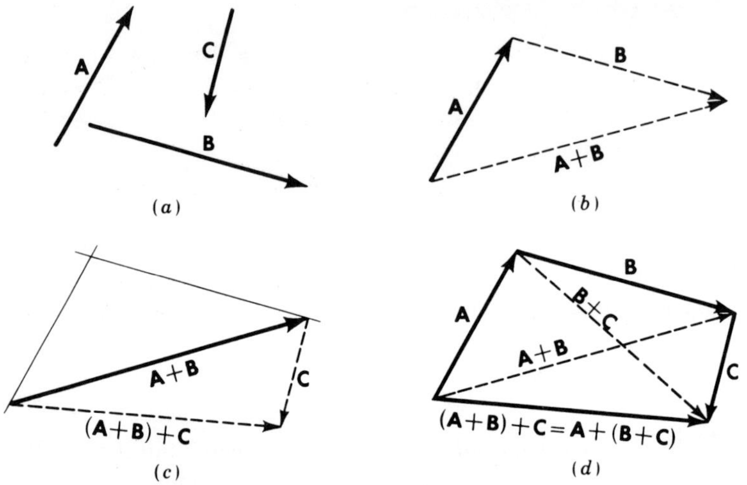

Fig. 2-6 Finding the sum of three vectors: (*a*) The given vectors; (*b*) first, find **A** + **B**; (*c*) second, add **C** to **A** + **B**; (*d*) the result is the same whether **C** is added to **A** + **B** or **A** is added to **B** + **C**.

tor is called *composition of vectors*. The single vector which results from the addition of the group is called the *resultant vector*.

2-5 VECTOR SUBTRACTION

The operation of vector subtraction is defined by the equation

$$\mathbf{A} - \mathbf{B} = \mathbf{A} + (-\mathbf{B}) \qquad (a)$$

Equation (*a*) states that we are to form the vector **A** − **B** by finding the sum of **A** and a vector opposite to **B**. The operation is illustrated in Fig. 2-7.

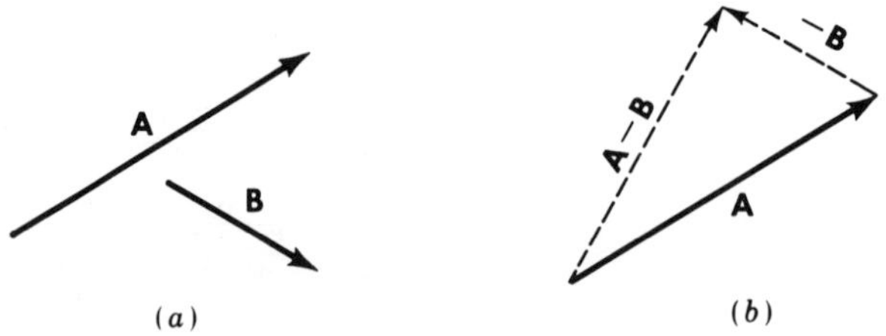

Fig. 2-7 Vector subtraction.

2-6 PRODUCTS OF VECTORS AND SCALARS

Consider the vector sum $\mathbf{A} + \mathbf{A}$ or $\mathbf{A} + \mathbf{A} + \mathbf{A}$. These sums may be abbreviated by writing $2\mathbf{A}$ and $3\mathbf{A}$, which is the product of a vector and a scalar. In the product $a\mathbf{R}$ the scalar a may have any positive or negative value. If a is negative, then $a\mathbf{R}$ is a vector opposite in sense to $\mathbf{R}$. If a is zero, then the product $a\mathbf{R}$ is called a *null vector* in order to retain a vector meaning.

Division of a vector by a scalar as indicated by the operation $\mathbf{R}/a$ may be interpreted in a similar manner to the product because the quantity $1/a$ is also a scalar.

2-7 RESOLUTION OF VECTORS

We have seen that the equation

$$\mathbf{C} = \mathbf{A} + \mathbf{B} \tag{a}$$

represents the addition of vector **B** to vector **A** to produce the resultant vector **C** and that this operation is called *composition of vectors*. Let us now consider the reverse operation in which vector **C** is specified and we are required to determine the vectors **A** and **B**. This is an important operation in vector mathematics and one which is extremely useful in kinematic analysis.

Suppose vector **C** represents a displacement, say, from R to T (Fig. 2-8). The same displacement could have occurred in two steps, first from R to S and then from S to T. If we designate the displacement from R to S as vector **A** and the displacement from S to T as **B**, then **A** and **B** satisfy the requirements of our problem.

Examination of Fig. 2-9, however, will reveal that the previous solution is not unique. For example, the points S' and S'' result in vectors $\mathbf{A}'$, $\mathbf{B}'$ and $\mathbf{A}''$, $\mathbf{B}''$, which also satisfy Eq. (a). Evidently, in order to

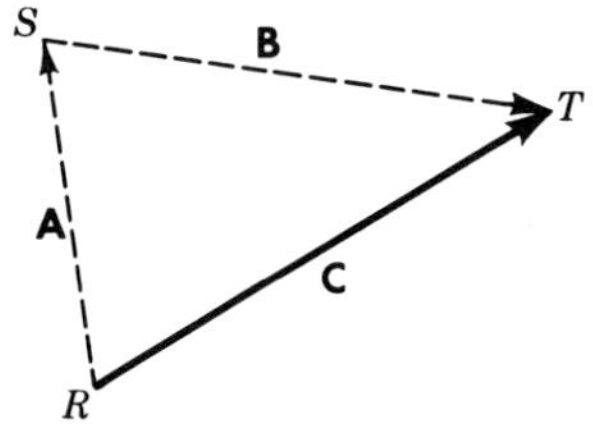

Fig. 2-8

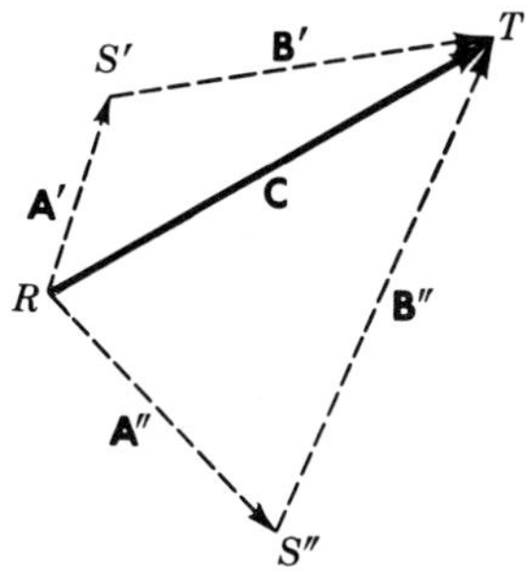

Fig. 2-9

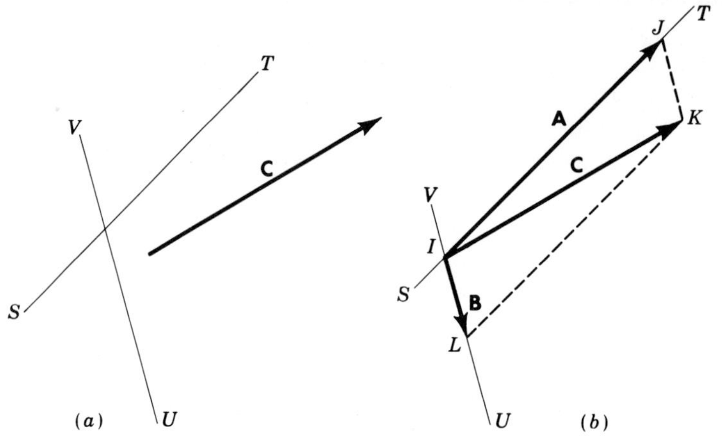

Fig. 2-10 Resolution of a vector.

obtain a unique solution we need to specify additional requirements to be fulfilled. Suppose now that we specify that vectors **A** and **B** are also to be directed along given lines of action. In Fig. 2-10*a* vector **C** is given as before, and we wish to find a vector **A** directed along line ST and a vector **B** along UV such that $\mathbf{C} = \mathbf{A} + \mathbf{B}$. In defining this problem we are not attempting to specify the sense of the desired vectors. The solution is shown in Fig. 2-10*b*. The vector **C** is placed with its origin at the intersection of lines ST and UV, and lines parallel to the desired directions of **A** and **B** are drawn through the terminus of **C**. This produces **A** and **B** in the desired directions as shown.

We note very particularly that line JK is *not* perpendicular to ST; neither is KL perpendicular to UV. In fact, this is the significance of the parallelogram law. The reader is cautioned not to fall into the trap of dropping a perpendicular, say, from K to line ST to get **A**.

A particular case of the parallelogram law occurs when the compo-

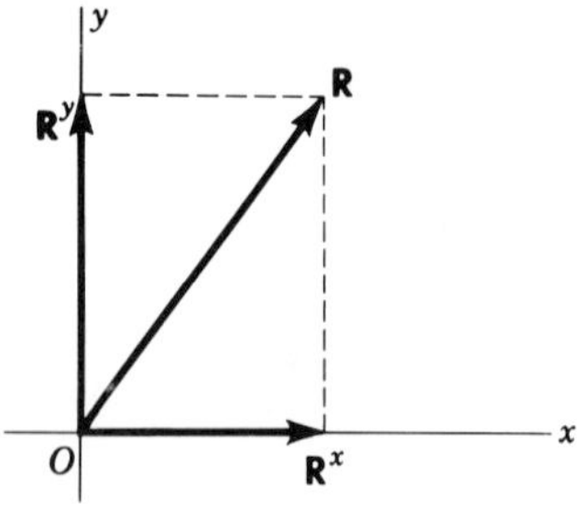

Fig. 2-11 Rectangular components of a vector.

nent vectors are at right angles to each other. The parallelogram becomes a rectangle, and the component vectors are then called the *rectangular components* of a vector. Figure 2-11 shows that the rectangular components of vector **R** are $\mathbf{R}^x$ and $\mathbf{R}^y$.

The operation which we have just concluded is called *resolution of vectors,* and the vectors obtained from this operation are called *component vectors.*

2-8 UNIT VECTORS

We have already seen that division of a vector by a scalar is the same as multiplication of the vector by the reciprocal of the scalar. If the scalar happens to be the magnitude of the vector itself, then the result of this operation is called a *unit vector.* Thus, for every vector **r** there is a corresponding unit vector defined by the equation

$$\hat{\mathbf{r}} = \frac{\mathbf{r}}{r} \qquad (a)$$

The use of the caret to designate unit vectors is particularly convenient because it enables us to use the same character for the vector, its magnitude, and its direction. When speaking of the unit vector $\hat{\mathbf{r}}$, call it *unit r* or *r-hat.*

When dealing with particular reference systems, accepted practice dictates that we employ particular symbols to designate unit vectors associated with the reference axes. In the rectangular cartesian system of Fig. 2-12 the three unit vectors $\hat{\mathbf{i}}$, $\hat{\mathbf{j}}$, and $\hat{\mathbf{k}}$, called a *triad,* define the x, y, and z directions, respectively. We shall reserve these three letters for this purpose; hence the carets may be omitted, if desired, and the triad of unit vectors designated simply as **i**, **j**, and **k**.

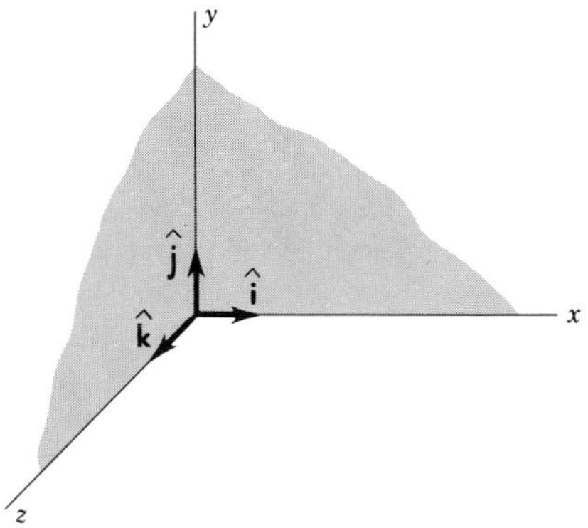

Fig. 2-12

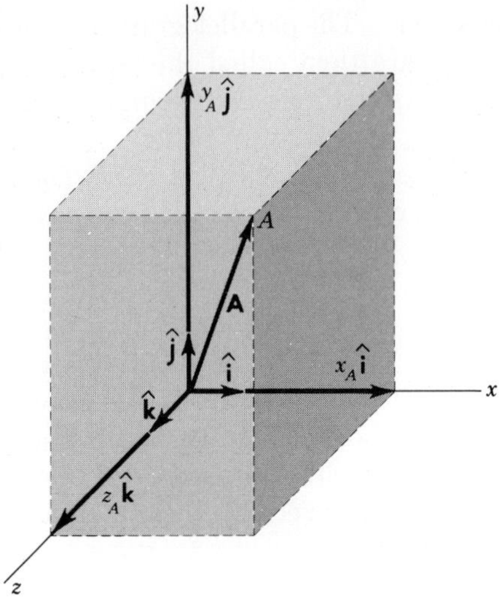

Fig. 2-13

Any point A in space having the coordinates x_A, y_A, and z_A is defined by its *position vector*

$$\mathbf{A} = x_A\hat{\mathbf{i}} + y_A\hat{\mathbf{j}} + z_A\hat{\mathbf{k}} \tag{b}$$

as shown in Fig. 2-13.

If another point in space is defined by the vector

$$\mathbf{B} = x_B\hat{\mathbf{i}} + y_B\hat{\mathbf{j}} + z_B\hat{\mathbf{k}} \tag{c}$$

then the sum of the two vectors **A** and **B** is a third vector

$$\mathbf{C} = \mathbf{A} + \mathbf{B} = (x_A + x_B)\hat{\mathbf{i}} + (y_A + y_B)\hat{\mathbf{j}} + (z_A + z_B)\hat{\mathbf{k}} \tag{d}$$

The magnitude of any vector

$$\mathbf{A} = x\hat{\mathbf{i}} + y\hat{\mathbf{j}} + z\hat{\mathbf{k}} \tag{e}$$

is

$$A = \sqrt{x^2 + y^2 + z^2} \tag{f}$$

The direction cosines of **A** are

$$\cos\alpha = \frac{x}{A} \qquad \cos\beta = \frac{y}{A} \qquad \cos\gamma = \frac{z}{A} \tag{g}$$

where the angles α, β, and γ are, respectively, the angles measured from the coordinate axes x, y, and z to the vector **A**.

2-9 PLANE NOTATION

When the paths of various points of a mechanism in motion lie in a single plane or in parallel planes, such a mechanism is called a *plane mechanism.* A very substantial portion of our investigations in this book will deal with such mechanisms; hence the development of special methods for such problems will be useful.

In Fig. 2-14 the vector

$$\mathbf{r} = r\hat{\mathbf{r}} = \mathbf{r}^x + \mathbf{r}^y \qquad (a)$$

has two rectangular components of magnitude

$$r^x = r\cos\theta \qquad r^y = r\sin\theta \qquad (b)$$

with

$$r = \sqrt{(r^x)^2 + (r^y)^2} \qquad \theta = \tan^{-1}\frac{r^y}{r^x} \qquad (c)$$

We also note that $\mathbf{r}$ can be expressed in the form

$$\mathbf{r} = r^x\hat{\mathbf{\imath}} + r^y\hat{\mathbf{\jmath}} = r\cos\theta\,\hat{\mathbf{\imath}} + r\sin\theta\,\hat{\mathbf{\jmath}} \qquad (2\text{-}1)$$

When plane vector problems are to be solved by graphical means, it is often desirable to express a vector by specifying its magnitude and direction in the form

$$\mathbf{r} = r\underline{/\theta} \qquad (2\text{-}2)$$

Example 2-1 Express the vectors $\mathbf{C} = 10\underline{/30^\circ}$ and $\mathbf{D} = 8\underline{/-15^\circ}$ in rectangular notation and find the sum $\mathbf{C} + \mathbf{D}$.

Solution The vectors are shown in Fig. 2-15.

$$\mathbf{C} = 10\cos 30^\circ\,\hat{\imath} + 10\sin 30^\circ\,\hat{\jmath} = 8.66\hat{\imath} + 5\hat{\jmath}$$

$$\mathbf{D} = 8\cos 15^\circ\,\hat{\imath} - 8\sin 15^\circ\,\hat{\jmath} = 7.73\hat{\imath} - 2.07\hat{\jmath}$$

$$\mathbf{E} = \mathbf{C} + \mathbf{D} = (8.66 + 7.73)\hat{\imath} + (5 - 2.07)\hat{\jmath} = 16.39\hat{\imath} + 2.93\hat{\jmath}$$

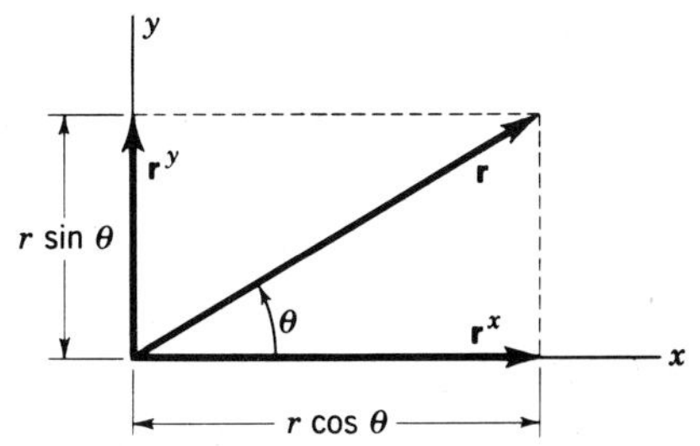

Fig. 2-14

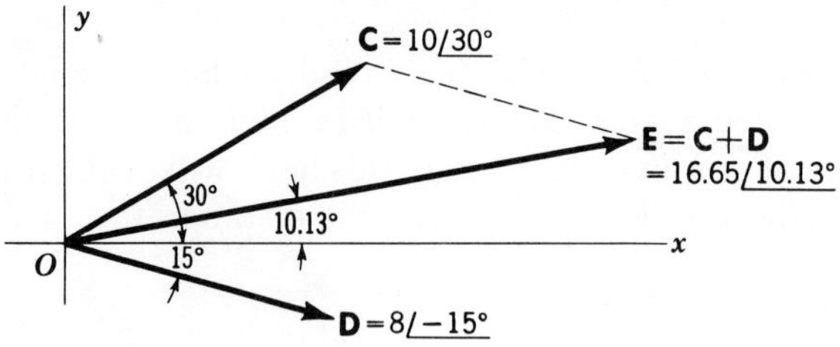

Fig. 2-15

The magnitude is

$$E = \sqrt{(16.39)^2 + (2.93)^2} = 16.65$$

and the angle is

$$\theta = \tan^{-1}\frac{2.93}{16.39} = 10.13°$$

and so the result is

$$\mathbf{E} = 16.65\underline{/10.13°}$$

The reader should be familiar with the use of the slide rule in performing vector operations. Example 2-1 may be solved on the slide rule as follows:

1. Given $C\underline{/\theta}$, find $C^x\hat{\imath} + C^y\hat{\jmath}$. Scale D gives $\sin\theta$ for values of θ on scale S (black) and $\cos\theta$ for values of θ on scale S (red), so the solution may be obtained by a single setting. *Set an index of the slide to C on scale D. Read the x component at θ on scale S (red) and the y component at θ on scale S (black).*

   Example. Find $x\hat{\imath} + y\hat{\jmath}$ for the vector $8\underline{/-15°}$. Set the right slide index to 8 on scale D. At $\theta = 15°$ on S (red), read $x = 7.73$. At $\theta = 15°$ (black), read $y = 2.07$. Therefore

   $$\mathbf{C} = 8\underline{/-15°} = 7.73\hat{\imath} - 2.07\hat{\jmath}.$$

2. Given $x\hat{\imath} + y\hat{\jmath}$, find $C\underline{/\theta}$. *Set x or y, whichever is larger, on scale D and move the slide index to this setting. Opposite the smaller, y or x, on scale D read θ on scale T.* If x is larger than y, then θ is less than 45° and the black numbering of scale T is used. If x is smaller than y, then θ is larger than 45° and the red numbering of scale T is used.

   Example. Find $E\underline{/\theta}$ for $16.39\hat{\imath} + 2.93\hat{\jmath}$. Set the left slide index at 16.39 on scale D. At 2.93 on scale D, read 10.13° on scale T.

Move 10.13° on scale S to 2.93 on scale D. At slide index, read E = 16.65 on scale D. Therefore $16.39\hat{\imath} + 2.93\hat{\jmath} = 16.65\underline{/10.13°}$.

2-10 COMPLEX NUMBERS

Though complex numbers are not vectors, they may be treated as such provided we are willing to refrain from using certain operations which are defined for complex numbers but not for vectors. If we adopt the convention that a vector is always to originate at the origin of the coordinate system, then the sense of a vector is always outward, or away from the origin. Then any system of mathematics which specifies the length and direction will completely define the vector. In Fig. 2-16 let the x and y coordinates of the head of the vector **R** be a and b. Then **R** may be expressed as

$$\mathrm{R} = a + jb \tag{2-3}$$

where j is prefixed to b to indicate 90° counterclockwise rotation. When used in this manner j is an operator which indicates that b is to be taken in the y direction.

In this notation the numbers a, 2, 6, -4 are real numbers (Fig. 2-17). They are always taken from the origin along the x axis, to the right if the sign is positive and to the left if it is negative.

When the symbol j is prefixed to a number it means that the vector is to be rotated through an angle of 90° from the positive x direction in a *counterclockwise* direction. If the sign of j is negative, then the vector is to be rotated 90° in a clockwise direction. These are called *imaginary* quantities and are always along the y axis. In Fig. 2-17, $+j4$ is shown on the positive y axis and $-j3$ on the negative y axis. Therefore j is plotted above x when it is positive and below x when it is negative.

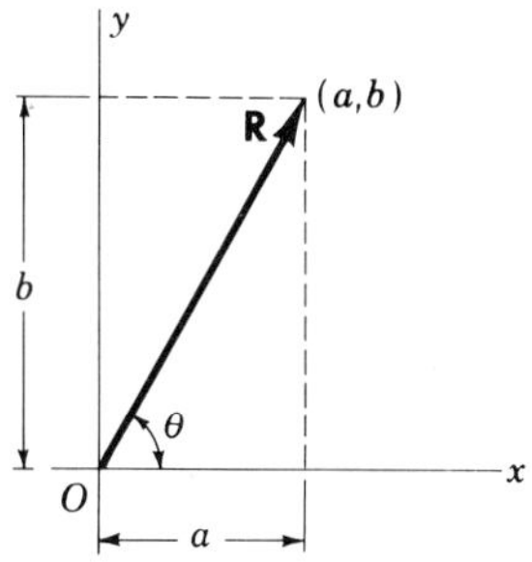

Fig. 2-16

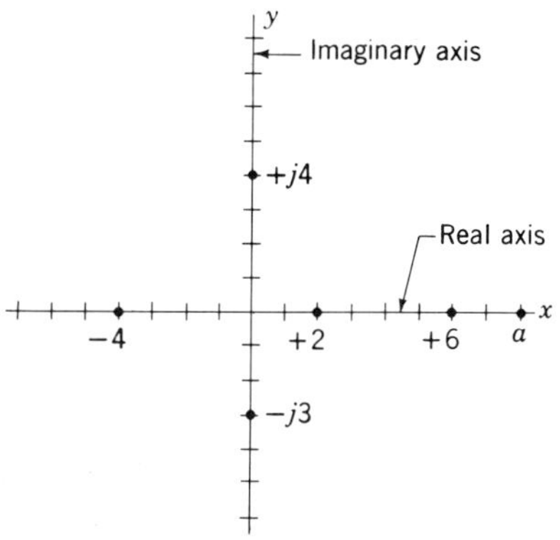

Fig. 2-17

The operator j may be used repeatedly, but each use represents a 90° rotation in the counterclockwise direction. For example, $j^2 3$ requires the rotation of 3 through 180° and must produce -3. Therefore $j^2 = -1$ and $j = \sqrt{-1}$. Using the same reasoning we find:

$$j = 90° \text{ counterclockwise} = \sqrt{-1} = j$$
$$j^2 = 180° \text{ counterclockwise} = (\sqrt{-1})^2 = -1$$
$$j^3 = 270° \text{ counterclockwise} = (\sqrt{-1})^3 = -\sqrt{-1} = -j$$
$$j^4 = 360° \text{ counterclockwise} = (\sqrt{-1})^4 = +1$$

Complex numbers in polar form appear frequently in the literature of mechanism analysis and synthesis. To illustrate, take the well-known transformation from trigonometry

$$e^{\pm j\theta} = \cos\theta \pm j\sin\theta \tag{2-4}$$

Employing complex rectangular notation for the vector $\mathbf{R} = R\underline{/\theta}$ yields

$$\mathbf{R} = R\cos\theta + jR\sin\theta \tag{a}$$

note that $R\sin\theta$ is not a vector; its just a real number

Therefore, from Eq. (2-4) $\mathbf{R}$ can also be written in the form

$$\mathbf{R} = Re^{j\theta} \tag{2-5}$$

which is the complex polar form. Expression of a vector in this form, as the product of its magnitude with its direction, is especially useful when a vector must be differentiated.

Soon we shall learn of two vector operations which bear the title "product." The reason for this, as we shall learn, is that these operations are similar to the multiplication of real numbers. But there is no vector operation which is similar to the division of real numbers.

On the other hand, it is possible to divide or to multiply one complex number by another. For example, let $A = ae^{j\theta}$ and $B = be^{j\phi}$. The real number a in the complex number A is called the *modulus* of A. And the angle θ is called the *argument* of A. Note very particularly that these complex numbers A and B are *not* being treated as vectors. Now

$$\frac{A}{B} = \frac{ae^{j\theta}}{be^{j\phi}} = \frac{a}{b}e^{j(\theta-\phi)} = \frac{a}{b}[\cos(\theta-\phi) + j\sin(\theta-\phi)] \tag{b}$$

Thus, to divide one complex number by another, divide their moduli and subtract their arguments. The reader should have no difficulty in developing a similar rule for the multiplication of two complex numbers.

The purpose in presenting this analysis is to show that certain operations are *not defined when such numbers are treated as vectors.* In particular, complex numbers can be multiplied or divided, but there is no corresponding operation for vectors.

2-11 ROTATION OF VECTOR AXES IN TWO DIMENSIONS

Sometimes the solution of a vector equation may be obtained very simply by referring the vector to a new reference system. This can be done quite easily by placing the vector in polar form. For example, suppose it is desired to transform the vector $\mathbf{R} = R/\theta$ from the xy system to the XY system. If we designate the positive orientation of the XY system as ϑ deg from the xy system, then the vector $\mathbf{R}$ referred to the new system is

$$\mathbf{R} = R/\theta - \vartheta \quad \text{from inspection of the diagram (Fig. 2-18)} \tag{2-6}$$

Note, particularly, as shown in Fig. 2-18, that the vector has not changed. The only change is in the coordinate system.

By rotating the axes the solution of vector equations is often simplified. Thus, if we refer the vector

$$\mathbf{A} = x_A \hat{\imath} + y_A \hat{\jmath} = A/\theta$$

to an XY system located counterclockwise at an angle θ from the xy system, then

$$X_A = A \quad \text{and} \quad Y_A = 0$$

With such a rotation one of the variables is automatically eliminated. After the solution is complete, the resultant vector may be referred back to the original system, by using Eq. (2-6), if this is desired.

2-12 TWO-DIMENSIONAL VECTOR EQUATIONS

The three-dimensional vector equation

$$\mathbf{A} + \mathbf{B} + \mathbf{C} = \mathbf{S} \tag{a}$$

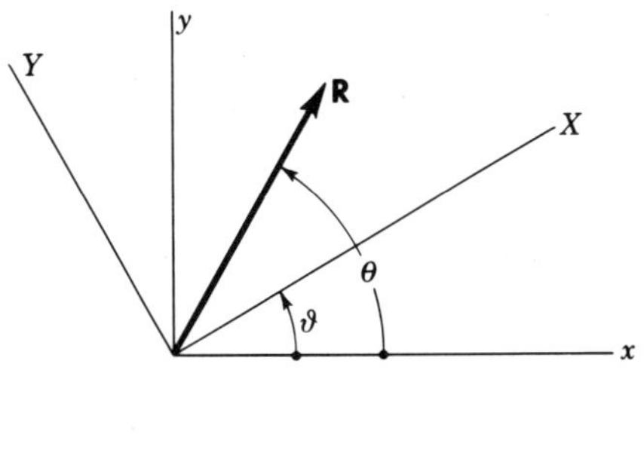

Fig. 2-18

with **S** known in direction and magnitude, may be transformed to

$$(x_A + x_B + x_C)\hat{i} + (y_A + y_B + y_C)\hat{j} + (z_A + z_B + z_C)\hat{k} = x_S\hat{i} + y_S\hat{j} + z_S\hat{k} \quad (b)$$

Equation (b) then leads to the following three scalar equations:

$$\begin{aligned} x_A + x_B + x_C &= x_S \\ y_A + y_B + y_C &= y_S \\ z_A + z_B + z_C &= z_S \end{aligned} \quad (c)$$

these relations may be used in analytical solutions

If these equations are *linearly independent* and *consistent*, they can be solved simultaneously for three unknowns. These may be three magnitudes, three directions, or any combination of magnitudes and directions. For some combinations the problem is indeed difficult to solve. Therefore, we shall delay our consideration of the general three-dimensional problem until it is needed.

A two-dimensional vector equation may have as unknowns two directions, two magnitudes, or one direction and one magnitude. Sometimes it is desirable to write the number of *known* quantities above each vector in an equation like this:

$$\overset{2}{\mathbf{R}} = \overset{2}{\mathbf{A}} + \overset{1}{\mathbf{B}} + \overset{2}{\mathbf{C}} + \overset{1}{\mathbf{D}} \quad (d)$$

This identifies the unknown terms and indicates whether a solution can be found. In Eq. (d) the vectors **R**, **A**, and **C** are <u>completely defined. They can be replaced by their sum</u>

$$\mathbf{S} = \mathbf{A} + \mathbf{C} - \mathbf{R} \quad (e)$$

giving

$$\overset{2}{\mathbf{S}} + \overset{1}{\mathbf{B}} + \overset{1}{\mathbf{D}} = 0 \quad (f)$$

Thus, any plane vector equation can be reduced to, at most, a three-term equation having one known vector.

If the two unknowns are the direction and magnitude of a single vector, then the problem is trivial. Thus, if the equation to be solved is

$$\overset{2}{\mathbf{S}} + \overset{0}{\mathbf{A}} = 0 \quad (g)$$

then the solution is

$$\mathbf{A} = -\mathbf{S} \quad (h)$$

In the equation

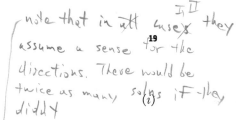

$$\overset{2}{\mathbf{S}} + \overset{1}{\mathbf{A}} + \overset{1}{\mathbf{B}} = 0 \qquad (i)$$

three distinct cases occur. For each of these cases the unknowns are:

Case 1. Magnitude of **A** and magnitude of **B**.
Case 2. Magnitude of **A** and direction of **B**.
Case 3. Direction of **A** and direction of **B**.

We shall solve these three cases graphically in this section, and analytically in the section which follows.

By writing Eq. (i) in the form

$$\mathbf{S} + A\hat{\mathbf{A}} + B\hat{\mathbf{B}} = 0 \qquad (j)$$

then, for case 1, **S**, $\hat{\mathbf{A}}$, and $\hat{\mathbf{B}}$ are given, and A and B are to be found. The solution is shown in Fig. 2-19b. The steps are as follows: (1) Draw vector **S**. (2) Construct line 1-1 through the terminus of **S** parallel to $\hat{\mathbf{A}}$. (3) Construct line 2-2 through the origin of **S** parallel to $\hat{\mathbf{B}}$. (4) The intersection of these two lines then defines the magnitudes A and B, and these may be either positive or negative. Note that case 1 has a unique solution.

Examination of Fig. 2-19a indicates that the directions of $\hat{\mathbf{A}}$ and $\hat{\mathbf{B}}$ are given but *not* the senses. Had we permitted solutions having only positive magnitudes and had the senses of $\hat{\mathbf{A}}$ and $\hat{\mathbf{B}}$ been given too, then it could turn out that there is no solution. But there is nothing wrong with

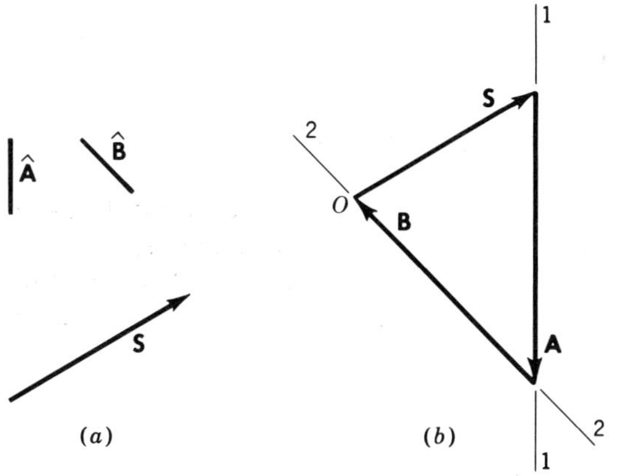

Fig. 2-19 Graphical solution of case 1. (a) Given: **S**, $\hat{\mathbf{A}}$, and $\hat{\mathbf{B}}$; (b) solution.

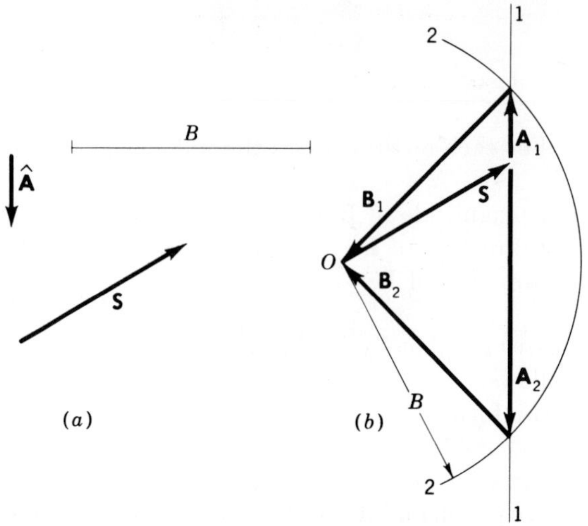

Fig. 2-20 Graphical solution of case 2. (*a*) Given: **S**, **Â**, and B; (*b*) solution.

a negative magnitude; hence it would have been perfectly appropriate to specify both the directions and the senses of **Â** and **B̂**.

For case 2, **S**, **Â**, and B are given, and A and **B̂** are to be found. The solution, shown in Fig. 2-20, is obtained as follows: (1) Construct line 1-1 through the terminus of **S** parallel to **Â**. (2) Adjust a compass to the distance B and construct arc 2-2 with center at the origin O. (3) The intersection of arc 2-2 with line 1-1 defines the *two* solutions $\mathbf{A}_1$, $\mathbf{B}_1$ and $\mathbf{A}_2$, $\mathbf{B}_2$.

Note, in Fig. 2-20*a*, that both the direction and the sense of **Â** are given in this case. This means that we are willing to accept negative magnitudes in the solution.

Finally, for case 3, **S**, A, and B are given, and **Â** and **B̂** are to be found, as shown in Fig. 2-21*a*. The steps in the solution are as follows: (1) Construct **S** and, with center at the terminus, construct arc 1-1 of radius A. (2) With center at origin O construct arc 2-2 of radius B. (3) The intersection of the two arcs defines the *two* solutions $\mathbf{A}_1$, $\mathbf{B}_1$ and $\mathbf{A}_2$, $\mathbf{B}_2$. A solution is obtained only if $A + B \geq S$.

it might be helpful to solve graphically first

2-13 ANALYTICAL SOLUTION OF VECTOR EQUATIONS

Turning now to the analytical solution for case 1 of Sec. 2-12, we have as given quantities the vector **S** and the directions of **A** and **B**. If we desig-

For all analytical problems, three vectors lie in one plane. 1) Fix a co-ordinate system in this plane, 2) find graphical solution in this plane, 3) resolve into unit vectors of this co-ordinate system by inspection 4) transfer back into original co-ordinate system

nate these directions as θ_A and θ_B, respectively, then four scalar equations can be written:

$$x_A + x_B = -x_S \tag{a}$$

$$y_A + y_B = -y_S \tag{b}$$

$$y_A = x_A \tan \theta_A \tag{c}$$

$$y_B = x_B \tan \theta_B \tag{d}$$

Since we wish to determine the magnitudes of **A** and **B**, the unknowns in these equations are x_A, y_A, x_B, and y_B. Eliminating all these except x_A yields

$$x_A = \frac{y_S - x_S \tan \theta_B}{\tan \theta_B - \tan \theta_A} \tag{2-7}$$

Then with x_A known, Eqs. (a), (c), and (d) can be used to determine the remaining components. The two magnitudes can then be obtained from

$$A = \sqrt{x_A{}^2 + y_A{}^2} \qquad B = \sqrt{x_B{}^2 + y_B{}^2}$$

Equation (2-7) fails, of course, if θ_A or θ_B should happen to be 90°. However, in this case either x_A or x_B is zero and the solution of Eqs. (a) through (d) is greatly simplified.

For case 2 the given quantities are the vector **S**, the direction θ_A of

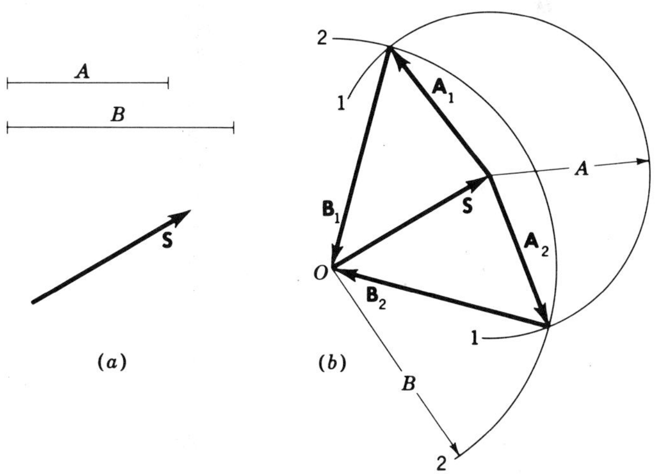

Fig. 2-21 Graphical solution of case 3. (a) Given: **S**, A, and B; (b) solution.

A, and the magnitude B of **B**. We begin by noting the four scalar relations

$$x_A + x_B = -x_S \qquad (e)$$

$$y_A + y_B = -y_S \qquad (f)$$

$$y_A = x_A \tan\theta_A \qquad (g)$$

$$y_B{}^2 = B^2 - x_B{}^2 \qquad (h)$$

The four unknowns in these equations are x_A, y_A, x_B, and y_B. If all these except x_B are eliminated, then the result is a quadratic equation. The solution is

$$x_B = \frac{(y_S - x_S\tan\theta_A)\tan\theta_A}{1+\tan^2\theta_A} \pm \frac{\sqrt{(y_S - x_S\tan\theta_A)^2 - (1+\tan^2\theta_A)(y_S{}^2 - B^2 - 2y_Sx_S\tan\theta_A + x_S{}^2\tan^2\theta_A}}{1+\tan^2\theta_A} \qquad (2\text{-}8)$$

With x_B known, Eqs. (*e*), (*f*), and (*g*) can be solved to obtain the remaining components. The magnitude of **A** and the direction of **B** are then readily found for each of the two solutions. Equation (2-8), of course, cannot be used if $\theta_A = 90°$.

Case 3 is much more complicated. As noted by Chace,[1] any solution of a vector equation in polynomial form must have a degree equal to or greater than the number of physically real solutions. When case 3 is solved in the general form, the solution is a higher-degree polynomial with two real solutions. Under these conditions Chace advises a reduction in the degree of the polynomial by taking advantage of whatever symmetry may exist. Figure 2-21 shows that the solution is symmetrical about the known vector **S**. We can use this symmetry by choosing a new axis system.

In case 3 the known quantities are **S**, A, and B, with **S** defined as

$$\mathbf{S} = S/\underline{\theta_S} \qquad (i)$$

in the xy system. We begin by defining the XY system such that

$$X_S = S \qquad Y_S = 0 \qquad (j)$$

As shown in Fig. 2-22, two solutions $\mathbf{A}_1$, $\mathbf{B}_1$ and $\mathbf{A}_2$, $\mathbf{B}_2$ exist, if we assume

[1] Milton A. Chace, "Development and Application of Vector Mathematics for Kinematic Analysis of Three-dimensional Mechanisms," Ph.D. thesis, The University of Michigan, Ann Arbor, Michigan, p. 19, 1964.

that $A + B > S$. From the geometry of this figure we write

$$S + X_A + X_B = 0 \qquad (k)$$
$$Y_A + Y_B = 0 \qquad (l)$$
$$X_A^2 + Y_A^2 = A^2 \qquad (m)$$
$$X_B^2 + Y_B^2 = B^2 \qquad (n)$$

for every known quantity you have, you should have an equation

These equations are now solved simultaneously for X_A. The result is

$$X_A = \frac{B^2 - A^2 - S^2}{2S} \qquad (2\text{-}9)$$

Note, from Fig. 2-22, that X_A and X_B are single-valued even though there are two solutions to case 3. With X_A known, Eq. (k) can be used to obtain X_B. Equations (m) and (n) are then used to obtain the pairs of values for Y_A and Y_B.

Now, with **A** and **B** in the XY reference system known, the final step is to transform these vectors back to the xy system. We designate **A** and **B** in the XY system as

$$\mathbf{A} = A\underline{/\vartheta_A} \qquad \mathbf{B} = B\underline{/\vartheta_B} \qquad (o)$$

To get to the xy system we must rotate the axes backward (clockwise) through the original angle θ_S. Therefore, **A** and **B** expressed in polar form are

$$\mathbf{A} = A\underline{/\vartheta_A + \theta_S} \qquad \mathbf{B} = B\underline{/\vartheta_B + \theta_S} \qquad (2\text{-}10)$$

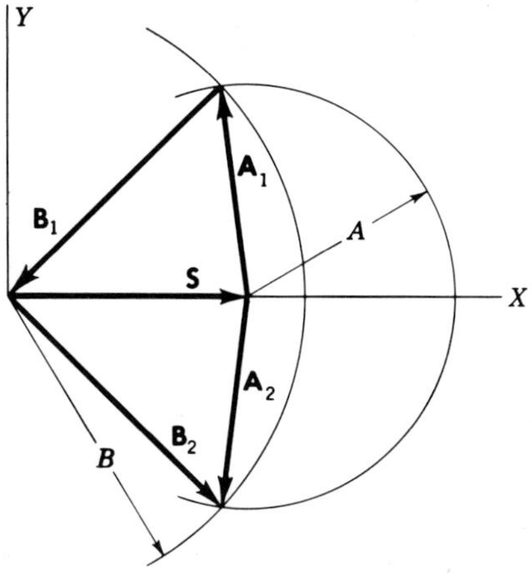

Fig. 2-22

The reader who attempts to solve case 3 without rotation of axes will gain a new respect for the graphical method.

2-14 VECTOR PRODUCTS

The operation to be defined here is called a vector product because it is similar to multiplication; yet we shall see that differences exist. The result of this operation is to be another vector having a direction and a magnitude. In Fig. 2-23 two nonparallel vectors **A** and **B** are separated by the angle θ. Define a unit vector $\hat{\mathbf{N}}$ perpendicular to the plane of **A** and **B** such that the triad **A**, **B**, and $\hat{\mathbf{N}}$ forms a right-handed system. Then the *vector product* of **A** and **B** is defined as

$$\mathbf{A} \times \mathbf{B} = AB \sin \theta \, \hat{\mathbf{N}} \quad \text{— a unit vector whose direction is defined by the right-hand rule} \tag{2-11}$$

where θ is the angle from **A** to **B** measured in the positive sense. The operation defined by Eq. (2-11) is frequently called a *cross product*, or sometimes a *vector cross product*. Note that the magnitude of the new vector, $AB \sin \theta$, is the area of a parallelogram formed by the vectors **A** and **B**.

Using the right-hand rule, note that the product **B** × **A** is

$$\mathbf{B} \times \mathbf{A} = -AB \sin \theta \, \hat{\mathbf{N}} \tag{a}$$

So the cross product is *not* commutative, since

$$\mathbf{A} \times \mathbf{B} = -\mathbf{B} \times \mathbf{A} \tag{b}$$

It is distributive, however, and for three vectors,

$$\mathbf{A} \times (\mathbf{B} + \mathbf{C}) = \mathbf{A} \times \mathbf{B} + \mathbf{A} \times \mathbf{C} \tag{c}$$

Since, when $\theta = 0$, $\sin \theta = 0$, if

$$\mathbf{A} \times \mathbf{B} = 0 \tag{d}$$

then the vectors **A** and **B** are parallel to each other.

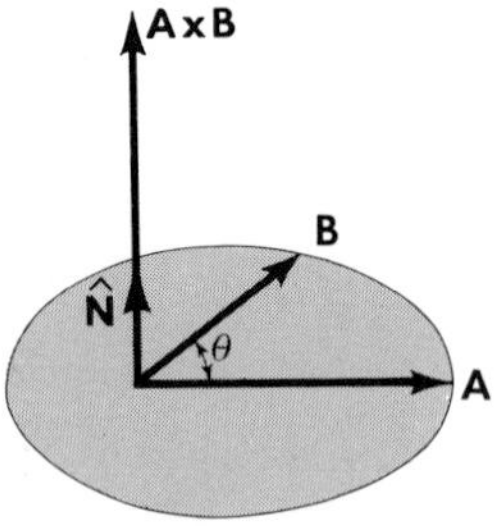

Fig. 2-23

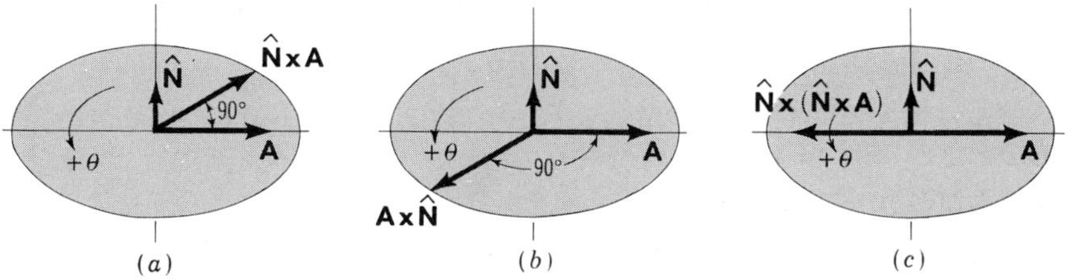

Fig. 2-24

Using the right-hand rule and the definition of a cross product we can obtain the following relations between the unit vectors $\hat{\mathbf{\imath}}$, $\hat{\mathbf{\jmath}}$, and $\hat{\mathbf{k}}$:

$$\begin{aligned} \hat{\mathbf{\imath}} \times \hat{\mathbf{\imath}} &= \hat{\mathbf{\jmath}} \times \hat{\mathbf{\jmath}} = \hat{\mathbf{k}} \times \hat{\mathbf{k}} = 0 \\ \hat{\mathbf{\imath}} \times \hat{\mathbf{\jmath}} &= -\hat{\mathbf{\jmath}} \times \hat{\mathbf{\imath}} = \hat{\mathbf{k}} \\ \hat{\mathbf{\jmath}} \times \hat{\mathbf{k}} &= -\hat{\mathbf{k}} \times \hat{\mathbf{\jmath}} = \hat{\mathbf{\imath}} \\ \hat{\mathbf{k}} \times \hat{\mathbf{\imath}} &= -\hat{\mathbf{\imath}} \times \hat{\mathbf{k}} = \hat{\mathbf{\jmath}} \end{aligned} \tag{2-12}$$

$\hat{\imath} \times \hat{\jmath} = \hat{k}$ defines the right-hand co-ord. system

Now consider the cross product of a vector **A** by a unit vector $\hat{\mathbf{N}}$ at right angles to it (Fig. 2-24*a*). The resulting vector $\hat{\mathbf{N}} \times \mathbf{A}$ is at right angles to **A**, and so the operation $\hat{\mathbf{N}} \times$ may be thought of as rotating the vector **A** through 90°. On the other hand, the vector $\mathbf{A} \times \hat{\mathbf{N}}$ is also at right angles to **A**, but in the opposite direction, as shown in Fig. 2-24*b*. So the operation $\times \hat{\mathbf{N}}$ rotates the vector **A** through −90°.

The triple cross product $\hat{\mathbf{N}} \times (\hat{\mathbf{N}} \times \mathbf{A}) = -\mathbf{A}$ rotates the vector **A** through 180°, as shown in Fig. 2-24*c*. It is equivalent to operating on **A** by $\hat{\mathbf{N}} \times$ two times in succession.

Now let us consider the cross product of the two vectors

$$\mathbf{A} = x_A\hat{\mathbf{\imath}} + y_A\hat{\mathbf{\jmath}} + z_A\hat{\mathbf{k}} \tag{e}$$

$$\mathbf{B} = x_B\hat{\mathbf{\imath}} + y_B\hat{\mathbf{\jmath}} + z_B\hat{\mathbf{k}} \tag{f}$$

By using Eq. (2-12), the cross product is

$$\mathbf{A} \times \mathbf{B} = (y_Az_B - z_Ay_B)\hat{\mathbf{\imath}} + (z_Ax_B - x_Az_B)\hat{\mathbf{\jmath}} + (x_Ay_B - y_Ax_B)\hat{\mathbf{k}} \tag{2-13}$$

It is convenient to write Eq. (2-13) as a determinant. Thus

$$\mathbf{A} \times \mathbf{B} = \begin{vmatrix} \hat{\mathbf{\imath}} & \hat{\mathbf{\jmath}} & \hat{\mathbf{k}} \\ x_A & y_A & z_A \\ x_B & y_B & z_B \end{vmatrix} \tag{2-14}$$

Example 2-2 Find **A** + **B**, **A** − **B**, and $\mathbf{A} \times \mathbf{B}$ for the vectors $\mathbf{A} = 2\hat{\mathbf{\imath}} - \hat{\mathbf{\jmath}} + \hat{\mathbf{k}}$ and $\mathbf{B} = \hat{\mathbf{\imath}} + 2\hat{\mathbf{\jmath}} - 3\hat{\mathbf{k}}$.

write this whole thing out for two-dimen. case and you won't get confused

Solution

$$\mathbf{A} + \mathbf{B} = (2 + 1)\hat{\imath} + (-1 + 2)\hat{\jmath} + (1 - 3)\hat{\mathbf{k}} = 3\hat{\imath} + \hat{\jmath} - 2\hat{\mathbf{k}}$$

$$\mathbf{A} - \mathbf{B} = (2 - 1)\hat{\imath} + (-1 - 2)\hat{\jmath} + (1 + 3)\hat{\mathbf{k}} = \hat{\imath} - 3\hat{\jmath} + 4\hat{\mathbf{k}}$$

$$\mathbf{A} \times \mathbf{B} = \begin{vmatrix} \hat{\imath} & \hat{\jmath} & \hat{\mathbf{k}} \\ 2 & -1 & 1 \\ 1 & 2 & -3 \end{vmatrix} = (3 - 2)\hat{\imath} + (1 + 6)\hat{\jmath} + (4 + 1)\hat{\mathbf{k}}$$

$$= \hat{\imath} + 7\hat{\jmath} + 5\hat{\mathbf{k}}$$

2-15 SCALAR MULTIPLICATION

In the preceding section two vectors were combined so as to produce a new third vector. In this section two vectors will be combined to produce a *scalar*. The result of this operation is called the *scalar product*, the *inner product*, or, more commonly, the *dot product*. It is defined by the equation

$$\mathbf{A} \cdot \mathbf{B} = AB \cos \theta \tag{2-15}$$

where A and B are the magnitudes of the two vectors, and θ is the angle between them.

Using Eq. (2-15) we can determine a number of interesting conclusions. Thus

$$\mathbf{A} \cdot \mathbf{A} = A^2 \tag{a}$$

since $\theta = 0$.

Also, if $\mathbf{A} \cdot \mathbf{B} = 0$, then one of the vectors may be a null vector. But in the more general case $\theta = 90°$ and $\mathbf{A}$ and $\mathbf{B}$ are perpendicular to each other.

Using the dot product with the unit vectors produces

$$\hat{\imath} \cdot \hat{\imath} = \hat{\jmath} \cdot \hat{\jmath} = \hat{\mathbf{k}} \cdot \hat{\mathbf{k}} = 1 \tag{b}$$

$$\hat{\imath} \cdot \hat{\jmath} = \hat{\jmath} \cdot \hat{\imath} = \hat{\jmath} \cdot \hat{\mathbf{k}} = \hat{\mathbf{k}} \cdot \hat{\jmath} = \hat{\imath} \cdot \hat{\mathbf{k}} = \hat{\mathbf{k}} \cdot \hat{\imath} = 0 \tag{c}$$

A very useful property is obtained by the dot product of any vector with a unit vector. Thus

$$\mathbf{A} \cdot \hat{\imath} = (A)(1) \cos \theta = A \cos \theta = x_A \tag{d}$$

This means that the dot product of any vector by a unit vector is the component of that vector in the direction of the unit vector. Therefore, the components of the vector $\mathbf{A}$ in the x, y, and z directions are

$$x_A = \mathbf{A} \cdot \hat{\imath} \qquad y_A = \mathbf{A} \cdot \hat{\jmath} \qquad z_A = \mathbf{A} \cdot \hat{\mathbf{k}} \tag{e}$$

Consider, now, the scalar product of the two vectors:

$$\mathbf{A} = x_A\hat{\mathbf{\imath}} + y_A\hat{\mathbf{\jmath}} + z_A\hat{\mathbf{k}} \tag{f}$$

$$\mathbf{B} = x_B\hat{\mathbf{\imath}} + y_B\hat{\mathbf{\jmath}} + z_B\hat{\mathbf{k}} \tag{g}$$

This gives

$$\mathbf{A} \cdot \mathbf{B} = (x_A\hat{\mathbf{\imath}} + y_A\hat{\mathbf{\jmath}} + z_A\hat{\mathbf{k}}) \cdot (x_B\hat{\mathbf{\imath}} + y_B\hat{\mathbf{\jmath}} + z_B\hat{\mathbf{k}}) \tag{h}$$

When the right side of this equation is expanded, six of the terms are like

$$x_Ay_B\hat{\mathbf{\imath}} \cdot \hat{\mathbf{\jmath}}, \ . \ . \ .$$

which, from Eq. (*c*), are all zero. The remaining three terms are

$$x_Ax_B\hat{\mathbf{\imath}} \cdot \hat{\mathbf{\imath}}, \ y_Ay_B\hat{\mathbf{\jmath}} \cdot \hat{\mathbf{\jmath}}, \ z_Az_B\hat{\mathbf{k}} \cdot \hat{\mathbf{k}}$$

So, from Eq. (*b*), we have

$$\mathbf{A} \cdot \mathbf{B} = x_Ax_B + y_Ay_B + z_Az_B \tag{2-16}$$

Example 2-3 A point P has the coordinates $x_P = 4$, $y_P = 8$. Find the component of the vector $\mathbf{A} = 6\hat{\imath} + 2\hat{\jmath}$ along a line from the origin to P.

Solution We shall solve this example by defining a unit vector $\hat{\mathbf{P}}$ in the direction of OP. The magnitude of the projection of $\mathbf{A}$ along OP is then $\mathbf{A} \cdot \hat{\mathbf{P}}$ and so the component vector of $\mathbf{A}$ in the direction of $\hat{\mathbf{P}}$ is $\hat{\mathbf{A}} \cdot \hat{\mathbf{P}}\hat{\mathbf{P}}$, a new vector. The graphical solution is shown in Fig. 2-25.

Analytically, we obtain the unit vector $\hat{\mathbf{P}}$ by writing

$$\mathbf{P} = 4\hat{\imath} + 8\hat{\jmath}$$

Then

$$P = \sqrt{(4)^2 + (8)^2} = 8.93$$

So the unit vector $\hat{\mathbf{P}}$ is

$$\hat{\mathbf{P}} = \frac{\mathbf{P}}{P} = \frac{4\hat{\imath} + 8\hat{\jmath}}{8.93} = 0.449\hat{\imath} + 0.897\hat{\jmath}$$

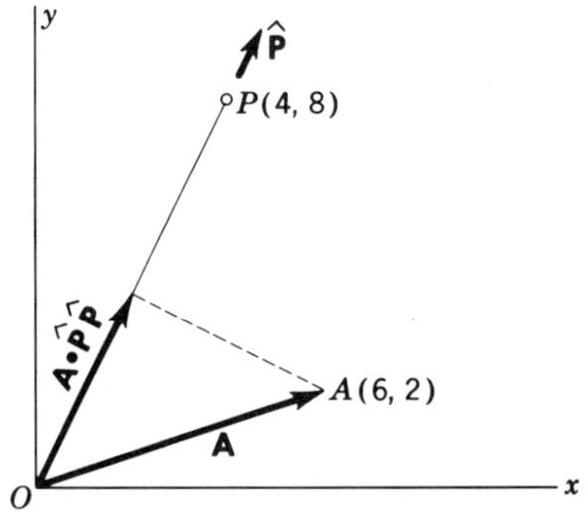

Fig. 2-25

Then

$$\mathbf{A} \cdot \hat{\mathbf{P}} = (6)(0.449)\hat{\imath} \cdot \hat{\imath} + (2)(0.897)\hat{\jmath} \cdot \hat{\jmath} = 4.48$$

hence, the component of **A** along *OP* is the new vector

$$\mathbf{A} \cdot \hat{\mathbf{P}}\hat{\mathbf{P}} = (4.48)(0.449\hat{\imath} + 0.897\hat{\jmath}) = 2.01\hat{\imath} + 4.02\hat{\jmath}$$

Example 2-4 What angle do the two vectors $\mathbf{A} = \hat{\imath} + 2\hat{\jmath} + 3\hat{\mathbf{k}}$ and $\mathbf{B} = -5\hat{\imath} + \hat{\jmath} - 2\hat{\mathbf{k}}$ make with each other?

Solution Since $\mathbf{A} \cdot \mathbf{B} = AB \cos \theta$,

$$\theta = \cos^{-1} \frac{\mathbf{A} \cdot \mathbf{B}}{AB}$$

Therefore, we successively calculate

$$\mathbf{A} \cdot \mathbf{B} = (1)(-5)\hat{\imath} \cdot \hat{\imath} + (2)(1)\hat{\jmath} \cdot \hat{\jmath} + (3)(-2)\hat{\mathbf{k}} \cdot \hat{\mathbf{k}} = -9$$

and

$$A = \sqrt{(1)^2 + (2)^2 + (3)^2} = 3.74 \qquad B = \sqrt{(-5)^2 + (1)^2 + (-2)^2} = 5.48$$

Therefore

$$\theta = \cos^{-1} \frac{-9}{(3.74)(5.48)} = 116°$$

2-16 IDENTITIES

There are a number of identities which will be useful in the studies which follow. Some of these we have already developed. We shall prove the remaining ones in this section and then tabulate them for ready reference.

The triple cross product If $\mathbf{R} = \mathbf{B} \times \mathbf{C}$ and $\mathbf{S} = \mathbf{A} \times \mathbf{R}$, then the triple cross product is

$$\mathbf{S} = \mathbf{A} \times \mathbf{R} = \mathbf{A} \times (\mathbf{B} \times \mathbf{C}) \tag{a}$$

which states that **S** is the cross product of the vector **A** with the vector $\mathbf{B} \times \mathbf{C}$. We note that the vector $\mathbf{B} \times \mathbf{C}$ is perpendicular to the plane of **B** and **C** and, therefore, that the vector **S** is in the plane of **B** and **C**.

On the other hand, the vector $\mathbf{A} \times \mathbf{B}$ is perpendicular to the plane containing **A** and **B**; hence, $(\mathbf{A} \times \mathbf{B}) \times \mathbf{C}$ is in the plane of **A** and **B**. We conclude that the parentheses *are* necessary and that

$$\mathbf{A} \times (\mathbf{B} \times \mathbf{C}) \neq (\mathbf{A} \times \mathbf{B}) \times \mathbf{C} \tag{b}$$

We next wish to prove the two identities

$$\mathbf{A} \times (\mathbf{B} \times \mathbf{C}) = (\mathbf{A} \cdot \mathbf{C})\mathbf{B} - (\mathbf{A} \cdot \mathbf{B})\mathbf{C} \tag{c}$$

$$(\mathbf{A} \times \mathbf{B}) \times \mathbf{C} = (\mathbf{C} \cdot \mathbf{A})\mathbf{B} - (\mathbf{C} \cdot \mathbf{B})\mathbf{A} \tag{d}$$

which are called *reduction formulas.* The proof consists in expanding both sides of Eqs. (*c*) and (*d*) and comparing the results term by term. Using Eq. (*c*) as an example and expanding the left side gives

$$\mathbf{A} \times (\mathbf{B} \times \mathbf{C}) = (x_A\hat{\mathbf{i}} + y_A\hat{\mathbf{j}} + z_A\hat{\mathbf{k}}) \times \begin{vmatrix} \hat{\mathbf{i}} & \hat{\mathbf{j}} & \hat{\mathbf{k}} \\ x_B & y_B & z_B \\ x_C & y_C & z_C \end{vmatrix}$$

$$\begin{aligned} &= (y_Ay_Cx_B - y_Ay_Bx_C - z_Az_Bx_C + z_Ax_Bz_C)\hat{\mathbf{i}} \\ &\quad + (z_Ay_Bz_C - z_Ay_Cz_B - x_Ay_Cx_B + x_Ay_Bx_C)\hat{\mathbf{j}} \\ &\quad + (x_Az_Bx_C - x_Ax_Bz_C - y_Ay_Bz_C + y_Ay_Cz_B)\hat{\mathbf{k}} \end{aligned} \qquad (e)$$

For the right-hand side we have

$$\begin{aligned} (\mathbf{A} \cdot \mathbf{C})\mathbf{B} - (\mathbf{A} \cdot \mathbf{B})\mathbf{C} &= (x_Ax_C + y_Ay_C + z_Az_C)(x_B\hat{\mathbf{i}} + y_B\hat{\mathbf{j}} + z_B\hat{\mathbf{k}}) \\ &\quad - (x_Ax_B + y_Ay_B + z_Az_B)(x_C\hat{\mathbf{i}} + y_C\hat{\mathbf{j}} + z_C\hat{\mathbf{k}}) \end{aligned} \qquad (f)$$

When Eq. (*f*) is expanded, terms like

$$x_Ax_Bx_C\hat{\mathbf{i}}$$

will cancel, and the remaining terms will be identical with Eq. (*e*).

Scalar products Consider the dot product

$$\mathbf{A} \cdot (\mathbf{A} \times \mathbf{B})$$

Here $\mathbf{A} \times \mathbf{B}$ is perpendicular to the plane containing $\mathbf{A}$ and $\mathbf{B}$ and hence perpendicular to the vector $\mathbf{A}$. Therefore, from the definition of the scalar product

$$\mathbf{A} \cdot (\mathbf{A} \times \mathbf{B}) = A|\mathbf{A} \times \mathbf{B}| \cos 90° = 0 \qquad (g)$$

where $|\mathbf{A} \times \mathbf{B}|$ is used to designate the magnitude of the vector $\mathbf{A} \times \mathbf{B}$. In a similar manner it follows that

$$(\mathbf{A} \times \mathbf{B}) \cdot \mathbf{B} = 0 \qquad (h)$$

Still another useful identity involves the dot product of one vector with the cross product of two others. In particular, we wish to prove that

$$\mathbf{A} \cdot (\mathbf{B} \times \mathbf{C}) = \mathbf{B} \cdot (\mathbf{C} \times \mathbf{A}) = \mathbf{C} \cdot (\mathbf{A} \times \mathbf{B}) \qquad (i)$$

To prove this, expand each side and then compare the results term by term. The left-hand side may be expanded to

$$\mathbf{A} \cdot (\mathbf{B} \times \mathbf{C}) = (x_A\hat{\mathbf{i}} + y_A\hat{\mathbf{j}} + z_A\hat{\mathbf{k}}) \cdot \begin{vmatrix} \hat{\mathbf{i}} & \hat{\mathbf{j}} & \hat{\mathbf{k}} \\ x_B & y_B & z_B \\ x_C & y_C & z_C \end{vmatrix}$$

$$\begin{aligned} &= x_Ay_Bz_C - x_Ay_Cz_B + y_Az_Bx_C - y_Ax_Bz_C \\ &\quad + z_Ay_Cx_B - z_Ay_Bx_C \end{aligned} \qquad (j)$$

The terms $\mathbf{B} \cdot (\mathbf{C} \times \mathbf{A})$ and $\mathbf{C} \cdot (\mathbf{A} \times \mathbf{B})$ will yield the same result when expanded.

The term $\mathbf{A} \cdot (\mathbf{B} \times \mathbf{C})$ can be easily interpreted geometrically. We have already noted that the magnitude $|\mathbf{B} \times \mathbf{C}|$ is the area of a parallelogram formed by the edges $\mathbf{B}$ and $\mathbf{C}$. Thus if the vector $\mathbf{A}$ forms the third edge of a parallelepiped, then $\mathbf{A} \cdot (\mathbf{B} \times \mathbf{C})$ is the volume of the parallelepiped.

We note in Eq. (i) that the cyclic order, ABC, BCA, CAB, is preserved. If this order is reversed, then the result changes sign. Thus

$$\mathbf{A} \cdot (\mathbf{B} \times \mathbf{C}) = -\mathbf{A} \cdot (\mathbf{C} \times \mathbf{B}) \tag{k}$$

Next, consider a dot product of two cross products:

$$(\mathbf{A} \times \mathbf{B}) \cdot (\mathbf{C} \times \mathbf{D}) = (\mathbf{A} \cdot \mathbf{C})(\mathbf{B} \cdot \mathbf{D}) - (\mathbf{A} \cdot \mathbf{D})(\mathbf{B} \cdot \mathbf{C}) \tag{l}$$

From Eq. (i) we can write

$$(\mathbf{A} \times \mathbf{B}) \cdot (\mathbf{C} \times \mathbf{D}) = [\mathbf{D} \times (\mathbf{A} \times \mathbf{B})] \cdot \mathbf{C}$$

Then, from Eq. (c),

$$\mathbf{D} \times (\mathbf{A} \times \mathbf{B}) = (\mathbf{B} \cdot \mathbf{D})\mathbf{A} - (\mathbf{A} \cdot \mathbf{D})\mathbf{B} \tag{m}$$

Dotting Eq. (m) through with $\mathbf{C}$ then yields Eq. (l).

Table 2-1 summarizes many of the important relations developed in this chapter.[1]

[1] For a thorough discussion of vector algebra, see Lawrence E. Goodman and William H. Warner, "Statics," chap. 1, Wadsworth Publishing Company, Inc., Belmont, Calif., 1964. See also Wilfred Kaplan, "Advanced Calculus," chap. 1, Addison-Wesley Publishing Company, Inc., Reading, Mass., 1952.

Table 2-1 Identities

Identity	*Equation number*
$\mathbf{A} \times \mathbf{B} = AB \sin\theta\, \hat{\mathbf{N}}$	(2-11)
$\mathbf{A} \times \mathbf{B} = \begin{vmatrix} \hat{\mathbf{i}} & \hat{\mathbf{j}} & \hat{\mathbf{k}} \\ x_A & y_A & z_A \\ x_B & y_B & z_B \end{vmatrix}$	(2-14)
$\mathbf{A} \cdot \mathbf{B} = AB \cos\theta$	(2-15)
$\mathbf{A} \cdot \mathbf{A} = A^2$	(2-17)
$\mathbf{A} \cdot \mathbf{B} = x_A x_B + y_A y_B + z_A z_B$	(2-16)
$\mathbf{A} \times (\mathbf{B} \times \mathbf{C}) = (\mathbf{A} \cdot \mathbf{C})\mathbf{B} - (\mathbf{A} \cdot \mathbf{B})\mathbf{C}$	(2-18)
$(\mathbf{A} \times \mathbf{B}) \times \mathbf{C} = (\mathbf{C} \cdot \mathbf{A})\mathbf{B} - (\mathbf{C} \cdot \mathbf{B})\mathbf{A}$	(2-19)
$\mathbf{A} \cdot (\mathbf{A} \times \mathbf{B}) = (\mathbf{A} \times \mathbf{B}) \cdot \mathbf{B} = 0$	(2-20)
$\mathbf{A} \cdot (\mathbf{B} \times \mathbf{C}) = \mathbf{B} \cdot (\mathbf{C} \times \mathbf{A}) = \mathbf{C} \cdot (\mathbf{A} \times \mathbf{B})$	(2-21)
$\mathbf{A} \cdot (\mathbf{B} \times \mathbf{C}) = -\mathbf{A} \cdot (\mathbf{C} \times \mathbf{B})$	(2-22)
$(\mathbf{A} \times \mathbf{B}) \cdot (\mathbf{C} \times \mathbf{D}) = (\mathbf{A} \cdot \mathbf{C})(\mathbf{B} \cdot \mathbf{D}) - (\mathbf{A} \cdot \mathbf{D})(\mathbf{B} \cdot \mathbf{C})$	(2-23)

2-17 THE CHACE SOLUTIONS TO THE PLANE VECTOR EQUATION

Chace[1] was the first to obtain explicit solutions in closed form to both the two- and the three-dimensional vector equations. We have already seen that the algebra involved, even in solving the plane vector equation, can become extremely difficult. Chace's approach takes advantage of the brevity of vector notation and yields results which are expressed vectorially.

Since many readers may wish to refer to the original work, we shall employ Chace's nomenclature. The equation to be solved is

$$\boxed{s\hat{\mathbf{s}} + r\hat{\mathbf{r}} + \mathbf{C} = 0} \tag{2-24}$$

where

$$\mathbf{C} = \sum_{i=1}^{n-2} \mathbf{C}_i \tag{2-25}$$

Here $\mathbf{C}$ is the sum of the $n - 2$ vectors in the equation whose magnitudes and directions are known. The remaining two vectors $\mathbf{s} = s\hat{\mathbf{s}}$ and $\mathbf{r} = r\hat{\mathbf{r}}$ have two unknowns consisting of two magnitudes, two directions, or one magnitude and one direction.

If both of the unknowns are in the same vector, the case is trivial, and so we shall proceed at once to case 1 of Sec. 2-12. For this case the two magnitudes s and r are unknown, and so $\hat{\mathbf{s}}$, $\hat{\mathbf{r}}$, and $\mathbf{C}$ are given. The procedure consists in eliminating one of the unknowns by taking the dot product of every vector in Eq. (2-17) by a new vector such that an unknown becomes zero. We can eliminate the vector $\mathbf{r}$ by taking the dot product of every term in Eq. (2-24) with $\hat{\mathbf{r}} \times \hat{\mathbf{k}}$. Thus

$$s\hat{\mathbf{s}} \cdot (\hat{\mathbf{r}} \times \hat{\mathbf{k}}) + r\hat{\mathbf{r}} \cdot (\hat{\mathbf{r}} \times \hat{\mathbf{k}}) + \mathbf{C} \cdot (\hat{\mathbf{r}} \times \hat{\mathbf{k}}) = 0 \tag{a}$$

Then, from Eq. (2-20), $\hat{\mathbf{r}} \cdot (\hat{\mathbf{r}} \times \hat{\mathbf{k}}) = 0$; hence

$$s = -\frac{\mathbf{C} \cdot (\hat{\mathbf{r}} \times \hat{\mathbf{k}})}{\hat{\mathbf{s}} \cdot (\hat{\mathbf{r}} \times \hat{\mathbf{k}})} \tag{b}$$

The sign of this expression can be changed by interchanging the cyclic order of the vectors in the denominator according to Eq. (2-22). Since Eq. (*b*) gives the magnitude of $\mathbf{s}$, we then have

$$\boxed{\mathbf{s} = \frac{\mathbf{C} \cdot (\hat{\mathbf{r}} \times \hat{\mathbf{k}})}{\hat{\mathbf{r}} \cdot (\hat{\mathbf{s}} \times \hat{\mathbf{k}})}\hat{\mathbf{s}}} \tag{2-26}$$

[1] *Op. cit.*, chaps. 2 and 3. See also Milton A. Chace, Mechanism Analysis by Vector Mathematics, *Trans. 7th Conf. Mech.*, Purdue University, Lafayette, Ind., 1962; Vector Analysis of Linkages, *J. Eng. Ind.*, ser. B, vol. 85, pp. 289–297, August, 1963.

In a similar manner, we obtain the unknown vector $\mathbf{r}$ as

$$\mathbf{r} = \frac{\mathbf{C} \cdot (\hat{\mathbf{s}} \times \hat{\mathbf{k}})}{\hat{\mathbf{s}} \cdot (\hat{\mathbf{r}} \times \hat{\mathbf{k}})} \hat{\mathbf{r}} \tag{2-27}$$

For case 2 the unknowns are s and $\hat{\mathbf{r}}$. We begin by taking the dot product of every term in Eq. (2-24) with $(\hat{\mathbf{s}} \times \hat{\mathbf{k}})$, which can be found. The product $s\hat{\mathbf{s}} \cdot (\hat{\mathbf{s}} \times \hat{\mathbf{k}}) = 0$, and we have left

$$r\hat{\mathbf{r}} \cdot (\hat{\mathbf{s}} \times \hat{\mathbf{k}}) + \mathbf{C} \cdot (\hat{\mathbf{s}} \times \hat{\mathbf{k}}) = 0 \tag{c}$$

Now, from the definition of a dot product [Eq. (2-15)],

$$\mathbf{A} \cdot \mathbf{B} = AB \cos \theta$$

we note that

$$r\hat{\mathbf{r}} \cdot (\hat{\mathbf{s}} \times \hat{\mathbf{k}}) = r \cos \theta \tag{d}$$

where θ is the angle between $\hat{\mathbf{r}}$ and $(\hat{\mathbf{s}} \times \hat{\mathbf{k}})$. Thus

$$\cos \theta = \hat{\mathbf{r}} \cdot (\hat{\mathbf{s}} \times \hat{\mathbf{k}}) \tag{e}$$

The vectors $\hat{\mathbf{s}}$ and $(\hat{\mathbf{s}} \times \hat{\mathbf{k}})$ are perpendicular to one another; hence we are free to choose another coordinate system $\hat{\boldsymbol{\lambda}}\hat{\boldsymbol{\mu}}\hat{\boldsymbol{\nu}}$ having the directions

$$\hat{\boldsymbol{\lambda}} = \hat{\mathbf{s}} \times \hat{\mathbf{k}} \qquad \hat{\boldsymbol{\mu}} = \hat{\mathbf{s}}$$

In this reference system, the unknown unit vector $\hat{\mathbf{r}}$ can be written as

$$\hat{\mathbf{r}} = \cos \theta\, (\hat{\mathbf{s}} \times \hat{\mathbf{k}}) + \sin \theta\, \hat{\mathbf{s}} \tag{f}$$

If we now substitute Eq. (e) into (c) and solve for $\cos \theta$, we obtain

$$\cos \theta = -\frac{\mathbf{C} \cdot (\hat{\mathbf{s}} \times \hat{\mathbf{k}})}{r} \tag{g}$$

Then

$$\sin \theta = \pm \sqrt{1 - \cos^2 \theta} = \pm \frac{1}{r} \sqrt{r^2 - [\mathbf{C} \cdot (\hat{\mathbf{s}} \times \hat{\mathbf{k}})]^2} \tag{h}$$

Substituting Eqs. (g) and (h) into (f) and multiplying both sides by the known magnitude r gives

$$\mathbf{r} = -[\mathbf{C} \cdot (\hat{\mathbf{s}} \times \hat{\mathbf{k}})](\hat{\mathbf{s}} \times \hat{\mathbf{k}}) \pm \sqrt{r^2 - [\mathbf{C} \cdot (\hat{\mathbf{s}} \times \hat{\mathbf{k}})]^2}\, \hat{\mathbf{s}} \tag{2-28}$$

To obtain the vector $\mathbf{s}$, we write

$$\mathbf{s} = -(\mathbf{C} + \mathbf{r}) \tag{i}$$

which is sufficient. However, if we substitute $\mathbf{r}$ from Eq. (2-28) into

★ trickey but understandable and helpful

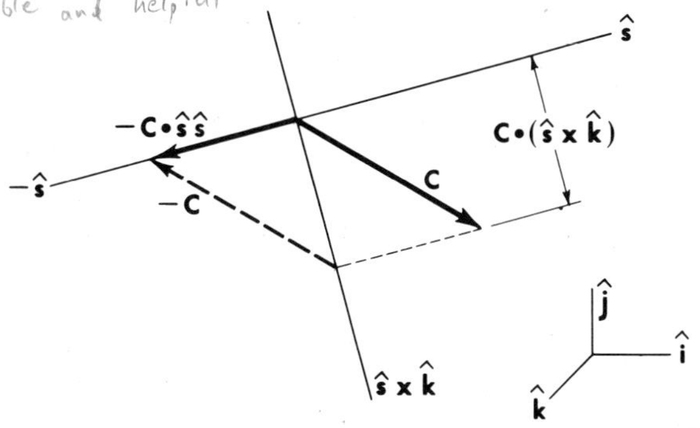

Fig. 2-26

Eq. (i), the first two terms are

$$-\mathbf{C} + [\mathbf{C} \cdot (\hat{\mathbf{s}} \times \hat{\mathbf{k}})](\hat{\mathbf{s}} \times \hat{\mathbf{k}})$$

note that $\bar{C} \cdot (\hat{s} \times \hat{k})$ is not a vector

This expression can be simplified as follows: As shown in Fig. 2-26, the $\hat{\mathbf{s}} \times \hat{\mathbf{k}}$ direction is measured clockwise 90° from the $\hat{\mathbf{s}}$ direction. The magnitude $\mathbf{C} \cdot (\hat{\mathbf{s}} \times \hat{\mathbf{k}})$ is the projection of $\mathbf{C}$ in the $\hat{\mathbf{s}} \times \hat{\mathbf{k}}$ direction. Therefore, when $-\mathbf{C}$ is added to the vector $[\mathbf{C} \cdot (\hat{\mathbf{s}} \times \hat{\mathbf{k}})](\hat{\mathbf{s}} \times \hat{\mathbf{k}})$, the result is a vector of magnitude $\mathbf{C} \cdot \hat{\mathbf{s}}$ in the $-\hat{\mathbf{s}}$ direction. With this substitution, Eq. (i) becomes

s and $\hat{r}$ unknown

★
$$\mathbf{s} = \{-\mathbf{C} \cdot \hat{\mathbf{s}} \mp \sqrt{r^2 - [\mathbf{C} \cdot (\hat{\mathbf{s}} \times \hat{\mathbf{k}})]^2}\}\hat{\mathbf{s}} \qquad (2\text{-}29)$$

Finally, in case 3 the unknowns are the two directions $\hat{\mathbf{r}}$ and $\hat{\mathbf{s}}$. This case is illustrated in Fig. 2-27, where the vector $\mathbf{C}$ is given, together with the magnitudes r and s. The problem is solved by finding the points of

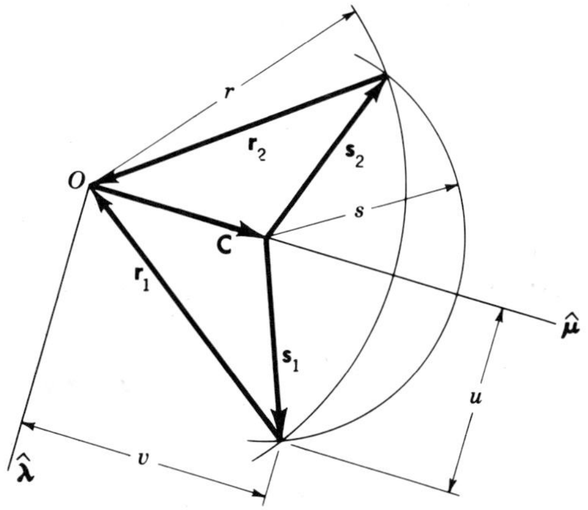

Fig. 2-27

intersection of two circles of radii r and s. We begin by defining a new reference system whose directions are specified by the triad $\hat{\lambda}\hat{u}\hat{v}$ ($\hat{v}$ is not used). As shown in Fig. 2-27, this system is oriented so that

$$\hat{\lambda} = \hat{C} \times \hat{k} \qquad \hat{u} = \hat{C} \qquad \hat{k} \text{ always comes out of the paper} \qquad (j)$$

If the coordinates of the points of intersection in this system are designated as u and v and if we select the first solution, then

$$\mathbf{r} = -u\hat{\lambda} - v\hat{u}$$
$$\mathbf{s} = u\hat{\lambda} + (v - C)\hat{u} \qquad (k)$$

The procedure is quite similar to that of Sec. 2-13. The equation of the circle of radius r is

$$u^2 + v^2 = r^2 \qquad (l)$$

The circle of radius s has the equation

$$u^2 + (v - C)^2 = s^2$$

or

$$u^2 + v^2 - 2vC + C^2 = s^2 \qquad (m)$$

Solving Eq. (l) for u^2 and substituting this into Eq. (m) yields

$$r^2 - 2vC + C^2 = s^2 \qquad (n)$$

and so

$$v = \frac{r^2 - s^2 + C^2}{2C} \qquad (o)$$

Substituting Eq. (o) into (l) and solving for u gives

$$u = \pm \sqrt{r^2 - \left(\frac{r^2 - s^2 + C^2}{2C}\right)^2} \qquad (p)$$

The final step is to substitute these values of u and v into Eqs. (k) and replace $\hat{\lambda}$ and $\hat{u}$ with the directions given by Eq. (j). The results are

$$\mathbf{r} = \mp \sqrt{r^2 - \left(\frac{r^2 - s^2 + C^2}{2C}\right)^2} (\hat{C} \times \hat{k}) - \left(\frac{r^2 - s^2 + C^2}{2C}\right)\hat{C} \qquad (2\text{-}30)$$

$$\mathbf{s} = \pm \sqrt{r^2 - \left(\frac{r^2 - s^2 + C^2}{2C}\right)^2} (\hat{C} \times \hat{k}) + \left(\frac{r^2 - s^2 + C^2}{2C} - C\right)\hat{C}$$
$$(2\text{-}31)$$

This completes the solution to the three cases of the plane vector triangle. The uses of these formulas will be demonstrated many times in the chapters which follow.

$\hat{s}$ and $\hat{r}$ unknown

PROBLEMS

2-1. Find the sum of $\mathbf{A} = 8\underline{/45^\circ}$ and $\mathbf{B} = 6\underline{/-30^\circ}$.

2-2. Find the sum of $\mathbf{A} = 10\underline{/120^\circ}$ and $\mathbf{B} = 6\underline{/-90^\circ}$.

2-3. Find the resultant of $\mathbf{A} = 12\underline{/60^\circ}$, $\mathbf{B} = 10\underline{/150^\circ}$, and $\mathbf{C} = 8\underline{/90^\circ}$.

2-4. Find the resultant of $\mathbf{A} = 6\underline{/-120^\circ}$, $\mathbf{B} = 4\underline{/-135^\circ}$, and $\mathbf{C} = 2\underline{/30^\circ}$.

2-5. Add the following vectors and find the resultant: $\mathbf{A} = 9\underline{/30^\circ}$, $\mathbf{B} = 2\underline{/45^\circ}$, $\mathbf{C} = 6\underline{/90^\circ}$, and $\mathbf{D} = 4\underline{/180^\circ}$.

2-6. Add the vectors $\mathbf{A} = 8\underline{/0^\circ}$, $\mathbf{B} = 10\underline{/150^\circ}$, $\mathbf{C} = 10\underline{/210^\circ}$, and $\mathbf{D} = 6\underline{/315^\circ}$.

2-7. Find the sum of the vectors shown in the figure.

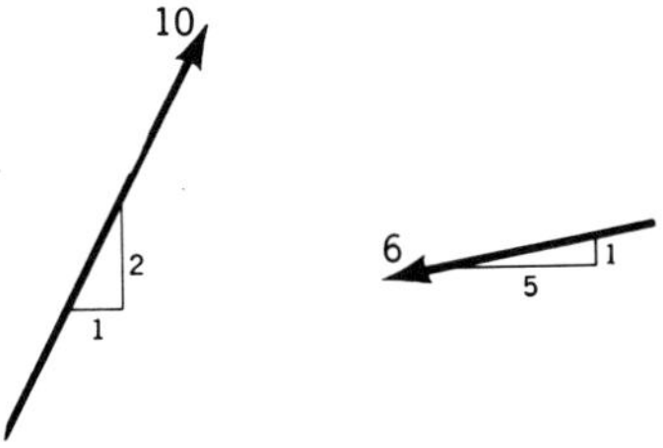

Prob. 2-7

2-8. Add the vectors shown in the figure.

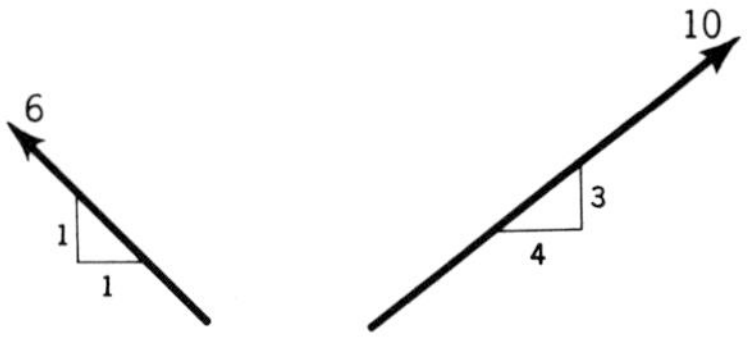

Prob. 2-8

2-9 to **2-12.** Using graphical means, resolve each vector shown in the figure into x and y components. Check the results analytically.

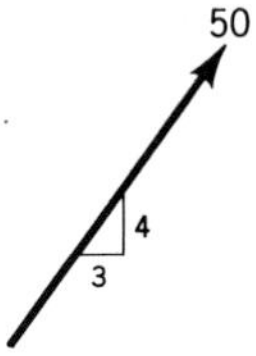

Prob. 2-9

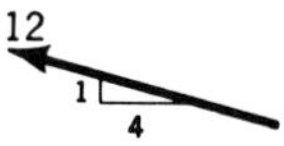

Prob. 2-10

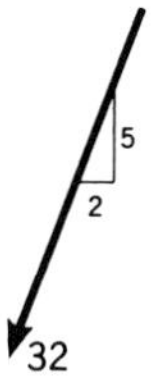

Prob. 2-11

Prob. 2-12

2-13 to **2-16.** Resolve each vector into components in the directions shown, using graphical methods. Check each result, using analytical methods.

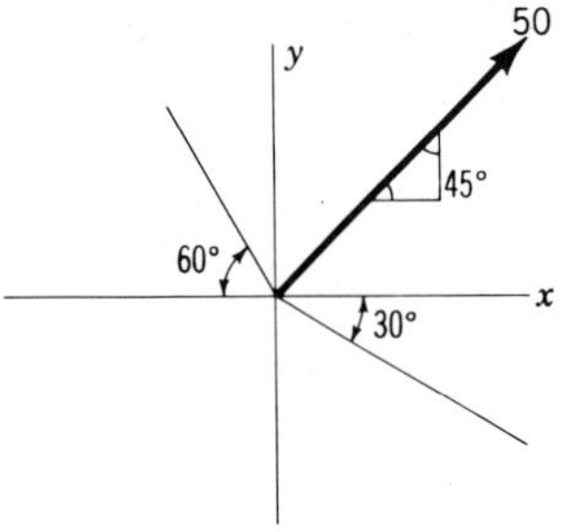

Prob. 2-13

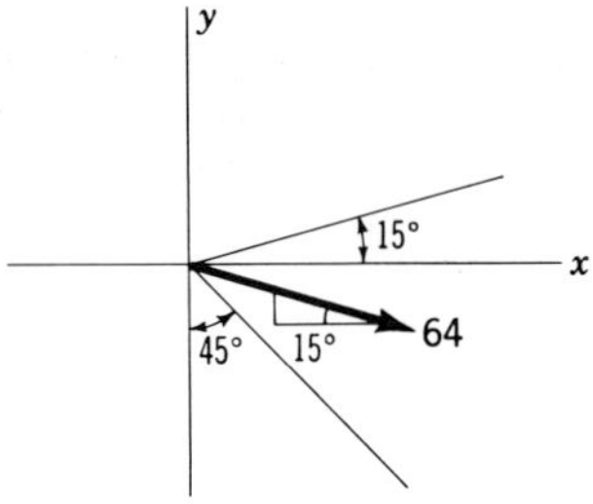

Prob. 2-14

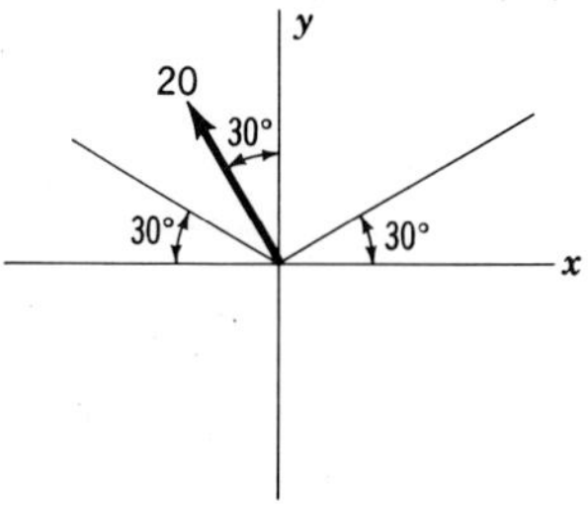

Prob. 2-15

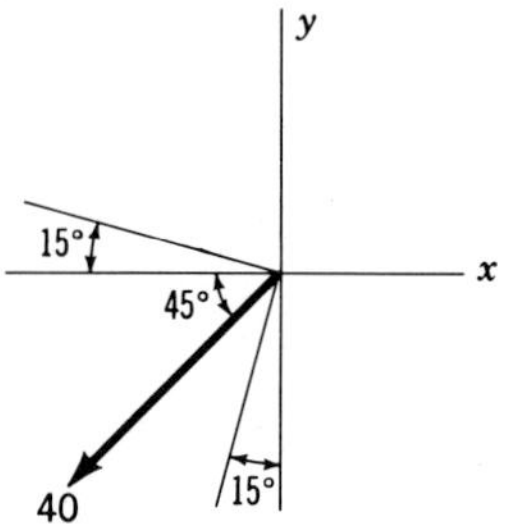

Prob. 2-16

2-17 to **2-20.** Find A − B.

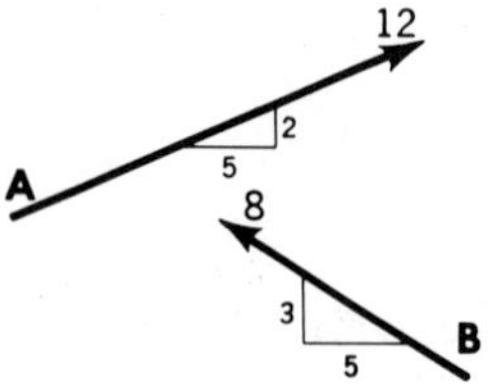

Prob. 2-17

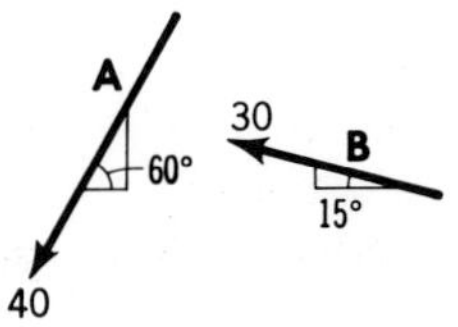

Prob. 2-18

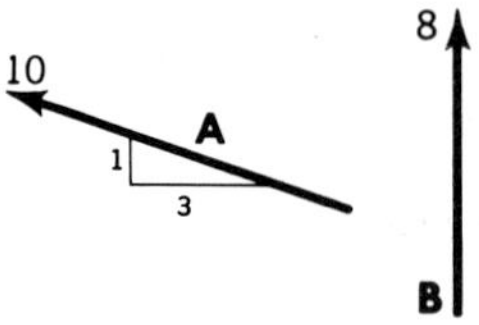

Prob. 2-19

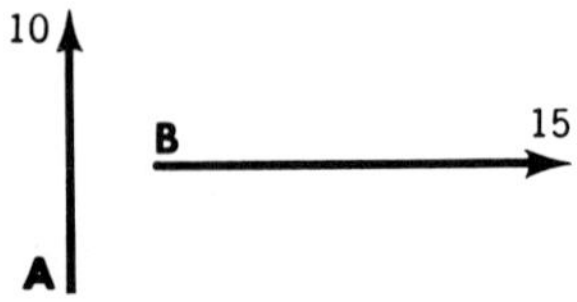

Prob. 2-20

2-21. Express the vector $\mathbf{A} = 19.0\underline{/60^\circ}$ in $\hat{\mathbf{i}}\hat{\mathbf{j}}\hat{\mathbf{k}}$ notation.

2-22. Express the vector $\mathbf{A} = 500\underline{/120^\circ}$ in $\hat{\mathbf{i}}\hat{\mathbf{j}}\hat{\mathbf{k}}$ notation.

2-23. Find the x and y components of the vector $\mathbf{A} = 40\underline{/-50^\circ}$.

2-24. A vector **B** has a magnitude of 16 and passes through $P_1(1,0)$ and $P_2(3,2)$. Express the vector in polar form and also in $\hat{\mathbf{i}}\hat{\mathbf{j}}\hat{\mathbf{k}}$ notation.

2-25. A vector **B** has a magnitude of 64 and passes through two points whose coordinates are $(4,-2,0)$ and $(1,4,0)$. Express the vector in $\hat{\mathbf{i}}\hat{\mathbf{j}}\hat{\mathbf{k}}$ notation and in polar notation.

2-26. Vector **R**, shown in the figure, has a slope of 3 on 7 and an x component whose magnitude is 84. Express **R** in polar form.

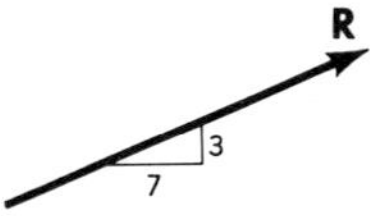

Prob. 2-26

2-27. Resolve the vector $\mathbf{R} = 120\underline{/15^\circ}$ into two components, one passing through the origin of the reference system and point $P_1(2,-1)$, the other passing through the origin and point $P_2(1,4)$.

2-28. If $\mathbf{A} = 10\underline{/-30^\circ}$ and $\mathbf{B} = 8\underline{/45^\circ}$, find $\mathbf{A} + \mathbf{B}$.

2-29. Find the sum of $\mathbf{A} = -4\hat{\mathbf{i}} + 8\hat{\mathbf{j}}$, $\mathbf{B} = 2\hat{\mathbf{i}} - 6\hat{\mathbf{j}}$, and $\mathbf{C} = -5\hat{\mathbf{i}} - 3\hat{\mathbf{j}}$. Express the result in ordinary polar form.

2-30. If $\mathbf{A} = 120\underline{/-150^\circ}$ and $\mathbf{B} = 80\underline{/75^\circ}$, find $\mathbf{B} - \mathbf{A}$.

2-31. Solve analytically the equation $\mathbf{R} = \mathbf{A} + \mathbf{B} + \mathbf{C}$, where **R** has a slope of 3 on 1 in the third quadrant, **A** has a slope of 1 on 2 in the second quadrant, $\mathbf{B} = 6\hat{\mathbf{i}} + \hat{\mathbf{j}}$, and $\mathbf{C} = -4\hat{\mathbf{i}} - 4\hat{\mathbf{j}}$.

2-32. Solve the equation $\mathbf{R} = \mathbf{A} + \mathbf{B} + \mathbf{C}$, where $\mathbf{R} = 8\underline{/-45^\circ}$, $\mathbf{A} = 10\underline{/120^\circ}$, **B** has a slope of 4 on 1 in the fourth quadrant, and **C** has a slope of 1 on 3 in the fourth quadrant.

2-33. If $\mathbf{r} = \mathbf{a} + \mathbf{b}$, with $\mathbf{a} = 2\hat{\mathbf{j}} + 4\hat{\mathbf{k}}$ and $\mathbf{b} = 3\hat{\mathbf{i}} - \hat{\mathbf{k}}$, find r and $\hat{\mathbf{r}}$.

2-34. If $\hat{\mathbf{r}} = l\hat{\mathbf{i}} + m\hat{\mathbf{j}} + n\hat{\mathbf{k}}$, prove that $l^2 + m^2 + n^2 = 1$.

2-35. Find r if $r = (\mathbf{a} \times \mathbf{b}) \cdot \mathbf{c}$ with $\mathbf{a} = \hat{\mathbf{i}} - 2\hat{\mathbf{j}} + \hat{\mathbf{k}}$, $\mathbf{b} = 2\hat{\mathbf{i}} - \hat{\mathbf{k}}$, and $\mathbf{c} = 3\hat{\mathbf{i}} - \hat{\mathbf{j}}$.

2-36. Let $\mathbf{A} = 3\hat{\mathbf{i}} + 4\hat{\mathbf{k}}$ and $\mathbf{B} = -2\hat{\mathbf{j}} - \hat{\mathbf{k}}$. Find A, $\hat{\mathbf{A}}$, B, $\hat{\mathbf{B}}$, $\mathbf{A} \cdot \mathbf{B}$, $\mathbf{A} \times \mathbf{B}$, $3\mathbf{A} + 4\mathbf{B}$, and the angle between **A** and **B**.

2-37. If $\mathbf{A} = \hat{\mathbf{i}} + \hat{\mathbf{j}} + \hat{\mathbf{k}}$ and $\mathbf{B} = -\hat{\mathbf{i}} + 2\hat{\mathbf{j}} - \hat{\mathbf{k}}$, find the projection of each vector on the other.

3
Position and Displacement

3-1 MOTION OF A PARTICLE

The motion of a particle may be defined by the relations

$$x = x(t) \qquad y = y(t) \qquad z = z(t) \tag{a}$$

where x, y, and z are the position coordinates of the point corresponding to the time t. If $x(t)$, $y(t)$, and $z(t)$ are specified, then the position may be found for any time t. This is the general case of the motion of a particle in space and is illustrated in Fig. 3-1.

Example 3-1 Describe the motion of a point whose position with respect to time is given by the equations $x = a \cos 2\pi t$, $y = a \sin 2\pi t$, $z = bt$.

Solution Substitution of values for t from 0 to 2 gives x, y, and z as shown in the following table.

t	x	y	z
0	a	0	0
¼	0	a	$b/4$
½	$-a$	0	$b/2$
¾	0	$-a$	$3b/4$
1	a	0	b
5⁄4	0	a	$5b/4$
3⁄2	$-a$	0	$3b/2$
7⁄4	0	$-a$	$7b/4$
2	a	0	$2b$

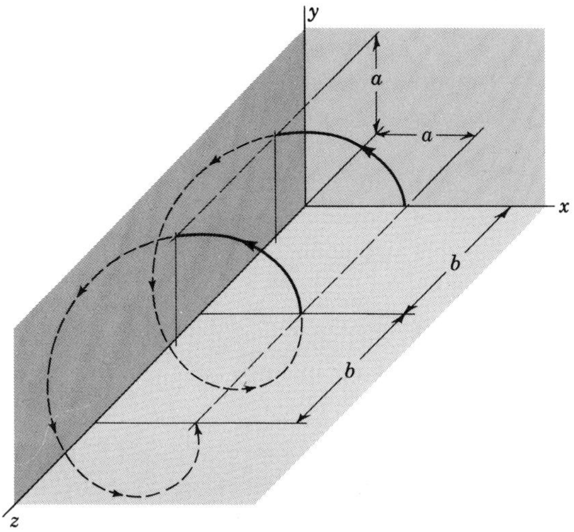

Fig. 3-1 Helical motion of a particle.

As shown in Fig. 3-1, the point moves in *helical motion* with radius a around the positive z axis. Note that if $z = z(t) = 0$, the moving point is confined to the xy plane and the motion is a circle with its center at the origin.

In the preceding paragraph we have used the word *particle* and the word *point* as if they were interchangeable. When the word point is used we have in mind something whose dimensions are zero—something which has no length, no width, and no thickness. When the word particle is used we have in mind something whose dimensions are not important, that is, a body with size and therefore dimensions; but we choose to ignore these because they will have no effect upon the results of the analysis which we are about to make.

If the successive positions of a moving point are connected, a line is obtained. The resulting line has no width because the point has no dimensions. However, the line does have length because the point does occupy successive positions as time changes. The line, representing the successive positions of the point, is called the *path*, or *locus*, of the point.

If three coordinates are necessary to describe the path of a point, then the point is said to move in space and it has *spatial motion*. If the path can be described by only two coordinates, that is, if one of the coordinates is always zero or constant, then the path may be contained by a single plane and the point is said to have *plane motion*. Sometimes it happens that the path of the point may be described by a single coordinate. This means that two of the position coordinates are always zero or constant. In this case the point moves in a straight line and its motion is called *straight-line motion*, or *rectilinear motion*, and it is parallel to the coordinate axis.

In each of the three cases described, it is assumed that the coordinate system is chosen so as to obtain the least number of coordinates necessary to describe the movement of the point. Thus the description of rectilinear motion requires one coordinate. A point whose path is a *plane curve* requires two coordinates. And a point whose path is a *space curve*, sometimes called a *skew curve*, requires three position coordinates.

3-2 MOTION OF A RIGID BODY

When the dimensions of an object become important in studying its motion, then it is necessary to regard the object as a body containing many particles or points. For example, in studying the motion of a missile, under certain conditions we might regard it as a particle, but if its spinning motion became important in the analysis, then it would be necessary to consider the missile as a body made up of a great many particles.

In kinematic analysis it is customary to make one very important assumption in studying the motion of bodies. Here we shall assume that any two particles in a body always remain the same distance apart no matter how large the forces may be which act to change their spacing. This is equivalent to saying that the body is incapable of experiencing deformation—that it is absolutely rigid.

Translation If we imagine the twisting and spiraling of a rigid body in three-dimensional space, it is easy to see that its motion can be very complex. For the present we shall restrict our study to motions which are simple though very important. Let us consider the double-crank-and-slider mechanism of Fig. 3-2. This is composed of the cranks 2 and 4, connecting rods 3 and 5, and the slider 6. All these members may be treated as rigid bodies. If we look at slider 6 we see that, when the mechanism is operating, all particles on 6 have exactly the same motion. This is called *translation*. Translation of a rigid body occurs when each

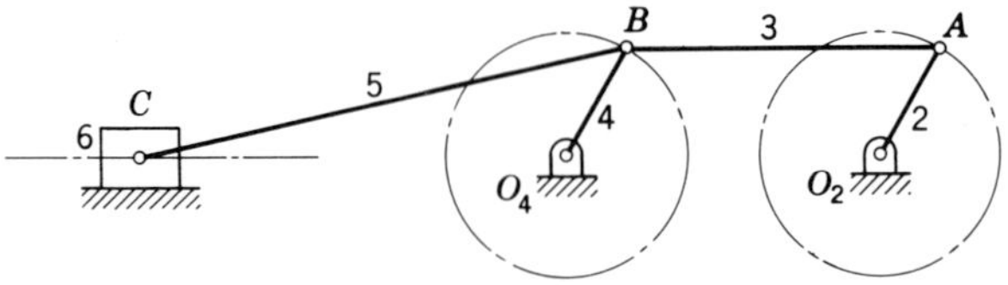

Fig. 3-2 A double-crank-and-slider mechanism.

particle on that body has exactly the same motion as every other particle of which it is composed. Now consider the connecting rod 3. Points A and B on it move in circles both of which have the same radius. As the connecting rod moves, it is always constrained to occupy a position which is parallel to a previous position. Thus all points have exactly the same motion, that is, circular with equal radii. Therefore the connecting rod 3 moves with translation, too, because all particles have exactly the same motion. This enables us to define two kinds of translation. *Rectilinear translation* occurs when all particles of a rigid body have exactly the same motion and that motion is a straight line. The slider 6 has rectilinear translation. On the other hand, if the motion of each particle is exactly the same as every other particle but the motion is curved, which is true of the connecting rod 3, then the body is said to have *curvilinear translation*.

In the translation of a rigid body the motion of a single particle describes the motion of all other particles of the body. This means that when a rigid body translates we need consider only the motion of a single particle in that body.

Rotation If there exists a straight line in a moving body such that all particles of the body which are coincident with this line have a zero velocity with respect to some reference, then the body is said to be in *rotation* about this reference, and the straight line is the *axis of rotation.*

The motion of cranks 2 and 4 about the fixed points O_2 and O_4 is a special case of rotation about a *fixed reference.*

When a rigid body rotates, a straight line constructed between two arbitrary points in the body does not remain parallel to itself. For this reason we must consider the angular changes in position.

Plane motion Connecting rod 5 has a motion which consists of both rotation and translation and is more complex. The path of B_5 is a circle, but that of C_5 is a straight line. Each particle of the rod in between these describes a different path during its motion. However, the path of a single particle of the rod may be contained by a single plane, and the paths of all particles describe planes which are parallel to each other. So we can describe the motion of 5 as *plane motion.* A rigid body moves with plane motion, then, if the paths of particles composing the body form parallel planes. Rectilinear and curvilinear translation are special cases of plane motion.

3-3 DEGREES OF FREEDOM

It is convenient in the description of mechanical systems to utilize the term *degrees of freedom* to express the number of dimensions, or coordi-

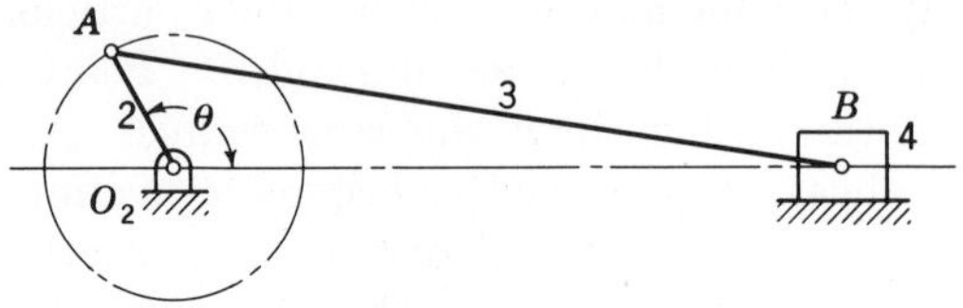

Fig. 3-3 Slider-crank mechanism.

nates, required to specify the position of all its parts; that is, if three coordinates are necessary to describe the position of a mechanical system, then that system is said to have three degrees of freedom or three freedoms.

As an example of degrees of freedom consider the slider-crank mechanism shown in Fig. 3-3. If the crank angle θ is specified, then the position of every part can be found. Thus only one coordinate, θ in this case, is necessary to define the position of all the elements of the slider-crank mechanism. So this mechanism is said to have a single freedom.

When we are dealing with particles, rectilinear motion of the particle represents a single freedom because only one coordinate is necessary to define its position. If the particle moves with plane motion in a circle about a fixed reference point, then the motion is still a single degree of freedom because only one coordinate, an angle, is necessary to define the position.

If the particle is restricted to a plane, but can move anywhere in that plane, then it has two degrees of freedom because two coordinates are necessary to locate its position. A particle free to move anywhere in space would then require three coordinates and would have three degrees of freedom.

A rigid body which moves in space requires six coordinates to describe its position. The body may rotate about three mutually perpendicular axes, and it may translate in the direction of each of these axes. So six coordinates are required to define its position, and consequently there are six degrees of freedom.

A flexible system, such as a belt or rope, has an infinite number of degrees of freedom because there are an infinite number of particles, each of which may have a different motion.

In specifying the position of a system we usually have a choice of coordinates. In Fig. 3-4 the rotating wheel 2 has a pendulum 3 mounted upon it at point A. Two coordinates are necessary to describe the position of this system. We can use θ_2 to define the position of wheel 2 relative to a fixed line on the frame. Then the position of pendulum 3 may be described either by the angle θ_3, between the pendulum axis and a fixed line on the frame, or by the angle θ_3', between the pendulum axis and a line

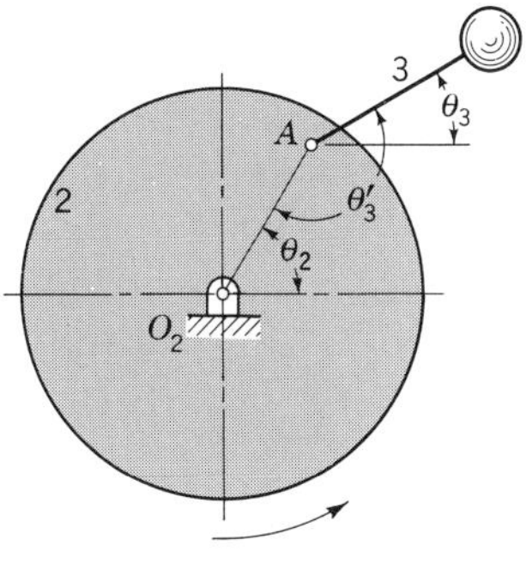

Fig. 3-4

fixed to the wheel. Either pair of coordinates describes the position of the system.

3-4 DISPLACEMENT OF A PARTICLE

A moving particle generates a path. The distance measured along the path between any two positions of the particle is the distance the particle has moved during the time interval. The net change in the location of the particle in the time interval is called the *displacement.*

In Fig. 3-5*a*, a particle R is initially located at R_1 and later at R_2. The displacement of R in this time interval is the directed line segment from R_1 to R_2. It is a vector quantity because it has both magnitude and direction.

It is important to note that displacement is the net position change and ignores the manner in which the change is accomplished. So it does not matter how the particle got from R_1 to R_2. For example, Fig. 3-5*b* shows that the particle R may have traveled from R_1 to R_2 by a very circuitous route. The displacement of R is the vector from R_1 to R_2, designated as $\mathbf{\Delta R}$, regardless of the path taken or of the total distance traveled.

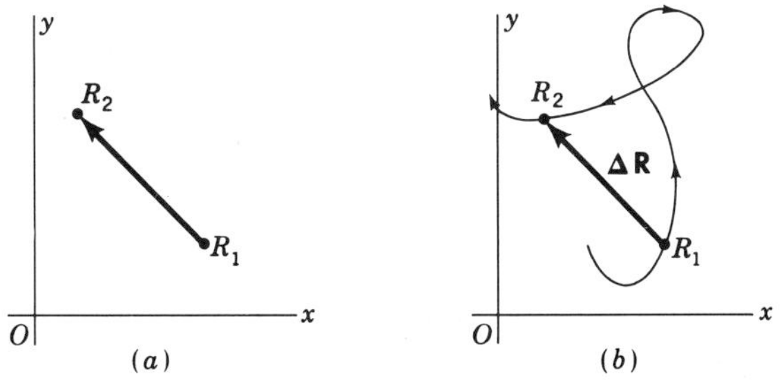

Fig. 3-5 Displacement of a particle.

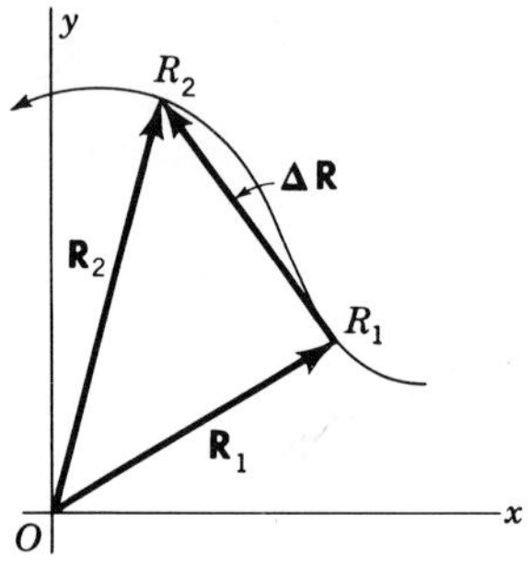

Fig. 3-6

Let us now express the displacement of a particle in quantitative form. Figure 3-6 illustrates the general case of the plane motion of a particle. R_1 and R_2 are two positions of R as it moves on the path. These points are completely defined by the rectangular coordinates $R_1(x_1,y_1)$ and $R_2(x_2,y_2)$ or the polar coordinates $R_1(r_1,\theta_1)$ and $R_2(r_2,\theta_2)$. Or the points may be defined by the position vectors $\mathbf{R}_1$ and $\mathbf{R}_2$. If the position vectors are now placed on the figure, then the vector equation

$$\mathbf{R}_2 = \mathbf{R}_1 + \Delta\mathbf{R} \qquad \text{or} \qquad \Delta\mathbf{R} = \mathbf{R}_2 - \mathbf{R}_1 \tag{3-1}$$

defines the displacement $\Delta\mathbf{R}$.

3-5 DISPLACEMENT OF A RIGID BODY

When a rigid body translates, the motion of a single particle describes the motion of every other particle composing the body. Therefore the displacement of a particle which moves in translation also describes the translation of a rigid body.

We have seen that when a rigid body rotates, a straight line constructed between two separated particles in the body does not remain parallel to itself. Displacement has been defined as the net change in the position of something regardless of the path traveled. It is logical to extend this definition to include angular displacement. Since the measurement of an angle requires the existence of two lines, it is reasonable to define *angular displacement* as the net angular change in the position of a line. Again, notice that the definition includes nothing concerning the route which the line takes in getting from one position to another. Thus in Fig. 3-7, suppose the line PQ is initially located at P_1Q_1 with an angular location of θ_1 with respect to the x axis. Then, at some later instant in time, suppose the line is located at P_2Q_2 and that it makes an angle of θ_2 with the x axis. The line has moved with both translation and rotation. Considering only the rotation, for the moment, we see

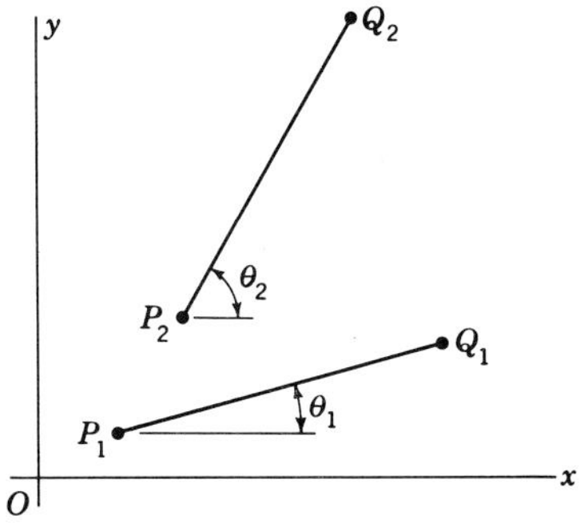

Fig. 3-7 Displacement of a line.

that its angular position has changed from θ_1 to θ_2. Therefore the angular displacement is

$$\Delta\theta = \Delta_2 - \theta_1 \tag{3-2}$$

The magnitude $\Delta\theta$ of this displacement is in radians, and the positive direction is taken as counterclockwise in accordance with the right-hand rule. In the next chapter we shall find that angular velocity, which is the time rate of change of angular displacement, can be treated as a vector. But angular displacements cannot be treated as vectors unless they happen to be infinitesimal ones. The reason for this is that they do not obey the commutative law of vector addition. This means that if several gross displacements are added vectorially, the result will depend upon the order in which they are added.

To illustrate, in Fig. 3-8*a* a plane $ABCO$ is rotated, first, 90° about

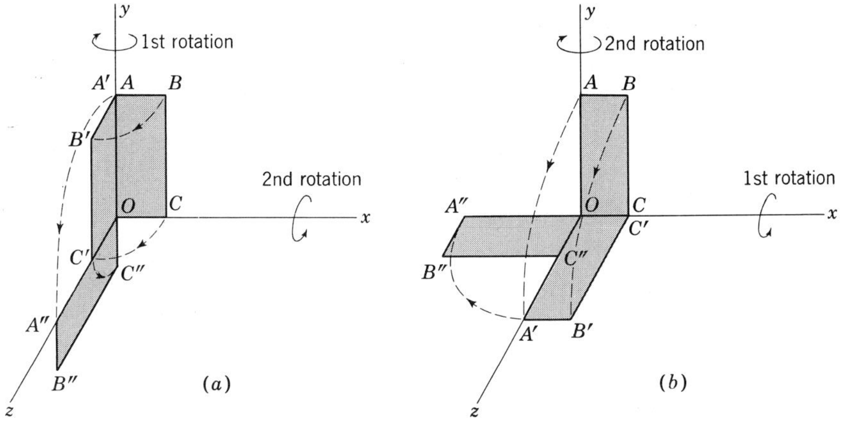

Fig. 3-8 Angular displacements cannot be added vectorially because the result depends upon the order in which they are added.

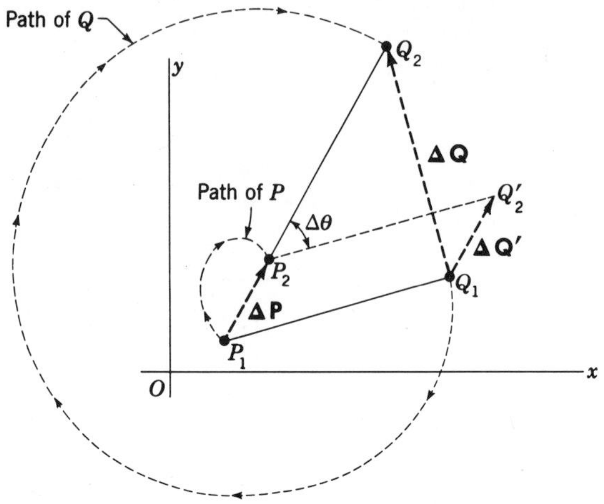

Fig. 3-9 Combined translation and rotation of a line.

the y axis and, second, 90° about the x axis. The final position is $A''B''C''O$, which is in the yz plane. In Fig. 3-8b the plane occupies the same starting position and again is rotated about each axis through an angle of 90° in the same direction. But the first rotation is about the x axis and the second about the y axis. The order of the rotations has been reversed, and the final position of the plane is seen to be in the xz plane instead of in the yz plane as it was after the original rotations.

Infinitesimal angular displacements are commutative; and, in fact, angular displacements which occur about the same axis are also commutative. To avoid confusion, however, we shall usually treat all such motions as scalar quantities.

Figure 3-9 is another illustration of the motion of a line. Here the motion has been separated into the translational component and the rotational component by constructing the line in a fictitious intermediate position. The initial position is P_1Q_1, the final position P_2Q_2. There are an infinite number of paths which P and Q could have taken to arrive at P_2Q_2. One of these possible paths is shown. But no matter which path is taken, the displacement is the net position change. So, in Fig. 3-9, the displacement is the translational component $\mathbf{\Delta P}$ or $\mathbf{\Delta Q'}$, and the rotational component is $\Delta\theta$.

Notice that the total displacement of the line may also be defined by the *two* displacement vectors $\mathbf{\Delta P}$ and $\mathbf{\Delta Q}$.

Example 3-2 A point P moves along the path $y = x^2 - 4$. If it starts from P_1 where $x_1 = 1$, what is its displacement when it goes through the point P_2 ($x_2 = 3$)? What is the angular displacement of the position vector?

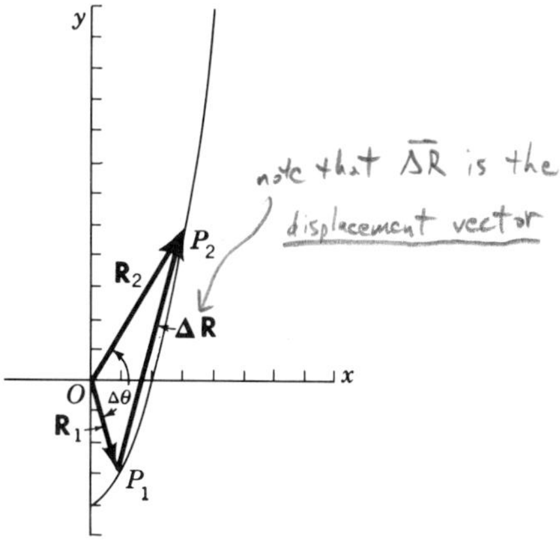

Fig. 3-10

Solution The path is shown in Fig. 3-10. By substituting x_1 and x_2 in the equation, the y coordinates are found to be

$$y_1 = 1^2 - 4 = -3 \qquad y_2 = 3^2 - 4 = 5$$

So the position vectors are

$$\mathbf{R}_1 = \hat{\imath} - 3\hat{\jmath} \qquad \mathbf{R}_2 = 3\hat{\imath} + 5\hat{\jmath}$$

Then

$$\Delta\mathbf{R} = \mathbf{R}_2 - \mathbf{R}_1 = (3\hat{\imath} + 5\hat{\jmath}) - (\hat{\imath} - 3\hat{\jmath}) = 2\hat{\imath} + 8\hat{\jmath}$$

Therefore the displacement vector is

$$\Delta\mathbf{R} = 8.24\underline{/75.95^\circ}$$

Since $\mathbf{R}_1 = \hat{\imath} - 3\hat{\jmath}$ and $\mathbf{R}_2 = 3\hat{\imath} + 5\hat{\jmath}$, we find

$$\theta_1 = -71.55^\circ \qquad \theta_2 = 59.05^\circ$$

Therefore the angular displacement of the position vector is

$$\Delta\theta = \theta_2 - \theta_1 = 59.05 - (-71.55) = 130.60^\circ$$

3-6 RELATIVE DISPLACEMENT

The vectors shown in Fig. 3-11*a* are *position* vectors. $\mathbf{R}_A$ and $\mathbf{R}_B$ describe the positions of A and B in the xy coordinate system. The vector $\mathbf{R}_{BA}$ is a *relative-position* vector; it describes the position of B relative to A in the *same xy* system. These three vectors are related by the equation

$$\mathbf{R}_B = \mathbf{R}_A + \mathbf{R}_{BA} \tag{3-3}$$

Vectors can be arranged in either polygon or polar form. In Fig. 3-11*b* the same three vectors are shown in polar form; it is evident here that $\mathbf{R}_{BA}$ describes the position of B relative to A in the *same reference system*.

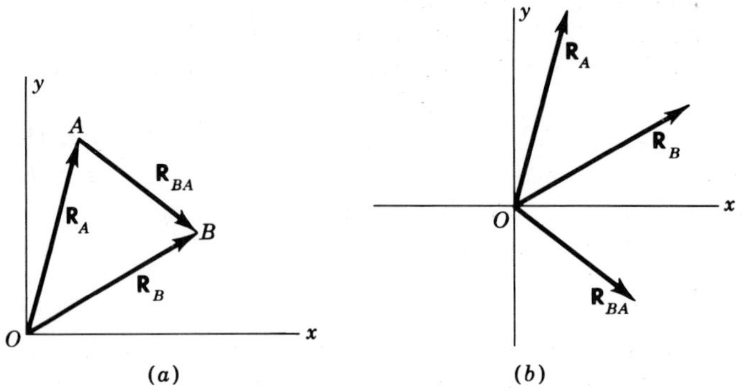

Fig. 3-11

Suppose, now, that A and B undergo simultaneous displacements $\Delta\mathbf{R}_A$ and $\Delta\mathbf{R}_B$ in the same interval of time. How are these displacements related?

If A and B are fixed in the same rigid body then the body may translate, it may rotate, or it may have a motion consisting of both translation and rotation. These three cases are shown in Fig. 3-12. For convenience we fasten a moving $x'y'$ system to the body with origin at A.

For the case of pure translation the $x'y'$ system moves from position 1 to position 2 (Fig. 3-12a). The displacement of A and the displacement of B are identical; hence

$$\Delta\mathbf{R}_B = \Delta\mathbf{R}_A \tag{a}$$

Also $\mathbf{R}_{B_2A} = \mathbf{R}_{B_1A}$ and so $\Delta\mathbf{R}_{BA}$, the displacement of B relative to A, is zero. The equation

$$\Delta\mathbf{R}_B = \Delta\mathbf{R}_A + \Delta\mathbf{R}_{BA} \tag{3-4}$$

applies, but $\Delta\mathbf{R}_{BA} = 0$.

Now let the $x'y'$ system have a motion of pure rotation (Fig. 3-12b). It is evident that point B has an absolute displacement $\Delta\mathbf{R}_B$ due to the fact that B is fixed in the rotating $x'y'$ system. However, the relative-position vector $\mathbf{R}_{BA}$ is also fixed in the $x'y'$ system; hence

$$\mathbf{R}_{B_2A} = \mathbf{R}_{B_1A} \tag{b}$$

which means that the displacement of B relative to A is again zero. How, then, can we describe the displacement of B? The answer lies in describing the motion of B relative to a *nonrotating coordinate system*. Thus, if the positions of B_1 and B_2 are *both* referred to the $x_1'y_1'$ system, then

$$\Delta\mathbf{R}_B = \mathbf{R}_{B_2A} - \mathbf{R}_{B_1A} = \Delta\mathbf{R}_{BA} \tag{c}$$

Equation (3-4) still holds, but now $\Delta\mathbf{R}_A = 0$.

Equation (c) demonstrates a fundamental proposition: *When both A and B are fixed in the same rigid body, the displacement of B relative to A, $\Delta\mathbf{R}_{BA}$, means the displacement of B in a nonrotating coordinate system whose origin is at A.* and whose origin moves with A

The case in which the $x'y'$ system containing A and B has both translation and rotation is shown in Fig. 3-12c. The displacement of B is made up of the two components $\Delta\mathbf{R}_A$, the translational component, and $\Delta\mathbf{R}_{BA}$, the rotational component. Thus duh!!!

$$\Delta\mathbf{R}_B = \Delta\mathbf{R}_A + \Delta\mathbf{R}_{BA} \tag{3-4}$$

Let us proceed now to the more general case in which we have a point with its motion defined relative to another moving body. Let B and A, as before, be fixed to the same rigid body and define a third point P instantaneously coincident with point B in the B_1 position. In Fig. 3-13a the $x'y'$ system with both A and B fixed to it moves with pure translation from position 1 to position 2. At the same time point P moves along the

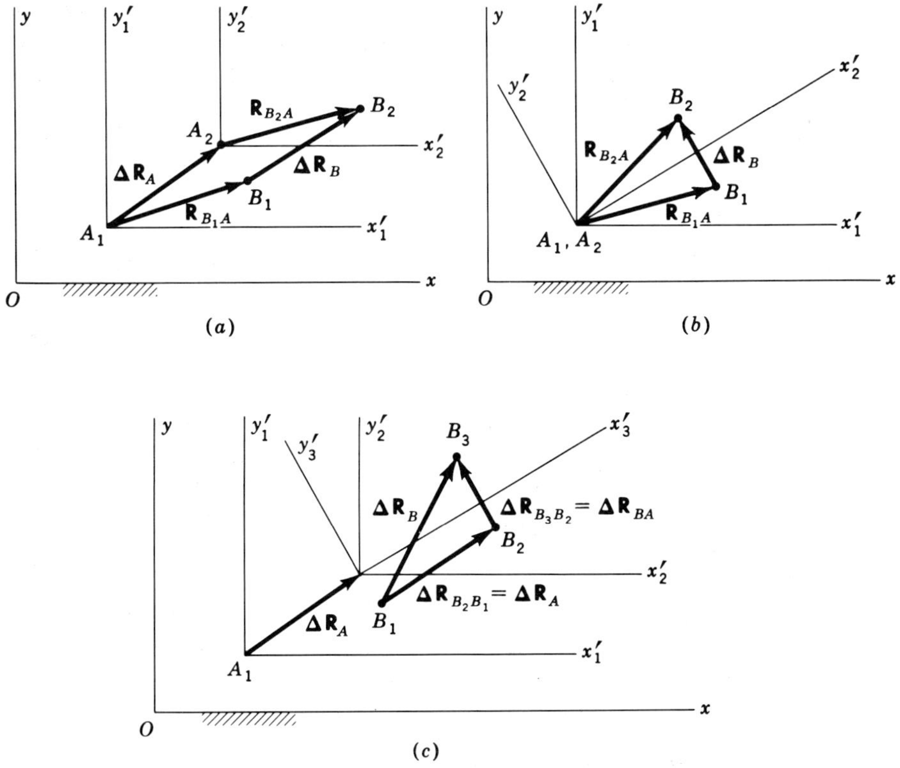

Fig. 3-12 (a) Translation; (b) rotation; (c) translation and rotation.

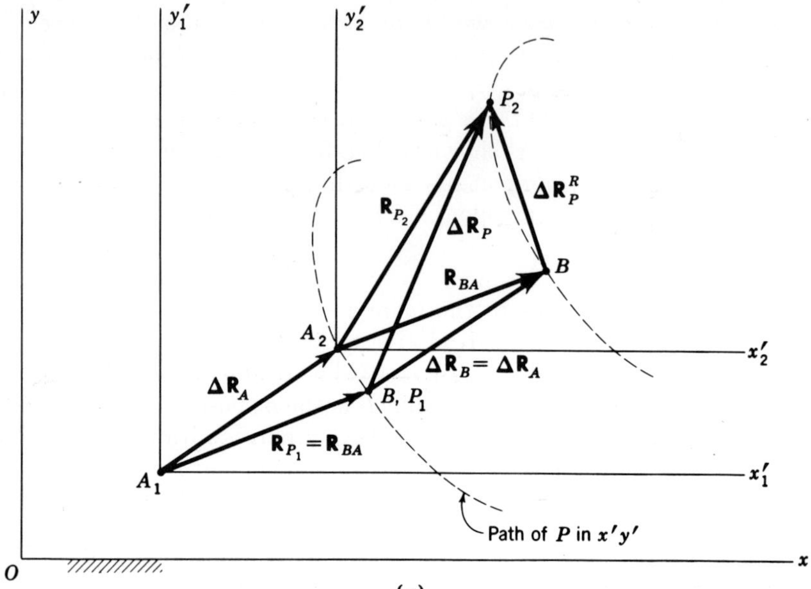

y
y_1'
y_2'
P_2
$\Delta \mathbf{R}_P^R$
$\mathbf{R}_{P_2}$
$\Delta \mathbf{R}_P$
B
$\mathbf{R}_{BA}$
A_2
x_2'
$\Delta \mathbf{R}_B = \Delta \mathbf{R}_A$
$\Delta \mathbf{R}_A$
B, P_1
$\mathbf{R}_{P_1} = \mathbf{R}_{BA}$
A_1
x_1'
Path of P in $x'y'$
O
x

(a)

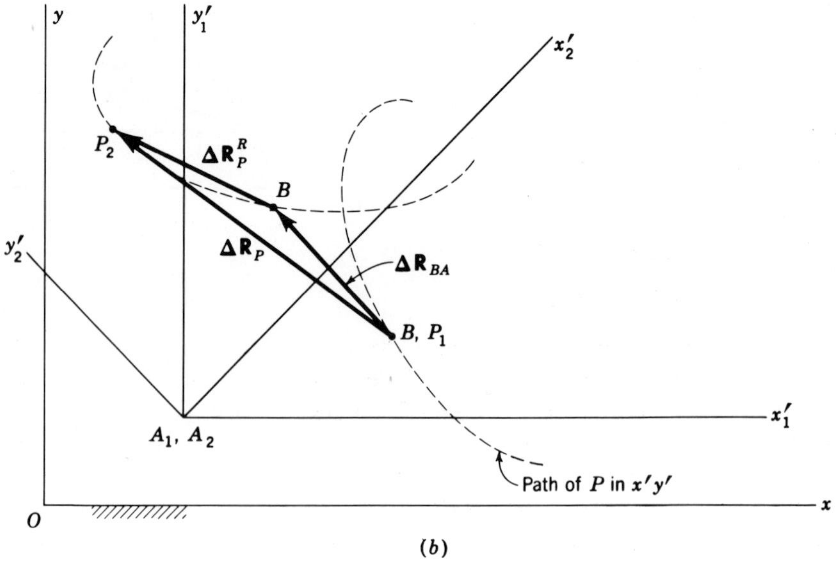

y
y_1'
x_2'
P_2
$\Delta \mathbf{R}_P^R$
B
y_2'
$\Delta \mathbf{R}_P$
$\Delta \mathbf{R}_{BA}$
B, P_1
A_1, A_2
x_1'
Path of P in $x'y'$
O
x

(b)

Fig. 3-13

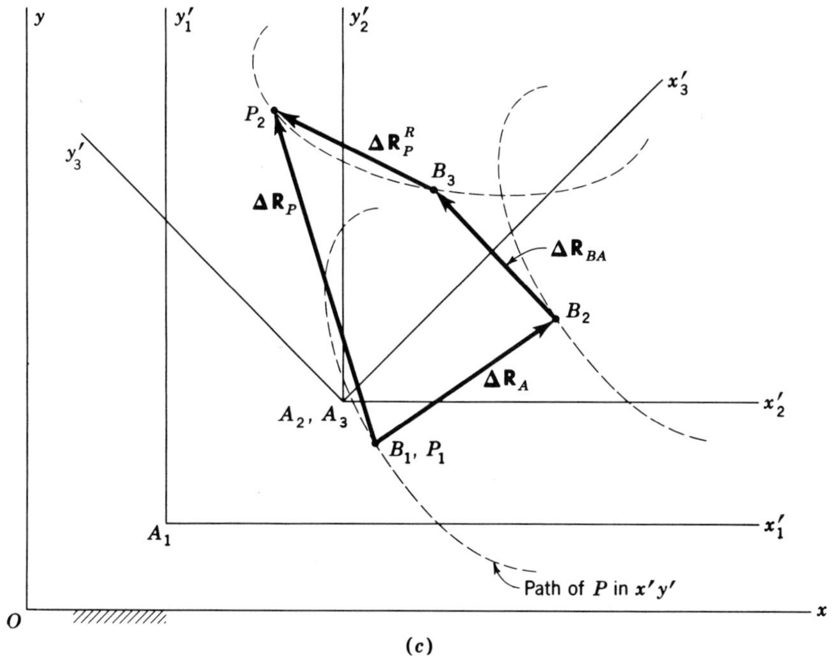

Fig. 3-13 *(Continued)*

path from P_1 to P_2. An observer attached to the $x'y'$ system would report the displacement of P as

$$\Delta\mathbf{R}_P{}^R = \mathbf{R}_{P_2} - \mathbf{R}_{P_1} \tag{d}$$

where the superscript R indicates that this is a *true relative displacement.* The absolute displacement of P is then

$$\Delta\mathbf{R}_P = \Delta\mathbf{R}_A + \Delta\mathbf{R}_P{}^R \tag{e}$$

where P moves relative to a translating system.

Next, consider the motion of P in a reference frame having pure rotation (Fig. 3-13*b*). In this case $\Delta\mathbf{R}_A$ is zero but the absolute displacement of P still is made up of two components. One of these, $\Delta\mathbf{R}_{BA}$, is the displacement of the coincident point B due to the rotation of the system containing B; and the other, $\Delta\mathbf{R}_P{}^R$, is the true relative displacement of P in the rotating $x'y'$ system. Thus, the absolute displacement of P is

$$\Delta\mathbf{R}_P = \Delta\mathbf{R}_{BA} + \Delta\mathbf{R}_P{}^R \tag{f}$$

The most general problem of all is that in which P moves relative to a system having both translation and rotation (Fig. 3-13*c*). The figure shows the $x'y'$ system translating from position 1 to position 2 and rotating from position 2 to position 3. The figure also shows that the displacement of P now consists of three components:

$\mathbf{\Delta R}_A$ = translational component, due to translation of $x'y'$ from position 1 to position 2

$\mathbf{\Delta R}_{BA}$ = rotational component, relative displacement of coincident point B fixed in $x'y'$

$\mathbf{\Delta R}_{P^R}$ = true relative displacement of P in $x'y'$

Thus

$$\mathbf{\Delta R}_P = \mathbf{\Delta R}_A + \mathbf{\Delta R}_{BA} + \mathbf{\Delta R}_{P^R} \tag{3-5}$$

If we substitute from Eq. (3-4) for the displacement of the coincident point B, in this equation, we obtain

$$\mathbf{\Delta R}_P = \mathbf{\Delta R}_B + \mathbf{\Delta R}_{P^R} \tag{3-6}$$

Equation (3-6) states that *the absolute displacement of a point P, which moves relative to a moving reference frame, is equal to the absolute displacement of a point B, fixed to the reference frame and coincident with P at its initial position, plus the displacement of P relative to the moving frame.*

3-7 POSITION

One of the first problems which arises in the analysis of mechanisms is that of locating the various points of interest. If the analysis is to be carried out by graphical means, then the problem is trivial. But it is not so trivial if the solution is to be obtained by analytical or computer methods.

To illustrate, let us begin with the slider-crank mechanism shown in Fig. 3-14. Here, link 2 is a crank which is constrained to rotate about a fixed pivot at O_2. Link 3 is the connecting rod and link 4 the slider. The problem is to determine the locations of A and B when the link lengths r_2 and r_3 are given together with the position θ_2 of the crank or the location

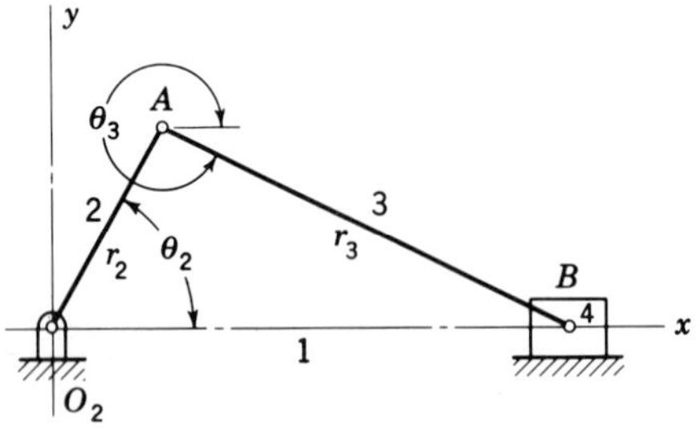

Fig. 3-14

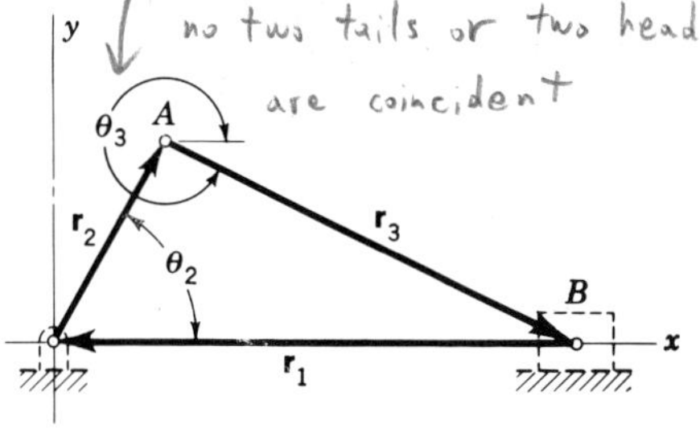

Fig. 3-15

For Chace soln, draw vectors so that no two tails or two heads are coincident

of the slider x_B. The problem can be solved by using a vector or an analytical approach.

If we assume that θ_2 is given, then the analytical solution is started by writing

$$y_A = r_2 \sin \theta_2 \qquad x_A = r_2 \cos \theta_2 \tag{3-7}$$

which defines the position of point A. Next, note that

$$r_2 \sin \theta_2 = r_3 \sin \theta_3$$

so that

$$\sin \theta_3 = \frac{r_2}{r_3} \sin \theta_2 \Rightarrow \Theta_3 \tag{a}$$

From the geometry of Fig. 3-14 we see that

$$x_B = r_2 \cos \theta_2 + r_3 \cos \theta_3 \tag{b}$$

where the minus sign in the second term results because of our choice of the origin for θ_3. Next, since

$$\cos \theta_3 = \pm \sqrt{1 - \sin^2 \theta_3}$$

we have, from Eq. (a),

$$\cos \theta_3 = -\frac{1}{r_3} \sqrt{r_3^2 - r_2^2 \sin^2 \theta_2} \Rightarrow \Theta_3 \tag{c}$$

From Eqs. (b) and (c), the position of B is

$$x_B = r_2 \cos \theta_2 + \sqrt{r_3^2 - r_2^2 \sin^2 \theta_2} \Rightarrow x_B, \ y_B = 0 \tag{3-8}$$

NOW In the vector approach, the links are replaced by vectors, as shown in Fig. 3-15. We can then write

$$r_1 \hat{r}_1 + r_2 \hat{r}_2 + r_3 \hat{r}_3 = 0 \tag{d}$$

With θ_2 given, the unknowns in this equation are the magnitude r_1 and the direction $\hat{r}_3$. This is a good example of the need for the Chace equations.

The solution of Eq. (*d*) corresponds to case 2 of Chap. 2 [Eqs. (2-28) and (2-29)]. By making the appropriate substitutions, the solution to Eq. (*d*) is

$$\mathbf{r}_3 = -[\mathbf{r}_2 \cdot (\hat{\mathbf{r}}_1 \times \hat{\mathbf{k}})](\hat{\mathbf{r}}_1 \times \hat{\mathbf{k}}) - \sqrt{r_3^2 - [\mathbf{r}_2 \cdot (\hat{\mathbf{r}}_1 \times \hat{\mathbf{k}})]^2}\, \hat{\mathbf{r}}_1 \tag{3-9}$$

$$\mathbf{r}_1 = \{-\mathbf{r}_2 \cdot \hat{\mathbf{r}}_1 + \sqrt{r_3^2 - [\mathbf{r}_2 \cdot (\hat{\mathbf{r}}_1 \times \hat{\mathbf{k}})]^2}\}\, \hat{\mathbf{r}}_1 \tag{3-10}$$

Example 3-3 Use the Chace equations to find the position of the slider of Fig. 3-15 with $r_2 = 1$ in., $r_3 = 3$ in., and $\theta_2 = 150°$.

Solution Placing the given information in vector form, we have

$$\hat{\mathbf{r}}_1 = -\hat{\mathbf{i}} \qquad r_3 = 3$$

$$\mathbf{r}_2 = 1 \cos(150)\hat{\mathbf{i}} + 1 \sin(150)\hat{\mathbf{j}} = -0.866\hat{\mathbf{i}} + 0.5\hat{\mathbf{j}}$$

It is also convenient to calculate the unit vector

$$\hat{\mathbf{r}}_1 \times \hat{\mathbf{k}} = \begin{vmatrix} \hat{\mathbf{i}} & \hat{\mathbf{j}} & \hat{\mathbf{k}} \\ -1 & 0 & 0 \\ 0 & 0 & 1 \end{vmatrix} = \hat{\mathbf{j}}$$

Substituting into Eq. (3-9) gives

$$\mathbf{r}_3 = -[(-0.866\hat{\mathbf{i}} + 0.5\hat{\mathbf{j}}) \cdot \hat{\mathbf{j}}]\hat{\mathbf{j}} - \sqrt{(3)^2 - [(-0.866\hat{\mathbf{i}} + 0.5\hat{\mathbf{j}}) \cdot \hat{\mathbf{j}}]^2}\,(-\hat{\mathbf{i}})$$
$$= 2.96\hat{\mathbf{i}} - 0.5\hat{\mathbf{j}}$$

Hence

$$r_3 = \sqrt{(2.96)^2 + (0.5)^2} = 3 \text{ in.}$$

which checks. The direction is

$$\theta_3 = \tan^{-1}\frac{-0.5}{2.96} = -9.6°$$

To obtain the position of the slider, use Eq. (3-10). Thus

$$\mathbf{r}_1 = \{-(-0.866\hat{\mathbf{i}} + 0.5\hat{\mathbf{j}}) \cdot (-\hat{\mathbf{i}}) + \sqrt{(3)^2 - [(-0.866\hat{\mathbf{i}} + 0.5\hat{\mathbf{j}}) \cdot \hat{\mathbf{j}}]^2}\}(-\hat{\mathbf{i}})$$
$$= -2.094\hat{\mathbf{i}}$$

From Fig. 3-15 note that the magnitude of this vector is x_B; hence

$$x_B = 2.094 \text{ in.}$$

The four-bar linkage shown in Fig. 3-16 is called the *crank-and-rocker* mechanism. Thus link 2, which is the crank, can rotate in a full circle; but the rocker, link 4, can only oscillate. We shall generally follow the accepted practice of designating the frame or fixed link as link 1. Link 3, in Fig. 3-16, is called the *coupler* or *connecting rod*. With the four-bar linkage the position problem generally consists in finding the positions of the coupler and output link or rocker when the dimensions of all the members are given together with the crank position.

To obtain the analytical solution we designate l as the distance

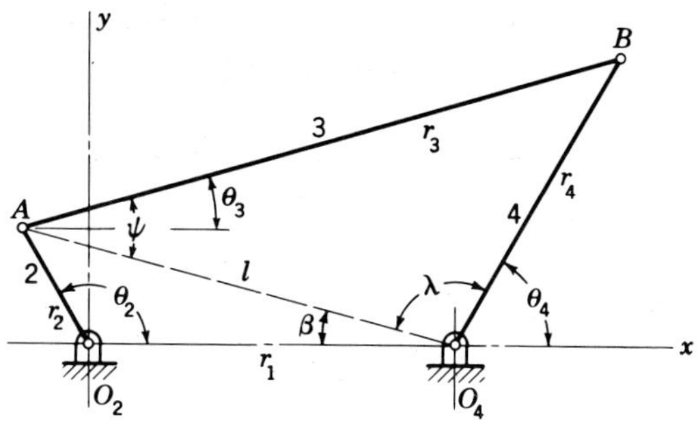

Fig. 3-16

AO_4 in Fig. 3-16. The cosine law can then be written twice for each of the two triangles O_2AO_4 and ABO_4. In terms of the angles and link lengths shown in the figure we then have

$$l = \sqrt{r_1^2 + r_2^2 - 2r_1r_2 \cos \theta_2} \tag{3-11}$$

$$\beta = \cos^{-1} \frac{r_1^2 + l^2 - r_2^2}{2r_1 l} \tag{3-12}$$

$$\psi = \cos^{-1} \frac{r_3^2 + l^2 - r_4^2}{2r_3 l} \tag{3-13}$$

$$\lambda = \cos^{-1} \frac{r_4^2 + l^2 - r_3^2}{2r_4 l} \tag{3-14}$$

There will generally be two values of λ corresponding to each value of θ_2. If θ_2 is in the range $0 \leq \theta_2 \leq \pi$, the unknown directions are taken as

$$\theta_3 = \psi - \beta \tag{3-15}$$

$$\theta_4 = \pi - \lambda - \beta \tag{3-16}$$

But if θ_2 is in the range $\pi \leq \theta_2 \leq 2\pi$, then

$$\theta_3 = \psi + \beta \tag{3-17}$$

$$\theta_4 = \pi - \lambda + \beta \tag{3-18}$$

To solve the same problem by using the Chace approach, we again replace each link with a vector, as shown in Fig. 3-17. If θ_2 is given, we first form the resultant vector

$$\mathbf{R} = \mathbf{r}_1 + \mathbf{r}_2 \tag{e}$$

The vector equation can then be written as

$$\mathbf{R} + r_3\hat{\mathbf{r}}_3 + r_4\hat{\mathbf{r}}_4 = 0 \tag{f}$$

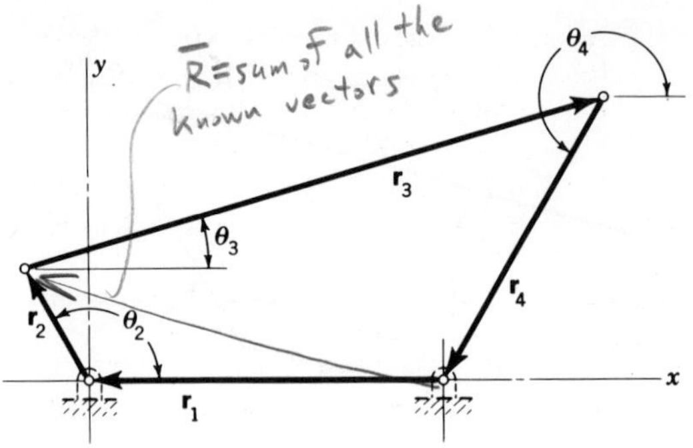

Fig. 3-17

where the two directions $\hat{\mathbf{r}}_3$ and $\hat{\mathbf{r}}_4$ are unknown. This is case 3 of Chap. 2, and the solutions are given by Eqs. (2-30) and (2-31). Substituting the nomenclature of Fig. 3-17 gives

$$\mathbf{r}_3 = \mp\sqrt{r_3{}^2 - \left(\frac{r_3{}^2 - r_4{}^2 + R^2}{2R}\right)^2}\,(\hat{\mathbf{R}} \times \hat{\mathbf{k}}) - \left(\frac{r_3{}^2 - r_4{}^2 + R^2}{2R}\right)\hat{\mathbf{R}} \tag{3-19}$$

$$\mathbf{r}_4 = \pm\sqrt{r_3{}^2 - \left(\frac{r_3{}^2 - r_4{}^2 + R^2}{2R}\right)^2}\,(\hat{\mathbf{R}} \times \hat{\mathbf{k}}) + \left(\frac{r_3{}^2 - r_4{}^2 + R^2}{2R} - R\right)\hat{\mathbf{R}} \tag{3-20}$$

The upper set of signs gives the solution for the crossed linkage. The lower set therefore applies to the linkage of Fig. 3-17.

PROBLEMS

3-1. Describe the path of a point which is known to move according to the equations $x = at \cos 2\pi t$, $y = at \sin 2\pi t$, $z = 0$.

3-2. If point A moves on the path of Prob. 3-1, find its displacement between $t = 2$ and $t = 2.5$.

3-3. A point moves along the path $y = x^2 + x - 16$. Find the displacement between the positions $x_1 = 2$ and $x_2 = 4$. What is the angular displacement of the position vector?

3-4. The path of a point is defined by the equation $y = 60 - x^3/3$. What is the displacement of a point whose motion begins at $x = 0$ and ends at $x = 3$?

3-5. A point moves from P to Q on the path $y = 2x^2 - 28$. Find the distance from P to Q and the direction if P is located at $x = 4$ and Q at $x = -3$.

3-6. The position vector of a point is given by the equation $\mathbf{R} = 100e^{i2\pi t}$. What is the path of the point? Determine the displacement of the point between the positions defined by $t = 0.10$ and $t = 0.40$. What is the angular displacement of the position vector between these points?

3-7. The equation $\mathbf{R} = (t^2 + 4)e^{-j\pi t/10}$ defines the path of a point. In which direction is the position vector rotating? Where is the point located when $t = 0$? What other value can t have in order that the direction of the position vector be the same as it is when $t = 0$? What is the displacement corresponding to these two values of t?

3-8. Using the equation of Prob. 3-7 to express the path of a point, find the displacement of the point and the angular displacement of the position vector corresponding to the positions $t_1 = 12$ and $t_2 = 16$.

3-9. The location of a point P is defined by the equation $\mathbf{R} = (4t + 2)e^{j\pi t^2/30}$ where t is the time in seconds. The point initiated its motion when $t = 0$.

(*a*) Find the position vector describing the initial position of the point.
(*b*) Find the position vector describing the location of the point 3 sec later.
(*c*) What is the displacement after 3 sec?
(*d*) Find the angular displacement of the position vector after 3 sec.

3-10. Link 2 in the figure moves according to the relation $\theta = \pi t/4$. A block 3 slides outward on the link according to the equation $r = t^2 + 2$. Using the equation for relative motion, find the displacement of the block between the positions defined by $t = 0$ and $t = 2$.

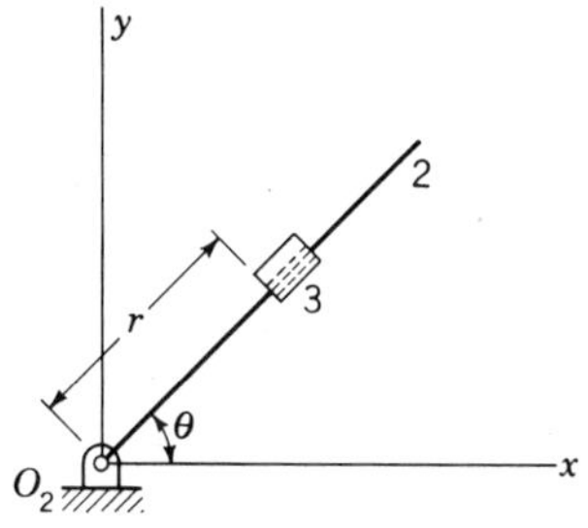

Prob. 3-10

3-11. A point P moves from A to B along link 3 while link 2 rotates from $\theta = 30°$ to $\theta = 120°$. Find the displacement of P using the equation for relative motion.

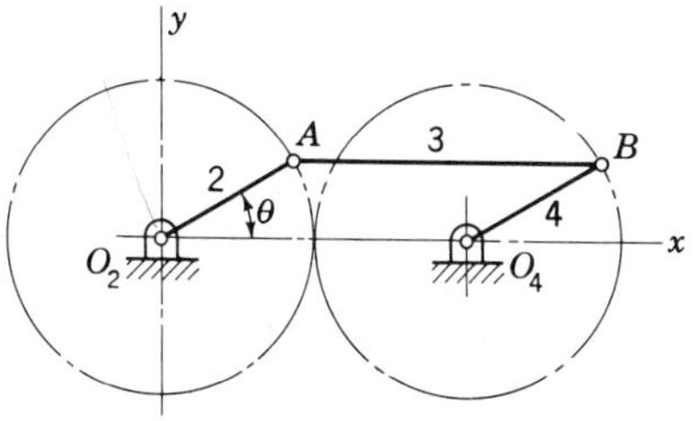

Prob. 3-11 $O_2A = 3$ in., $AB = 6$ in., $O_2O_4 = 6$ in., $O_4B = 3$ in.

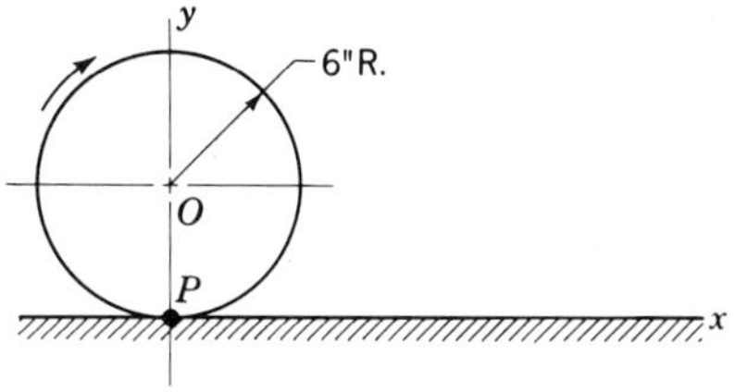

Prob. 3-12

3-12. A wheel with center at O rolls without slipping. The wheel rolls so that the center O is displaced 10 in. to the right. What is the displacement of point P on the periphery during this motion? Use the equation for relative motion.

4
Velocity

4-1 DEFINITION OF VELOCITY

In Fig. 4-1 a point is first observed at R_1 and later at R_2. During this time interval the displacement of the point is

$$\Delta \mathbf{r} = \mathbf{r}_2 - \mathbf{r}_1$$

where $\mathbf{r}_1$ and $\mathbf{r}_2$ are the position vectors which define the location of the point at the beginning and end of the time interval considered. The *average velocity* of the point during the time interval Δt is $\Delta \mathbf{r}/\Delta t$. The *instantaneous velocity* (hereafter called simply *velocity*) is the limit of this ratio, and is given by

$$\mathbf{v} = \lim_{\Delta t \to 0} \frac{\Delta \mathbf{r}}{\Delta t} = \frac{d\mathbf{r}}{dt} = \dot{\mathbf{r}} \tag{4-1}$$

Since $\Delta \mathbf{r}$ is a vector, there are *two* convergencies in taking this limit—the *magnitude* and the *direction* of $\Delta \mathbf{r}$. Of course we have assumed that a limit does, in fact, exist. Velocity is, therefore, the *time rate of change of position*.

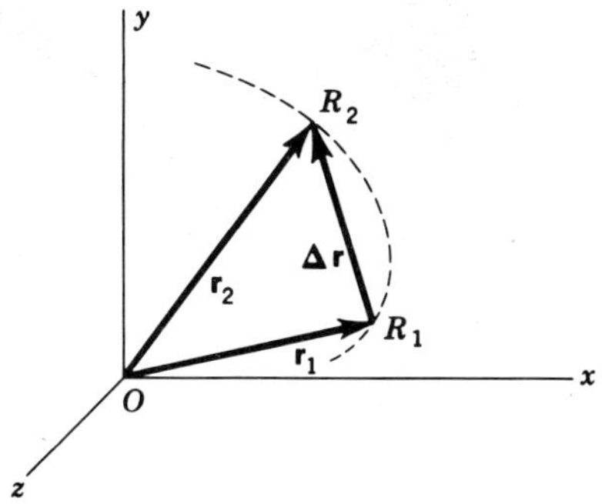

Fig. 4-1

We might observe, from Fig. 4-1, that the position vectors $\mathbf{r}_1$ and $\mathbf{r}_2$ depend upon the location of the coordinate system. But the displacement vector $\Delta\mathbf{r}$ and hence the velocity $\mathbf{v}$ are independent of the choice of the reference system.

The differentiation of vectors follows the same rules as the differentiation of scalars. Thus

$$\frac{d}{dt}(r\hat{\mathbf{r}}) = \dot{r}\hat{\mathbf{r}} + r\dot{\hat{\mathbf{r}}} \tag{4-2}$$

$$\frac{d}{dt}(\mathbf{r} + \mathbf{s}) = \dot{\mathbf{r}} + \dot{\mathbf{s}} \tag{4-3}$$

$$\frac{d}{dt}(\mathbf{r} \times \mathbf{s}) = \dot{\mathbf{r}} \times \mathbf{s} + \mathbf{r} \times \dot{\mathbf{s}} \tag{4-4}$$

$$\frac{d}{dt}(\mathbf{r} \cdot \mathbf{s}) = \dot{\mathbf{r}} \cdot \mathbf{s} + \mathbf{r} \cdot \dot{\mathbf{s}} \tag{4-5}$$

The dependency of the velocity on the path of motion of a particle can be illustrated in another manner. In Fig. 4-2 we wish to describe the

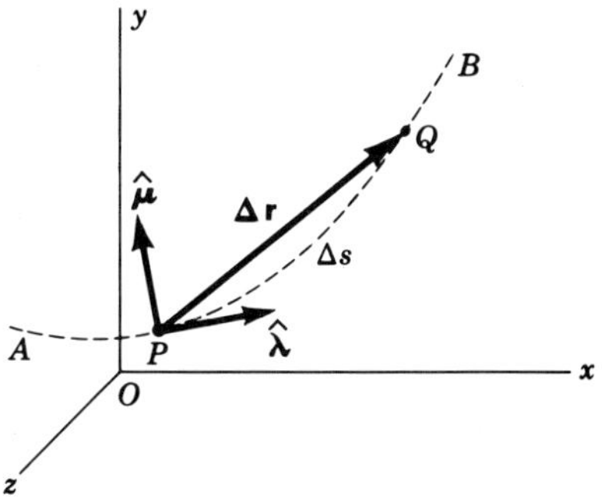

Fig. 4-2

velocity of P as it moves along the path AB. We visualize ourselves as observers moving with P; hence it is natural to choose a coordinate system with an origin at P. The $\hat{\boldsymbol{\lambda}}\hat{\boldsymbol{\mu}}\hat{\boldsymbol{\nu}}$ system, which is illustrated, has been chosen so that $\hat{\boldsymbol{\lambda}}$ is tangent to the path, $\hat{\boldsymbol{\mu}}$ is normal to the path, and $\hat{\boldsymbol{\nu}}$ is related by the equation

$$\hat{\boldsymbol{\nu}} = \hat{\boldsymbol{\lambda}} \times \hat{\boldsymbol{\mu}} \tag{a}$$

Now define Δs, a scalar, as the arc from P to some point Q, and $\Delta \mathbf{r}$ as the chord. Then the limit of $\Delta \mathbf{r}/\Delta s$, as Δs approaches zero, is unity. Hence

$$\lim_{\Delta s \to 0} \frac{\Delta \mathbf{r}}{\Delta s} = \frac{d\mathbf{r}}{ds} = \hat{\boldsymbol{\lambda}} \tag{4-6}$$

Here both $\Delta \mathbf{r}$ and Δs are considered as functions of time, and so we have

$$\lim_{\Delta t \to 0} \frac{\Delta \mathbf{r}}{\Delta t} = \lim_{\Delta t \to 0} \frac{\Delta \mathbf{r}}{\Delta s}\frac{\Delta s}{\Delta t} = \frac{d\mathbf{r}}{ds}\frac{ds}{dt} \tag{b}$$

Therefore the velocity referred to the $\hat{\boldsymbol{\lambda}}\hat{\boldsymbol{\mu}}\hat{\boldsymbol{\nu}}$ system is

$$\boxed{\mathbf{v} = \frac{ds}{dt}\hat{\boldsymbol{\lambda}} = \dot{s}\hat{\boldsymbol{\lambda}}} \tag{4-7}$$

where $\dot{s}$ is the *speed* of P along the path and $\hat{\boldsymbol{\lambda}}$ is a unit vector tangent to the path at P. The importance of this result is that the velocity is always tangent to the path.

If

$$\mathbf{r} = x\hat{\mathbf{i}} + y\hat{\mathbf{j}} + z\hat{\mathbf{k}}$$

then the velocity in rectangular coordinates is

$$\dot{\mathbf{r}} = \dot{x}\hat{\mathbf{i}} + \dot{y}\hat{\mathbf{j}} + \dot{z}\hat{\mathbf{k}} \tag{4-8}$$

4-2 ANGULAR VELOCITY

In Fig. 4-3 we picture a rigid body rotating about some axis OA. This means that all points in the body, such as point P, move in circular paths about OA at the instant under consideration. The angular velocity of the body is given by the vector $\boldsymbol{\omega}$ having a direction OA and a sense according to the right-hand rule. The magnitude of the angular velocity is the time rate of change of *any* line in the body whose direction is normal to the axis of rotation. If we designate the angular displacement of any of these lines as $\Delta\theta$ and the time interval as Δt, then the magnitude of the

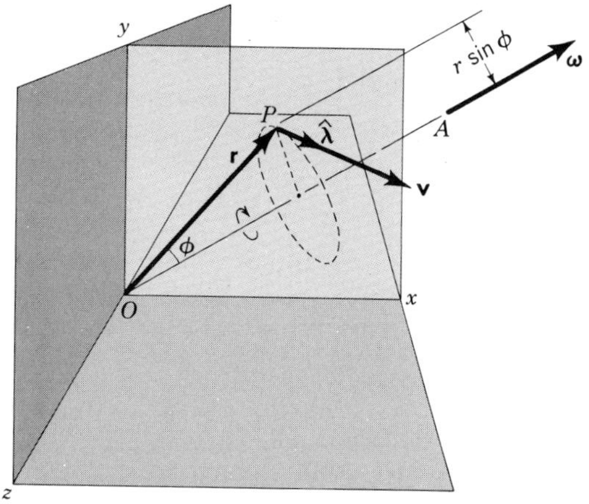

Fig. 4-3

angular velocity is

$$\omega = \lim_{\Delta t \to 0} \frac{\Delta \theta}{\Delta t} = \dot{\theta} \qquad (4\text{-}9)$$

Suppose, now, we let the axis of rotation OA of the rigid body be fixed. Let the position vector **r** define the position of a point P fixed in the rotating body. Then consider the vector product $\boldsymbol{\omega} \times \mathbf{r}$. According to Eq. (2-11) its magnitude is $\omega r \sin \phi$, where ϕ is the angle between $\boldsymbol{\omega}$ and **r**, as shown in Fig. 4-3. The direction is that of the unit tangent $\hat{\lambda}$, perpendicular to the plane of $\boldsymbol{\omega}$ and **r**. Therefore

$$\dot{\mathbf{r}} = \mathbf{v} = \boldsymbol{\omega} \times \mathbf{r} \qquad (4\text{-}10)$$

applies to any vector fixed in the rigid body; where $|\vec{r}|=0$

We emphasize that Eq. (4-10) refers to the velocity of a point in a rigid body rotating about a fixed axis.

Figure 4-4 shows a body rotating about a fixed axis with an angular velocity ω. Points A and B are fixed in the body and are defined by position vectors **A** and **B** from the fixed axis of rotation. **A** and **B** also define a vector **C**, also fixed in the body, such that

$$\mathbf{C} = \mathbf{B} - \mathbf{A} \qquad (a)$$

remember this, I'll definitely like this

The rate of change of **C** due to its rotation is

$$\dot{\mathbf{C}} = \dot{\mathbf{B}} - \dot{\mathbf{A}} \qquad (b)$$

simple things

But both **A** and **B** originate from the axis of rotation. Equation (4-10) applies, and so

$$\dot{\mathbf{C}} = \boldsymbol{\omega} \times \mathbf{B} - \boldsymbol{\omega} \times \mathbf{A} \tag{c}$$

or

$$\dot{\mathbf{C}} = \boldsymbol{\omega} \times (\mathbf{B} - \mathbf{A}) = \boldsymbol{\omega} \times \mathbf{C} \tag{d}$$

and therefore Eq. (4-10) applies to *any* vector fixed in the rotating body. Note that the operation $\boldsymbol{\omega} \times$ serves as a method of differentiation, provided the magnitude of the vector, upon which the operation is performed, is constant. Since a unit vector has a constant magnitude, Eq. (4-2) can be written

$$\dot{\mathbf{r}} = \dot{r}\hat{\mathbf{r}} + r\dot{\hat{\mathbf{r}}} = \dot{r}\hat{\mathbf{r}} + r(\boldsymbol{\omega} \times \hat{\mathbf{r}}) \tag{4-11}$$

where $\boldsymbol{\omega}$ is the angular velocity of $\mathbf{r}$. The reader should prove for himself that the two components of $\dot{\mathbf{r}}$ in Eq. (4-11) are at right angles to each other.

If we restrict our investigation of velocity to two dimensions, then complex polar notation is helpful in visualizing the directions of the two components of Eq. (4-11). Thus, if

$$\mathbf{r} = re^{j\theta} \tag{e}$$

then

$$\begin{aligned}\dot{\mathbf{r}} &= \dot{r}e^{j\theta} + jr\dot{\theta}e^{j\theta} \\ &= \dot{r}e^{j\theta} + jr\omega e^{j\theta}\end{aligned} \tag{4-12}$$

Equations (4-11) and (4-12) are now identical except for the fact that (4-12) applies only to two-dimensional space. To explain, refer to Fig. 4-5, where a position vector $\mathbf{r}$ having an instantaneous magnitude r and direction θ varies with time. Equation (4-12) is shown in Fig. 4-5*a*, and the terms $\dot{r}$ in the $e^{j\theta}$ direction and $r\omega$ in the $je^{j\theta}$ direction are added vec-

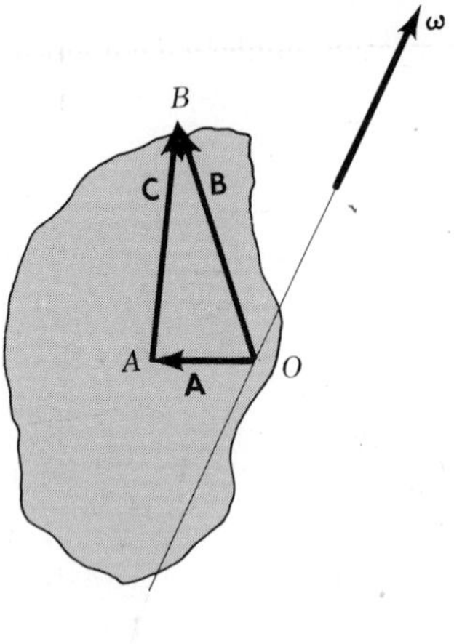

Fig. 4-4

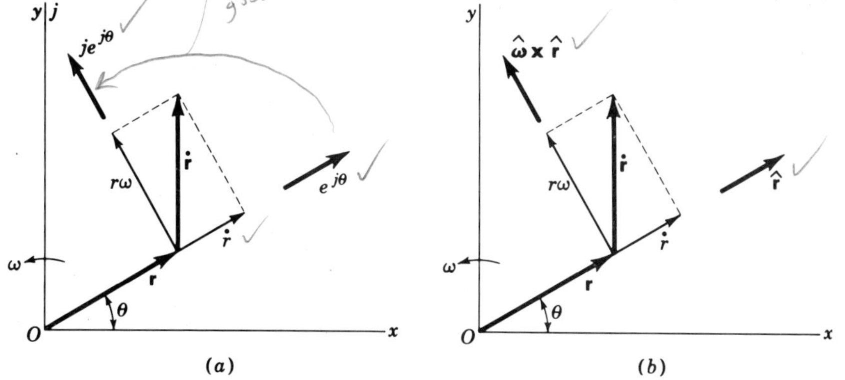

Fig. 4-5

torially to form the velocity $\dot{\mathbf{r}}$. Figure 4-5*b* illustrates Eq. (4-11). Here the component $\dot{r}$ due only to magnitude change is in the $\hat{\mathbf{r}}$ direction. If we note that

$$r(\omega \times \hat{\mathbf{r}}) = r\omega(\hat{\omega} \times \hat{\mathbf{r}}) \qquad (f)$$

then we see that the second component $r\omega$ is perpendicular to the first and in the direction of $\hat{\omega} \times \hat{\mathbf{r}}$. Since we are dealing with plane motion, the direction $\hat{\omega} \times \hat{\mathbf{r}}$ is obtained by rotating the vector $\hat{\mathbf{r}}$ 90° in the sense of ω.

The figure also shows that

$$\hat{\mathbf{r}} = e^{j\theta} \qquad \hat{\omega} \times \hat{\mathbf{r}} = je^{j\theta} \qquad (4\text{-}13)$$

for plane motion.

4-3 THE EULERIAN ANGLES

We have already seen that angular velocity is a vector; hence we conclude that, like all vectors, it can be resolved into the rectangular components

$$\omega = \omega^x\hat{\mathbf{i}} + \omega^y\hat{\mathbf{j}} + \omega^z\hat{\mathbf{k}} \qquad (a)$$

Unfortunately, one cannot find a set of three angles which specify the orientation of a rigid body and which also have ω^x, ω^y, and ω^z as their time derivatives.

To clarify the problem further, we visualize a rigid body rotating in space about a fixed point O at the origin of a ground or *absolute reference system xyz*. A moving reference system $x'y'z'$ is then defined so that it is fixed to the rotating body. The axes of the $x'y'z'$ system are called *body-fixed axes*. We might define the orientation of $x'y'z'$ by using direction cosines, but nine of these would be required; and these would be related by six orthogonality relations.

Three angles, called the *eulerian angles*, can be used to specify the orientation of the body-fixed axes. To illustrate the eulerian angles, we shall begin with the body-fixed axes coincident with the absolute reference axes. We then specify three successive rotations, which must occur in the specified order, to arrive at the $x'y'z'$ orientation.[1] A pictorial three-dimensional description of these rotations is most unsatisfactory. Consequently we shall employ the three orthographic views of Fig. 4-6. These views are all arranged so that the axes are in the plane of the paper or are directed positively out of the paper to the reader.

The first rotation is through the angle ϕ about the z axis and in the positive direction, as shown in view *a*. This rotation yields the $x_1y_1z_1$ system. Thus x rotates through ϕ to x_1, y to y_1, and z and z_1 are coincident. We visualize an angular-velocity vector $\dot{\boldsymbol{\phi}}$ coincident with z and z_1.

The next step is to construct view *b* by projecting orthographically along the positive y_1 axis. The second rotation is through the angle θ about the y_1 axis and in the positive direction, as shown. This rotation yields the $x_2y_2z_2$ system with z_1 rotating through θ to z_2 and x_1 to x_2. Note that y_1 and y_2 are coincident and that we can visualize another angular-velocity vector $\dot{\boldsymbol{\theta}}$ directed along the positive y_2 axis. Note also that the vector $\dot{\boldsymbol{\phi}}$ has been resolved into components along the x_2 and z_2 axes.

The last step is started by projecting orthographically from the positive z_2 axis in view *b* to obtain view *c*. This makes the vector $\dot{\boldsymbol{\theta}}$ appear on the positive y_2 axis, and the z_2 axis directed positively out of the figure. The third rotation is through the angle ψ about the z_2 axis. This yields the desired orientation and the $x'y'z'$ axes. The angular velocities are then resolved again into components along the $x'y'z'$ axes. By using views *b* and *c*, the components can be summed to yield

$$\omega^{x'} = \dot{\theta} \sin \psi - \dot{\phi} \sin \theta \cos \psi \tag{4-14}$$

$$\omega^{y'} = \dot{\theta} \cos \psi + \dot{\phi} \sin \theta \sin \psi \tag{4-15}$$

$$\omega^{z'} = \dot{\psi} + \dot{\phi} \cos \theta \tag{4-16}$$

[1] There appears to be little agreement among writers as to how these angles should be defined. In this book we employ the definition given by Hsuan Yeh and Joel I. Abrams, "Principles of Mechanics of Solids and Fluids," vol. 1, pp. 131–133, McGraw-Hill Book Company, New York, 1960, and John L. Synge and Byron A. Griffith, "Principles of Mechanics," 3d ed., pp. 259–261, McGraw-Hill Book Company, New York, 1959. A different method is presented in each of the following three books: Lawrence E. Goodman and William H. Warner, "Dynamics," pp. 451–456, Wadsworth Publishing Company, Inc., Belmont, Calif., 1964; Irving H. Shames, "Engineering Mechanics: Dynamics," pp. 565–568, Prentice-Hall, Inc., Englewood Cliffs, N.J., 1960; and Donald T. Greenwood, "Principles of Dynamics," pp. 332–336 and 380–383, Prentice-Hall, Inc., Englewood Cliffs, N.J., 1965.

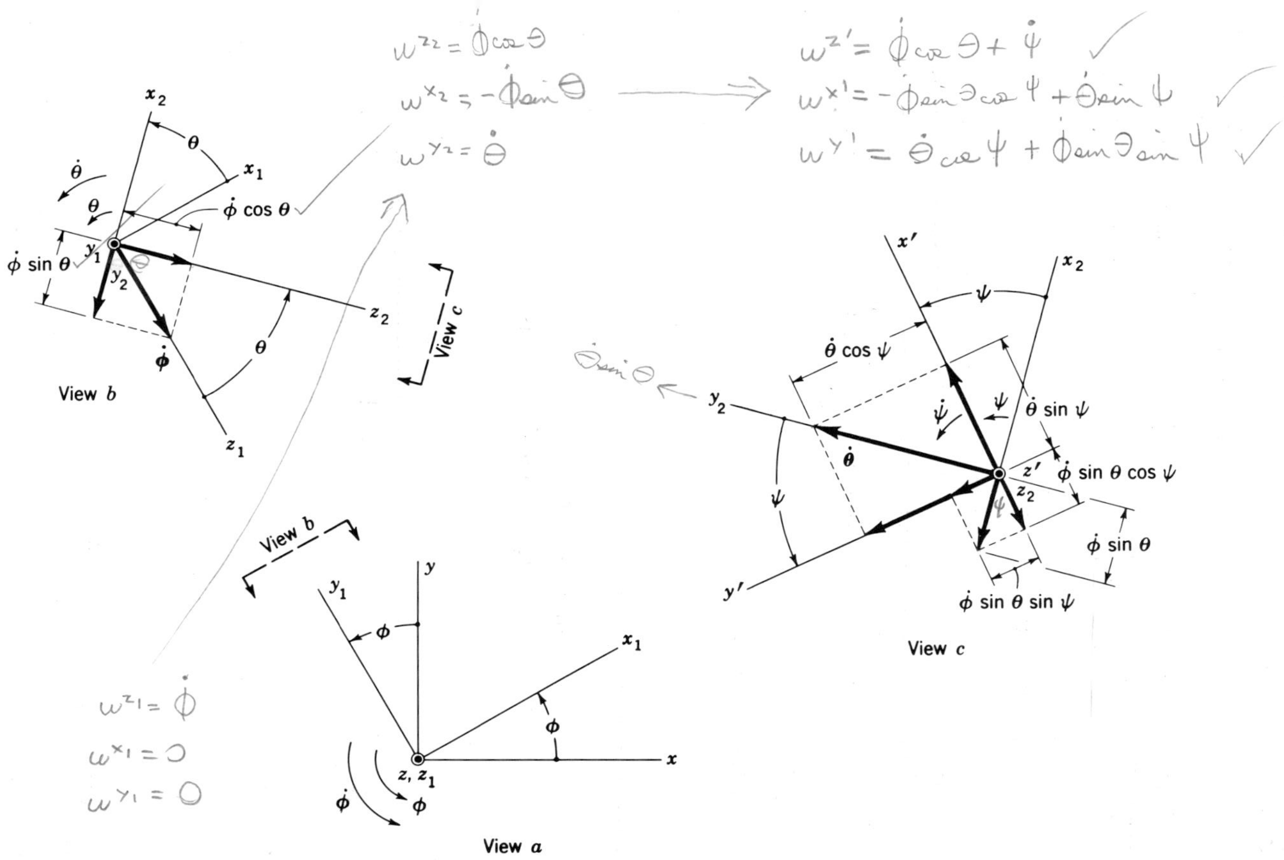

Fig. 4-6 Orthographic views of the three successive rotations required to define the eulerian angles.

Defining $\hat{\lambda}\hat{u}\hat{v}$ as the unit-vector triad of the moving system then gives

$$\omega = \omega^{x'}\hat{\lambda} + \omega^{y'}\hat{u} + \omega^{z'}\hat{v} \tag{4-17}$$

4-4 THE VELOCITY OF A RIGID BODY

We now wish to study the velocity of points in a rigid body which moves with both translation and rotation. If, in Fig. 4-7, we specify an angular velocity ω for the body and we also state that point A has a velocity $\mathbf{v}_A$, then the body has a combined motion of translation and rotation. The position of any other point B in the body is defined by the equation

$$\mathbf{r}_B = \mathbf{r}_A + \mathbf{r}_{BA} \tag{a}$$

Then the velocity of B is

$$\dot{\mathbf{r}}_B = \dot{\mathbf{r}}_A + \dot{\mathbf{r}}_{BA} \tag{b}$$

But $\dot{\mathbf{r}}_A = \mathbf{v}_A$, which was given. Since $\mathbf{r}_{BA}$ is a position vector in a rigid body, its length cannot change; hence

$$\dot{\mathbf{r}}_{BA} = \omega \times \mathbf{r}_{BA} \tag{c}$$

note that this is true because the length of $\mathbf{r}_{BA}$ cannot change

We therefore have

$$\mathbf{v}_B = \mathbf{v}_A + \omega \times \mathbf{r}_{BA} \tag{4-18}$$

where $\mathbf{v}_A$ is the absolute velocity of A and is the translation part of the motion.

We shall often write Eq. (4-18) in the form

$$\mathbf{v}_B = \mathbf{v}_A + \mathbf{v}_{BA} \tag{4-19}$$

which is read, *the velocity of B equals the velocity of A plus the velocity of B*

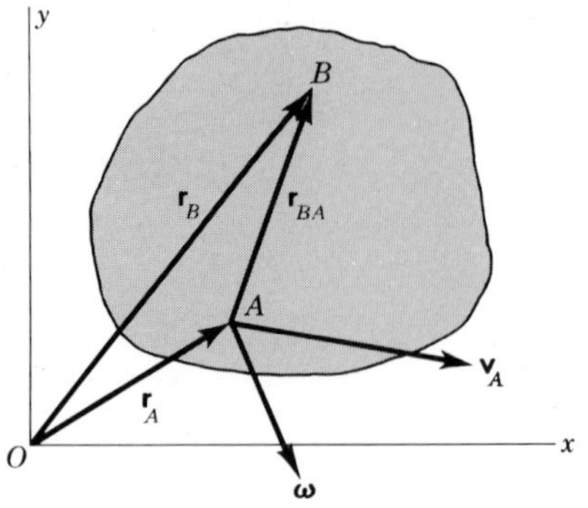

Fig. 4-7

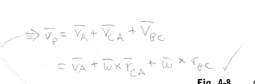

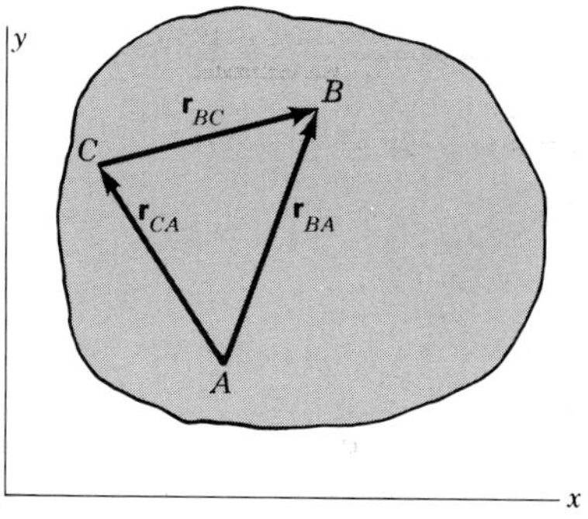

Fig. 4-8

relative to A. This is called the *relative-velocity equation*. It is easy to remember by noting the order of the subscripts: B, A, B, A.

It is important to note, in speaking of Eq. (4-19), that velocity is a vector and *cannot* be referred to a point, because vectors have directional properties. Thus, when we speak of the velocity of B relative to A, we really mean *the velocity of B in a reference system having A as the origin*.

Now, in Fig. 4-8, suppose we had chosen to compute the velocity of B, instead, by employing point C. We should then have begun with, say, $\boldsymbol{\omega}'$ as the angular velocity of the body and $\mathbf{v}_C$ as the velocity of point C, to which B is to be referred. Equation (4-18) would then have been written

$$\mathbf{v}_B = \mathbf{v}_C + \boldsymbol{\omega}' \times \mathbf{r}_{BC} \tag{d}$$

But

$$\mathbf{v}_C = \mathbf{v}_A + \boldsymbol{\omega} \times \mathbf{r}_{CA} \tag{e}$$

Now, since $\mathbf{r}_{BA} = \mathbf{r}_{CA} + \mathbf{r}_{BC}$ (Fig. 4-8), Eq. (4-18) can be written

$$\mathbf{v}_B = \mathbf{v}_A + \boldsymbol{\omega} \times (\mathbf{r}_{CA} + \mathbf{r}_{BC}) \tag{f}$$

But upon substituting Eq. (e) into Eq. (d) we get

$$\mathbf{v}_B = \mathbf{v}_A + \boldsymbol{\omega} \times \mathbf{r}_{CA} + \boldsymbol{\omega}' \times \mathbf{r}_{BC} \tag{g}$$

Then when we compare Eqs. (f) and (g) term by term we see that

$$\boldsymbol{\omega} \times \mathbf{r}_{BC} = \boldsymbol{\omega}' \times \mathbf{r}_{BC}$$

and hence that

$$\boldsymbol{\omega} = \boldsymbol{\omega}' \tag{h}$$

We are thus led to the conclusion that the velocity of a rigid body is obtained as the sum of a rotational velocity $\boldsymbol{\omega}$ about *any* reference axis in the body plus the velocity of that reference axis. The velocity of the reference is the translational component of the total velocity, and this will be *different* for each reference. The angular velocity ω is the same, however; it does *not* depend upon the choice of the reference axis.

4-5 VELOCITY POLYGONS

The graphic method using velocity polygons is a fast way of solving for the kinematics of problems in plane mechanisms. As an example of this approach let us consider the wheel of Fig. 4-9*a*. We shall give point A a velocity $\mathbf{V}_A$ and specify a second point B in the wheel whose velocity we wish to find. We begin the vector diagram by constructing $\mathbf{V}_A$ in the proper length and direction at another place on the paper (Fig. 4-9*b*).

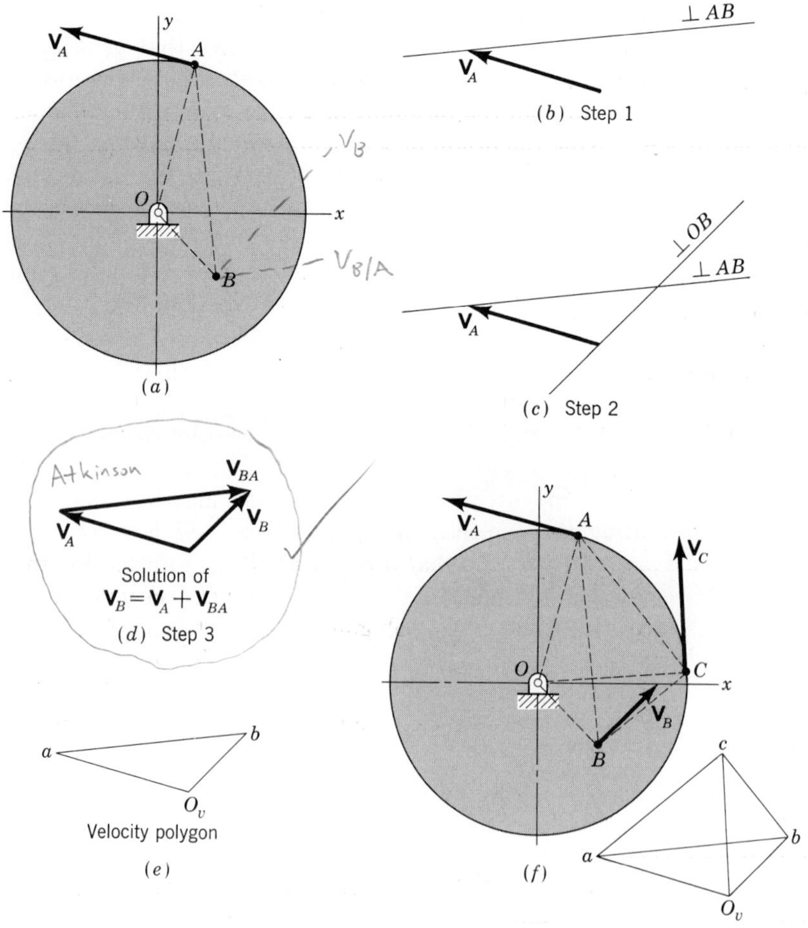

Fig. 4-9 Construction of velocity polygons.

Now, in order to find the velocity of B we write the relative-velocity equation

$$\overset{1}{\mathbf{V}_B} = \overset{2}{\mathbf{V}_A} + \overset{1}{\mathbf{V}_{BA}} \tag{a}$$

The numbers above the equation indicate the number of *known* quantities. We know that the direction of $\mathbf{V}_B$ is perpendicular to OB, but we do not know its magnitude. Both the magnitude and the direction of $\mathbf{V}_A$ are known. Since A and B are both points in the same rigid body they cannot get closer together or farther apart. Therefore the only way that B can move relative to A is in a direction perpendicular to AB, the line that joins them. So we know that the direction of $\mathbf{V}_{BA}$ is perpendicular to AB.

Equation (a) furnishes all the instructions to construct the vector diagram. The equation states that another vector $\mathbf{V}_{BA}$ must be added to $\mathbf{V}_A$ to form $\mathbf{V}_B$. So, through the terminus of $\mathbf{V}_A$, we construct a line perpendicular to AB. Now the origin of $\mathbf{V}_{BA}$ will be at the terminus of $\mathbf{V}_A$, and $\mathbf{V}_{BA}$ will lie along this line (Fig. 4-9b). The next step is to construct on the drawing a line representing the direction of $\mathbf{V}_B$. This step is done in Fig. 4-9c, where the figure has been reproduced to show each step more clearly. The intersection of the lines constructed in steps 1 and 2 gives the magnitudes of $\mathbf{V}_B$ and $\mathbf{V}_{BA}$, as shown in Fig. 4-9d. This completes the vector polygon.

The practice of constructing vector diagrams with thick black lines makes them easy to read, but when the diagram is the graphical solution of an equation, it is not very accurate. For this reason it is customary to construct a polygon, made with thin sharp lines, as a substitute for the diagram. In this way the directions and magnitudes may be determined with more accuracy. The velocity polygon of Fig. 4-9e has been constructed in this manner. Notice that it contains all the information that the vector diagram contains. The pole is designated O_v, the v indicating that it is a velocity polygon. The polygon shows that O_va is $\mathbf{V}_A$, O_vb is $\mathbf{V}_B$, and the directed line segment from a to b is $\mathbf{V}_{BA}$. Notice also that the line from b to a represents $\mathbf{V}_{AB}$.

In Fig. 4-9f the wheel has been drawn again and a third point C added. The resulting polygon is shown. This was obtained by writing two vector equations, one relating the velocity of B to A and the other relating the velocity of C to A. The reader should be able to identify $\mathbf{V}_C$, $\mathbf{V}_{CA}$, and $\mathbf{V}_{CB}$.

It is evident that vectors which originate from the pole represent absolute velocities and that vectors which do not originate at the pole represent relative velocities.

When vector problems are solved analytically, the polygon should be constructed as a freehand sketch in order to better visualize the solution. In this case it is often convenient to add the location of the coordinate axes to the sketch.

4-6 CONVENTIONS

The mechanism shown in Fig. 4-10 is called a *plane six-bar linkage.* It consists of a *fixed frame* 1, a *crank* 2, two *couplers* or *connecting rods* numbered 3 and 5, a *rocker* 4, and a *slider* 6. The slider is constrained to move only in the x direction; it is customary to show only one of the two required constraints for plane motion.

We shall always number the links of a mechanism beginning with the frame. If the mechanism is a part of a machine, then it will have an *input* or *driving* member. This we shall call the second link in the chain. The remaining links are then usually numbered successively.

It is also customary to describe the angular positions of the members of a kinematic chain from the x axis with the counterclockwise direction taken as the positive direction, as shown in Fig. 4-10. Thus θ_2 is the angular position of link 2, θ_3 the position of link 3, etc. Similarly, we shall designate the angular velocities of the various links as ω_2, ω_3, etc., with the counterclockwise direction designated as positive.

Pin connections are designated as A, B, C, etc., as shown in Fig. 4-10. One could also use the notation, say, A_{23}, indicating that links 2 and 3 are paired at A. Frame connections will be indicated by using the capital letter O with a numbered subscript indicating the movable link.

Example 4-1 The four-bar linkage shown in Fig. 4-11 is driven by crank 2 at 900 rpm counterclockwise, as shown. Find the velocities of point C and point D and the angular velocities of links 3 and 4.

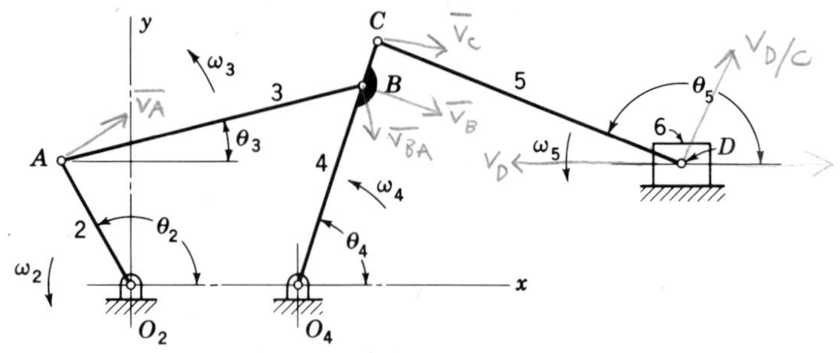

Fig. 4-10 Conventional notation for linkages. O_4BC is link 4, a single rigid link.

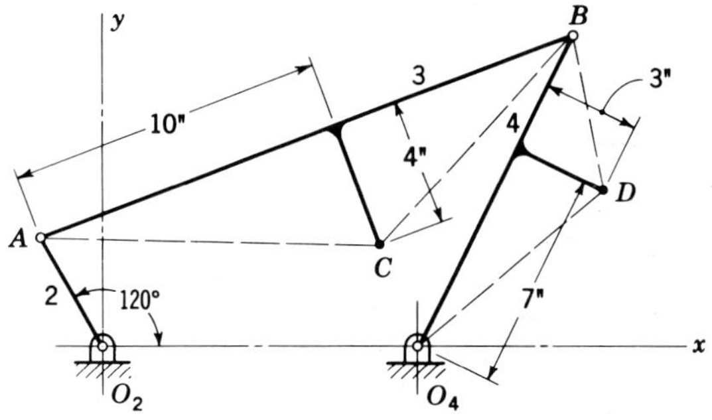

Fig. 4-11 $O_2O_4 = 10$ in., $O_2A = 4$ in., $AB = 18$ in., $O_4B = 11$ in., $n_2 = 900$ rpm ccw.

Graphic solution To obtain a graphic solution we first calculate the angular velocity of link 2 in radians per second. This is

$$\omega_2 = \frac{2\pi n_2}{60} = \frac{2\pi(900)}{60} = 94.2 \text{ rad/sec}$$

Then the velocity of A is

$$V_A = r_2\omega_2 = (\tfrac{4}{12})(94.2) = 31.4 \text{ fps}$$

In Fig. 4-12 we choose a pole O_v and construct the line O_va perpendicular to link 2. The length of this line is V_A to an appropriate velocity scale.

The next step is to obtain $\mathbf{V}_B$, and so we write the equation

$$\overset{1}{\mathbf{V}_B} = \overset{2}{\mathbf{V}_A} + \overset{1}{\mathbf{V}_{BA}} \tag{1}$$

The direction of $\mathbf{V}_B$ is perpendicular to O_4B; the direction of $\mathbf{V}_{BA}$ is perpendicular to AB; both magnitudes are unknown. In Fig. 4-12 construct O_vb per-

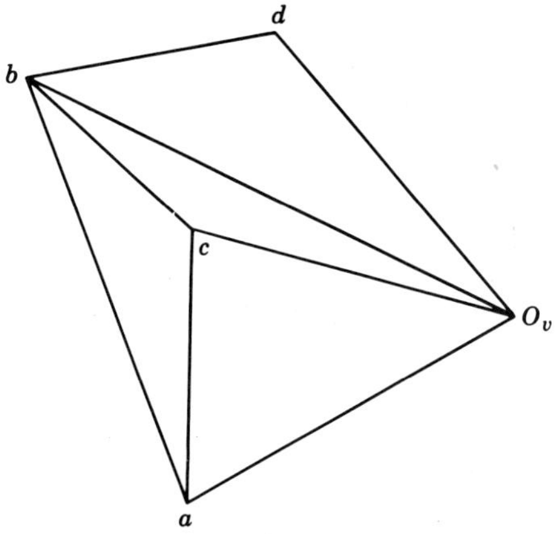

Fig. 4-12

pendicular to O_4B and ab perpendicular to AB. These lines intersect at b. The distance O_vb is V_B and the distance ba is V_{BA}. When these are scaled, we find $V_B = 45.6$ fps and $V_{BA} = 38.4$ fps. Since $V = r\omega$, the angular velocities of links 3 and 4 are

$$\omega_3 = \frac{V_{BA}}{r_3} = \frac{38.4}{\frac{18}{12}} = 25.6 \text{ rad/sec ccw}$$

$$\omega_4 = \frac{V_B}{r_4} = \frac{45.6}{\frac{11}{12}} = 49.7 \text{ rad/sec ccw}$$

There are several methods of obtaining the velocity of C. In one method we draw the line AC, as shown in Fig. 4-11. The distance AC can be measured or calculated. When it is calculated we find

$$r_{CA} = \sqrt{(10)^2 + (4)^2} = 10.78 \text{ in.}$$

Consequently,

$$V_{CA} = r_{CA}\omega_3 = \left(\frac{10.78}{12}\right)(25.6) = 22.7 \text{ fps}$$

Then, since

$$\overset{0}{\mathbf{V}_C} = \overset{2}{\mathbf{V}_A} + \overset{2}{\mathbf{V}_{CA}} \tag{2}$$

we draw a line ac perpendicular to AC. The length of this line, which is constructed on the velocity polygon, is made to correspond to 22.7 fps to the appropriate scale. The triangle O_vac is now the solution to Eq. (2); O_vc represents $\mathbf{V}_C$. When this line is scaled we find $V_C = 27.8$ fps.

An alternative method would be to use the distance CB in Fig. 4-11 and solve the vector equation

$$\mathbf{V}_C = \mathbf{V}_B + \mathbf{V}_{CB} \tag{3}$$

This solution would then produce the triangle O_vbc.

Suppose we wish to find $\mathbf{V}_C$ without going through the intermediate step of computing ω_3. In this case the equation

$$\overset{0}{\mathbf{V}_C} = \overset{2}{\mathbf{V}_A} + \overset{1}{\mathbf{V}_{CA}} \tag{4}$$

cannot be solved because it contains three unknowns. And, for the same reason, neither can the equation

$$\overset{0}{\mathbf{V}_C} = \overset{2}{\mathbf{V}_B} + \overset{1}{\mathbf{V}_{CB}} \tag{5}$$

be solved. However, Eqs. (4) and (5) have the same term on the left-hand sides, and so

$$\overset{2}{\mathbf{V}_A} + \overset{1}{\mathbf{V}_{CA}} = \overset{2}{\mathbf{V}_B} + \overset{1}{\mathbf{V}_{CB}} \tag{6}$$

Now, we have a vector equation with only two unknowns. To solve it, refer again to Fig. 4-12, where O_va is $\mathbf{V}_A$ and O_vb is $\mathbf{V}_B$. Since the direction of $\mathbf{V}_{CA}$ is known, through a draw ac perpendicular to AC. This takes care of the left-hand side of Eq. (6). For the right-hand side, draw a line bc through b perpendicular to BC. The intersection of ac and bc locates c. The velocity of C is then found by drawing the line $O_v c$.

The velocity of D can be found in the same manner. The result is

$$V_D = 31.8 \text{ fps}$$

Analytic solution In using vector analysis to obtain the velocities we must first replace all links by vectors. Using the link numbers as subscripts we first write the position equation

$$\mathbf{r}_1 + \mathbf{r}_2 + \mathbf{r}_3 - \mathbf{r}_4 = 0 \tag{7}$$

Here $\mathbf{r}_1$ is taken from O_4 to O_2, $\mathbf{r}_2$ from O_2 to A, $\mathbf{r}_3$ from A to B, and $\mathbf{r}_4$ from O_4 to B. Both $\mathbf{r}_1$ and $\mathbf{r}_2$ are known. Retaining the units of inches for a while, we have

$$\mathbf{r}_1 = 10\underline{/180°} = -10\hat{\imath}$$

$$\mathbf{r}_2 = 4\underline{/120°} = -2\hat{\imath} + 3.46\hat{\jmath}$$

If we designate $\mathbf{R} = \mathbf{r}_1 + \mathbf{r}_2$, then Eq. (7) becomes

$$\mathbf{R} + r_3\hat{\mathbf{r}}_3 - r_4\hat{\mathbf{r}}_4 = 0 \tag{8}$$

with the directions $\hat{\mathbf{r}}_3$ and $\hat{\mathbf{r}}_4$ unknown. Note that

$$\mathbf{R} = (-10 - 2)\hat{\imath} + 3.46\hat{\jmath} = -12\hat{\imath} + 3.46\hat{\jmath}$$

Hence

$$R = 12.46 \text{ in.}$$

and

$$\hat{\mathbf{R}} = \frac{-12\hat{\imath} + 3.46\hat{\jmath}}{12.46} = -0.963\hat{\imath} + 0.278\hat{\jmath}$$

Note also that

$$\hat{\mathbf{R}} \times \hat{\mathbf{k}} = \begin{vmatrix} \hat{\imath} & \hat{\jmath} & \hat{\mathbf{k}} \\ -0.963 & 0.278 & 0 \\ 0 & 0 & 1 \end{vmatrix} = 0.278\hat{\imath} + 0.968\hat{\jmath}$$

We now employ Eq. (3-19) to find $\mathbf{r}_3$ and use the lower sign. This gives

$$\mathbf{r}_3 = \sqrt{r_3{}^2 - \left(\frac{r_3{}^2 - r_4{}^2 + R^2}{2R}\right)^2}\,(\hat{\mathbf{R}} \times \hat{\mathbf{k}}) - \left(\frac{r_3{}^2 - r_4{}^2 + R^2}{2R}\right)\hat{\mathbf{R}}$$

$$= \sqrt{(18)^2 - \left[\frac{(18)^2 - (11)^2 + (12.46)^2}{(2)(12.46)}\right]^2}\,(0.278\hat{\imath} + 0.963\hat{\jmath})$$

$$- \left[\frac{(18)^2 - (11)^2 + 12.46)^2}{(2)(12.46)}\right](-0.963\hat{\imath} + 0.278\hat{\jmath})$$

$$= 16.84\hat{\imath} + 6.4\hat{\jmath} = 18\underline{/20.8°} \text{ in.}$$

Equation (3-20) is used to find $\mathbf{r}_4$. We should normally use the lower sign; however, we have reversed the direction $\mathbf{r}_4$ and so we must change the signs of both terms in Eq. (3-20). The result is

$$\mathbf{r}_4 = \sqrt{r_3^2 - \left(\frac{r_3^2 - r_4^2 + R^2}{2R}\right)^2}\,(\hat{\mathbf{R}} \times \hat{\mathbf{k}}) - \left(\frac{r_3^2 - r_4^2 + R^2}{2R} - R\right)\hat{\mathbf{R}} \tag{9}$$

When this equation is solved for $\mathbf{r}_4$ we obtain

$$\mathbf{r}_4 = 4.86\hat{\imath} + 9.86\hat{\jmath} = 11\underline{/63.7^\circ} \text{ in.}$$

From Eq. (4-10) we find the velocity of A as

$$\mathbf{V}_A = \boldsymbol{\omega}_2 \times \mathbf{r}_2 = \begin{vmatrix} \hat{\imath} & \hat{\jmath} & \hat{\mathbf{k}} \\ 0 & 0 & 94.2 \\ -2 & 3.46 & 0 \end{vmatrix} \tfrac{1}{12} = -27.2\hat{\imath} + 15.7\hat{\jmath} = 31.4\underline{/-150^\circ} \text{ fps}$$

We now visualize a nonrotating coordinate system $x'y'$ with origin at A. Using Eq. (4-19), we have

$$\mathbf{V}_B = \mathbf{V}_A + \mathbf{V}_{BA} \tag{10}$$

or

$$\boldsymbol{\omega}_4 \times \mathbf{r}_4 = \mathbf{V}_A + \boldsymbol{\omega}_3 \times \mathbf{r}_3 \tag{11}$$

Since $\boldsymbol{\omega}_3 = \omega_3\hat{\mathbf{k}}$ and $\boldsymbol{\omega}_4 = \omega_4\hat{\mathbf{k}}$, we have

$$\boldsymbol{\omega}_4 \times \mathbf{r}_4 = \begin{vmatrix} \hat{\imath} & \hat{\jmath} & \hat{\mathbf{k}} \\ 0 & 0 & \omega_4 \\ 4.86 & 9.86 & 0 \end{vmatrix} \tfrac{1}{12} = \omega_4(-0.822\hat{\imath} + 0.405\hat{\jmath})$$

$$\boldsymbol{\omega}_3 \times \mathbf{r}_3 = \begin{vmatrix} \hat{\imath} & \hat{\jmath} & \hat{\mathbf{k}} \\ 0 & 0 & \omega_3 \\ 16.84 & 6.4 & 0 \end{vmatrix} \tfrac{1}{12} = \omega_3(-0.533\hat{\imath} + 1.4\hat{\jmath})$$

Substituting these with $\mathbf{V}_A$ into Eq. (10) gives

$$\omega_4(-0.822\hat{\imath} + 0.405\hat{\jmath}) = -(27.2\hat{\imath} + 15.7\hat{\jmath}) + \omega_3(-0.533\hat{\imath} + 1.4\hat{\jmath})$$

This equation can be solved by separating the $\hat{\imath}$ and $\hat{\jmath}$ components and solving the two resulting algebraic equations for ω_3 and ω_4:

$$\boldsymbol{\omega}_3 = 25.6\hat{\mathbf{k}} \text{ rad/sec} \qquad \boldsymbol{\omega}_4 = 49.7\hat{\mathbf{k}} \text{ rad/sec}$$

Thus

$$\mathbf{V}_B = \boldsymbol{\omega}_4 \times \mathbf{r}_4 = 49.7(-0.822\hat{\imath} + 0.405\hat{\jmath}) = -40.8\hat{\imath} + 20.1\hat{\jmath} = 45.6\underline{/153.7^\circ} \text{ fps}$$

which is 90° ccw from $\mathbf{r}_4$. Also

$$\mathbf{V}_{BA} = \boldsymbol{\omega}_3 \times \mathbf{r}_3 = 25.6(-0.533\hat{\imath} + 1.4\hat{\jmath}) = -13.64\hat{\imath} + 35.8\hat{\jmath} = 38.4\underline{/110.8^\circ} \text{ fps}$$

which is 90° ccw from $\mathbf{r}_3$.

Now, to obtain a vector $\mathbf{r}_{CA}$ from A to C (see Fig. 4-11), we write

$$\mathbf{r}_{CA} = \tfrac{10}{18}\mathbf{r}_3 + \tfrac{4}{18}\mathbf{r}_3 \times \hat{\mathbf{k}} = 10.78\hat{\imath} - 0.19\hat{\jmath}$$

Then

$$\mathbf{V}_C = \mathbf{V}_A + \mathbf{V}_{CA} = \mathbf{V}_A + \boldsymbol{\omega}_3 \times \mathbf{r}_{CA} \tag{12}$$

The last term is

$$\boldsymbol{\omega}_3 \times \mathbf{r}_{CA} = \begin{vmatrix} \hat{\imath} & \hat{\jmath} & \hat{\mathbf{k}} \\ 0 & 0 & 25.6 \\ 10.78 & -0.19 & 0 \end{vmatrix} \tfrac{1}{12} = 0.4\hat{\imath} + 23\hat{\jmath}$$

We then have

$$\begin{aligned} \mathbf{V}_C &= (-27.2\hat{\imath} - 15.7\hat{\jmath}) + (0.4\hat{\imath} + 23\hat{\jmath}) \\ &= -26.8\hat{\imath} + 7.3\hat{\jmath} = 27.8\underline{/164.8^\circ} \text{ fps} \end{aligned}$$

The position vector for point D is

$$\mathbf{r}_D = \tfrac{7}{11}\mathbf{r}_4 + \tfrac{3}{11}\mathbf{r}_4 \times \hat{\mathbf{k}} = 5.78\hat{\imath} + 5.01\hat{\jmath}$$

Then

$$\begin{aligned} \mathbf{V}_D = \boldsymbol{\omega}_4 \times \mathbf{r}_D &= \begin{vmatrix} \hat{\imath} & \hat{\jmath} & \hat{\mathbf{k}} \\ 0 & 0 & 49.7 \\ 5.78 & 5.01 & 0 \end{vmatrix} \tfrac{1}{12} = -20.8\hat{\imath} + 24\hat{\jmath} \\ &= 31.8\underline{/130.9^\circ} \text{ fps} \end{aligned}$$

Before concluding this example, observe that the derivative of Eq. (7) is

$$\dot{\mathbf{r}}_1 + \dot{\mathbf{r}}_2 + \dot{\mathbf{r}}_3 - \dot{\mathbf{r}}_4 = 0 \tag{13}$$

Since link 1 is the frame, $\dot{\mathbf{r}}_1 = 0$. The other terms are $\dot{\mathbf{r}}_2 = \mathbf{V}_A$, $\dot{\mathbf{r}}_3 = \mathbf{V}_{BA}$, and $\dot{\mathbf{r}}_4 = \mathbf{V}_B$. Thus Eq. (13) *is* the relative-velocity equation.

4-7 VELOCITY IMAGES

A very helpful device in the velocity analysis of complicated linkages is the idea of an image of the link itself which gives the velocities to scale and their correct directions. Figure 4-13*a* illustrates a triangular-shaped link which is turning about the fixed point O_2. For convenience, the angular velocity is given as $\omega_2 = 1$ rad/sec. The dimensions of the sides of the link are given as r_A, r_B, and r_{BA}, as shown. The velocities of points A and B, and of B with respect to A, may now be calculated using the equation $V = r\omega$. Since the angular velocity is unity, the velocity magnitudes turn out to be equal, respectively, to the distances between the several points.

The velocity polygon is constructed in Fig. 4-13*b*. Since the sides of the polygon are equal in length to the sides of the original link, the

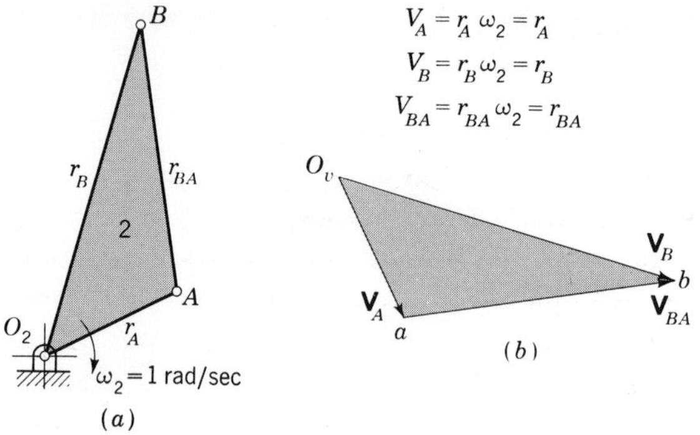

Fig. 4-13 The velocity polygon is the image of the link from which it was derived.

polygon is an exact duplicate of the link. However, each side has been rotated in the direction of ω_2 through an angle of 90°.

It is easy to see that if a value for ω_2 other than unity had been chosen and if, in addition, the velocity polygon had been drawn to some scale, then the polygon would have been larger or smaller than the original link, but still similar; that is, the corresponding angles of the polygon and link would have been equal. Also, the lengths of the corresponding sides would have been in the same proportion to each other. This leads to the following definition: *A velocity image of a link is a reproduction of the link to the same or to a different scale, rotated 90° in the direction of the angular velocity of the link.* This means that one can obtain the velocity image of a link by rotating the link 90° in the direction of the angular velocity and then by increasing or decreasing its size.

One of the important advantages of the idea of a velocity image of a link is the ease with which one can find the velocities of various points in the link. Thus, consider the link of Fig. 4-14*a*. If the velocity of one point in the link can be found, say that of A, then, by selecting a velocity scale, the velocity image can be drawn and the velocities of other points scaled off it.

As a demonstration let us select a velocity for A of 10 fps and plan to construct this as a vector 2 in. long. Then, in the velocity image (Fig. 4-14*c*), $\mathbf{V}_A$ is represented by the vector $O_v a$ with a length of 2 in. Our problem now is to construct the link on $O_v a$ with each side of the image increased in the same ratio as $O_2 A$ has been increased to obtain $O_v a$. A convenient way to do this is to use similar triangles, as shown in Fig. 4-14*b*. This is a standard construction for graphically increasing or decreasing the length of a group of lines in the same ratio and is described in any engineering-drawing book. Applying this method successively to each side of the link results in the image of Fig. 4-14*c* and the velocities $\mathbf{V}_B$, $\mathbf{V}_C$, and $\mathbf{V}_D$.

Scaling has been accomplished by similar triangles in this example, but any method that will accomplish the same result may be used.

In the preceding discussion the velocity image of a link rotating about a stationary, or ground, point was found. Let us now extend the discussion to include a connecting link not rotating about any fixed center. The connecting rod 3 of the slider-crank linkage in Fig. 4-15*a* is such a link. As before, we specify the angular velocity of link 2 as $\omega_2 = 1$ rad/sec. This results in a velocity image of link 2 which is identical in size with

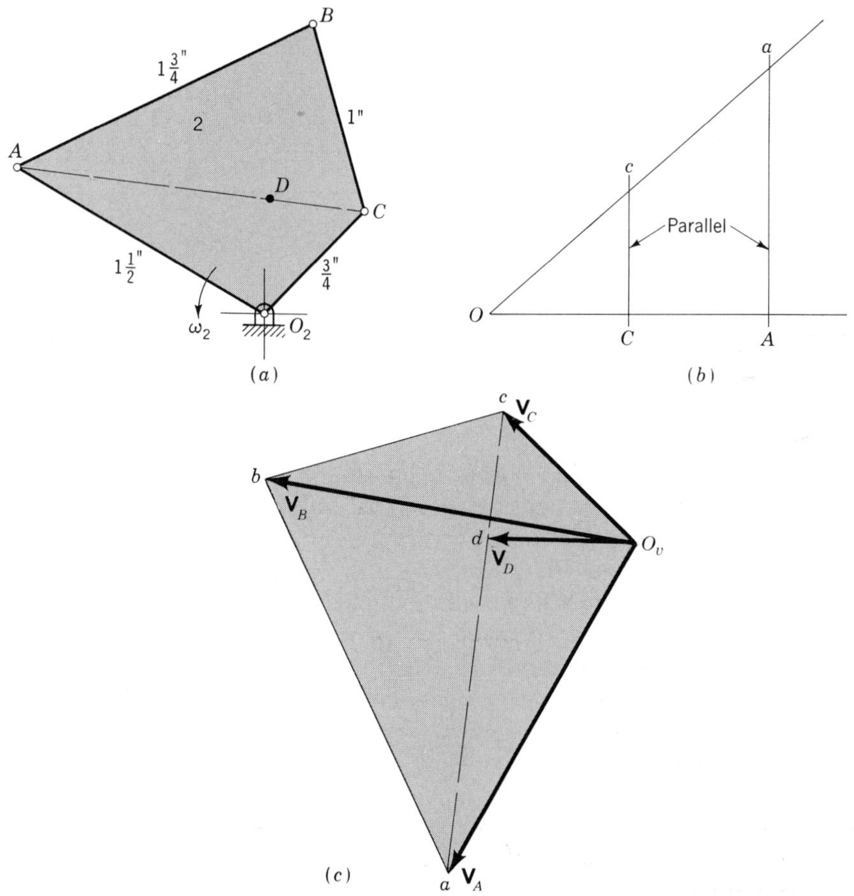

Fig. 4-14

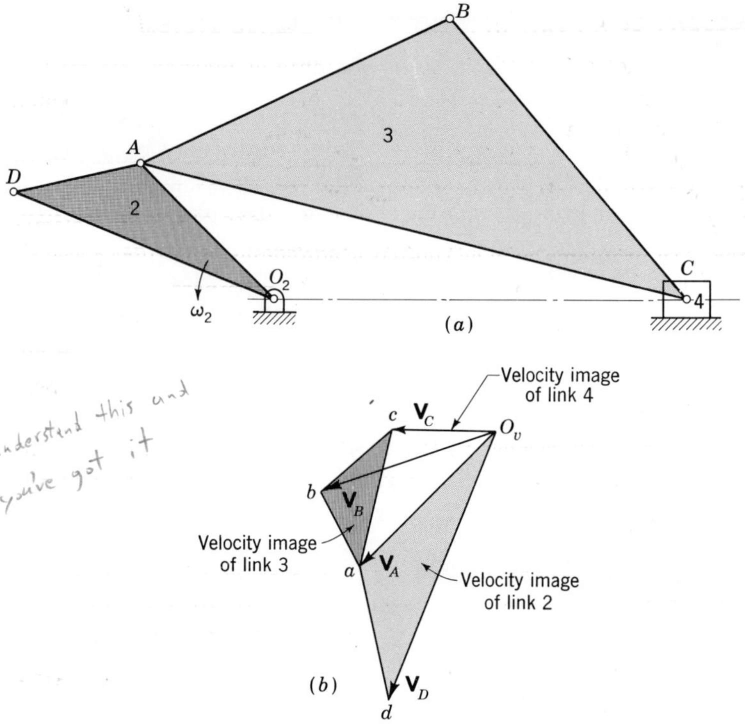

Fig. 4-15

link 2, but turned 90° in the direction of ω_2, as shown in Fig. 4-15*b*. The velocity polygon is completed using the methods discussed earlier. The reader can easily spot the velocity image of link 3 in the velocity polygon. The following points should be noted:

1. The velocity image of a link is a scale drawing of the link rotated 90° in the direction of the absolute angular velocity.
2. The letters identifying the end points of a link progress around the image in the same direction as around the link.
3. The ratio of the size of a velocity image to the size of a link is, in general, not the same for one link as it is for another link in the same mechanism.
4. The velocity image of a link moving in a straight line is the same for all points of the link, and for any link the velocity image shrinks to a point if the angular velocity becomes zero.
5. The absolute velocity of any point on any link is the vector from O_v to the image of the point.

$\hat{\lambda}, \hat{\mu}, \hat{\nu}$ ... remember this; you'll probably see it a lot

4-8 THE VELOCITY OF A POINT IN A MOVING REFERENCE SYSTEM

In analyzing the motions of the various elements of mechanisms we frequently encounter the problem of describing the motion of a point which moves relative to another moving system. A simple example of this is a rotating link along which another member has a sliding motion. With the angular velocity of the rotating link and the linear motion of the slider as the known quantities, we may wish to find the absolute velocity of the slider. In problems such as these it is convenient to utilize a second reference system attached to one of the moving members.

Let us consider, then, a moving $x'y'z'$ system having an origin A whose coordinates are known relative to the absolute reference system xyz (Fig. 4-16). We shall standardize on $\hat{\mathbf{i}}\hat{\mathbf{j}}\hat{\mathbf{k}}$ as the unit vectors for the absolute system and $\hat{\boldsymbol{\lambda}}\hat{\boldsymbol{\mu}}\hat{\boldsymbol{\nu}}$ for the moving system. The position of the origin of the moving system with respect to the fixed system is designated by the vector $\mathbf{r}$ in Fig. 4-16. Any point B can be defined as $\mathbf{R}$ relative to the fixed system or as $\boldsymbol{\rho}$ relative to the moving system.

If the $x'y'z'$ system moves in pure translation, then the time derivatives of the unit vectors $\hat{\boldsymbol{\lambda}}\hat{\boldsymbol{\mu}}\hat{\boldsymbol{\nu}}$ are zero, because their directions do not change. But if the moving system rotates with an angular velocity $\boldsymbol{\omega}$, as shown in Fig. 4-16, then, according to Eq. (4-10), the time derivatives of the unit vectors are

since the lengths of the unit vectors are const.

$$\dot{\hat{\boldsymbol{\lambda}}} = \boldsymbol{\omega} \times \hat{\boldsymbol{\lambda}} \qquad \dot{\hat{\boldsymbol{\mu}}} = \boldsymbol{\omega} \times \hat{\boldsymbol{\mu}} \qquad \dot{\hat{\boldsymbol{\nu}}} = \boldsymbol{\omega} \times \hat{\boldsymbol{\nu}} \tag{4-20}$$

because a unit vector has a fixed length. Note that Eq. (4-20) can also be written

$$\dot{\hat{\boldsymbol{\lambda}}} = \omega(\hat{\boldsymbol{\omega}} \times \hat{\boldsymbol{\lambda}}) \qquad \dot{\hat{\boldsymbol{\mu}}} = \omega(\hat{\boldsymbol{\omega}} \times \hat{\boldsymbol{\mu}}) \qquad \dot{\hat{\boldsymbol{\nu}}} = \omega(\hat{\boldsymbol{\omega}} \times \hat{\boldsymbol{\nu}}) \tag{4-21}$$

where $(\hat{\boldsymbol{\omega}} \times \hat{\boldsymbol{\lambda}})$, etc., are the directions of the new vectors.

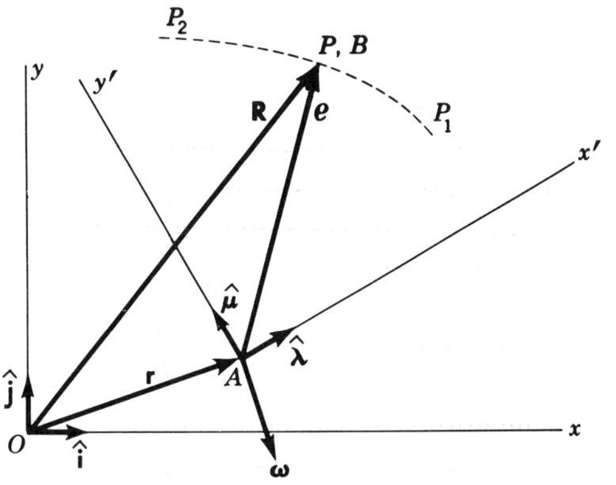

Fig. 4-16

If we regard $x'y'z'$ as a system which is fixed to a rigid body to which B is also fixed, then, according to Eq. (4-18), the velocity of B is

$$\mathbf{v}_B = \mathbf{v}_A + \boldsymbol{\omega} \times \boldsymbol{\rho} \tag{a}$$

where $\boldsymbol{\omega} \times \boldsymbol{\rho} = \mathbf{v}_{BA}$.

But now suppose we have another point P (Fig. 4-16) moving along the path P_1PP_2 relative to $x'y'z'$. Suppose also that P is instantaneously coincident with B. The position of P in xyz is given by the equation

$$\mathbf{R} = \mathbf{r} + \boldsymbol{\rho} \tag{b}$$

where

$$\boldsymbol{\rho} = x'\hat{\boldsymbol{\lambda}} + y'\hat{\boldsymbol{\mu}} + z'\hat{\boldsymbol{\nu}} \tag{c}$$

Substituting Eq. (c) in (b) and taking the derivative to get the velocity of P gives

$$\dot{\mathbf{R}} = \dot{\mathbf{r}} + (\dot{x}'\hat{\boldsymbol{\lambda}} + \dot{y}'\hat{\boldsymbol{\mu}} + \dot{z}'\hat{\boldsymbol{\nu}}) + (x'\dot{\hat{\boldsymbol{\lambda}}} + y'\dot{\hat{\boldsymbol{\mu}}} + z'\dot{\hat{\boldsymbol{\nu}}}) \tag{d}$$

The first term in this equation is the velocity of the origin of $x'y'z'$. The second term is the velocity that P has relative to the $x'y'z'$ system. The third term, by Eq. (4-20), is

$$x'\dot{\hat{\boldsymbol{\lambda}}} + y'\dot{\hat{\boldsymbol{\mu}}} + z'\dot{\hat{\boldsymbol{\nu}}} = \boldsymbol{\omega} \times \boldsymbol{\rho} \tag{e}$$

Therefore Eq. (d) becomes

$$\mathbf{v}_P = \mathbf{v}_A + \mathbf{v}^R + \boldsymbol{\omega} \times \boldsymbol{\rho} \tag{4-22}$$

where

$$\mathbf{v}^R = \dot{x}'\hat{\boldsymbol{\lambda}} + \dot{y}'\hat{\boldsymbol{\mu}} + \dot{z}'\hat{\boldsymbol{\nu}} \tag{4-23}$$

and is called the *relative velocity.* Note that $\mathbf{v}^R$ is the velocity that an observer attached to $x'y'z'$ would report for point P. If we go still further and substitute the terms for the velocity of B from Eq. (a), we get

$$\mathbf{v}_P = \mathbf{v}_B + \mathbf{v}^R \tag{4-24}$$

which states that the absolute velocity of a point P, moving with respect to a moving coordinate system, is equal to the absolute velocity of the coincident point B, fixed to the moving reference system, plus the velocity of P relative to the moving system.

The process can be extended indefinitely. Thus point P of Fig. 4-16 might be the origin of another reference system $x''y''z''$ which translates and rotates relative to $x'y'z'$. The velocity of a point Q whose motion is known relative to $x''y''z''$ is found by successively applying Eq. (4-23) until the fixed reference frame is reached.

you had the identical problem in 104A except that $\overline{AB}$ was zero

★ Note!!! that $\overline{V_B}$ is ⊥ $\overline{BO_4}$ and **not** to $\overline{BC}$

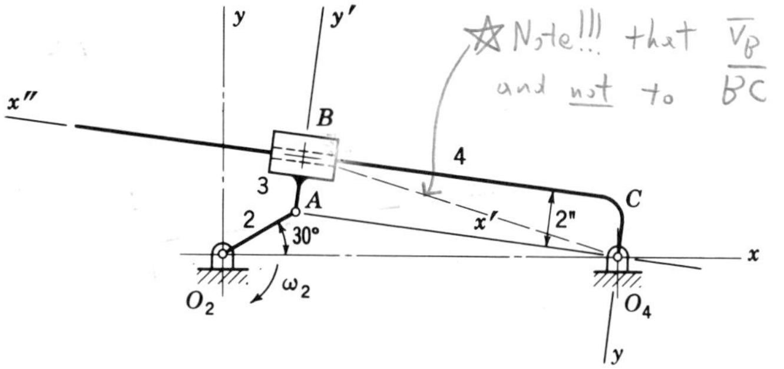

Fig. 4-17 $O_2A = 3$ in., $O_2O_4 = 14$ in., $\omega_2 = -36\mathbf{k}$ rad/sec.

★**Example 4-2** Shown in Fig. 4-17 is an inversion of the slider-crank mechanism. The crank is link 2 and is the driver. Link 3 slides on link 4 and is pivoted to the crank at A. Find the angular velocity of link 4.

Solution Choose an $x'y'$ system attached to link 3 with origin at A. Then B_3, a point fixed to link 3, and B_4, fixed to link 4, are coincident points. For these points Eq. (4-24) is

$$\mathbf{V}_{B_4} = \mathbf{V}_{B_3} + \mathbf{V}_{B_4A} \quad \checkmark \quad (1)$$

note that the slider is fixed with respect to the x'y' system

with

$$\mathbf{V}_{B_3} = \mathbf{V}_A + \mathbf{V}_{B_3A} \quad \checkmark \quad (2)$$

Therefore, Eq. (1) becomes *note*

$$\Rightarrow \quad \overset{1}{\mathbf{V}_{B_4}} = \overset{2}{\mathbf{V}_A} + \overset{1}{\mathbf{V}_{B_3A}} + \overset{1}{\mathbf{V}_{B_4A}} \quad \checkmark \quad (3)$$

Here, the magnitudes V_{B_4}, V_{B_3A}, and V_{B_4A} are unknown. The direction of $\mathbf{V}_{B_4}$ is perpendicular to O_4B. The direction of $\mathbf{V}_{B_3A}$ is perpendicular to AB. But $\mathbf{V}_{B_4A}$ also is directed perpendicular to AB. Consequently the sum $\mathbf{V}_{B_3A} + \mathbf{V}_{B_4A}$ can be considered as a single vector in obtaining the solution. ✓

We first calculate V_A:

$$V_A = (\tfrac{3}{12})(36) = 9 \text{ fps} \quad \checkmark$$

Beginning with O_v in Fig. 4-18 we draw a line $O_v a$ representing $\mathbf{V}_A$ to an appropriate scale. Then, through O_v construct a line perpendicular to O_4B. Finally, through a construct a third line parallel to BC, the direction of sliding. The point of intersection b_4 gives $O_v b_4$, the magnitude of $\mathbf{V}_{B_4}$, and ab_4, the magnitude of the sum $\mathbf{V}_{B_3A} + \mathbf{V}_{B_4A}$. Scaling the polygon gives

$$V_{B_4} = 7.3 \text{ fps} \qquad V_{B_3A} + V_{B_4A} = 6.8 \text{ fps}$$

The distance O_4B can be measured or computed. It is 11.6 in. Therefore

★ $\boxed{\omega_4 = \omega_3 = \dfrac{7.3}{11.6/12} = 7.55 \text{ rad/sec}}$

Another approach to this problem is to fasten the moving reference system to link 4. In Fig. 4-17 an $x''y''$ system with origin at C is shown fixed

since "3" and "4" can not rotate with respect to each other !!!!!

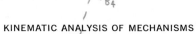

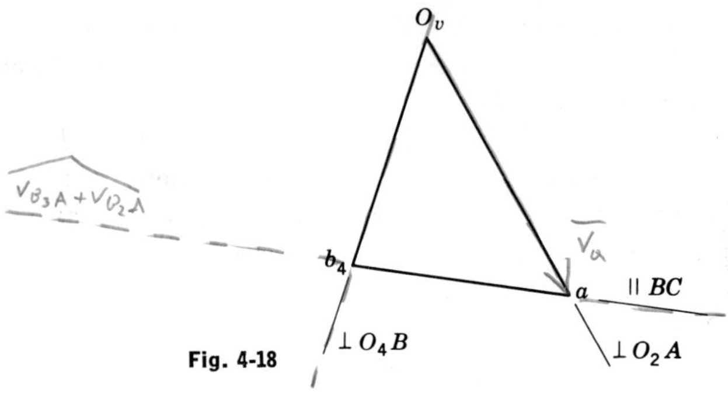

Fig. 4-18

to link 4. With this choice B_4 is fixed in the moving system and B_3 moves relative to $x''y''$. Equation (4-24) for this choice is

$$\mathbf{V}_{B_3} = \mathbf{V}_{B_4} + \mathbf{V}_{B_3C} \tag{4}$$

Then, since

$$\mathbf{V}_{B_3} = \mathbf{V}_A + \mathbf{V}_{B_3A} \tag{5}$$

we have

$$\overset{2}{\mathbf{V}_A} + \overset{1}{\mathbf{V}_{B_3A}} = \overset{1}{\mathbf{V}_{B_4}} + \overset{1}{\mathbf{V}_{B_3C}} \tag{6}$$

Again, we have too many unknowns to solve the equation. But the unknowns are all magnitudes since $\mathbf{V}_{B_3A}$ is perpendicular to AB, $\mathbf{V}_{B_4}$ is perpendicular to O_4B, and $\mathbf{V}_{B_3C}$ is parallel to BC. The components $\mathbf{V}_{B_3A}$ and $\mathbf{V}_{B_3C}$ have the same directions; hence the solution can be found by treating the difference in these two velocities as a single vector. When Eq. (6) is solved, we obtain the same polygon as in the first approach.

4-9 RELATIVE ANGULAR VELOCITY

When two bodies rotate with different velocities, then the vector difference between the two is defined as the *relative angular velocity*. Thus

$$\boldsymbol{\omega}_{21} = \boldsymbol{\omega}_2 - \boldsymbol{\omega}_1 \tag{4-25}$$

which may also be written

$$\boldsymbol{\omega}_2 = \boldsymbol{\omega}_1 + \boldsymbol{\omega}_{21} \tag{4-26}$$

where $\boldsymbol{\omega}_{21}$ is the angular velocity of body 2 relative to body 1.† The reader should compare this equation with Eq. (4-19) for linear velocities.

4-10 INSTANTANEOUS VELOCITY AXES

One of the more interesting concepts in kinematics is that of an instantaneous velocity axis for bodies which move relative to one another. In

† See Sec. 12-6.

particular, we shall find that an axis exists which is common to both bodies and about which either body can be considered as having rotation with respect to the other.

Our usage of these axes will be restricted to the study of plane motion; hence we shall usually refer to them as *centers* or as *poles*. These centers are regarded as rigidly attached to the bodies; hence they do have an acceleration and acquire a velocity. It is not correct, either, to speak of them as instantaneous axes of *pure rotation*, because they are not generally located at the center of curvature of the path which a point on one body generates with respect to the other. Even with these restrictions, however, we shall find that these velocity axes contribute substantially to the understanding of the kinematics of plane motion.

To begin, let us define a body 2 moving in the xy plane and specify that it may have any motion: translation, rotation, or a combination of the two (Fig. 4-19a). Any point A will have a velocity $\mathbf{V}_A$, and the body will have an angular velocity ω. With these two quantities known, the velocity of any other point in the body can be found. Locate, for example, P on a perpendicular to $\mathbf{V}_A$ and at radius $r_{PA} = V_A/\omega$, as shown in Fig. 4-19b. This perpendicular is erected counterclockwise from $\mathbf{V}_A$ if ω is positive. Then

$$\mathbf{V}_P = \mathbf{V}_A + \mathbf{V}_{PA} = 0$$

because $V_{PA} = \omega r_{PA} = \omega(V_A/\omega) = V_A$ and is opposite in direction to the vector $\mathbf{V}_A$. Thus, the body is instantaneously rotating about P. The velocity of any third point C is $V_C = \omega r_{PC}$.

This instantaneous axis may also be located when the velocities of two points in the body are given. In Fig. 4-20a points A and C have velocities $\mathbf{V}_A$ and $\mathbf{V}_C$. Perpendiculars to $\mathbf{V}_A$ and $\mathbf{V}_C$ intersect at P, the instantaneous center of rotation. Figure 4-20b shows how to locate P when the three points A, C, and P are on the same line.

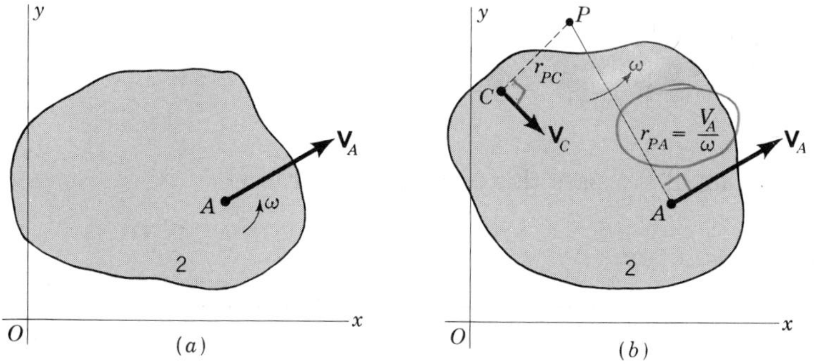

Fig. 4-19

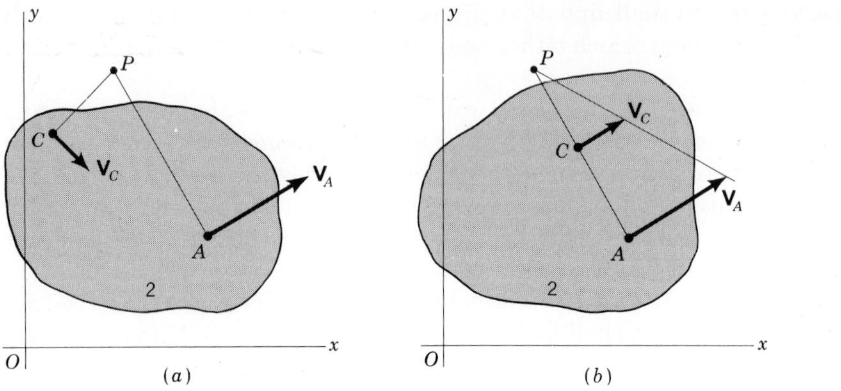

Fig. 4-20

In the analysis of plane-motion mechanisms many names have been used to designate these instantaneous axes. Some of them are *instantaneous centers, instant centers, IC's, centros,* and *poles.* In this book we shall refer to them either as instantaneous centers or as poles.

The paths of instantaneous centers are called *centrodes* or *polodes.* We shall call them polodes.

The velocity pole, in general, is *not* a stationary point, because it can have both a velocity and an acceleration. The velocity pole is defined as *the instantaneous location of a point common to two bodies which has the same velocity in each.* It may also be defined as *the instantaneous location of a point on one body about which another body is instantaneously rotating.*

Since we have adopted the convention of numbering the elements of a mechanism, it is convenient to designate an instantaneous center by using the numbers of the links associated with it. Thus P_{32}, or simply 32, identifies the pole common to links 3 and 2. (The order of the numbers has no significance.) A mechanism has as many poles as there are ways of pairing the members. Thus the number of poles in an n-link mechanism is

$$N = \frac{n(n-1)}{2} \tag{4-27}$$

4-11 THE ARONHOLD–KENNEDY THEOREM OF THREE CENTERS

According to Eq. (4-27) the number of poles in a four-bar linkage is six. In Fig. 4-21 we can identify four poles, A, B, C, and D, all of which satisfy the definition. Point B, for example, is a point in 2 about which 3

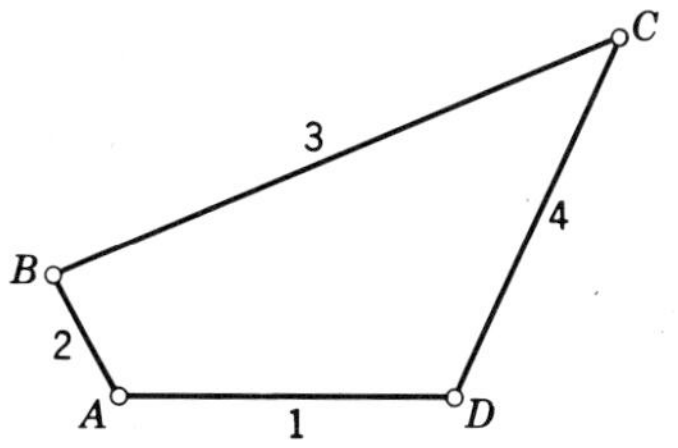

Fig. 4-21

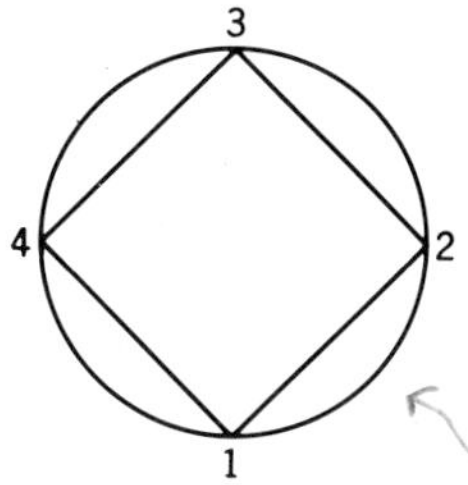

Fig. 4-22

rotates; it is a point in 3 about which 2 rotates; or it is a point common to 2 and 3 having the same velocity in each.

A good way to identify all the poles in an n-link mechanism is to space the link numbers around a circle, as shown in Fig. 4-22. Then each time a pole is found, a line is drawn connecting the pair of numbers which identify the pole. In Fig. 4-21 we have found P_{12}, P_{23}, P_{34}, and P_{14}, and the corresponding lines for these are shown in Fig. 4-22. The diagram shows missing lines for P_{13} and P_{24}. The location of these cannot be found from the definition of a pole.

After all possible poles have been found by using the definition of a velocity pole, the remaining ones are located by using the Aronhold-Kennedy theorem of three centers. This theorem, sometimes called Kennedy's rule, states: *When three bodies move relative to one another, they have three velocity poles all of which lie on the same straight line.*

To demonstrate the theorem, consider the three bodies of Fig. 4-23. Link 1 is the stationary frame containing pole 12, about which link 2 rotates, and pole 13, about which link 3 rotates. The shape of links 2 and

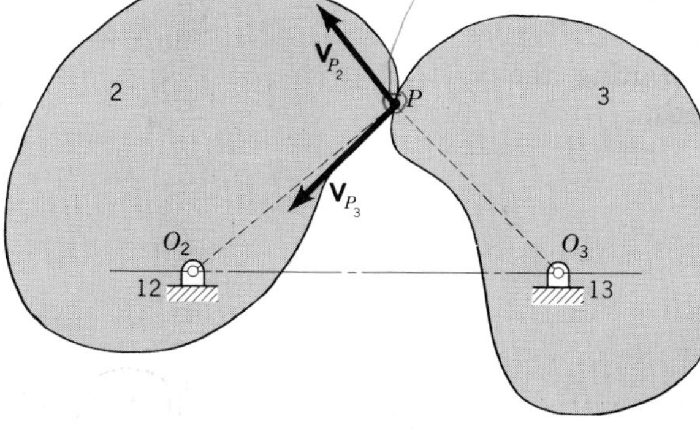

Fig. 4-23

3 is not defined, the only stipulation being that they each have different velocities; that is, links 2 and 3 have motion relative to each other. Now, Kennedy's rule states that the three poles 12, 13, and 23 must all lie on the same straight line. Suppose they are not on the same line; in fact, let us suppose that the third pole 23 is located at P in Fig. 4-23. Then the velocity of P as a point in link 2 would be in the direction $\mathbf{V}_{P_2}$ perpendicular to PO_2. But the velocity of P as a point in link 3 is $\mathbf{V}_{P_3}$ perpendicular to PO_3. These directions are not coincident. Yet the definition of a pole includes the statement that the point must be common to the two links with the same velocity in each. The point P, then, cannot be a velocity pole. In fact, the only location for P_{23} which will give the same direction for $\mathbf{V}_{P_2}$ and $\mathbf{V}_{P_3}$ is on the line joining the poles P_{12} and P_{13}. Thus all three poles must lie on the same straight line.

The examples which follow illustrate the methods of locating velocity poles.

Example 4-3 Find all the poles for the six-link mechanism of Fig. 4-24.

Solution The circle shown in the figure is used to find the poles. Those which can be found from the definition of a pole are connected on the circle diagram with

Note how useful the circle diagram is in simplifying things.

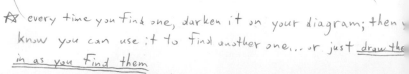

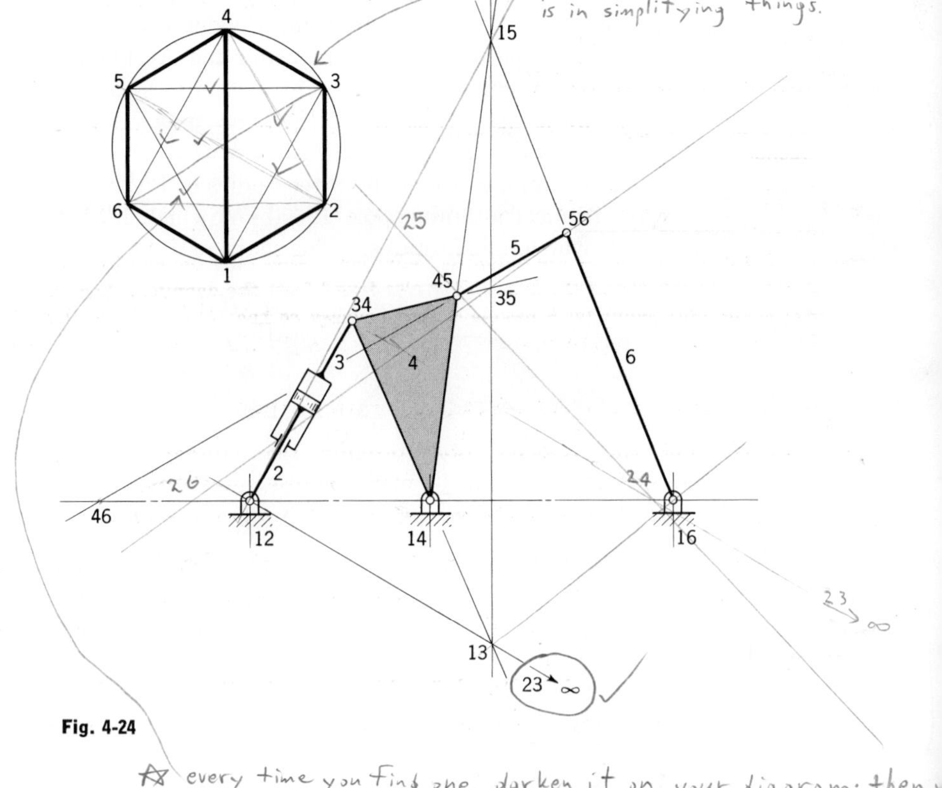

Fig. 4-24

every time you find one, darken it on your diagram; then you know you can use it to find another one... or just draw them in as you find them

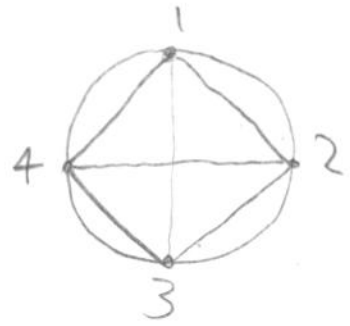

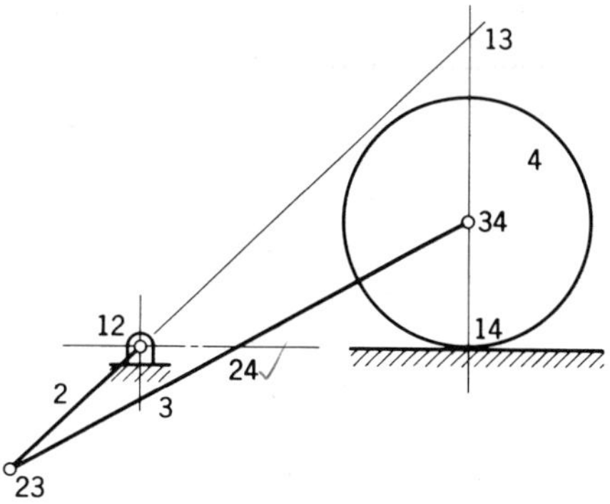

Fig. 4-25

heavy lines. Note that pole 23 is also obtained by definition. It is located at infinity on any line perpendicular to the direction of sliding of link 2 relative to link 3. This is equivalent to relative motion on a circle having an infinite radius.

The remaining poles are found by using Kennedy's rule, and these are indicated by thin lines on the circle diagram. To find a pole, locate any two triangles having a common side which identifies the desired pole. For example, to find 13, use 12 and 23 for the first straight line, and 14 and 43 for the second. Other poles can be found by using the following pairs of lines:

Pole 46: Use (45,56) and (14,16)
Pole 15: Use (16,56) and (14,45)
Pole 35: Use (34,45) and (13,15)

Not shown are poles 24, 25, 26, and 36; these are left as an exercise for the reader.

Example 4-4 Find all the poles for the mechanism in Fig. 4-25. Wheel 4 rolls on frame 1 without slipping.

Solution Poles 12, 23, and 34 are found by definition. Pole 14 is the point of tangency of the wheel and the frame; it is also found from the definition, because P_{14} is the only point on 4 having a zero velocity at the instant considered. P_{13} and P_{24} are then easily located by using Kennedy's rule.

4-12 VELOCITY ANALYSIS USING INSTANTANEOUS CENTERS

The properties of the velocity pole provide two simple graphical approaches for the velocity analysis of plane-motion mechanisms. We shall term these the *line-of-centers method* and the *link-to-link method.*

The line-of-centers method In Fig. 4-26*a* we assume that the angular velocity ω_2 of crank 2 of the four-bar linkage is given and we wish to find the velocities of points *B*, *D*, and *E*. The first step is to find all the poles; these are shown in the figure. Next, calculate $\mathbf{V}_4$.

Consider now the straight line defined by the poles P_{14}, P_{12}, and P_{24}. This is the line of Kennedy's theorem for links 1, 2, and 4 and is called

Use line of center ~~only~~ on test; not link-to-link

Line of centers on test.

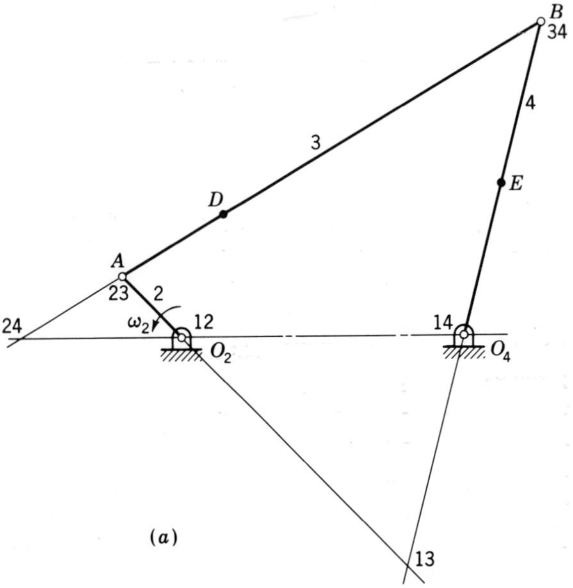

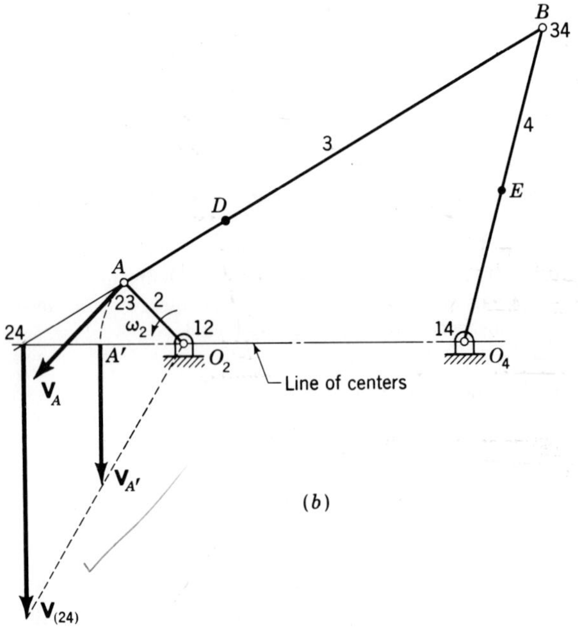

Fig. 4-26 Graphical velocity determination by the line-of-centers method.

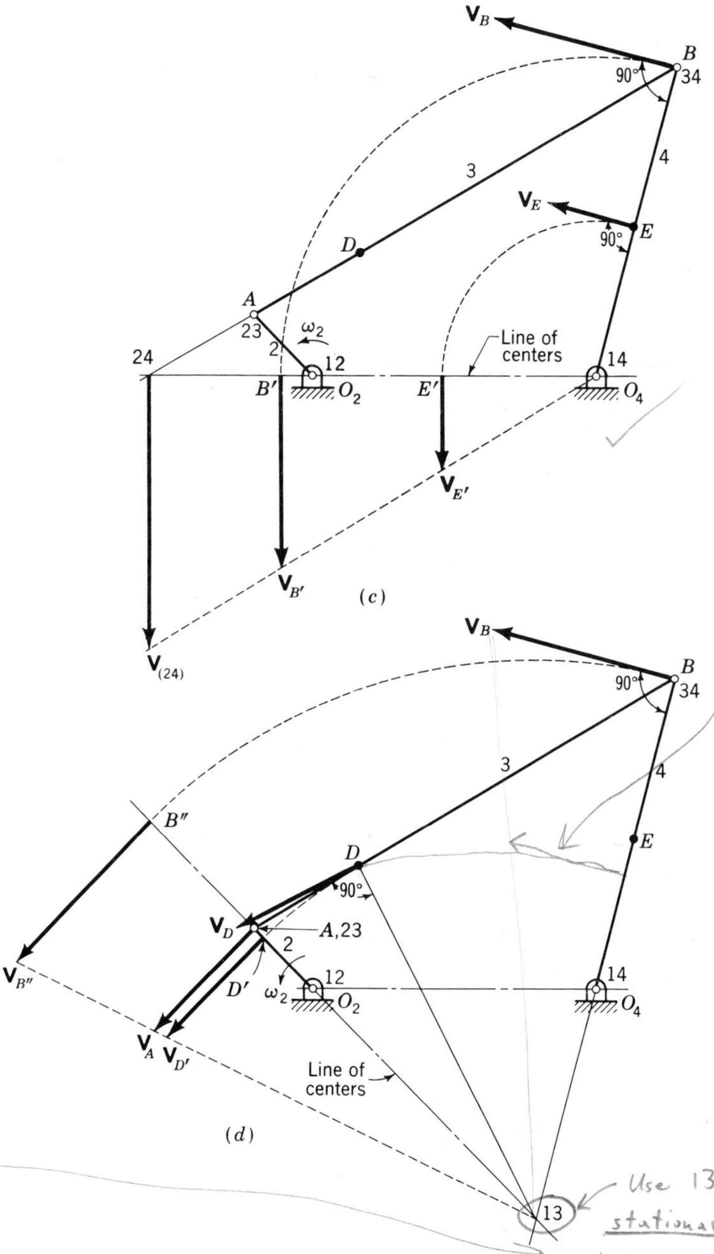

Fig. 4-26 *(Continued)*

note this other way of finding V_D

Use 13 because 1 is stationary. Since 1 is stationary, 12 is, and 13 is and, ..., 1N is.

right on...

the *line of centers*. According to the definition of a pole, P_{24} is common to links 2 and 4 and has the same velocity in each.

For the moment consider P_{24} as a point in 2. Since $\mathbf{V}_A$ is known, $\mathbf{V}_{(24)}$ can be found. The construction is shown in Fig. 4-26*b*. Rotate point A to A'. Both A and A' are points in 2 at the same radius from O_2; so their velocity magnitudes are equal. Using dividers, lay off $\mathbf{V}_{A'}$ as shown. Now P_{24} and A' are on the same line of centers and $\mathbf{V}_{(24)}$ is found by extending a line from O_2 through the terminus of $\mathbf{V}_{A'}$ as shown.

Next, consider P_{24} as a point in 4. Since $\mathbf{V}_{(24)}$ is known, we can find the velocity of any other point in 4, such as $\mathbf{V}_{B'}$ or $\mathbf{V}_{E'}$ (Fig. 4-26*c*). Since B' and E' have the same radii in 4 as B and E, we now have the magnitudes of $\mathbf{V}_B$ and $\mathbf{V}_E$, and these can be laid off in their proper direction, as shown in Fig. 4-26*c*.

To obtain $\mathbf{V}_D$ we note that D is in link 3; the known velocity $\mathbf{V}_A$ is in link 2, and link 1 is the reference link. Therefore links 1, 2, and 3 define the poles P_{12}, P_{13}, and P_{23} which lie on the line of centers for finding $\mathbf{V}_D$. P_{23} is the common pole and its velocity is given. Therefore, simply rotate D to D' on the line of centers using P_{13} as the center of the arc (Fig. 4-26*d*). The reason for this is that 3 is instantaneously rotating about P_{13}. $\mathbf{V}_{D'}$ is now found by using similar triangles, and its magnitude is transferred to D to give $\mathbf{V}_D$, which is perpendicular to the line from D to P_{13}. Since B may also be considered as a point in 3, its velocity may be found in a similar manner, as shown.

The line-of-centers method of velocity determination may be summarized as follows:

1. Identify the link containing the point whose velocity is known, the link whose velocity is to be found, and the reference or ground link.
2. Locate the three poles defined by step 1 and draw the line of centers through them. This line of centers now contains a common pole and two poles on the frame.
3. Find the velocity of the common pole, considering it as a point in the link whose velocity is known.
4. With the velocity of the common pole known, consider it as a point in the link whose velocity is unknown. The velocity of any other point in that link can now be found by using similar triangles.

The link-to-link method This method is applied by progressing from the point of known velocity through a connecting link or links and finding the velocities of points on each link along the way. The poles necessary for this method are usually easier to find than those for the line-of-centers method.

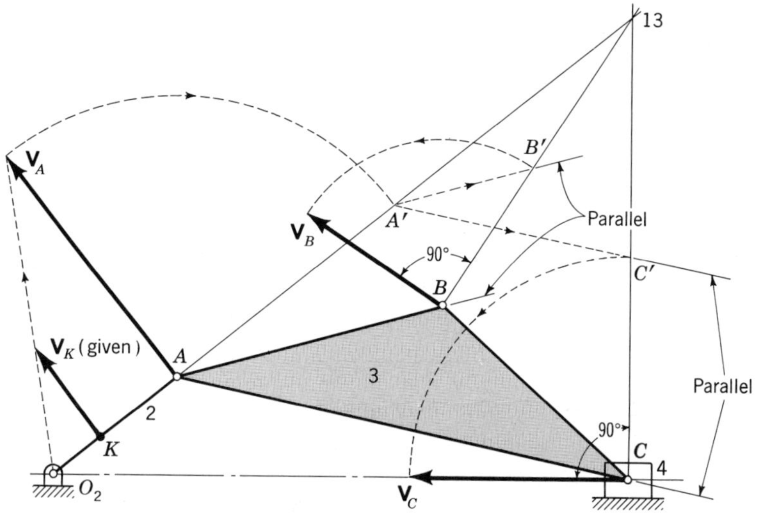

Fig. 4-27 Velocity determination by the link-to-link method.

The link-to-link solution of the slider-crank mechanism is illustrated in Fig. 4-27. The velocity of a point K on the crank is given. The velocities of points B and C on the connecting rod are to be found. Only one pole P_{13} in this example need be found; this is P_{13}, the point in 3 having an instantaneous zero velocity. Thus all points in 3 may be considered as instantaneously rotating about P_{13}. $\mathbf{V}_A$ is found first by using similar triangles and the known value of $\mathbf{V}_K$. Then the velocity of B is found as follows: Rotate $\mathbf{V}_A$ into its radius line, locating point A'. Through A' draw $A'B'$ parallel to AB. Then BB' is the magnitude of $\mathbf{V}_B$, and its direction is perpendicular to the radius line. For proof note that P_{13} with A' and B' and P_{13} with A and B form two similar triangles and that the segments AA' and BB' are proportional to the radii of A and B from the pole 13.

The velocity of C is found in a similar manner.

The directions of the velocity vectors may be found by imagining a small displacement of the linkage. Then the velocities must have a direction such that they all produce an angular velocity about P_{13} of the same sign.

Another link-to-link velocity analysis is shown in Fig. 4-28. Here we begin with the velocity of point A on link 2 and proceed successively through links 3, 4, 5, 6, and 7 to obtain the velocity of point E.

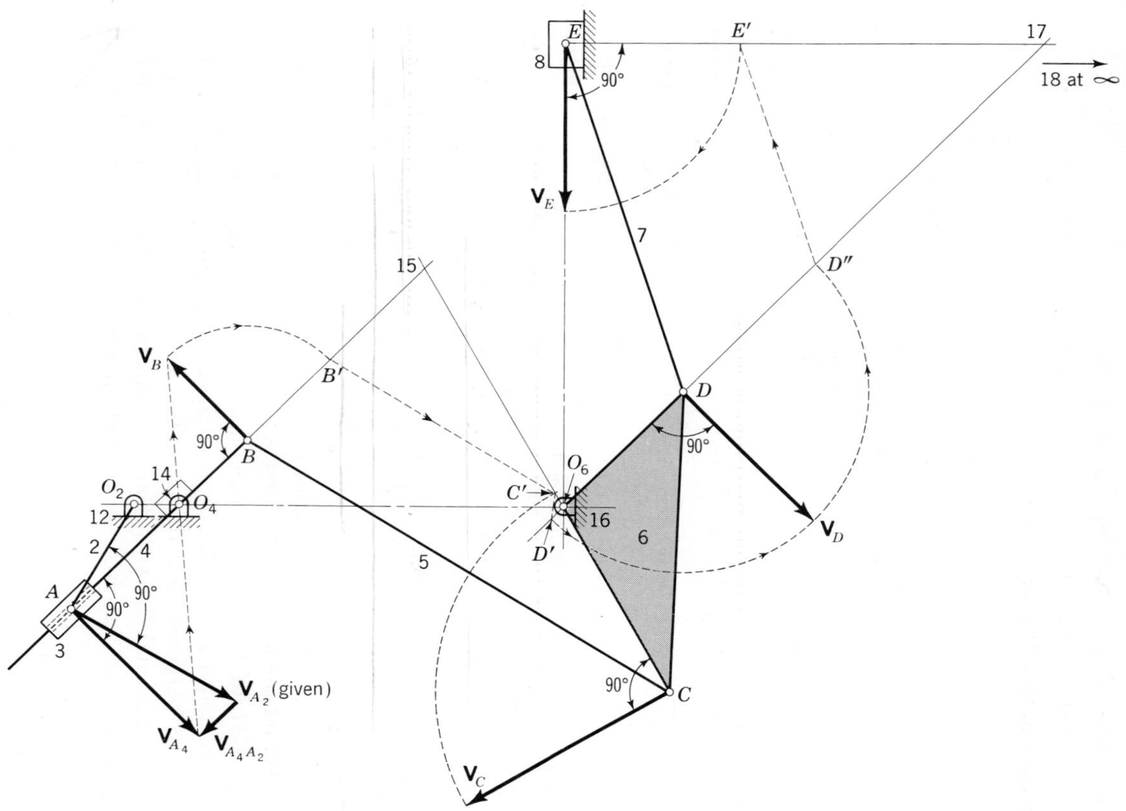

Fig. 4-28 Velocity determination by the link-to-link method.

4-13 THE ANGULAR-VELOCITY-RATIO THEOREM

In Fig. 4-29 P_{24} is the pole common to links 2 and 4. The velocity $\mathbf{V}_{(24)}$ is the same whether P_{24} is considered as a point in 2 or in 4. Let the magnitude of this velocity be

$$V_{(24)} = p_2\omega_{21} \tag{a}$$

where ω_{21} is the same as ω_2 but the additional subscript 1 has been added to emphasize the presence of the third link (the frame).

But, considering P_{24} as a point in link 4, we have

$$V_{(24)} = p_4\omega_{41} \tag{b}$$

Dividing Eq. (a) by (b) and rearranging gives

$$\frac{\omega_{21}}{\omega_{41}} = \frac{p_4}{p_2} \tag{4-28}$$

Equation (4-28) is called the angular-velocity-ratio theorem. This theorem relates the angular velocities of any two bodies, 2 and 4 in this instance, relative to a third body, frame 1. The theorem states that *the angular-velocity ratio of any two bodies relative to a third body is inversely proportional to the segments into which the common pole cuts the line of centers.*

The reader should prove for himself that the angular-velocity ratio is positive when the common pole lies outside the fixed poles (Fig. 4-29), and negative when it lies inside the fixed poles.

4-14 DIRECT CONTACT

Two elements of a mechanism which are in direct contact with each other may have a relative motion which is either pure rolling, pure sliding, or

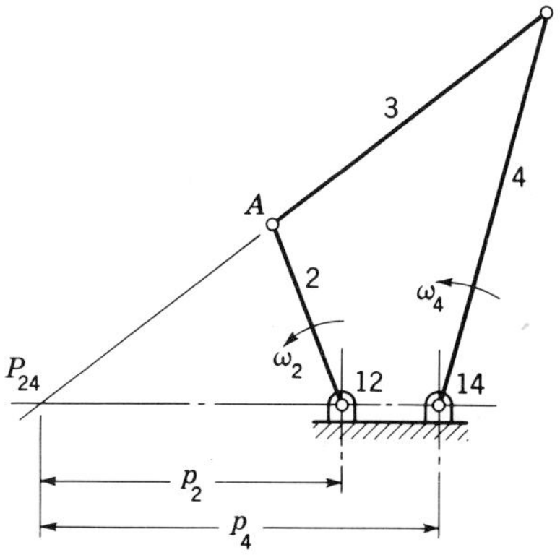

Fig. 4-29

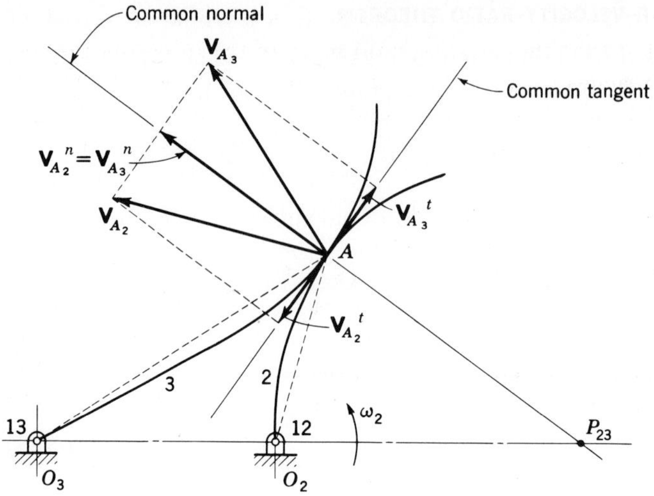

Fig. 4-30

both rolling and sliding. In Fig. 4-30 link 2 drives link 3 through direct contact, and the relative motion is combined rolling and sliding.

If the motion of 2 relative to 3 were pure rolling, the common pole P_{23} would be located at the point of contact A. If the motion were pure sliding, the common pole would be on the common normal at infinity. By Kennedy's rule P_{23} must lie on a line joining the fixed poles 12 and 13; hence the relative motion is mixed rolling and sliding.

Let us now refer to $\mathbf{V}_{A_2}$ as the velocity of point A on link 2 and $\mathbf{V}_{A_3}$ as the velocity of point A on link 3. Then $\mathbf{V}_{A_2}$ is perpendicular to O_2A and $\mathbf{V}_{A_3}$ is perpendicular to O_3A, as shown in Fig. 4-30. The components of these two velocities along the common normal are $\mathbf{V}_{A_2}{}^n$ and $\mathbf{V}_{A_3}{}^n$. These must be equal; otherwise link 2 will move out of contact or dig into link 3. The difference between the tangential components is the sliding velocity $\mathbf{V}_{A_3A_2}$, or

$$\mathbf{V}_{A_3A_2} = \mathbf{V}_{A_3}{}^t - \mathbf{V}_{A_2}{}^t = \mathbf{V}_{A_3} - \mathbf{V}_{A_2} \tag{4-29}$$

because $\mathbf{V}_{A_3}{}^n = \mathbf{V}_{A_2}{}^n$.

For pure rolling $\mathbf{V}_{A_3A_2} = 0$ and P_{23} must lie on the line of centers. Therefore the point of contact must also lie on the line of centers.

For pure sliding the common normal is parallel to the line of centers.

4-15 FREUDENSTEIN'S THEOREM

In the analysis and design of linkages it is often important to know the phases of the linkage at which the extreme values of the velocity occur.

The earliest work in determining extreme values is apparently that of Krause,[1] who stated that the velocity ratio ω_4/ω_2 of the drag-link mechanism (Fig. 4-31) reaches an extreme value when the connecting rod and driven link (links 3 and 4) become perpendicular to each other. Rosenauer,[2] however, has shown that this is not strictly true. Following Krause, Ferdinand Freudenstein of Columbia University developed a simple graphical method for determining the phases of the mechanism at which extreme values of the velocity ratio occur.[3]

Freudenstein's theorem makes use of the two moving poles of the four-bar linkage (Fig. 4-32). The two poles are 24 and 13 and are designated as P and Q, respectively. The line PQ is called the *collineation axis.* The theorem states: *At an extreme value of the velocity ratio in a four-bar linkage, the collineation axis PQ is perpendicular to the connecting link AB.*[4]

Using Eq. (4-28) we write

$$\frac{\omega_4}{\omega_2} = \frac{PO_2}{PO_2 + O_2O_4}$$

Since O_2O_4 is a fixed distance, the extremes of the velocity ratio occur when the distance PO_2 is maximum. Such positions may occur on either or both sides of O_2. Thus the problem reduces to that of finding the

[1] R. Krause, Die Doppelkurbel und Ihre Geschwindigkeitsgrenzen, *Maschinenbau/Getriebetechnik,* vol. 18, pp. 37–41, 1939; Zur Synthese der Doppelkurbel, *Maschinenbau/Getriebetechnik,* vol. 18, pp. 93–94, 1939.

[2] N. Rosenauer, Synthesis of Drag-link Mechanisms for Producing Nonuniform Rotational Motion within Prescribed Reduction Ratio Limits, *Australian J. Appl. Sci.,* vol. 8, pp. 1–6, 1957.

[3] Ferdinand Freudenstein, On the Maximum and Minimum Velocities and the Accelerations in Four-link Mechanisms, *Trans. ASME,* vol. 78, pp. 779–787, 1956.

[4] A. S. Hall of Purdue University contributed a rigorous proof of the theorem in the Appendix to Freudenstein's paper.

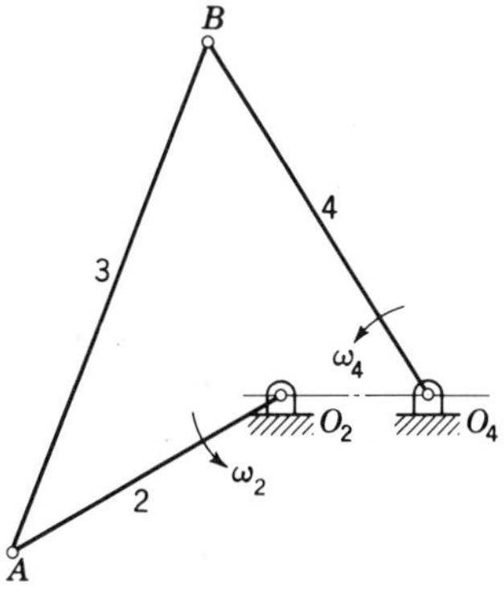

Fig. 4-31 The drag-link mechanism.

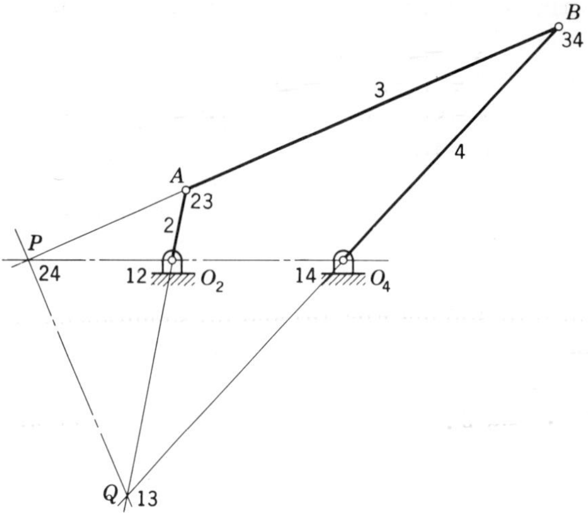

Fig. 4-32 The collineation axis.

geometry of the linkage when PO_2 is maximum. During motion P moves along O_2O_4 or its extension, but at an extreme value of the velocity ratio P must be instantaneously at rest. This occurs when the velocity of P, considered as a point in 3, is directed along AB. This will be true only when link 3 extended is perpendicular to the collineation axis because Q is the instantaneous center of rotation of link 3.

An inversion of the theorem states: *An extreme value of the velocity ratio ω_3/ω_2 of the four-bar linkage occurs when the collineation axis is perpendicular to the driven link (link 4).*

4-16 ANALYTICAL METHODS

When a velocity analysis of a mechanism must be performed without the aid of a computer, the graphic methods are undoubtedly the quickest to use. It is more economical, though, to use the digital or the analog computer when many mechanisms must be analyzed for all phases of their motions. For this reason, in this section we shall include algebraic and vector methods which are particularly useful for computer programming.

Consider first the slider-crank mechanism shown in Fig. 4-33. Because this is the engine mechanism, we have designated the crank radius as r and the connecting-rod length as l. From the geometry of the system

$$r \sin \theta = l \sin \phi \tag{a}$$

$$x = r \cos \theta + l \cos \phi \tag{b}$$

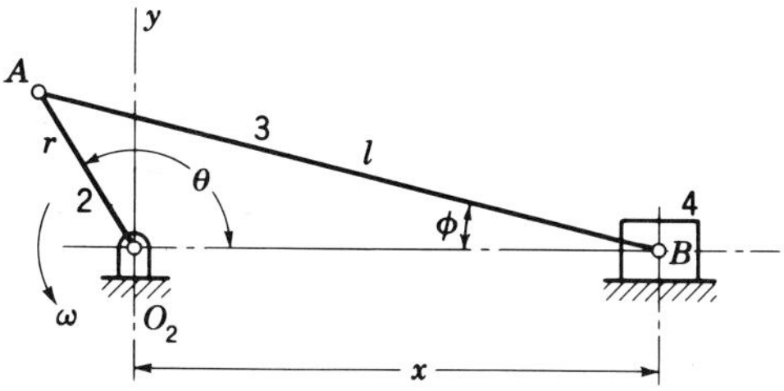

Fig. 4-33

The slider position is obtained by solving Eqs. (*a*) and (*b*) simultaneously to eliminate ϕ. This gives

$$x = r\cos\theta + l\sqrt{1 - \left(\frac{r}{l}\sin\theta\right)^2} \tag{4-30}$$

Differentiating Eq. (*b*) gives the slider velocity as

$$\dot{x} = -r\omega\left(\sin\theta + \frac{r\sin 2\theta}{2l\cos\phi}\right) \tag{4-31}$$

or, if ϕ is eliminated,

$$\dot{x} = -r\omega\left\{\sin\theta + \frac{r\sin 2\theta}{2l\sqrt{1 - [(r/l)\sin\theta]^2}}\right\} \tag{4-32}$$

The *Scotch yoke* shown in Fig. 4-34 is an interesting variation of the slider-crank mechanism. Here, link 3 rotates about a pole located at infinity. This has the effect of a connecting rod of infinite length, and the second terms of Eqs. (4-30) to (4-32) become zero. Thus the slider

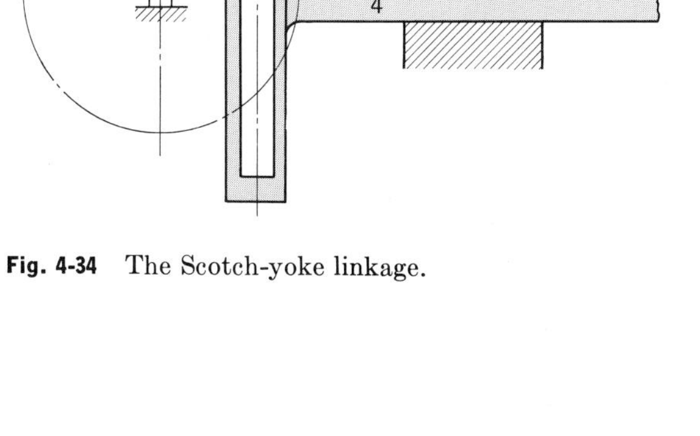

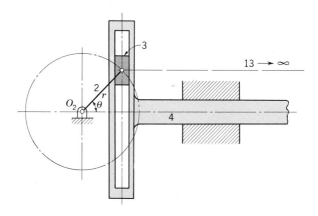

Fig. 4-34 The Scotch-yoke linkage.

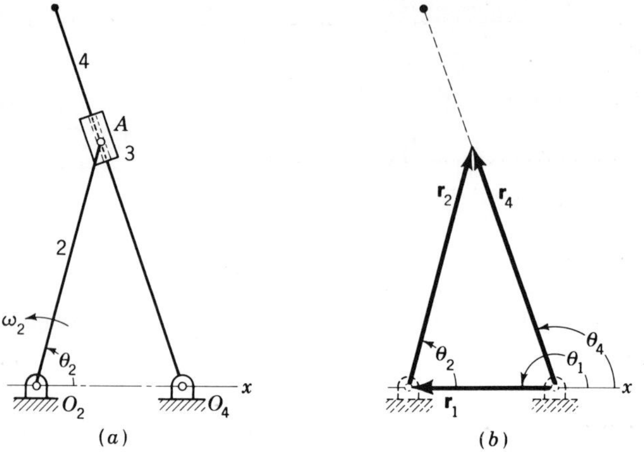

Fig. 4-35

moves with simple harmonic motion. It is for this reason that the deviation of the kinematics of the slider-crank motion from simple harmonic motion is sometimes said to be due to the *angularity of the connecting rod.*

A method of analyzing both plane and space mechanisms using complex polar notation has been developed by Raven.[1] The method consists in replacing each link of the mechanism by a vector and summing these vectors around every closed loop. The resulting equation or equations are then written by using complex numbers in the polar form and differentiated to obtain a velocity relation. Another transformation places the equations in complex rectangular notation and allows expression of the unknowns in terms of the knowns.

To illustrate Raven's approach let us analyze the inversion of the slider-crank linkage shown in Fig. 4-35*a*. We consider link 2 as the driver, with a constant angular velocity ω_2 and an angular position for the instant considered of θ_2. Replacing each link by its position vector, as shown in Fig. 4-35*b*, and summing around the loop gives

$$\mathbf{r}_1 + \mathbf{r}_2 - \mathbf{r}_4 = 0 \qquad (c)$$

Then transforming to complex notation yields

$$r_1 e^{j\theta_1} + r_2 e^{j\theta_2} - r_4 e^{j\theta_4} = 0 \qquad (d)$$

Since r_1, r_2, θ_1, and θ_2 are assumed to be given, Eq. (*d*) can be solved

[1] Francis H. Raven, Velocity and Acceleration Analysis of Plane and Space Mechanisms by Means of Independent-position Equations, *J. Appl. Mech.*, vol. 80, pp. 1–6, 1958.

directly for r_4 and θ_4. Remembering that r_1, r_2, and θ_1 are constants, we differentiate Eq. (d) to obtain

$$jr_2\dot{\theta}_2 e^{j\theta_2} - \dot{r}_4 e^{j\theta_4} - jr_4\dot{\theta}_4 e^{j\theta_4} = 0 \tag{e}$$

Since $\dot{\theta}_2$ and $\dot{\theta}_4$ are, respectively, the same as ω_2 and ω_4, Eq. (e) is, in fact, the equation of the velocity polygon, with

$$V_{A_2} = r_2\omega_2 \qquad V_{A_2A_4} = \dot{r}_4 \qquad V_{A_4} = r_4\omega_4$$

and is the complex form of the relative-velocity equation

$$\mathbf{V}_{A_2} = \mathbf{V}_{A_4} + \mathbf{V}_{A_2A_4} \tag{f}$$

Next, transform Eq. (e) to complex rectangular notation and separate the real and the imaginary terms. This gives

$$\begin{aligned} r_2\omega_2 \cos\theta_2 - \dot{r}_4 \sin\theta_4 - r_4\omega_4 \cos\theta_4 &= 0 \\ -r_2\omega_2 \sin\theta_2 - \dot{r}_4 \cos\theta_4 + r_4\omega_4 \sin\theta_4 &= 0 \end{aligned} \tag{g}$$

Solving these two equations simultaneously for $\dot{r}_4$ and ω_4 yields

$$\dot{r}_4 = r_2\omega_2 \sin(\theta_4 - \theta_2) \tag{4-33}$$

$$\omega_4 = \frac{r_2\omega_2}{r_4}\cos(\theta_4 - \theta_2) \tag{4-34}$$

The solution for the four-bar linkage is obtained in a similar manner. First, replace each link with a position vector, sum the vectors around a closed loop, then replace each vector in the resulting equation with a complex number in polar form. This gives

$$r_1 e^{j\theta_1} + r_2 e^{j\theta_2} + r_3 e^{j\theta_3} + r_4 e^{j\theta_4} = 0 \tag{h}$$

Since link 1 is the fixed link, the first time derivative is

$$jr_2\dot{\theta}_2 e^{j\theta_2} + jr_3\dot{\theta}_3 e^{j\theta_3} + jr_4\dot{\theta}_4 e^{j\theta_4} = 0 \tag{i}$$

As before, we transform the equation to rectangular form and then separate the real and the imaginary terms. With ω_2 given, the unknowns are ω_3 and ω_4. When the resulting two equations are solved simultaneously, the results are

$$\omega_3 = \frac{r_2\omega_2}{r_3}\,\frac{\sin(\theta_2 - \theta_4)}{\sin(\theta_4 - \theta_3)} \tag{4-35}$$

$$\omega_4 = \frac{r_2\omega_2}{r_4}\,\frac{\sin(\theta_2 - \theta_3)}{\sin(\theta_4 - \theta_3)} \tag{4-36}$$

In the Chace approach to the velocity analysis of the four-bar linkage the vectors are summed around a closed loop, as before.[1] Thus the position equation for the four-bar linkage is

$$\mathbf{r}_1 + \mathbf{r}_2 + \mathbf{r}_3 + \mathbf{r}_4 = 0 \tag{j}$$

Note that this is the same as Eq. (*h*). The given quantities are r_1, r_2, r_3, r_4, $\hat{\mathbf{r}}_1$, $\hat{\mathbf{r}}_2$, and ω_2. We assume that Eq. (*j*) has already been solved by using Eq. (3-19) or (3-20). Therefore, we begin the velocity analysis with $\hat{\mathbf{r}}_3$ and $\hat{\mathbf{r}}_4$ also known. The unknown quantities are ω_3 and ω_4; once these are known the velocities of all points of interest may be found.

Differentiating Eq. (*j*) with respect to time yields

$$\boldsymbol{\omega}_2 \times \mathbf{r}_2 + \boldsymbol{\omega}_3 \times \mathbf{r}_3 + \boldsymbol{\omega}_4 \times \mathbf{r}_4 = 0 \tag{k}$$

because r_2, r_3, and r_4 are constants. But in a plane-motion mechanism all the angular velocities are in the $\hat{\mathbf{k}}$ direction. Hence we can write

$$r_2\omega_2(\hat{\mathbf{k}} \times \hat{\mathbf{r}}_2) + r_3\omega_3(\hat{\mathbf{k}} \times \hat{\mathbf{r}}_3) + r_4\omega_4(\hat{\mathbf{k}} \times \hat{\mathbf{r}}_4) = 0 \tag{l}$$

To solve this equation, take the dot product with $\hat{\mathbf{r}}_4$ and then with $\hat{\mathbf{r}}_3$. Each of these operations reduces Eq. (*l*) to a scalar equation with only a single unknown. Thus, dotting through with $\hat{\mathbf{r}}_4$ gives

$$r_2\omega_2(\hat{\mathbf{k}} \times \hat{\mathbf{r}}_2) \cdot \hat{\mathbf{r}}_4 + r_3\omega_3(\hat{\mathbf{k}} \times \hat{\mathbf{r}}_3) \cdot \hat{\mathbf{r}}_4 = 0 \tag{m}$$

since

$$r_4\omega_4(\hat{\mathbf{k}} \times \hat{\mathbf{r}}_4) \cdot \hat{\mathbf{r}}_4 = 0 \tag{n}$$

Solving Eq. (*m*) for ω_3 then yields

$$\omega_3 = -\frac{r_2\omega_2}{r_3}\,\frac{(\hat{\mathbf{k}} \times \hat{\mathbf{r}}_2) \cdot \hat{\mathbf{r}}_4}{(\hat{\mathbf{k}} \times \hat{\mathbf{r}}_3) \cdot \hat{\mathbf{r}}_4} \tag{4-37}$$

If we then dot Eq. (*l*) through with $\hat{\mathbf{r}}_3$ in a similar manner and solve the result for ω_4, we get

$$\omega_4 = -\frac{r_2\omega_2}{r_4}\,\frac{(\hat{\mathbf{k}} \times \hat{\mathbf{r}}_2) \cdot \hat{\mathbf{r}}_3}{(\hat{\mathbf{k}} \times \hat{\mathbf{r}}_4) \cdot \hat{\mathbf{r}}_3} \tag{4-38}$$

That these equations are indeed the same as Raven's can be shown quite simply. From Eq. (4-38) we can write

$$\hat{\mathbf{k}} \times \hat{\mathbf{r}}_2 = \begin{vmatrix} \hat{\mathbf{i}} & \hat{\mathbf{j}} & \hat{\mathbf{k}} \\ 0 & 0 & 1 \\ \cos\theta_2 & \sin\theta_2 & 0 \end{vmatrix} = -\sin\theta_2\,\hat{\mathbf{i}} + \cos\theta_2\,\hat{\mathbf{j}}$$

[1] Milton A. Chace, Vector Analysis of Linkages, *J. Eng. Ind.*, ser. B, vol. 85, pp. 289–297, August, 1963.

Then

$$(\hat{\mathbf{k}} \times \hat{\mathbf{r}}_2) \cdot \hat{\mathbf{r}}_3 = (-\sin\theta_2\,\hat{\mathbf{i}} + \cos\theta_2\,\hat{\mathbf{j}}) \cdot (\cos\theta_3\,\hat{\mathbf{i}} + \sin\theta_3\,\hat{\mathbf{j}})$$

$$= -\sin\theta_2 \cos\theta_3 + \cos\theta_2 \sin\theta_3$$

$$= \sin(\theta_3 - \theta_2) \qquad (o)$$

In a similar manner we find that

$$(\hat{\mathbf{k}} \times \hat{\mathbf{r}}_4) \cdot \hat{\mathbf{r}}_3 = -\sin(\theta_4 - \theta_3) \qquad (p)$$

When the terms from Eqs. (o) and (p) are substituted into Eq. (4-38), the result is Eq. (4-36).

4-17 GRAPHICAL DIFFERENTIATION

The process of differentiating any function $x = f(t)$ by graphic means is illustrated in Fig. 4-36. Select some point on the curve, say point A, and draw the tangent BC. Since

$$\frac{dx}{dt} = \lim_{\Delta t \to 0} \frac{\Delta x}{\Delta t}$$

then the derivative at point A is approximately

$$\dot{x} = \frac{CD}{BD}$$

Now if we draw additional tangents at other points and construct triangles in a similar manner, and if, in addition, we construct all the triangles so that their abscissas are all the same and equal to BD, then all we must do to obtain the derived curve is to plot the ordinate of each of these triangles. Then the curve passing through these ordinate values will be the derivative of the original function.

Rather than construct these triangles on the given curve itself, it is expedient to use the same abscissa for all of them and to construct them adjacent to the axes of the derived curve. Thus, in Fig. 4-36b, triangle BCD of the function is represented by triangle PEO'. So the distance PO' is the common abscissa for all the triangles. Only lines parallel to the tangents need be drawn to complete the triangles. The distance PO' is called the *pole distance*, and its length is designated as h, having units of seconds for a velocity-time graph. The point A', measured from the t axis, thus represents the velocity of point A. With this method of construction the pole distance is first selected. It may have any convenient length. Then tangents are drawn at various points on the given curve and, with instruments, lines are drawn through the pole (point P)

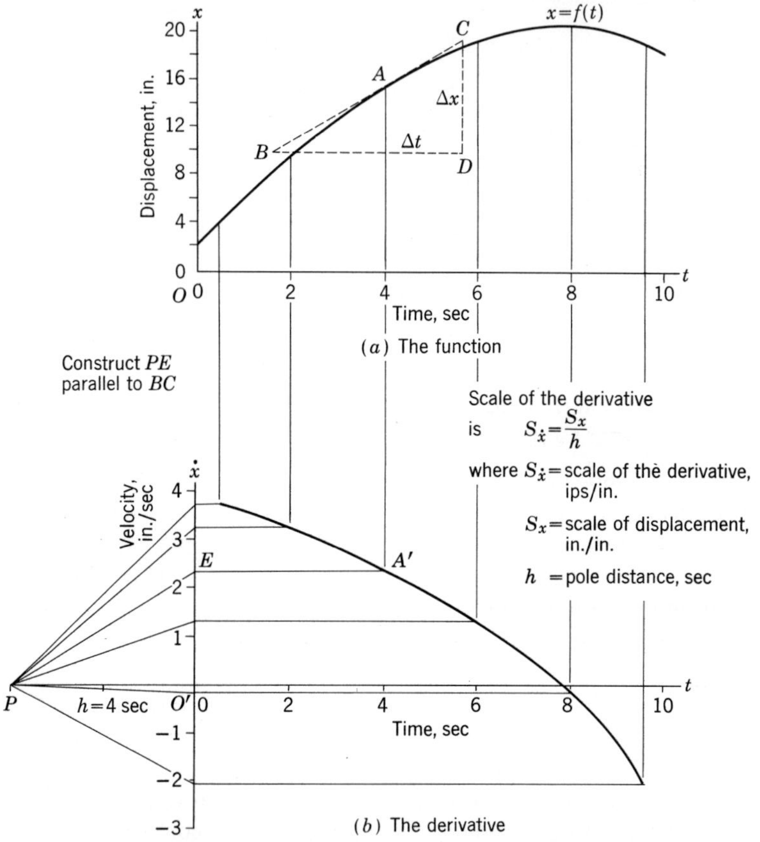

Fig. 4-36 Graphical differentiation.

parallel to the respective tangents. The intersection of these rays from the pole with the velocity axis gives the velocity.

The scale of the derivative is determined in the following manner:

$$S_{\dot{x}} = \frac{S_x}{h} \tag{4-39}$$

where $S_{\dot{x}}$ = scale of velocity, ips/in. or fps/in.
S_x = scale of displacement, in./in. or ft/in.
h = pole distance, sec

The displacement scale S_x means that so many inches (or feet) of displacement occupy 1 in. on the drawing. Similarly, the velocity scale $S_{\dot{x}}$ means that so many units of velocity (inches per second or feet per second) occupy 1 in. on the drawing. The time may also be in minutes.

It is desirable to employ the same time scale for both graphs. Thus the value of the time scale does not appear in Eq. (4-39).

The accuracy of the derived curve depends a great deal upon how well the tangents have been constructed. The number of points taken and the scale of the original curve also affect the accuracy. No general rules can be given for selecting the number of points, but if the curvature is rapid, then a great many should be employed. Of course, a steep curve can always be flattened out by increasing the length of the time axis.

Graphical differentiation deals only with magnitudes; that is, directions are not included. If the direction of the velocity vector is not constant, then the x and y components may be plotted separately and differentiated.

4-18 GRAPHICAL INTEGRATION

Integration is the reverse of differentiation. If the velocity-time relation is given, then the displacement-time relation may be obtained by integration; or if the acceleration-time relation is first defined, then the velocity-time graph may be obtained by integration. The method of graphical integration does not depend upon the construction of tangents for its accuracy, and it is inherently more accurate than graphical differentiation.

In order to demonstrate this method, a problem in simple integration will first be illustrated. In Fig. 4-37 a graph of some function $\dot{x} = f(t)$ has been plotted using suitable scales for both $\dot{x}$ and t. To integrate this function graphically means that we wish to derive a curve such that, for any value of t, say $t = t_1$, the ordinate of the derived curve will represent the area under the original curve between $t = 0$ and $t = t_1$. Between $t = 0$ and $t = 1$ there is a rectangular area of 10 square units. In Fig. 4-37*b* this point is plotted at the coordinates $t = 1$ and $x = 10$. Between $t = 0$ and $t = 1$ the area is directly proportional to t so that a straight line can be drawn between the origin and the first point. Now between $t = 1$ and $t = 2$ the area is 5 square units, and so the total area up to $t = 2$ is 15 square units. This value is plotted on the line $t = 2$, and a second straight line is drawn. In the same way an area of 30 square units is found between $t = 2$ and $t = 3$, making a total of 45 square units under the curve. This value is now plotted on the line $t = 3$, and the points connected. This completes the integration. It is noted that the inclination of the lines in the derived curve is proportional to the height of the rectangles in the original curve. For example, between $t = 1$ and $t = 2$ the rectangle has a small height and the derived curve a small inclination; on the other hand, between $t = 2$ and $t = 3$ the rectangle has a large height and the inclination of the derived curve is large.

In Fig. 4-38 the problem is reproduced. To integrate this function graphically we construct the pole distance OP a convenient length, and then draw the rays $P1$, $P2$, and $P3$. Since the slope of these rays is proportional to the height of the respective rectangles we may draw lines parallel to them to obtain the derived curve (integral). This has been done in Fig. 4-38*b*. The scale of the integral is

$$S_x = hS_{\dot{x}} \tag{4-40}$$

where S_x = scale of integral
h = pole distance, in time units
$S_{\dot{x}}$ = scale of given function

If the original function is a velocity curve, then its integral will be a displacement curve. The scale units for two such curves might be inches per second of velocity per inch of the drawing and inches of displacement per inch of drawing.

In the example above we have used rectangles to illustrate the integration process. When the original function is a curve, then it should be divided into narrow strips and the average height of each strip treated exactly as the rectangles have been treated above. The number of strips depends upon the accuracy desired. The strips need not be taken at equal time intervals.

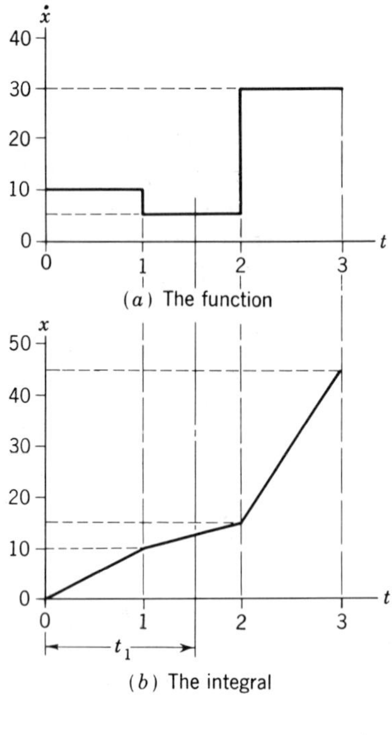

Fig. 4-37

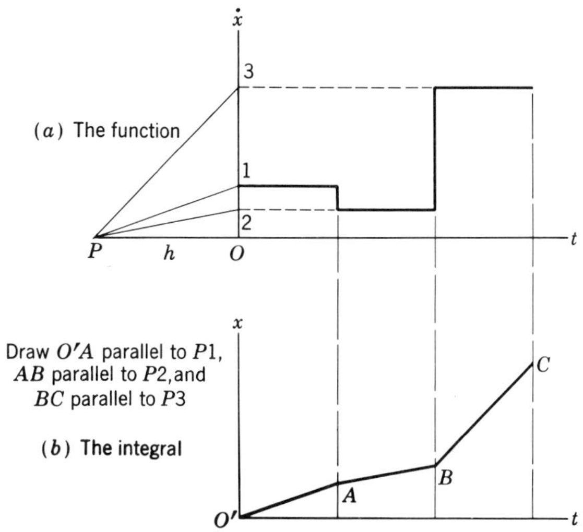

Fig. 4-38

PROBLEMS[1]

4-1. The position vector of a point is given by the equation $\mathbf{R} = 100e^{j\pi t}$, where $\mathbf{R}$ is in inches. Find the velocity of the point and the angular velocity of the position vector at $t = 0.40$ sec.

4-2. The equation $\mathbf{R} = (t^2 + 4)e^{-j\pi t/10}$ defines the path of a particle. If $\mathbf{R}$ is in inches, find the velocity of the particle at $t = 20$ sec.

4-3. If automobile A is traveling south at 55 mph and automobile B north 60° east at 40 mph, what is the velocity of B with respect to A?

4-4. In the figure, wheel 2 rotates at 600 rpm and drives wheel 3 without slipping. Find the velocity in feet per minute of point B on wheel 3 with respect to point A on wheel 2.

[1] In assigning problems, the instructor should specify the method of solution to be used, because a number of approaches have been presented in the text.

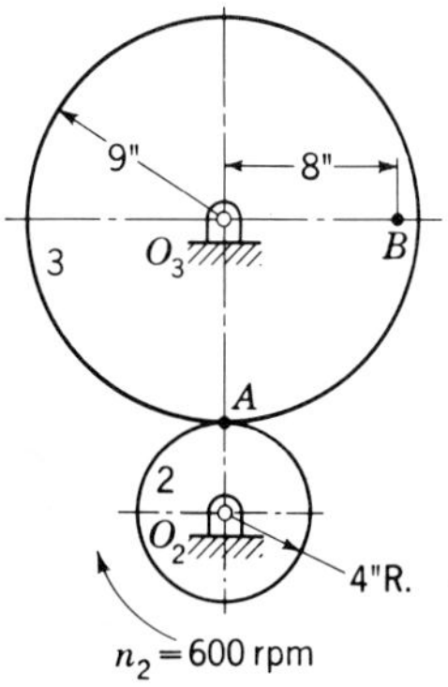

Prob. 4-4

4-5. Two points, A and B, located on the radius of a rotating wheel (see figure) have speeds of 800 and 1400 fpm, respectively.

(*a*) What is the distance between the points?

(*b*) Find $\mathbf{V}_{AB}$, $\mathbf{V}_{BA}$, and the rpm of the wheel.

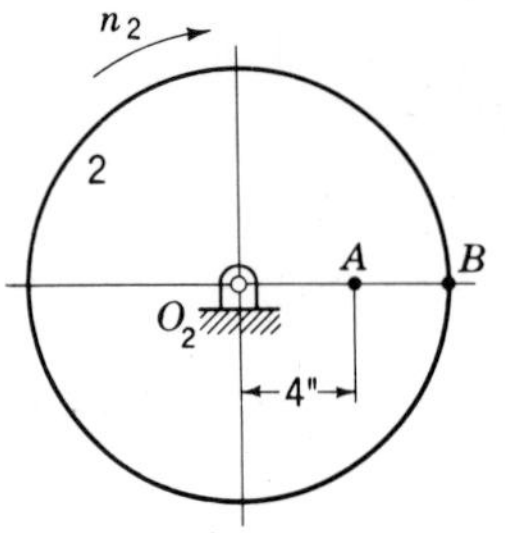

Prob. 4-5

60 mph = 88 ft/sec

4-6. A plane leaves point B and flies east at 350 mph. Simultaneously at point A, 200 miles southeast (see figure), a plane leaves and flies northeast at a speed of 390 mph.

(*a*) How close will the planes come to each other if they are flying at the same altitude?

(*b*) If they both leave at 6:00 PM, at what time will this occur?

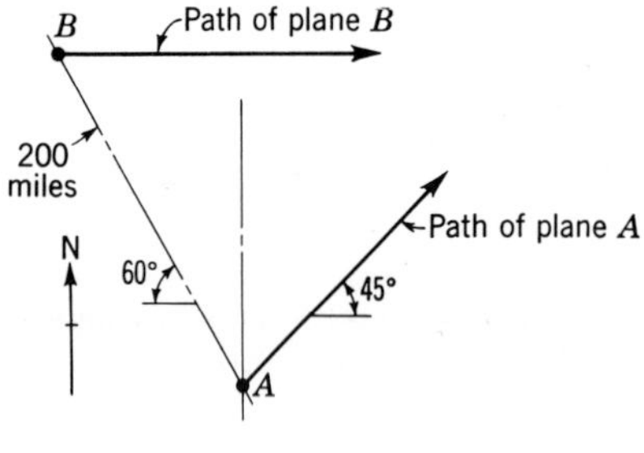

Prob. 4-6

4-7. To the data of Prob. 4-6 add a wind of 30 mph from the west.

(*a*) If A flies the same heading, what is its new path?

(*b*) What change will the wind make in the results of Prob. 4-6?

4-8. The velocity of point B for the linkage shown in the figure is 120 fps. Find $\mathbf{V}_A$ and ω_3.

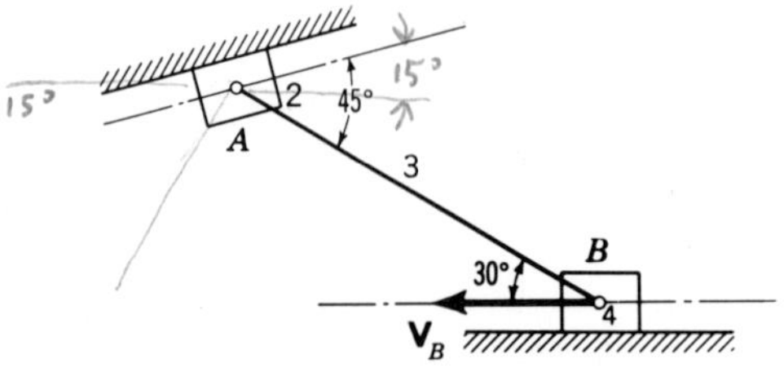

Prob. 4-8 $AB = 16$ in.

4-9. Find the velocity of B and the angular velocities of links 3 and 4 of the open four-bar linkage shown in the figure. Crank 2 is the driver.

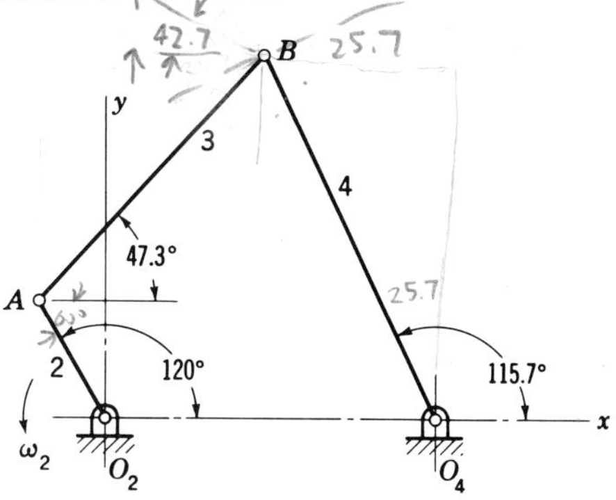

Prob. 4-9 $O_2A = 4$ in., $AB = 10$ in., $O_4B = 12$ in., $O_2O_4 = 10$ in., $\omega_2 = 45$ rad/sec; the directions shown for links 3 and 4 were computed graphically.

4-10. Crank 2 of the push-link mechanism shown in the figure drives the mechanism. Find V_B, V_C, ω_3, and ω_4.

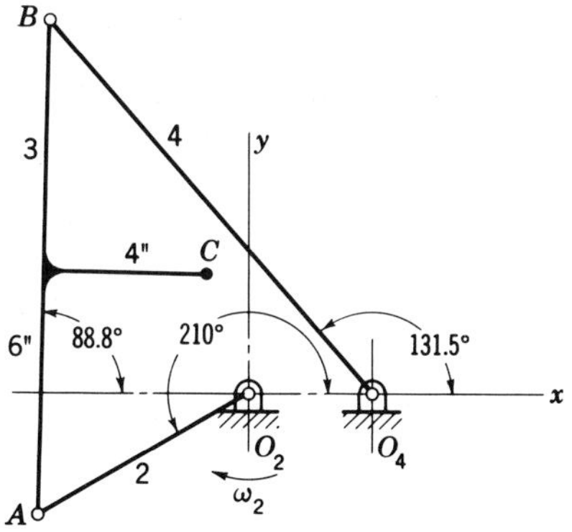

Prob. 4-10 $O_2A = 6$ in., $AB = 12$ in., $O_4B = 12$ in., $O_2O_4 = 3$ in., $\omega_2 = 60$ rad/sec; the directions of links 3 and 4 were computed graphically.

4-11. An offset slider-crank linkage is illustrated having the slider as the driver. Find the velocity of point C on the connecting rod and also the angular velocities of links 2 and 3.

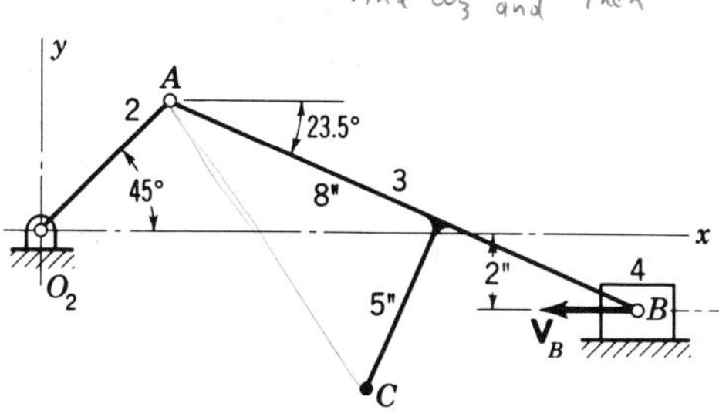

Prob. 4-11 $O_2A = 5$ in., $AB = 14$ in., $V_B = 10$ fps.

4-12. Find the velocity of point C on link 4 of the mechanism shown in the figure. What is the angular velocity of link 3?

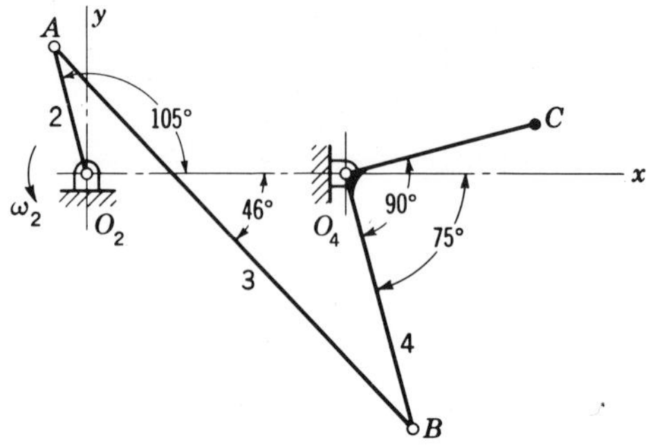

Prob. 4-12 $O_2O_4 = 16$ in., $O_2A = 8$ in., $AB = 32$ in., $O_4B = 16$ in., $O_4C = 12$ in., $\omega_2 = 48$ rad/sec.

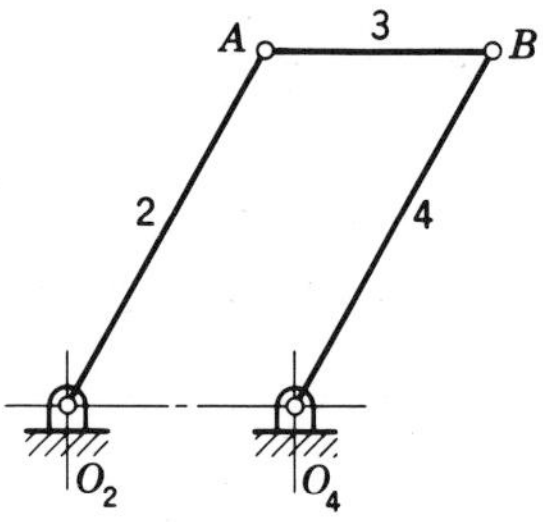

Prob. 4-13

4-13. The figure illustrates the parallel-bar linkage in which opposite links have equal lengths. For this linkage show that ω_3 is always zero, and that $\omega_2 = \omega_4$.

4-14. The figure illustrates the antiparallel or crossed-bar mechanism with links 2 and 4 having equal lengths and also with link 1 having the same length as the distance AB of link 3. Find the linear velocities of points C and D.

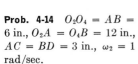

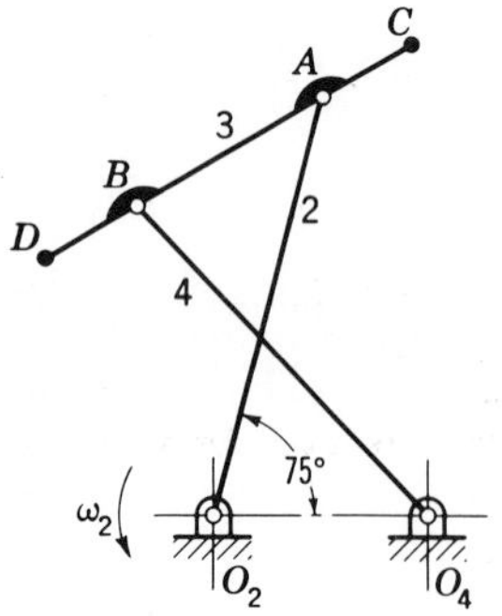

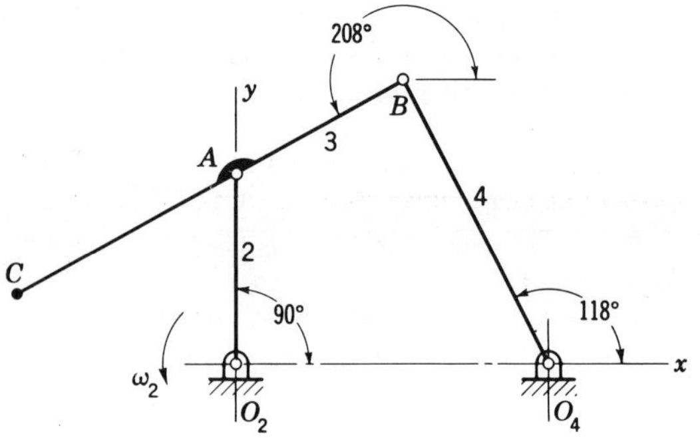

Prob. 4-14 $O_2O_4 = AB = 6$ in., $O_2A = O_4B = 12$ in., $AC = BD = 3$ in., $\omega_2 = 1$ rad/sec.

Prob. 4-15 $O_2A = AB = 6$ in., $AC = 8$ in., $O_4B = 10$ in., $O_2O_4 = 10$ in., $\omega_2 = 60$ rad/sec.

$\frac{360}{12} = 30$ ft/sec

4-15. Find the velocity of point C of the mechanism shown in the figure. Also find the angular velocities of links 3 and 4.

4-16. Find the velocity of point B and the angular velocities of links 3 and 4 of the inversion of the slider-crank mechanism shown in the figure.

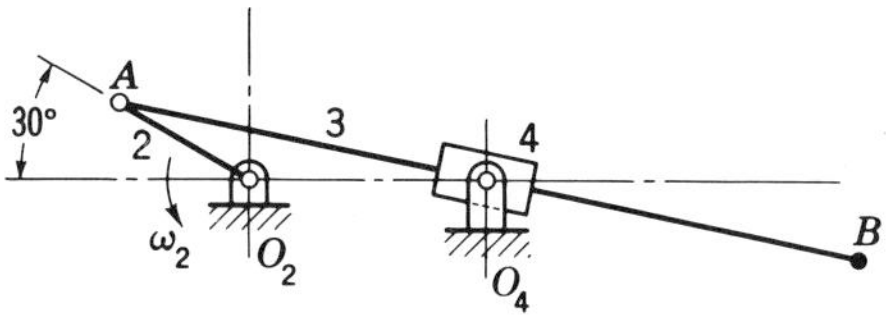

Prob. 4-16 $O_2A = 3$ in., $AB = 16$ in., $O_2O_4 = 5$ in., $\omega_2 = 60$ rad/sec.

4-17. Link 2 of the linkage shown in the figure has an angular velocity of 10 rad/sec. Find the angular velocity of link 6 and the velocities of points B and D.

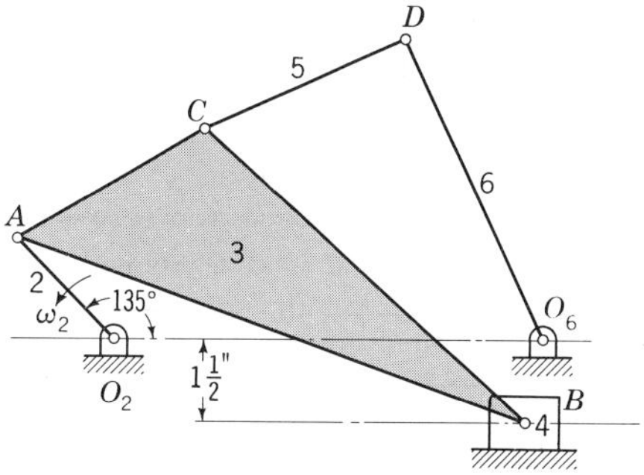

Prob. 4-17 $O_2A = 2.5$ in., $O_2O_6 = 8$ in., $AB = 10$ in., $AC = CD = 4$ in., $BC = 8$ in., $O_6D = 6$ in.

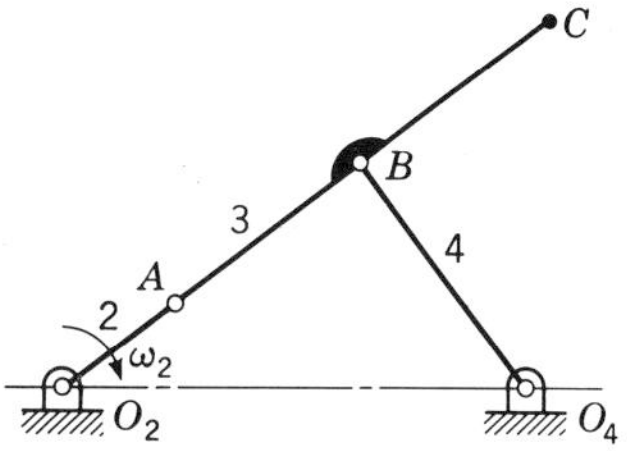

Prob. 4-18 $O_2O_4 = 10$ in., $O_2A = 3$ in., $AB = CB = 5$ in., $O_4B = 6$ in.

4-18. Find the velocity of the coupler point C and the angular velocity of link 4 of the mechanism illustrated in the figure if crank 2 has an angular velocity of 30 rad/sec.

4-19. The angular velocity of link 2 of the drag-link mechanism shown in the figure is 16 rad/sec. Plot a polar velocity diagram of the velocity of point B for all crank positions. Check the positions of maximum and minimum velocities by using Freudenstein's theorem.

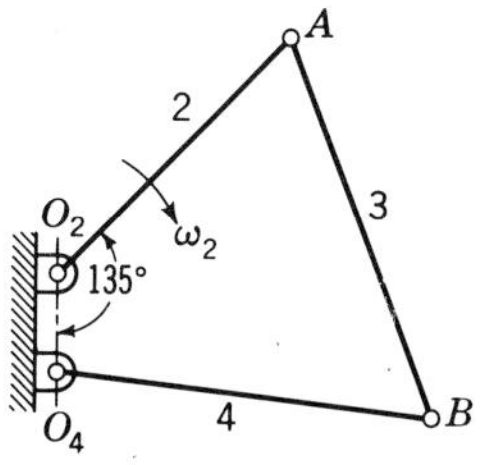

Prob. 4-19 $O_2O_4 = 4$ in., $O_2A = 14$ in., $AB = 17$ in., $O_4B = 16$ in.

4-20. Link 2 of the mechanism shown in the figure has an angular velocity of 36 rad/sec. Find the angular velocity of link 3 and the velocity of point B.

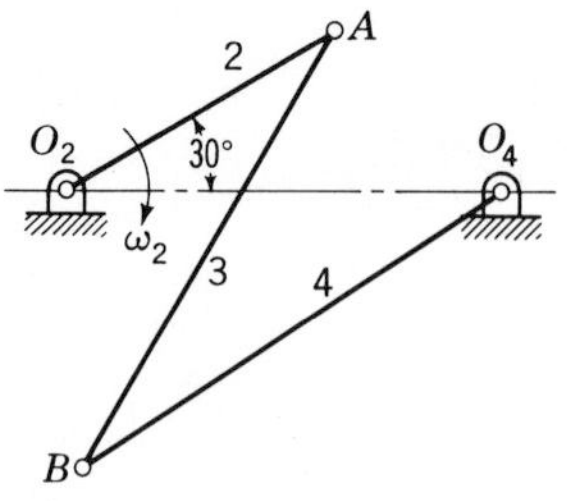

Prob. 4-20 $O_2A = 5$ in., $O_2O_4 = 7$ in., $AB = O_4B = 8$ in.

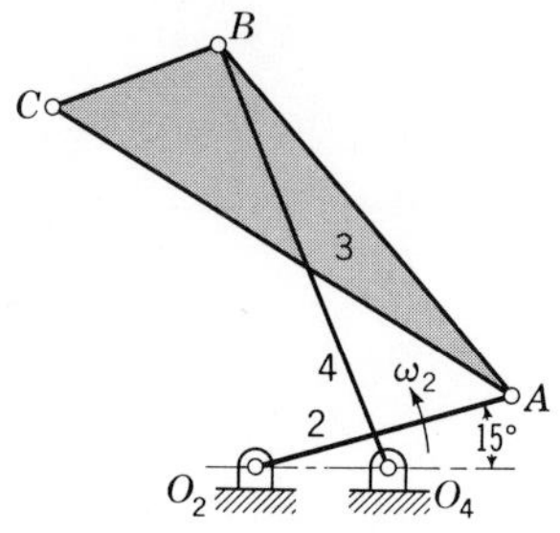

Prob. 4-21 $O_2O_4 = 3$ in., $O_2A = 6$ in., $AB = O_4B = 10$ in., $BC = 4$ in., $AC = 12$ in.

4-21. Find the velocity of point C and the angular velocity of link 3 of the push-link mechanism shown in the figure. Link 2 is the driver and rotates at 8 rad/sec.

4-22. Link 2 of the mechanism shown in the figure has an angular velocity of 56 rad/sec. Find V_C.

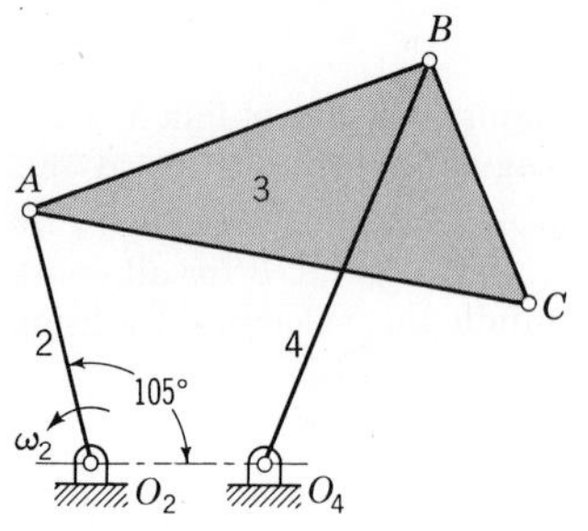

Prob. 4-22 $O_2O_4 = 4$ in., $O_2A = BC = 6$ in., $AB = O_4B = 10$ in., $AC = 12$ in.

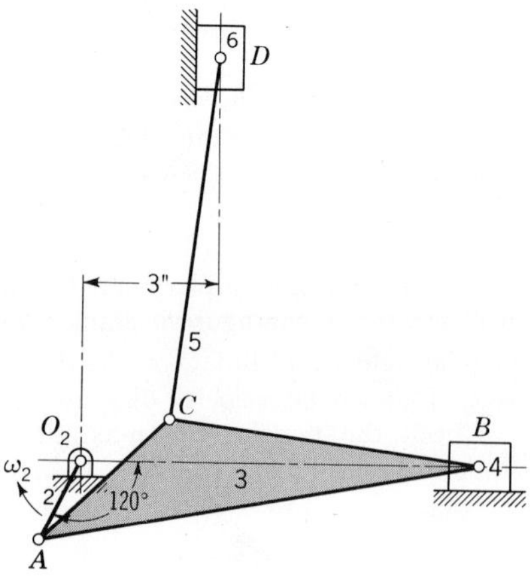

Prob. 4-23 $O_2A = 2$ in., $AB = 10$ in., $BC = 7$ in., $AC = 4$ in., $CD = 8$ in.

4-23. Find the velocities of points B, C, and D of the double-slider linkage shown in the figure if crank 2 rotates at 42 rad/sec.

4-24. The figure shows the mechanism used on a two-cylinder 60° V-engine consisting, in part, of an articulated connecting rod. Crank 2 rotates at 2000 rpm. Find the velocities of points B, C, and D.

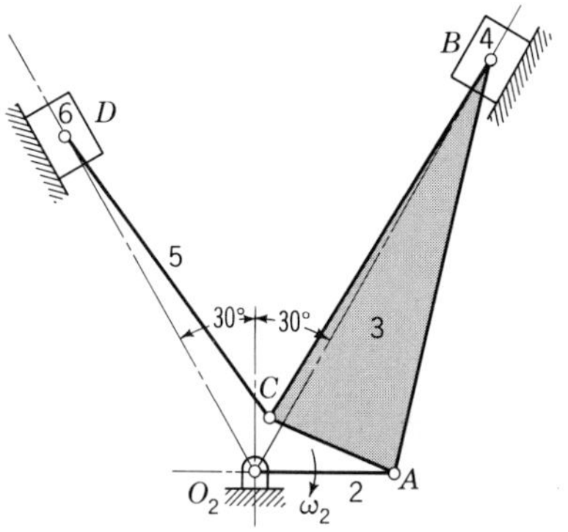

Prob. 4-24 $O_2A = 2$ in., $AB = 6$ in., $BC = 6$ in., $AC = 2$ in., $CD = 5$ in.

4-25. Make a complete velocity analysis of the linkage shown in the figure.

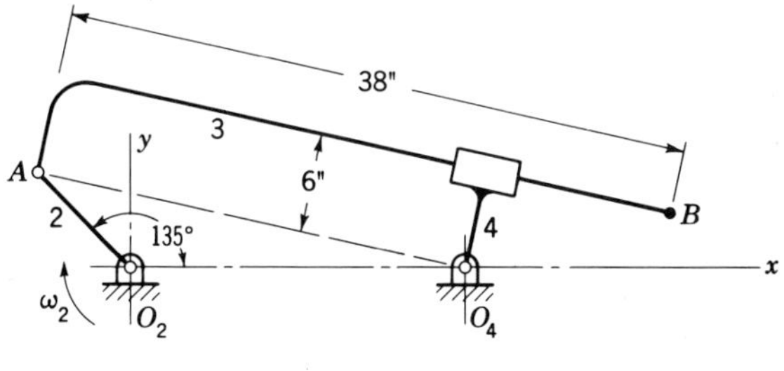

Prob. 4-25 $O_2A = 8$ in., $O_2O_4 = 20$ in., $\omega_2 = 24$ rad/sec.

4-26. Find V_B for the linkage shown in the figure if $V_A = 1$ fps.

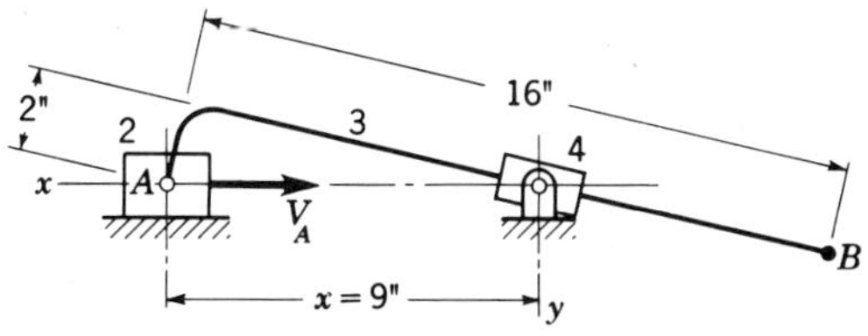

Prob. 4-26

4-27. The figure shows a variation of the Scotch-yoke mechanism with an inclined crosshead. Find the velocity of the crosshead.

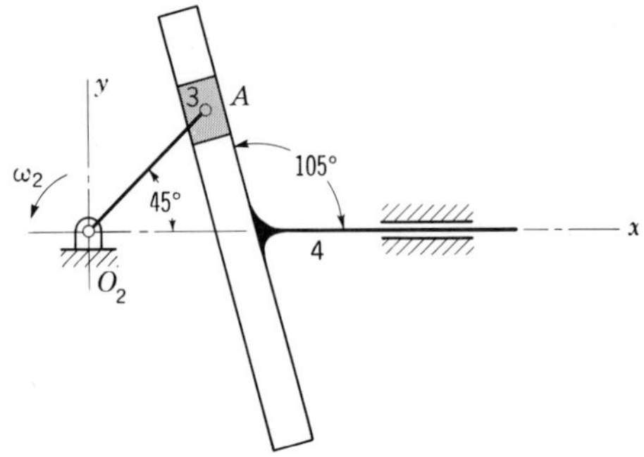

Prob. 4-27 $O_2A = 10$ in., $\omega_2 =$ 36 rad/sec.

4-28. Make a complete velocity analysis of the linkage shown in the figure.

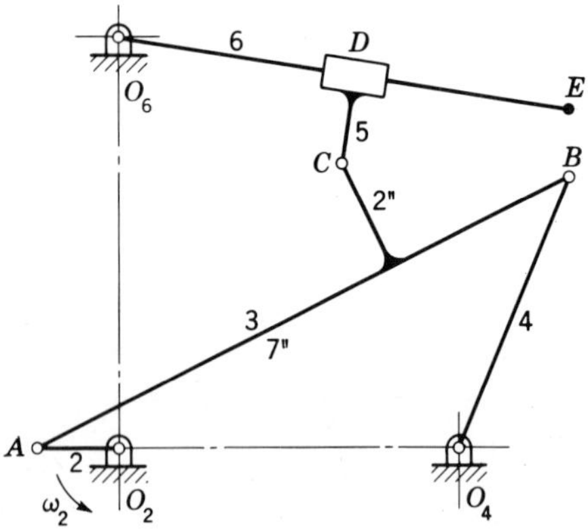

Prob. 4-28 $O_2A = CD = 1.5$ in., $AB = 10.5$ in., $O_4B = 5$ in., $O_2O_4 = 6$ in., $O_2O_6 = 7$ in., $O_6E = 8$ in., $\omega_2 = 72$ rad/sec.

4-29. Using the linkage of Prob. 4-23, locate all the velocity poles.

4-30. Find the velocity poles for the mechanism of Prob. 4-24.

4-31. Find the velocity poles for the mechanism of Prob. 4-25.

4-32. Use the mechanism of Prob. 4-28 and find all the poles.

5

Acceleration

see section 4-8

read ch. 5; 1-5

5-13, 5-15, 5-25, 5-31, 5-34

5-1 DEFINITION OF ACCELERATION

Instantaneous acceleration (hereafter called simply *acceleration*) is the time rate of change of velocity and is defined by the equation

$$\mathbf{a} = \lim_{\Delta t \to 0} \frac{\Delta \mathbf{v}}{\Delta t} = \frac{d\mathbf{v}}{dt} = \dot{\mathbf{v}} = \ddot{\mathbf{r}} \tag{5-1}$$

where $\Delta\mathbf{v}$ is the incremental change in $\mathbf{v}$ during the time Δt. In a similar manner, the angular acceleration of a rotating body is defined by the equation

$$\dot{\boldsymbol{\omega}} = \lim_{\Delta t \to 0} \frac{\Delta \boldsymbol{\omega}}{\Delta t} = \frac{d\boldsymbol{\omega}}{dt} \tag{5-2}$$

In Sec. 4-1 [Eq. (4-7)] we found the velocity of a point P moving along a path, such as the one of Fig. 5-1a, to be

$$\mathbf{v} = \dot{s}\boldsymbol{\hat{\lambda}} \tag{a}$$

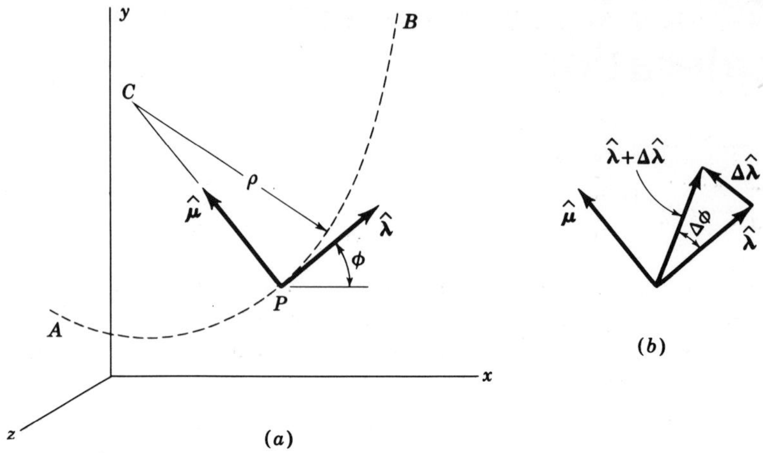

Fig. 5-1

where $\dot{s}$ is the speed of P along the path and where the right-handed triad $\hat{\lambda}\hat{\mu}\hat{\nu}$ has been chosen so that $\hat{\lambda}$ and $\hat{\mu}$ are tangent and normal, respectively, to the path at P. Differentiating Eq. (a) to obtain the acceleration yields

$$\mathbf{a} = \ddot{s}\hat{\lambda} + \dot{s}\dot{\hat{\lambda}} \qquad (b)$$

The second term of this equation requires additional clarification.

In Fig. 5-1a let ϕ represent the inclination of $\hat{\lambda}$ to any axis selected in the plane containing $\hat{\lambda}$, $\hat{\mu}$, and point C, the instantaneous center of curvature of the path at P. As P moves along the line AB, both $\hat{\lambda}$ and $\hat{\mu}$ are functions of ϕ. In Fig. 5-1b the vectors $\hat{\lambda}$ and $\hat{\lambda} + \Delta\hat{\lambda}$, both of which are unit vectors, have been transferred to the same origin. Then the limit of the ratio $\Delta\hat{\lambda}/\Delta\phi$ is

$$\frac{d\hat{\lambda}}{d\phi} = \lim_{\Delta\phi\to 0} \frac{\Delta\hat{\lambda}}{\Delta\phi} = \lim_{\Delta\phi\to 0} \frac{2\sin(\Delta\phi/2)\hat{\mu}}{\Delta\phi} = \hat{\mu} \qquad (c)$$

The second term of Eq. (b) can now be arranged in the form

$$\dot{s}\dot{\hat{\lambda}} = \frac{ds}{dt}\frac{d\hat{\lambda}}{d\phi}\frac{d\phi}{ds}\frac{ds}{dt} \qquad (d)$$

The term $d\phi/ds$ is the rate with respect to the distance s along the path with which a tangent to the path changes. This is called the *curvature*, and its reciprocal is the *radius of curvature* ρ. Thus

$$\frac{1}{\rho} = \frac{d\phi}{ds} \qquad (e)$$

With the help of Eq. (*c*), Eq. (*d*) now can be written

$$\dot{s}\dot{\hat{\boldsymbol{\lambda}}} = \frac{\dot{s}^2}{\rho}\hat{\mathbf{u}} \qquad (f)$$

Hence Eq. (*b*) becomes

$$\boxed{\mathbf{a} = \ddot{s}\hat{\boldsymbol{\lambda}} + \frac{\dot{s}^2}{\rho}\hat{\mathbf{u}}} \qquad (5\text{-}3)$$

Thus the acceleration vector **a** has two perpendicular components, a tangential component of magnitude $\ddot{s}$ and a normal component of magnitude $\dot{s}^2/\rho$ because it is directed toward the center of curvature C. Equation (5-3) can also be written

$$\mathbf{a} = \mathbf{a}^t + \mathbf{a}^n \qquad (5\text{-}4)$$

5-2 THE ACCELERATION OF A RIGID BODY

In Sec. 4-4 the velocity of points in a rigid body moving with both translation and rotation was determined. It was found that the velocity of any point in a rigid body could be obtained as the sum of a rotational component, due to the angular velocity $\boldsymbol{\omega}$ of the body, about *any* reference axis in the body plus the velocity of that reference. Thus, the velocity of any point B in a rigid body can be obtained from the equation

$$\mathbf{v}_B = \mathbf{v}_A + \boldsymbol{\omega} \times \mathbf{r}_{BA} \qquad (a)$$

where $\boldsymbol{\omega}$ is the angular velocity of the body, $\mathbf{v}_A$ is the velocity of the reference and is the translational component of $\mathbf{v}_B$, and $\mathbf{r}_{BA}$ is a position vector fixed in the body which defines the position of B relative to the reference.

We employ the same nomenclature in Fig. 5-2 and also specify that the reference axis has an acceleration $\mathbf{a}_A$ and that the body has an

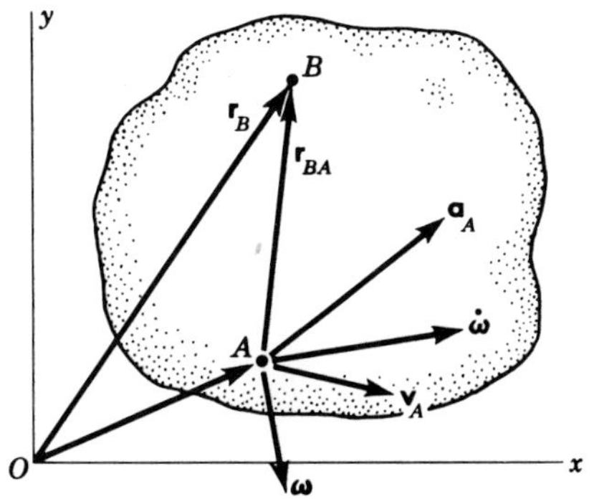

Fig. 5-2

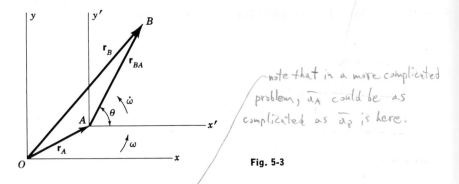

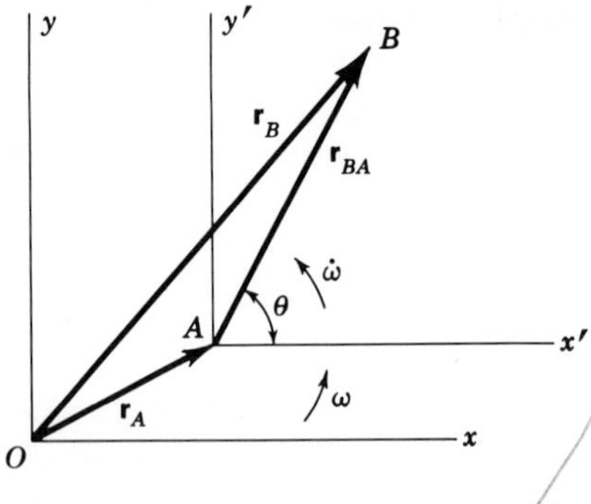

Fig. 5-3

angular acceleration $\dot{\boldsymbol{\omega}}$, as shown. Note that $\dot{\boldsymbol{\omega}}$ generally does *not* have the same direction as $\boldsymbol{\omega}$. The acceleration of B is obtained by taking the derivative of Eq. (*a*):

$$\dot{\mathbf{v}}_B = \dot{\mathbf{v}}_A + \dot{\boldsymbol{\omega}} \times \mathbf{r}_{BA} + \boldsymbol{\omega} \times \dot{\mathbf{r}}_{BA} \tag{b}$$

But $\dot{\mathbf{v}}_B = \mathbf{a}_B$, $\dot{\mathbf{v}}_A = \mathbf{a}_A$, and $\dot{\mathbf{r}}_{BA} = \boldsymbol{\omega} \times \mathbf{r}_{BA}$; so Eq. (*b*) can be written

$$\boxed{\mathbf{a}_B = \mathbf{a}_A + \dot{\boldsymbol{\omega}} \times \mathbf{r}_{BA} + \boldsymbol{\omega} \times (\boldsymbol{\omega} \times \mathbf{r}_{BA})} \tag{5-5}$$

In this expression $\mathbf{a}_A$ is the acceleration of the reference and is the translational component of the total acceleration of B. The other two components are due to the rotation of the body. In order to picture the directions of these, let us first study them in terms of a two-dimensional problem.

In Fig. 5-3 let A and B be two points fixed in a rigid body which has a combined motion of translation and rotation in the xy plane. Define, also, a moving $x'y'$ system with origin at A, but restrict this system to pure translation. Thus x' must remain parallel to x. As given quantities we specify the velocity and acceleration of A and the angular velocity ω and angular acceleration $\dot{\omega}$ of the rigid body. For plane motion these angular quantities can be treated as scalars because the corresponding vectors always have the same direction.[1]

The location of point B can now be specified by the vector equation

$$\mathbf{r}_B = \mathbf{r}_A + \mathbf{r}_{BA}$$

Transforming the second term on the right to the complex polar form yields

$$\mathbf{r}_B = \mathbf{r}_A + r_{BA}e^{j\theta} \tag{c}$$

[1] They may have different senses, however, since a scalar quantity can be either positive or negative.

The first derivative of Eq. (*c*) is the velocity. Thus

$$\mathbf{v}_B = \mathbf{v}_A + j\omega r_{BA}e^{j\theta} \tag{d}$$

This is the complex polar form of Eq. (*a*). The second term on the right is $\mathbf{v}_{BA}$. Its magnitude is ωr_{BA} and its direction $je^{j\theta}$ is perpendicular to $\mathbf{r}_{BA}$ in the sense of ω, as shown in Fig. 5-4.

The acceleration is the derivative of Eq. (*d*) and is

$$\mathbf{a}_B = \mathbf{a}_A + j\dot{\omega} r_{BA}e^{j\theta} - \omega^2 r_{BA}e^{j\theta} \tag{e}$$

The second two components in Eq. (*e*) correspond exactly with the second two components in Eq. (5-5). The term $j\dot{\omega} r_{BA}e^{j\theta}$ is due to the angular acceleration of the body. The magnitude is $\dot{\omega} r_{BA}$ and the direction is $je^{j\theta}$, which is in the same direction as v_{BA}. Note that B traces out a circle in its motion relative to A. Since this acceleration component is perpendicular to $\mathbf{r}_{BA}$ and hence tangent to the circle, it is convenient to call it a *tangential* component and designate it by the equation

$$a_{BA}{}^t = \dot{\omega} r_{BA} \tag{5-6}$$

Figure 5-4 also shows this acceleration component.

The third component of acceleration in Eq. (*e*) is called the *radial* or the *centripetal* component. Its magnitude is $\omega^2 r_{BA}$, and its direction $-e^{j\theta}$ is opposite to the vector $\mathbf{r}_{BA}$. This component is also shown in Fig. 5-4. The superscript r is used to denote the radial component as follows:

$$a_{BA}{}^r = \omega^2 r_{BA} \tag{5-7}$$

Let us now examine the last two terms of Eq. (5-5) in three-dimensional space. The term

$$\mathbf{a}_{BA}{}^t = \dot{\boldsymbol{\omega}} \times \mathbf{r}_{BA} \tag{5-8}$$

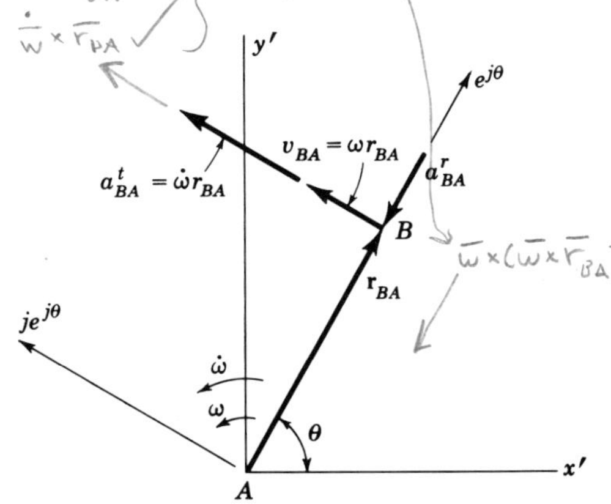

Fig. 5-4

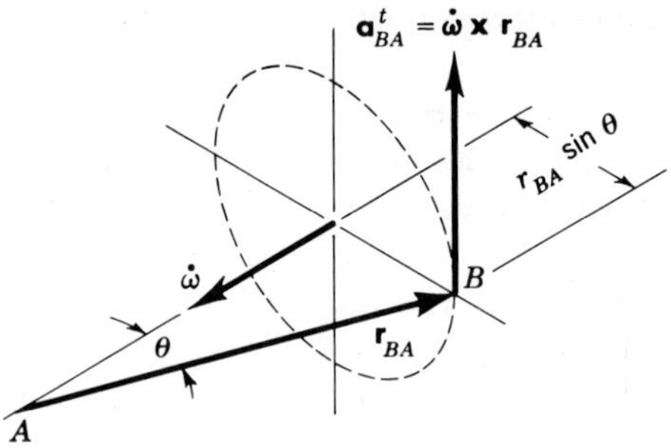

Fig. 5-5

according to the definition of the cross product, is perpendicular to the plane containing $\dot{\boldsymbol{\omega}}$ and $\mathbf{r}_{BA}$ with a sense according to the right-hand rule. Because of $\dot{\boldsymbol{\omega}}$ we visualize B as accelerating around a circle, as shown in Fig. 5-5. The plane of this circle is normal to the plane containing $\dot{\boldsymbol{\omega}}$ and $\mathbf{r}_{BA}$. Using the definition of the cross product, we find that the magnitude of this tangential acceleration component is

$$|\dot{\boldsymbol{\omega}} \times \mathbf{r}_{BA}| = \dot{\omega} r_{BA} \sin \theta$$

where $r_{BA} \sin \theta$, as shown, is the radius of the circle.

The direction of the radial acceleration component

$$\mathbf{a}_{BA}{}^r = \boldsymbol{\omega} \times (\boldsymbol{\omega} \times \mathbf{r}_{BA}) \tag{5-9}$$

is shown in Fig. 5-6. This component is in the plane containing $\boldsymbol{\omega}$ and $\mathbf{r}_{BA}$ and it is perpendicular to $\boldsymbol{\omega}$. The magnitude is

$$|\boldsymbol{\omega} \times (\boldsymbol{\omega} \times \mathbf{r}_{BA})| = \omega^2 r_{BA} \sin \phi$$

where $r_{BA} \sin \phi$ is the radius of the circle in Fig. 5-6.

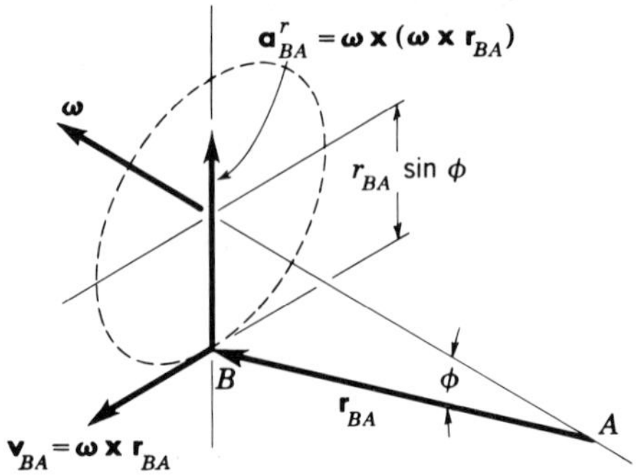

Fig. 5-6

Again we emphasize that $\dot{\boldsymbol{\omega}}$ and $\boldsymbol{\omega}$ do not generally have the same directions in three-dimensional space.

Let us now summarize the results of this section. A point fixed in a rigid body has three components of acceleration. The first of these is the acceleration of the reference (point A in Fig. 5-2). This is the translational component of the acceleration, and its value depends upon the particular reference selected for the analysis. There are two components of acceleration due to the rotation of the body. One of these is the tangential component, and it is due to the change in the angular velocity of the body. The other is the radial component.

Equation (5-5) can now be written as

$$\boxed{\mathbf{a}_B = \mathbf{a}_A + \mathbf{a}_{BA}} \tag{5-10}$$

which is called the *relative-acceleration equation*. It is also convenient to designate the components of the relative acceleration as

$$\boxed{\mathbf{a}_{BA} = \mathbf{a}_{BA}{}^r + \mathbf{a}_{BA}{}^t} \tag{5-11}$$

The relative-acceleration equation can be solved by any of the methods which were used in Chap. 4. The next several examples will serve to illustrate the acceleration-polygon approach and the direct analytical method.

Example 5-1 For the four-bar linkage shown in Fig. 5-7a, find the acceleration of A and B and the angular acceleration of links 3 and 4. Crank 2 has a positive angular velocity of 200 rad/sec.

Solution Figure 5-7 shows both the velocity polygon and the acceleration polygon with the circled numbers indicating the order of construction; the method of finding the directions of each vector is also indicated. The velocity polygon must, of course, be found first, because the angular velocities of links 3 and 4 are needed in the acceleration analysis.

The velocity of A is

$$V_A = O_2A\omega_2 = (\tfrac{6}{12})(200) = 100 \text{ fps}$$

Now the velocity polygon can be drawn. From it we obtain

$$V_{BA} = 128 \text{ fps} \qquad V_B = 129 \text{ fps}$$

and so the angular velocities of links 3 and 4 are

$$\omega_3 = \frac{V_{BA}}{AB} = \frac{128}{\frac{18}{12}} = 85.3 \text{ rad/sec ccw}$$

$$\omega_4 = \frac{V_B}{O_4B} = \frac{129}{\frac{12}{12}} = 129 \text{ rad/sec ccw}$$

where the directions are obtained from an examination of the velocity polygon.

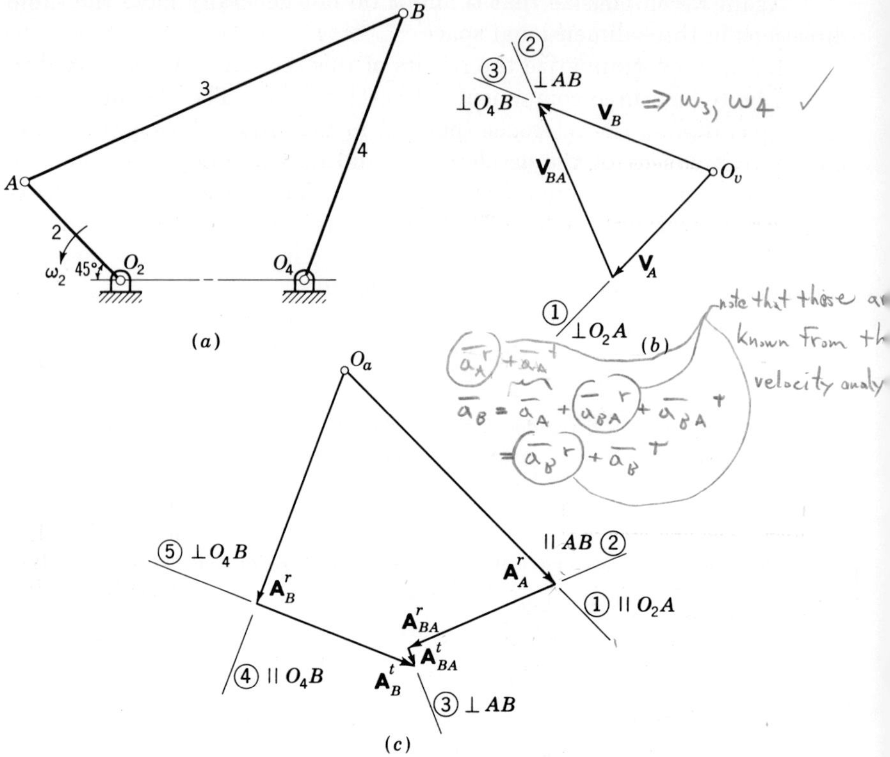

Fig. 5-7 (*a*) The linkage; $O_2A = 6$ in., $AB = 18$ in., $O_4B = 12$ in., $O_2O_4 = 8$ in. (*b*) The velocity polygon. (*c*) The acceleration polygon.

We now have enough information to determine some of the acceleration components. These are calculated as follows:

$$A_A{}^t = O_2A\dot{\omega}_2 = 0$$

$$A_A{}^r = O_2A\omega_2{}^2 = (\tfrac{6}{12})(200)^2 = 20{,}000 \text{ fps}^2$$

$$A_{BA}{}^r = BA\omega_3{}^2 = (\tfrac{18}{12})(85.3)^2 = 10{,}900 \text{ fps}^2$$

$$A_B{}^r = O_4B\omega_4{}^2 = (\tfrac{12}{12})(129)^2 = 16{,}650 \text{ fps}^2$$

The next step is to write the relative-acceleration equation in terms of its components. Thus, since

$$\mathbf{A}_B = \mathbf{A}_A + \mathbf{A}_{BA}$$

we have

$$\overset{1}{\mathbf{A}_B{}^t} + \overset{2}{\mathbf{A}_B{}^r} = \overset{2}{\mathbf{A}_A{}^r} + \overset{2}{\mathbf{A}_{BA}{}^r} + \overset{1}{\mathbf{A}_{BA}{}^t}$$

The numbers above each term indicate the number of known values for each vector. Beginning with the right side of the equation, the acceleration polygon is drawn by constructing $\mathbf{A}_A{}^r$, then $\mathbf{A}_{BA}{}^r$, and lastly $\mathbf{A}_{BA}{}^t$, because its magnitude is unknown. Note that the acceleration pole is designated O_a. Beginning at the pole again and using the left side of the equation, construct $\mathbf{A}_B{}^r$, and then $\mathbf{A}_B{}^t$ because its magnitude is unknown. This completes the polygon and we now label it, as shown, and scale off the following results:

$$A_B{}^t = 11{,}600 \text{ fps}^2 \qquad A_B = 20{,}600 \text{ fps}^2 \qquad A_{BA}{}^t = 1550 \text{ fps}^2$$

The angular accelerations are then computed as follows:

$$\dot{\omega}_3 = \frac{A_{BA}{}^t}{AB} = \frac{1550}{\frac{18}{12}} = 1033 \text{ rad/sec}^2 \text{ cw}$$

$$\dot{\omega}_4 = \frac{A_B{}^t}{O_4B} = \frac{11{,}600}{\frac{12}{12}} = 11{,}600 \text{ rad/sec}^2 \text{ cw}$$

Example 5-2 Solve Example 5-1 by using the direct analytical approach.

Solution The first step is to analyze the linkage for position. Since this procedure has already been demonstrated in Chaps. 3 and 4, we shall omit the analysis and display only the results; see Fig. 5-8. In vector form, the position vectors of the links are then as follows:

$$\mathbf{r}_2 = \left(\tfrac{6}{12}\right)\underline{/135^\circ} = -0.354\hat{\imath} + 0.354\hat{\jmath}$$

$$\mathbf{r}_3 = \left(\tfrac{18}{12}\right)\underline{/22.4^\circ} = 1.385\hat{\imath} + 0.571\hat{\jmath}$$

$$\mathbf{r}_4 = \left(\tfrac{12}{12}\right)\underline{/68.4^\circ} = 0.368\hat{\imath} + 0.931\hat{\jmath}$$

where $\mathbf{r}_3$ is a vector from A to B, and $\mathbf{r}_4$ a vector from O_4 to B.

The velocity analysis is then performed by using the methods of Chap. 4. The results are

$$\boldsymbol{\omega}_2 = 200\hat{\mathbf{k}} \qquad \boldsymbol{\omega}_3 = 85.3\hat{\mathbf{k}} \qquad \boldsymbol{\omega}_4 = 129\hat{\mathbf{k}}$$

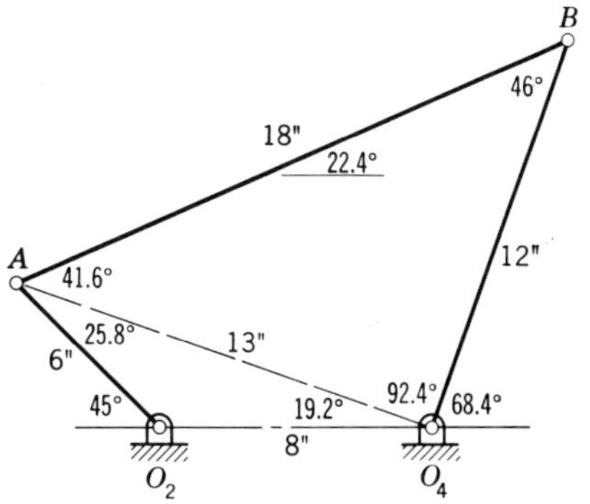

Fig. 5-8

The known acceleration components can now be computed. Thus

$$\mathbf{A}_A{}^r = \boldsymbol{\omega}_2 \times (\boldsymbol{\omega}_2 \times \mathbf{r}_2) = 200\hat{\mathbf{k}} \times \begin{vmatrix} \hat{\mathbf{i}} & \hat{\mathbf{j}} & \hat{\mathbf{k}} \\ 0 & 0 & 200 \\ -0.354 & 0.354 & 0 \end{vmatrix}$$

$$= \begin{vmatrix} \hat{\mathbf{i}} & \hat{\mathbf{j}} & \hat{\mathbf{k}} \\ 0 & 0 & 200 \\ -70.8 & -70.8 & 0 \end{vmatrix} = 14{,}140\hat{\mathbf{i}} - 14{,}140\hat{\mathbf{j}} \tag{1}$$

$$\mathbf{A}_{BA}{}^r = \boldsymbol{\omega}_3 \times (\boldsymbol{\omega}_3 \times \mathbf{r}_3) = 85.3\hat{\mathbf{k}} \times \begin{vmatrix} \hat{\mathbf{i}} & \hat{\mathbf{j}} & \hat{\mathbf{k}} \\ 0 & 0 & 85.3 \\ 1.387 & 0.571 & 0 \end{vmatrix}$$

$$= \begin{vmatrix} \hat{\mathbf{i}} & \hat{\mathbf{j}} & \hat{\mathbf{k}} \\ 0 & 0 & 85.3 \\ -48.8 & 118 & 0 \end{vmatrix} = -10{,}080\hat{\mathbf{i}} - 4160\hat{\mathbf{j}} \tag{2}$$

$$\mathbf{A}_B{}^r = \boldsymbol{\omega}_4 \times (\boldsymbol{\omega}_4 \times \mathbf{r}_4) = 129\hat{\mathbf{k}} \times \begin{vmatrix} \hat{\mathbf{i}} & \hat{\mathbf{j}} & \hat{\mathbf{k}} \\ 0 & 0 & 129 \\ 0.368 & 0.931 & 0 \end{vmatrix}$$

$$= \begin{vmatrix} \hat{\mathbf{i}} & \hat{\mathbf{j}} & \hat{\mathbf{k}} \\ 0 & 0 & 129 \\ -120 & 47.5 & 0 \end{vmatrix} = -6130\hat{\mathbf{i}} - 15{,}500\hat{\mathbf{j}} \tag{3}$$

Though $\dot{\omega}_3$ and $\dot{\omega}_4$ are both unknown, we can incorporate them into the solution in the following manner:

$$\mathbf{A}_{BA}{}^t = \dot{\boldsymbol{\omega}}_3 \times \mathbf{r}_3 = \begin{vmatrix} \hat{\mathbf{i}} & \hat{\mathbf{j}} & \hat{\mathbf{k}} \\ 0 & 0 & \dot{\omega}_3 \\ 1.387 & 0.571 & 0 \end{vmatrix} = -0.571\dot{\omega}_3\hat{\mathbf{i}} + 1.387\dot{\omega}_3\hat{\mathbf{j}} \tag{4}$$

$$\mathbf{A}_B{}^t = \dot{\boldsymbol{\omega}}_4 \times \mathbf{r}_4 = \begin{vmatrix} \hat{\mathbf{i}} & \hat{\mathbf{j}} & \hat{\mathbf{k}} \\ 0 & 0 & \dot{\omega}_4 \\ 0.368 & 0.931 & 0 \end{vmatrix} = -0.931\dot{\omega}_4\hat{\mathbf{i}} + 0.368\dot{\omega}_4\hat{\mathbf{j}} \tag{5}$$

Expanding Eq. (5-10) into components and noting that $\mathbf{A}_A{}^t = 0$ gives

$$\mathbf{A}_B{}^t + \mathbf{A}_B{}^r = \mathbf{A}_A{}^r + \mathbf{A}_{BA}{}^r + \mathbf{A}_{BA}{}^t \tag{6}$$

If we now substitute Eqs. (1) to (5) in (6) and separate the $\hat{\mathbf{i}}$ and $\hat{\mathbf{j}}$ components, we obtain the following pair of simultaneous algebraic equations:

$$0.571\dot{\omega}_3 - 0.931\dot{\omega}_4 = 10{,}190 \tag{7}$$

$$-1.387\dot{\omega}_3 + 0.368\dot{\omega}_4 = -2800 \tag{8}$$

When these equations are solved simultaneously, the results are found to be

$$\dot{\boldsymbol{\omega}}_3 = -1033\hat{\mathbf{k}} \text{ rad/sec}^2 \qquad \dot{\boldsymbol{\omega}}_4 = -11{,}600\hat{\mathbf{k}} \text{ rad/sec}^2$$

Example 5-3 The linkage shown in Fig. 5-9 was analyzed for velocities in Example 4-1. The results were $\omega_3 = 25.6$ rad/sec ccw and $\omega_4 = 49.7$ rad/sec ccw. Determine the angular acceleration of links 3 and 4 and the linear acceleration of points C and D.

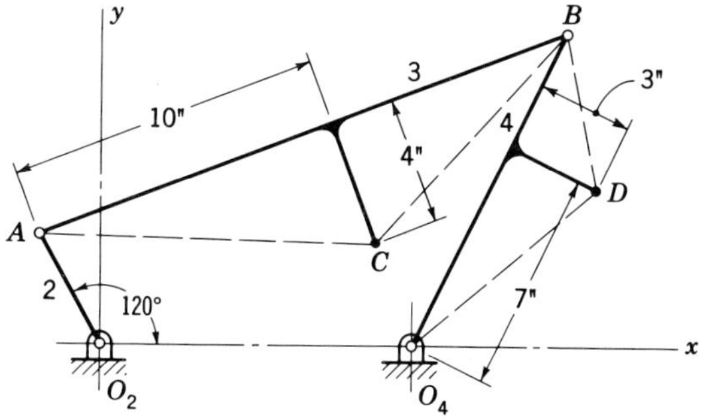

Fig. 5-9 $O_2O_4 = 10$ in., $O_2A = 4$ in., $AB = 18$ in., $O_4B = 11$ in., $\omega_2 = 94.2$ rad/sec ccw.

Solution Using the information from the velocity analysis, we first calculate the following scalar acceleration components:

$$A_A{}^r = O_2A\omega_2{}^2 = (\tfrac{4}{12})(94.2)^2 = 2960 \text{ fps}^2$$

$$A_{BA}{}^r = AB\omega_3{}^2 = (\tfrac{18}{12})(25.6)^2 = 983 \text{ fps}^2$$

$$A_B{}^r = O_4B\omega_4{}^2 = (\tfrac{11}{12})(49.7)^2 = 2270 \text{ fps}^2$$

Since $\dot{\omega}_2 = 0$, the relative-acceleration equation is written as

$$\overset{2}{\mathbf{A}_B{}^r} + \overset{1}{\mathbf{A}_B{}^t} = \overset{2}{\mathbf{A}_A{}^r} + \overset{2}{\mathbf{A}_{BA}{}^r} + \overset{1}{\mathbf{A}_{BA}{}^t}$$

Here, $\mathbf{A}_B{}^t$ is perpendicular to link 4 and $\mathbf{A}_{BA}{}^t$ is perpendicular to link 3. The acceleration polygon for this equation is then constructed as shown in Fig. 5-10.

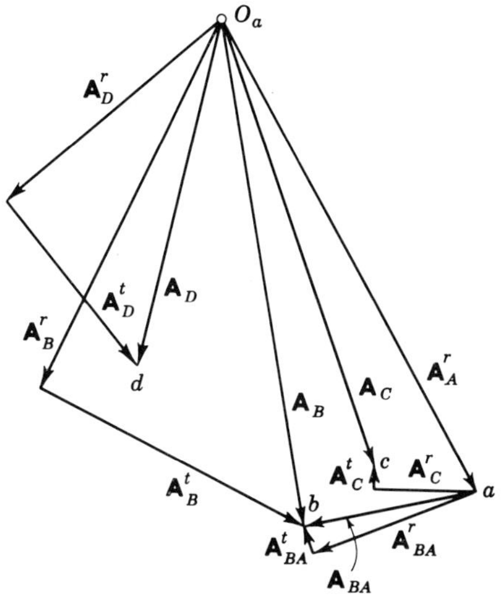

Fig. 5-10

Having first located point b, we can scale off the following unknown magnitudes:

$$A_B = 2820 \text{ fps}^2 \qquad A_B{}^t = 1670 \text{ fps}^2 \qquad A_{BA}{}^t = 160 \text{ fps}^2$$

The angular accelerations are found to be

$$\dot{\omega}_3 = \frac{A_{BA}{}^t}{AB} = \frac{160}{\frac{18}{12}} = 108 \text{ rad/sec}^2 \text{ ccw}$$

$$\dot{\omega}_4 = \frac{A_B{}^t}{O_4B} = \frac{1670}{\frac{11}{12}} = 1820 \text{ rad/sec}^2 \text{ cw}$$

To obtain the acceleration of D we can write the equation

$$\overset{0}{\mathbf{A}_D} = \overset{2}{\mathbf{A}_D{}^r} + \overset{2}{\mathbf{A}_D{}^t}$$

As indicated by the numbers above each component, all the terms on the right are known or can be computed. From Fig. 5-9 the distance AD is

$$AD = \sqrt{(7)^2 + (3)^2} = 7.62 \text{ in.}$$

Therefore

$$A_D{}^r = O_4D\omega_4{}^2 = (7.62/12)(49.7)^2 = 1570 \text{ fps}^2$$

and this component is parallel to the line O_4D. Next,

$$A_D{}^t = O_4D\dot{\omega}_4 = (7.62/12)(1820) = 1160 \text{ fps}^2$$

which is perpendicular to O_4D. These two components are added to the polygon of Fig. 5-10. The acceleration of D is then scaled and found to be

$$A_D = 1960 \text{ fps}^2$$

Finally, we have yet to find the acceleration of point C on the coupler. The equation for the acceleration of C is

$$\overset{0}{\mathbf{A}_C} = \overset{2}{\mathbf{A}_A} + \overset{2}{\mathbf{A}_{CA}{}^r} + \overset{2}{\mathbf{A}_{CA}{}^t}$$

though we could just as well refer its motion to point B. The distance from A to C, from Fig. 5-9, is

$$AC = \sqrt{(10)^2 + (4)^2} = 10.78 \text{ in.}$$

Therefore

$$A_{CA}{}^r = AC\omega_3{}^2 = (10.78/12)(25.6)^2 = 588 \text{ fps}^2$$

$$A_{CA}{}^t = AC\dot{\omega}_3 = (10.78/12)(108) = 97 \text{ fps}^2$$

When these two components are constructed on the polygon and added, the acceleration of C is found to be

$$A_C = 2580 \text{ fps}^2$$

in the direction shown in Fig. 5-10.

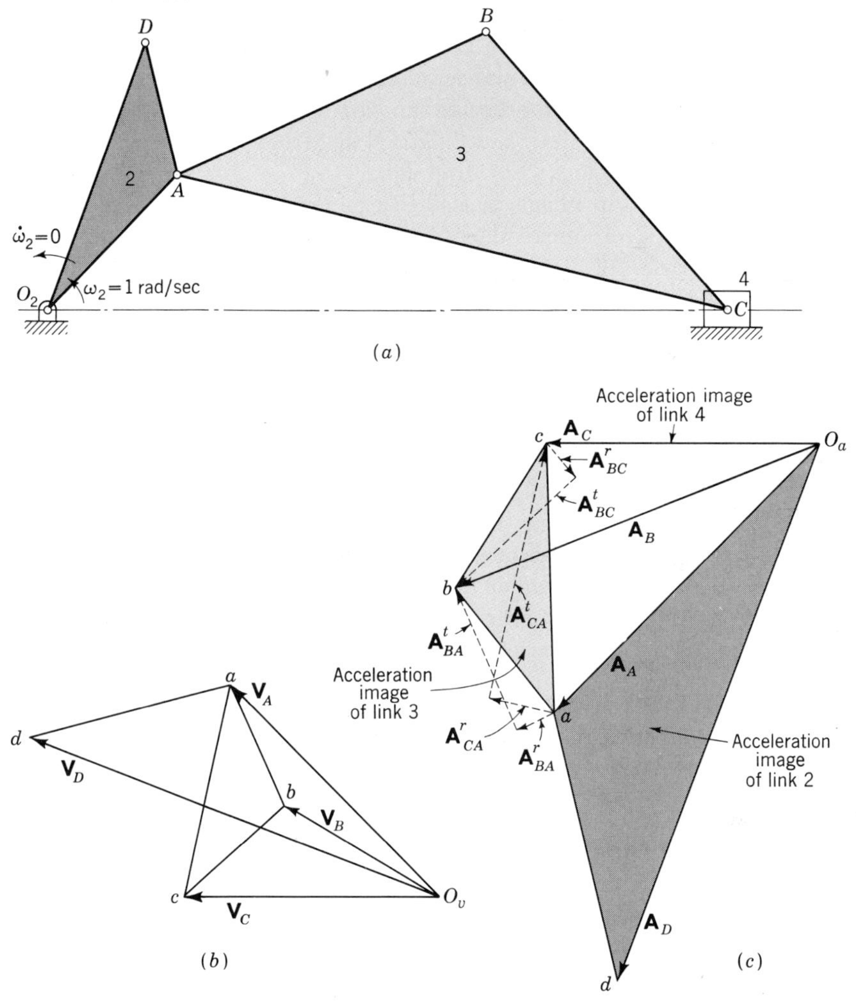

Fig. 5-11

5-3 ACCELERATION IMAGES

The acceleration image of a link is obtained in much the same manner as a velocity image. Figure 5-11*a* is a slider-crank mechanism in which links 2 and 3 have been given triangular shapes in order to illustrate the images. Figure 5-11*b* is the velocity image, and the acceleration image is shown in *c*. The angular acceleration of the crank (link 2) is zero, and notice that the corresponding acceleration image is turned 180° from the orientation of the link itself. On the other hand, notice that link 3 has a counterclockwise angular acceleration and that its image is oriented less than 180° from the link. Thus the orientation of the acceleration image depends upon the angular acceleration of the link in question. It can be shown from the geometry of the figure that the orientation of the acceleration image is $[180 - \tan^{-1}(\dot{\omega}/\omega^2)]$ deg from the link, measured in the positive (counterclockwise) direction.

The acceleration image is used in exactly the same manner as the velocity image. Note particularly that the acceleration image is formed by the *total* acceleration vectors, not the component vectors.

Really doesn't save any time if you aren't given the position diagram to scale (on have to construct it anyway)

5-4 COMPLETE GRAPHICAL ACCELERATION ANALYSIS

Consider the link AB, in Fig. 5-12a, rotating at ω rad/sec. The velocity of point B with reference to A is $\mathbf{V}_{BA}$, and its magnitude is obtained by multiplying the length of the link by the angular velocity. Its direction is at right angles to the link, as shown. In this section we shall demonstrate that by constructing the triangle AMN of Fig. 5-12b the radial acceleration of B with reference to A is obtained if the acceleration scale is related in a specified manner to the velocity and space scales.

Graphical methods of acceleration analysis have already been developed, but all of them have required calculations in order to obtain some of the acceleration components. An analysis in which these intermediate steps are eliminated not only is a time-saver but also greatly reduces the opportunity for error. The method to be developed is easy to understand and is quite general.

The magnitudes of the velocity and of the radial component of acceleration of point B of the link of Fig. 5-12 are related by the equation

$$A_{BA}{}^r = BA\omega^2 = \frac{V_{BA}{}^2}{BA} \qquad (a)$$

Fig. 5-12   (a)   (b)

Use this construction and use $\left[\dfrac{S_s S_a}{S_v{}^2} = 1\right]$ for scales

Note that you can only find radial components with this, but tangential magnitudes are usually unknowns anyway.

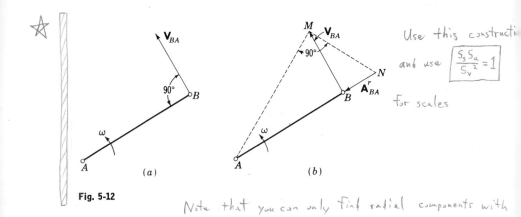

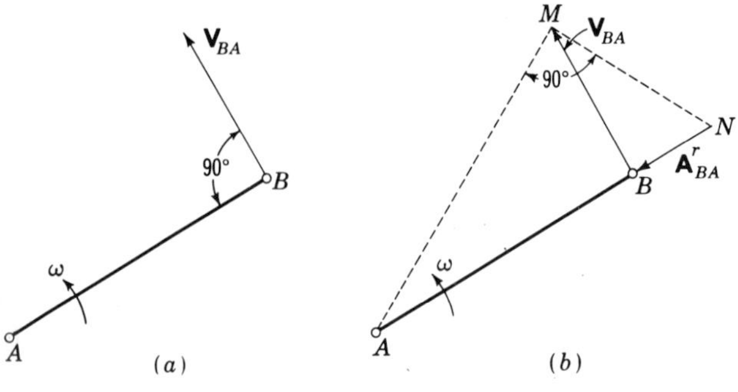

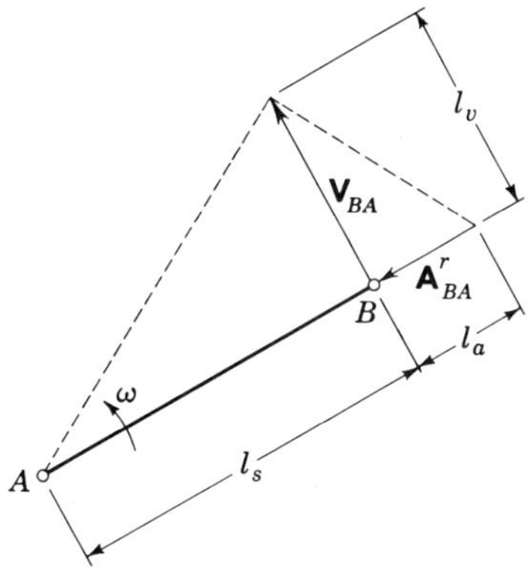

Fig. 5-13

since

$$V_{BA} = BA\omega \tag{b}$$

Equation (*a*) can also be written

$$\frac{A_{BA}{}^r}{V_{BA}} = \frac{V_{BA}}{BA} \tag{c}$$

Now notice that Eq. (*c*) expresses the relation between two similar triangles in Fig. 5-12*b*:

$$\frac{NB}{BM} = \frac{BM}{BA} \tag{d}$$

Thus, if the space scale, the velocity scale, and the acceleration scale are chosen in accordance with the relationship expressed by Eq. (*c*), then the construction of Fig. 5-12 may be used to find the radial acceleration component when the velocity is known. Or, alternatively, a similar construction can be used to find the velocity when the radial acceleration component is given.

The scales are defined as follows:

S_s = space scale, ft/in.
S_v = velocity scale, fps/in.
S_a = acceleration scale, fps²/in.

where the units of inches are inches of drawing; thus if the velocity scale is given as $S_v = 100$ fps/in., this means that 100 fps is represented by a line 1 in. long on the drawing. Similarly, if $S_a = 800$ fps²/in., then an acceleration of 1200 fps² would be represented by a line $1\frac{1}{2}$ in. long on the drawing.

The relationship of the scales is demonstrated by reference to Fig. 5-13. Here the link AB has been drawn a distance l_s in. long, the velocity vector $\mathbf{V}_{BA}$ a distance l_v in. long, and the radial acceleration component $\mathbf{A}_{BA}{}^r$ a distance l_a in. long. Then, by similar triangles,

$$\frac{l_a}{l_v} = \frac{l_v}{l_s} \tag{e}$$

Applying the scales,

$$l_s = \frac{BA}{S_s} \qquad l_v = \frac{V_{BA}}{S_v} \qquad l_a = \frac{A_{BA}{}^r}{S_a}$$

Substitution of these equations into Eq. (e) gives

$$A_{BA}{}^r = \frac{V_{BA}{}^2}{BA}\frac{S_sS_a}{S_v{}^2} \tag{f}$$

And so the scales are related by the equation

$$\frac{S_sS_a}{S_v{}^2} = 1 \qquad \text{or} \qquad S_v = \sqrt{S_sS_a} \tag{5-12}$$

Ss and Sv are known, not Ss and Sa

In making an analysis, first choose two of the scales; then use Eq. (5-12) to find the third.

Example 5-4 Determine the velocities and accelerations of the four-bar linkage shown in Fig. 5-14. The angular velocity of the driver, together with the dimensions, is specified in the figure caption. The indicated scales have been chosen to agree with the relation expressed by Eq. (5-12).

Solution Using the angular velocity of the crank, we first calculate the velocity of point A. Thus

$$V_A = O_2A\omega_2 = (0.333)(24) = 8 \text{ fps}$$

Starting with this vector, we now construct the velocity polygon by using the S_v scale. Now take the three vectors $\mathbf{V}_A$, $\mathbf{V}_B$, and $\mathbf{V}_{BA}$ from the velocity polygon and construct them on the linkage. $\mathbf{V}_A$ is drawn from A perpendicular to O_2A; $\mathbf{V}_{BA}$ from B perpendicular to BA; and $\mathbf{V}_B$ from B perpendicular to O_4B. Next find $\mathbf{A}_A{}^r$, $\mathbf{A}_{BA}{}^r$, and $\mathbf{A}_B{}^r$, respectively, using the graphical construction just described. This completes the construction on the linkage, and the last step is to draw the acceleration polygon. The equation is

$$\overset{1}{\mathbf{A}_B{}^t} + \overset{2}{\mathbf{A}_B{}^r} = \overset{2}{\mathbf{A}_A{}^r} + \overset{1}{\mathbf{A}_{BA}{}^t} + \overset{2}{\mathbf{A}_{BA}{}^r}$$

This equation is now solved by constructing the acceleration polygon as described previously. The results may be scaled directly from the polygon, and the angular accelerations obtained by calculation.

Figure 5-15 shows a method of obtaining the velocity when the radial acceleration is given.

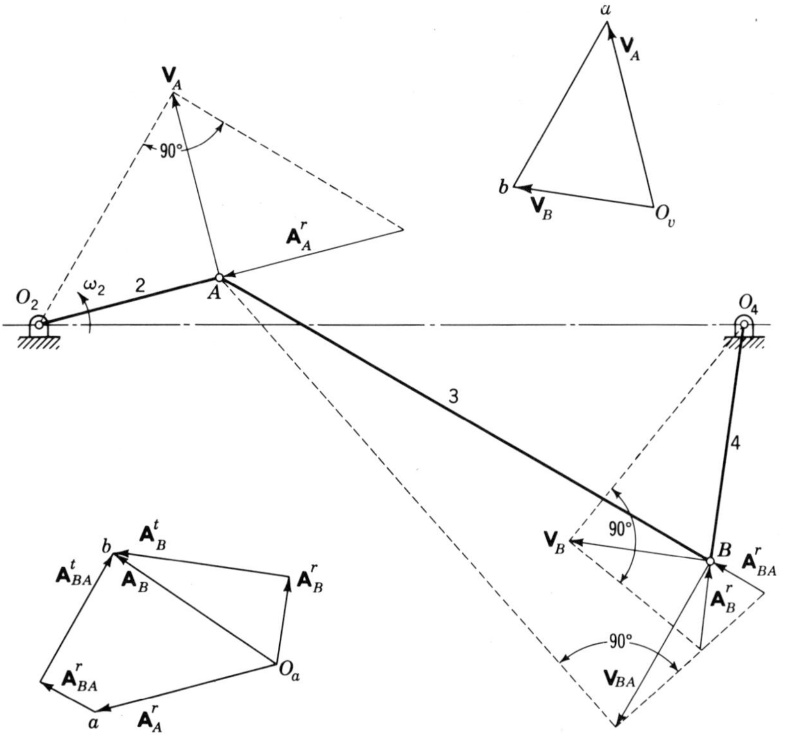

Fig. 5-14 A complete graphical acceleration analysis. $O_2A = 0.333$ ft, $AB = 1.00$ ft, $O_4B = 0.416$ ft, $O_2O_4 = 1.25$ ft, $\omega_2 = 24$ rad/sec, $S_s = 0.333$ ft/in., $S_v = 8$ fps/in., $S_a = 192$ fps²/in.

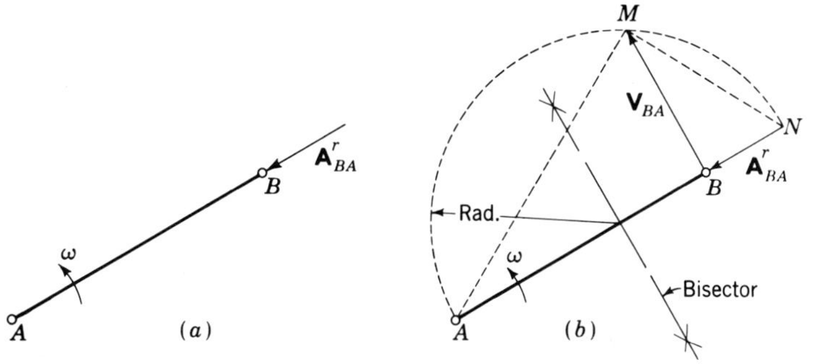

Fig. 5-15 (*a*) Given the radial acceleration vector $\mathbf{A}_{BA}{}^r$, find the velocity $\mathbf{V}_{BA}$. (*b*) Solution: Bisect AN and construct a semicircle on AN as the diameter. Then BM is $\mathbf{V}_{BA}$ because AMN is a right triangle.

5-5 THE ACCELERATION OF A POINT IN A MOVING REFERENCE SYSTEM

In Sec. 4-8 we found it necessary to develop an expression for the velocity of a point whose motion is known relative to a moving coordinate system. Let us now investigate the acceleration of that point.

To review the problem, Fig. 5-16 illustrates a point P which moves relative to the moving $x'y'z'$ reference system. Point B is fixed to the system and is instantaneously coincident with P. Since we are now interested in the acceleration of P, we specify that the origin A of $x'y'z'$ has a velocity $\mathbf{v}_A$ and an acceleration $\mathbf{a}_A$ and also that $x'y'z'$ has an angular velocity $\boldsymbol{\omega}$ and an angular acceleration $\dot{\boldsymbol{\omega}}$.

Point B is fixed in the $x'y'z'$ system. Equation (4-18) gives the velocity of B as

$$\mathbf{v}_B = \mathbf{v}_A + \boldsymbol{\omega} \times \boldsymbol{\rho} \tag{a}$$

and, from Eq. (5-5), the acceleration of B is

$$\mathbf{a}_B = \mathbf{a}_A + \dot{\boldsymbol{\omega}} \times \boldsymbol{\rho} + \boldsymbol{\omega} \times (\boldsymbol{\omega} \times \boldsymbol{\rho}) \tag{b}$$

Finally, from Eq. (4-22) of Sec. 4-8, we learned that the velocity of point P may be expressed as

$$\mathbf{v}_P = \mathbf{v}_A + \mathbf{v}^R + \boldsymbol{\omega} \times \boldsymbol{\rho} \tag{c}$$

where

$$\mathbf{v}^R = \dot{x}'\hat{\boldsymbol{\lambda}} + \dot{y}'\hat{\boldsymbol{\mu}} + \dot{z}'\hat{\boldsymbol{\nu}} \tag{d}$$

and

$$\boldsymbol{\rho} = x'\hat{\boldsymbol{\lambda}} + y'\hat{\boldsymbol{\mu}} + z'\hat{\boldsymbol{\nu}} \tag{e}$$

With this brief review out of the way and the equations before us, we now differentiate Eq. (c) to obtain the acceleration of P. The result is

$$\dot{\mathbf{v}}_P = \dot{\mathbf{v}}_A + \dot{\mathbf{v}}^R + \dot{\boldsymbol{\omega}} \times \boldsymbol{\rho} + \boldsymbol{\omega} \times \dot{\boldsymbol{\rho}} \tag{f}$$

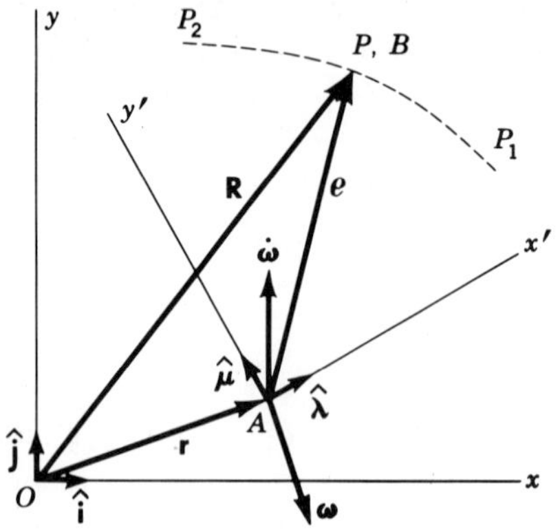

Fig. 5-16

Let us now consider the terms in Eq. (*f*) one at a time. We first note that $\mathbf{a}_P = \dot{\mathbf{v}}_P$ and $\mathbf{a}_A = \dot{\mathbf{v}}_A$. To get the term $\dot{\mathbf{v}}^R$, we must differentiate Eq. (*d*). The result is

$$\dot{\mathbf{v}}^R = (\ddot{x}'\hat{\boldsymbol{\lambda}} + \ddot{y}'\hat{\boldsymbol{\mu}} + \ddot{z}'\hat{\boldsymbol{\nu}}) + (\dot{x}'\dot{\hat{\boldsymbol{\lambda}}} + \dot{y}'\dot{\hat{\boldsymbol{\mu}}} + \dot{z}'\dot{\hat{\boldsymbol{\nu}}})$$

$$= \mathbf{a}^R + \boldsymbol{\omega} \times \mathbf{v}^R \qquad (g)$$

because the unit vectors $\hat{\boldsymbol{\lambda}}\hat{\boldsymbol{\mu}}\hat{\boldsymbol{\nu}}$ have a fixed length and rotate at angular velocity $\boldsymbol{\omega}$.

The third term in Eq. (*f*) is the same as Eq. (5-8) and needs no clarification.

The last term may be expanded to

$$\boldsymbol{\omega} \times \dot{\boldsymbol{\rho}} = \boldsymbol{\omega} \times \frac{d}{dt}(x'\hat{\boldsymbol{\lambda}} + y'\hat{\boldsymbol{\mu}} + z'\hat{\boldsymbol{\nu}})$$

$$= \boldsymbol{\omega} \times [(\dot{x}'\hat{\boldsymbol{\lambda}} + \dot{y}'\hat{\boldsymbol{\mu}} + \dot{z}'\hat{\boldsymbol{\nu}}) + (x'\dot{\hat{\boldsymbol{\lambda}}} + y'\dot{\hat{\boldsymbol{\mu}}} + z'\dot{\hat{\boldsymbol{\nu}}})]$$

$$= \boldsymbol{\omega} \times \mathbf{v}^R + \boldsymbol{\omega} \times (\boldsymbol{\omega} \times \boldsymbol{\rho}) \qquad (h)$$

note that ω and ω̇ are associated with the moving frame!!!

Substituting all of these back into Eq. (*f*) yields

$$\mathbf{a}_P = \mathbf{a}_A + \mathbf{a}^R + 2\boldsymbol{\omega} \times \mathbf{v}^R + \dot{\boldsymbol{\omega}} \times \boldsymbol{\rho} + \boldsymbol{\omega} \times (\boldsymbol{\omega} \times \boldsymbol{\rho}) \qquad (5\text{-}13)$$

Now, if we substitute the acceleration of *B* from Eq. (*b*) into (5-13) we get

$$\mathbf{a}_P = \mathbf{a}_B + \mathbf{a}^R + \mathbf{a}^C \qquad (5\text{-}14)$$

where $\mathbf{a}^R$ is the acceleration of *P* relative to $x'y'z'$, and

$$\mathbf{a}^C = 2\boldsymbol{\omega} \times \mathbf{v}^R \qquad (5\text{-}15)$$

and is called the *Coriolis component of acceleration*. The numeral 2 in Eq. (5-15) indicates that the Coriolis component comes into existence for *two* reasons. The first is suggested by Eq. (*g*), which indicates the existence of the part $\boldsymbol{\omega} \times \mathbf{v}^R$ because the vector $\mathbf{v}^R$ is rotating and hence *changing its direction*. The second reason is indicated in Eq. (*h*), where the other half of the Coriolis acceleration arises because the relative-velocity vector $\mathbf{v}^R$ is *changing its magnitude*.

Equation (5-15) can also be expressed in the form

$$\mathbf{a}^C = 2\omega\hat{\boldsymbol{\omega}} \times v^R\hat{\mathbf{v}}^R = 2\omega v^R(\hat{\boldsymbol{\omega}} \times \hat{\mathbf{v}}^R) \qquad (5\text{-}16)$$

Using the definition of the cross product we see that the magnitude of $\mathbf{a}^C$ is

$$a^C = 2\omega v^R \sin\theta \qquad (5\text{-}17)$$

Note that, in a complicated case, these can be found by first doing an analysis considering the x'y'z' system as fixed. This could be carried on to an x''y''z'' system to handle any problem complexity.

where θ is the angle between $\boldsymbol{\omega}$ and $\mathbf{v}^R$. The direction is normal to the plane formed by the vectors $\boldsymbol{\omega}$ and $\mathbf{v}^R$ and with a sense in accordance with the right-hand rule.

With plane motion $\boldsymbol{\omega}$ and $\mathbf{v}^R$ are perpendicular to each other, and so $a^C = 2\omega v^R$ and the direction is perpendicular to $\mathbf{v}^R$ in the sense of $\boldsymbol{\omega}$.

Equation (5-14) states that the acceleration of a point P in a moving reference system is equal to the acceleration of a coincident point B fixed to the moving system, plus the acceleration of P relative to the moving system, plus the Coriolis acceleration. The Coriolis acceleration is twice the cross product of the angular velocity of the moving reference system with the velocity of P relative to that system.

The unit-vector method provides an alternative approach to the derivation of Eq. (5-15). We first express the position of P in the form

$$\mathbf{R} = \mathbf{r} + \rho\hat{\boldsymbol{\rho}} \tag{i}$$

Then the velocity of P is

$$\begin{aligned}\dot{\mathbf{R}} &= \dot{\mathbf{r}} + \dot{\rho}\hat{\boldsymbol{\rho}} + \rho\dot{\hat{\boldsymbol{\rho}}} \\ &= \dot{\mathbf{r}} + \dot{\rho}\hat{\boldsymbol{\rho}} + \rho(\boldsymbol{\omega} \times \hat{\boldsymbol{\rho}})\end{aligned} \tag{j}$$

But the last term of Eq. (j) is

$$\rho(\boldsymbol{\omega} \times \hat{\boldsymbol{\rho}}) = \boldsymbol{\omega} \times \boldsymbol{\rho} = \mathbf{v}_{BA}$$

Equation (j) can therefore be written

$$\mathbf{v}_P = \mathbf{v}_B + \mathbf{v}^R \tag{k}$$

where $\mathbf{v}^R = \dot{\rho}\hat{\boldsymbol{\rho}}$. Note that Eq. (k) is the same as (4-24) in Sec. 4-8.

The next derivative of Eq. (j) is

$$\ddot{\mathbf{R}} = \ddot{\mathbf{r}} + \ddot{\rho}\hat{\boldsymbol{\rho}} + \dot{\rho}\dot{\hat{\boldsymbol{\rho}}} + \dot{\rho}(\boldsymbol{\omega} \times \hat{\boldsymbol{\rho}}) + \rho(\dot{\boldsymbol{\omega}} \times \hat{\boldsymbol{\rho}}) + \rho(\boldsymbol{\omega} \times \dot{\hat{\boldsymbol{\rho}}}) \tag{l}$$

The third and fourth terms of Eq. (l) are identical, since

$$\dot{\rho}\dot{\hat{\boldsymbol{\rho}}} = \dot{\rho}(\boldsymbol{\omega} \times \hat{\boldsymbol{\rho}})$$

The last term may be expanded to

$$\rho(\boldsymbol{\omega} \times \dot{\hat{\boldsymbol{\rho}}}) = \rho[\boldsymbol{\omega} \times (\boldsymbol{\omega} \times \hat{\boldsymbol{\rho}})]$$

With only slight manipulations Eq. (l) can now be written

$$\ddot{\mathbf{R}} = \ddot{\mathbf{r}} + \ddot{\rho}\hat{\boldsymbol{\rho}} + 2\boldsymbol{\omega} \times (\dot{\rho}\hat{\boldsymbol{\rho}}) + \dot{\boldsymbol{\omega}} \times \boldsymbol{\rho} + \boldsymbol{\omega} \times (\boldsymbol{\omega} \times \boldsymbol{\rho}) \tag{5-18}$$

The terms in this equation can be identified as follows:

$$\begin{aligned}
\mathbf{a}_P &= \ddot{\mathbf{R}}, \text{ absolute acceleration of } P \\
\mathbf{a}_A &= \ddot{\mathbf{r}}, \text{ absolute acceleration of } A \\
\mathbf{a}^R &= \ddot{\rho}\hat{\boldsymbol{\rho}}, \text{ acceleration of } P \text{ relative to moving system} \\
\mathbf{v}^R &= \dot{\rho}\hat{\boldsymbol{\rho}}, \text{ velocity of } P \text{ relative to the moving system} \\
\mathbf{a}^C &= 2\boldsymbol{\omega} \times (\dot{\rho}\hat{\boldsymbol{\rho}}) = 2\boldsymbol{\omega} \times \mathbf{v}^R, \text{ Coriolis acceleration} \\
\mathbf{a}_{BA}{}^t &= \dot{\boldsymbol{\omega}} \times \boldsymbol{\rho}, \text{ transverse component of acceleration of } B \text{ relative to } A \\
\mathbf{a}_{BA}{}^r &= \boldsymbol{\omega} \times (\boldsymbol{\omega} \times \boldsymbol{\rho}), \text{ radial component of acceleration of } B \text{ relative to } A
\end{aligned}$$

With these substitutions Eq. (5-18) is seen to be identical with (5-14).

Example 5-5 In Fig. 5-17 a block 3 slides outward on link 2 at a uniform velocity of 30 fps, while link 2 is rotating counterclockwise at a constant angular velocity of 50 rad/sec. Find the absolute acceleration of the block.

Solution As shown in Fig. 5-17, we choose a moving $x'y'$ system fixed to link 2. Then A_2 is fixed in the moving system and A_3 has motion relative to it. In terms of this nomenclature, Eq. (5-14) is written

$$\mathbf{A}_{A_3} = \mathbf{A}_{A_2} + \mathbf{A}^R + \mathbf{A}^C \tag{1}$$

The terms for this equation are computed as follows:

$$\mathbf{A}_{A_2}{}^t = O_2A\dot{\omega}_2\underline{/\theta} = 0$$

$$\mathbf{A}_{A_2}{}^r = O_2A\omega_2{}^2\underline{/\theta} = (\tfrac{6}{12})(50)^2\underline{/45^\circ - 180^\circ} = 1250\underline{/-135^\circ} \text{ fps}^2$$

$$\mathbf{A}^R = 0$$

$$\mathbf{A}^C = 2\omega_2V^R\underline{/\theta} = (2)(50)(30)\underline{/45^\circ + 90^\circ} = 3000\underline{/135^\circ} \text{ fps}^2$$

Transforming into components and substituting into Eq. (1) gives

$$\begin{aligned}
\mathbf{A}_{A_3} = \mathbf{A}_{A_2}{}^r + \mathbf{A}^C &= (-885\hat{\imath} - 885\hat{\jmath}) + (-2120\hat{\imath} + 2120\hat{\jmath}) \\
&= -3005\hat{\imath} + 1235\hat{\jmath} \\
&= 3250\underline{/157.6^\circ} \text{ fps}^2
\end{aligned}$$

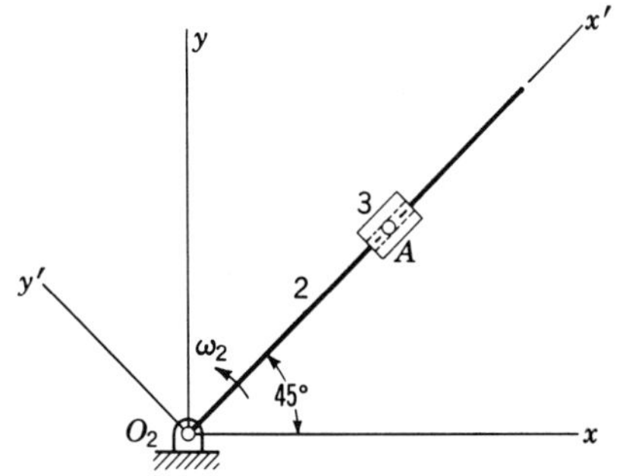

Fig. 5-17 $O_2A = 6$ in.

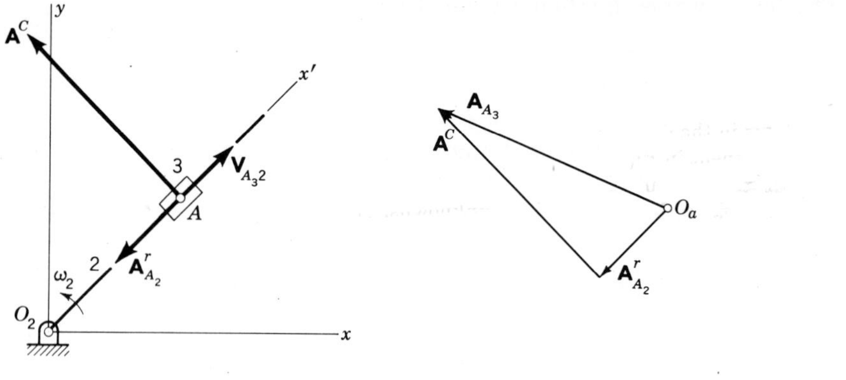

Fig. 5-18

Figure 5-18 shows the acceleration polygon and the directions of the two components.

Example 5-6 Do an acceleration analysis of the linkage shown in Fig. 5-19.
Solution A complete velocity analysis, using the polygon shown in the figure, yields

$$V_A = 12 \text{ fps} \qquad V_{B_3A} = 10.1 \text{ fps} \qquad V^R = V_{B_3} = 6.5 \text{ fps}$$

$$\omega_3 = \omega_4 = 7.77 \text{ rad/sec cw}$$

where the subscript B_3 refers to point B on link 3. Note that $V_{B_4} = 0$.

To solve for the accelerations we first write the equation

$$\overset{0}{\mathbf{A}_{B_3}} = \overset{2}{\mathbf{A}_A} + \overset{2}{\mathbf{A}_{BA}{}^r} + \overset{1}{\mathbf{A}_{BA}{}^t} \tag{1}$$

Neither the direction nor the magnitude of $\mathbf{A}_{B_3}$ is known, as indicated in Eq. (1).

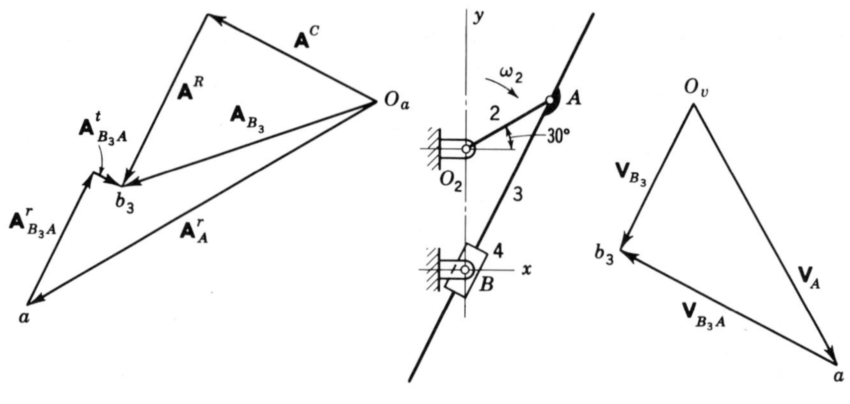

Fig. 5-19 $O_2B = 10$ in., $O_2A = 8$ in., $AB = 15.6$ in., $\omega_2 = 18$ rad/sec.

The first two components on the right are

$$A_A = A_A{}^r = O_2A\omega_2{}^2 = (\tfrac{8}{12})(18)^2 = 216 \text{ fps}^2$$

$$A_{BA}{}^r = AB\omega_3{}^2 = (15.6/12)(7.77)^2 = 78.5 \text{ fps}^2$$

$\mathbf{A}_A{}^r$ is in the direction of AO_2; $\mathbf{A}_{BA}{}^r$ is a vector directed from B to A. The last component in Eq. (1) is $\mathbf{A}_{BA}{}^t$; we know this vector is perpendicular to line AB, but its magnitude is unknown.

Since Eq. (1) has three unknowns, it cannot be solved. We therefore define a moving $x'y'$ system attached to link 4 with origin at B. Equation (5-14) applies, and, for this situation, it is written

$$\overset{0}{\mathbf{A}_{B_3}} = \overset{2}{\mathbf{A}_{B_4}} + \overset{1}{\mathbf{A}^R} + \overset{2}{\mathbf{A}^C} \tag{2}$$

In this equation $\mathbf{A}_{B_4} = 0$. The term $\mathbf{A}^R$ is in the direction of AB, but its magnitude is unknown. The Coriolis acceleration is

$$A^C = 2\omega_4 V^R = (2)(7.77)(6.5) = 101 \text{ fps}^2$$

and its direction is in the direction that $\mathbf{V}^R$ would have if it were rotated 90° in the sense of ω_4. Therefore $\mathbf{A}^C$ is perpendicular to AB and directed to the left because ω_4 is in the clockwise direction.

Solving Eqs. (1) and (2) simultaneously to eliminate $\mathbf{A}_{B_3}$ gives

$$\overset{1}{\mathbf{A}^R} + \overset{2}{\mathbf{A}^C} = \overset{2}{\mathbf{A}_A} + \overset{2}{\mathbf{A}_{BA}{}^r} + \overset{1}{\mathbf{A}_{BA}{}^t} \tag{3}$$

The polygon for this equation is shown in Fig. 5-19 too. When this polygon is scaled, the results are found to be

$$A^R = 103 \text{ fps}^2 \qquad A_{BA}{}^t = 16 \text{ fps}^2 \qquad A_{B_3} = 144 \text{ fps}^2$$

The angular accelerations of links 3 and 4 are

$$\dot\omega_3 = \dot\omega_4 = \frac{A_{BA}{}^t}{BA} = \frac{16}{15.6/12} = 12.3 \text{ rad/sec}^2 \text{ ccw}$$

duh!!!!! Just consider "3" as fixed and use $\omega_{4/3}$ and $\dot\omega_{4/3}$ with $\bar r = \overline{BC}$ in eq. (5-5)

Example 5-7 The linkage of Fig. 5-20*a* furnishes an excellent means of obtaining still more insight into the meaning of Eq. (5-14). Given the coincident points C_4, attached to link 4, and C_3, attached to link 3, find the acceleration of C_4 relative to link 3.

Solution The velocity polygon is shown in Fig. 5-20*b*. The results of a velocity analysis are

$$\omega_3 = 1.90 \text{ rad/sec ccw} \qquad \omega_4 = 2.29 \text{ rad/sec cw} \qquad V_{C_4C_3} = 2.24 \text{ fps}$$

Note that $\mathbf{V}_{C_4C_3}$ is a vector extending from c_3 to c_4 in Fig. 5-20*b*.

To obtain the accelerations, we first compute

$$A_A{}^r = O_2A\omega_2{}^2 = (\tfrac{4}{12})(10)^2 = 33.3 \text{ fps}^2$$

$$A_{BA}{}^r = AB\omega_3{}^2 = (\tfrac{14}{12})(1.90)^2 = 4.21 \text{ fps}^2$$

$$A_B{}^r = O_4B\omega_4{}^2 = (\tfrac{10}{12})(2.29)^2 = 4.37 \text{ fps}^2$$

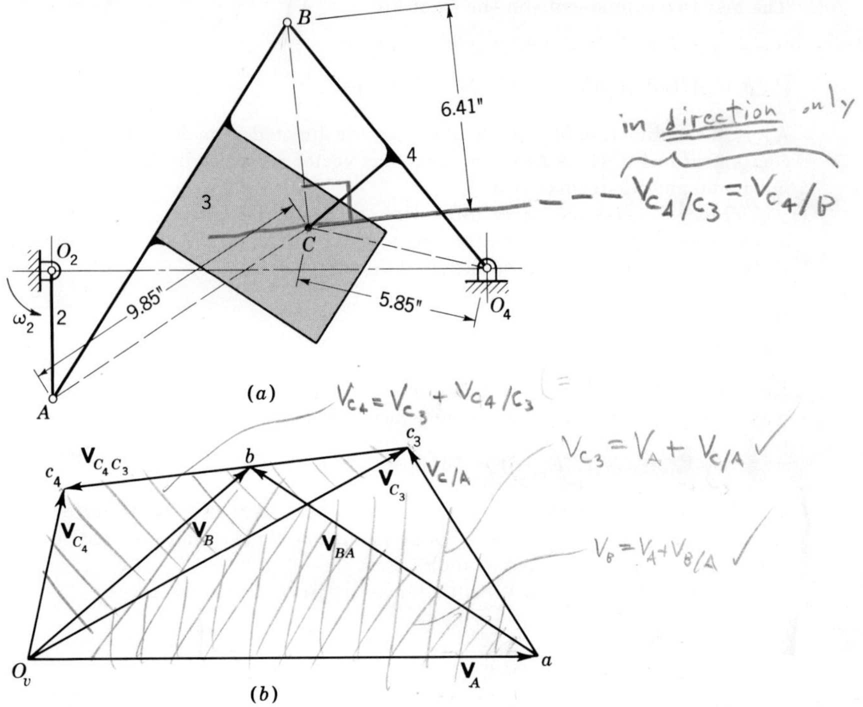

Fig. 5-20 $O_2A = 4$ in., $AB = 14$ in., $O_4B = 10$ in., $O_2O_4 = 14$ in., $\omega_2 = 10$ rad/sec ccw.

The acceleration polygon (Fig. 5-21) is the solution to the equation

$$\overset{1}{\mathbf{A}_B^t} + \overset{2}{\mathbf{A}_B^r} = \overset{2}{\mathbf{A}_A^r} + \overset{1}{\mathbf{A}_{BA}^t} + \overset{2}{\mathbf{A}_{BA}^r} \tag{1}$$

and yields the portion of the polygon defined by O_aab. When the polygon is scaled, we find

$$A_B^t = 26.1 \text{ fps}^2 \qquad A_{BA}^t = 30.3 \text{ fps}^2$$

and so

$$\dot{\omega}_4 = \frac{A_B^t}{O_4B} = \frac{26.1}{\frac{10}{12}} = 31.3 \text{ rad/sec}^2 \text{ cw}$$

$$\dot{\omega}_3 = \frac{A_{BA}^t}{AB} = \frac{30.3}{\frac{14}{12}} = 26 \text{ rad/sec}^2 \text{ cw}$$

We now have enough information to compute the accelerations of C_3 and C_4. For C_3, we write

$$\overset{0}{\mathbf{A}_{C_3}} = \overset{2}{\mathbf{A}_A} + \overset{2}{\mathbf{A}_{C_3A}^r} + \overset{2}{\mathbf{A}_{C_3A}^t} \tag{2}$$

$\mathbf{A}_A$ is known. The magnitudes of the two relative acceleration components for this equation are

$$A^r_{C_3A} = C_3A\omega_3^2 = \left(\frac{9.85}{12}\right)(1.90)^2 = 2.96 \text{ fps}^2$$

$$A^t_{C_3A} = C_3A\dot{\omega}_3 = \left(\frac{9.85}{12}\right)(26) = 21.4 \text{ fps}^2$$

Equation (2) can now be solved to locate point c_3 in Fig. 5-21. The total acceleration of C_3 would be a vector from O_a to c_3, but this has been omitted from the polygon for the sake of clarity.

The next step is to obtain the acceleration of C_4. The equation to be solved is

$$\overset{0}{\mathbf{A}_{C_4}} = \overset{2}{\mathbf{A}_{C_4}^r} + \overset{2}{\mathbf{A}_{C_4}^t} \qquad (3)$$

The two components on the right have magnitudes of

$$A_{C_4}^r = O_4C_4\omega_4^2 = \left(\frac{5.85}{12}\right)(2.29)^2 = 2.56 \text{ fps}^2$$

$$A_{C_4}^t = O_4C_4\dot{\omega}_4 = \left(\frac{5.85}{12}\right)(31.3) = 15.3 \text{ fps}^2$$

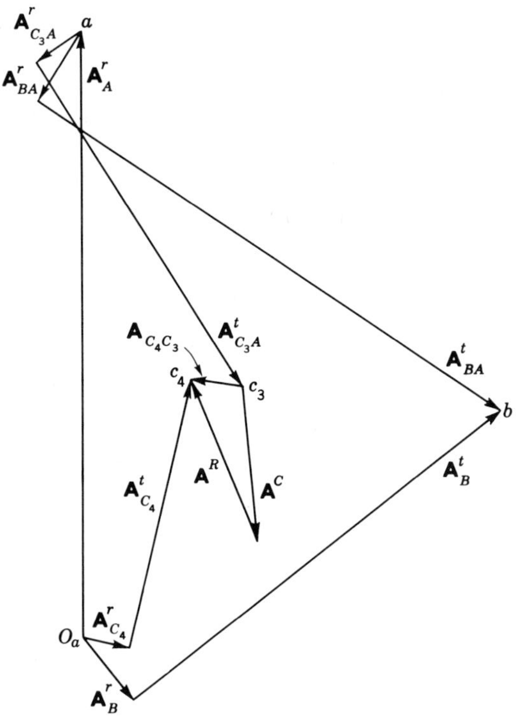

Fig. 5-21

These two vectors are added to the polygon beginning at O_a in Fig. 5-21. This locates point c_4, and the total velocity of C_4 (not shown) would be a vector from O_a to c_4.

At this point we are tempted to construct a vector from c_3 to c_4, label it $\mathbf{A}_{C_4C_3}$, and call it the acceleration of C_4 relative to C_3. This is all perfectly appropriate, but $\mathbf{A}_{C_4C_3}$ *is not the acceleration of* C_4 *relative to link* 3! By this we mean it is not the acceleration that an observer attached to link 3 would report as the acceleration of C_4. We are thus led to the use of the equations for the motion of a point in a moving reference system.

To begin this investigation we attach an $x'y'$ system to link 3. Then C_4 is the moving point and C_3 is the coincident point fixed to the moving system. Equation (4-24), for the velocity of C_4, is written

$$\mathbf{V}_{C_4} = \mathbf{V}_{C_3} + \mathbf{V}^R$$

Thus

$$\mathbf{V}^R = \mathbf{V}_{C_4} - \mathbf{V}_{C_3} = \mathbf{V}_{C_4C_3} \tag{4}$$

so that the relative velocity $\mathbf{V}^R$ *is* the same as $\mathbf{V}_{C_4C_3}$.

Next, using the nomenclature of Fig. 5-20, we write Eq. (5-14) as

$$\overset{2}{\mathbf{A}_{C_4}} = \overset{2}{\mathbf{A}_{C_3}} + \overset{0}{\mathbf{A}^R} + \overset{2}{\mathbf{A}^C} \tag{5}$$

Both $\mathbf{A}_{C_4}$ and $\mathbf{A}_{C_3}$ are known. From Eq. (5-15) the magnitude of the Coriolis acceleration is

$$A^C = 2\omega_3 V^R = (2)(1.90)(2.24) = 8.66 \text{ fps}^2$$

and its direction is perpendicular to $\mathbf{V}^R$ in the sense of ω_3. When Eq. (5) is solved, we find

$$A^R = 10.1 \text{ fps}^2$$

and in the direction shown in Fig. 5-21. This is the relative acceleration, the acceleration of C_4 that would be reported by an observer attached to and moving with link 3.

We might note that both $\mathbf{V}^R$ and $\mathbf{A}^R$ are independent of the choice of the origin of the $x'y'$ system.

Another approach to the problem would have been to choose a moving $x''y''$ system attached to link 4 instead of link 3. The velocity equation would then have given the velocity as

$$\mathbf{V}^R = \mathbf{V}_{C_3C_4} = -\mathbf{V}_{C_4C_3}$$

and so the direction of the relative-velocity vector would have been reversed and the magnitude would have been unchanged. The relative acceleration, however, would change in magnitude as well as in direction because the Coriolis component would be affected by ω_4, which is different from ω_3. We are thus led to the conclusion that the relative acceleration depends upon the reference system from which it is observed.

The reader may wish to demonstrate for himself that $\mathbf{A}^R$ is indeed the relative acceleration. Figure 5-22 is an inversion of the mechanism of Fig. 5-20 with link 3 as the fixed link. If link 1 in Fig. 5-22 is driven with

$$\omega_1 = -\omega_3 \qquad \dot{\omega}_1 = -\dot{\omega}_3$$

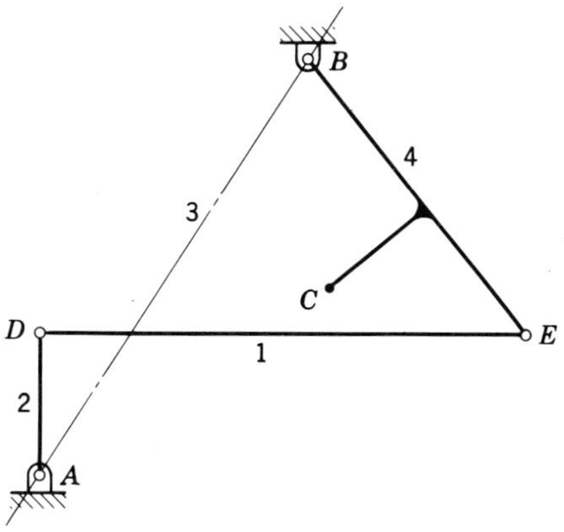

Fig. 5-22

then the acceleration of C_4 can easily be determined. The result, since link 3 is fixed, is the acceleration of C_4 relative to link 3 and hence should be the same as the value found for $\mathbf{A}^R$ in Fig. 5-21.

STOP!!!

5-6 THE ANALYSIS OF PLANE-MOTION DIRECT-CONTACT MECHANISMS[1]

The mechanism shown in Fig. 5-23 consists of three bodies or links. Link 2 is called the *cam;* it is the driving member and rotates at a constant angular velocity ω_2. Link 3 is the driven member and is called the *follower*. To maintain contact, the follower must be held against the cam by some external means such as a spring. The problem is to determine the angular velocity and acceleration of the follower.

In Fig. 5-23 we designate the point of contact as point B at the intersection of the *common normal* and *common tangent*. But there are really three coincident points here, each of which has a unique motion. Point B_2 is fixed to the cam; hence its motion is perpendicular to the line O_2B_2 in the sense of ω_2. Point B_3 is fixed to the follower, and so its motion is perpendicular to the line O_3B_3.

To obtain the motion of B it is necessary to define an *equivalent mechanism*. Two mechanisms are said to be *kinematically equivalent* when the motions of the drivers and the followers are identical. For the mechanism of Fig. 5-23 we first note that the cam and follower profiles are both made of circle arcs for the phase of the motion under study. The centers of these arcs are points A and C. An equivalent four-bar mechanism is formed by pinning a fictitious link X to A and C. Thus

[1] In the English language the book Alexander Cowie, "Kinematics and Design of Mechanisms," International Textbook Company, Scranton, Pa., 1961, is, without doubt, the best and most complete reference on this subject. See pp. 47–50, 102–104, and 163–191.

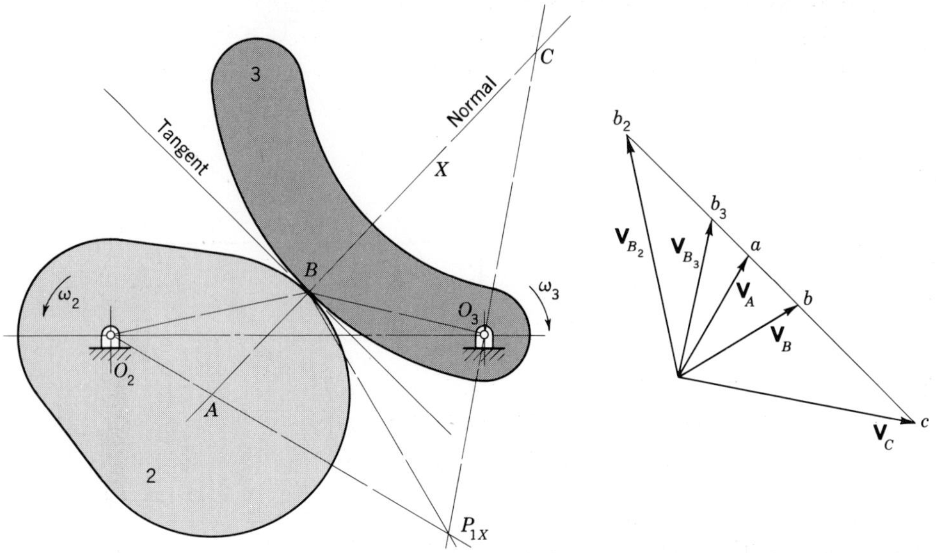

Fig. 5-23

the equivalent mechanism has a crank O_2A, a coupler AC, and a follower O_3C. In the general case the radii of curvature of the cam and follower vary, and so the dimensions of the equivalent mechanism will vary too. The motion of the point of contact B may be found by considering it as fixed to link X and constructing the velocity and acceleration polygons in the usual manner. The velocity polygon for such an analysis is shown in Fig. 5-23 as O_vabc. Note that point b can be located by using the image approach or by computing $\mathbf{V}_B$ directly after the angular velocity ω_X of link X has been found.

The direction of $\mathbf{V}_B$ can also be found by employing velocity poles. The pole P_{1X} for link X is located at the intersection of a line through O_2A and another through O_3C. The velocity of B, then, is a vector perpendicular to $P_{1X}B$.

In the special case in which one of the contacting surfaces is flat, the center of curvature is on the common normal at infinity, and the fictitious link X then is infinitely long. The equivalent mechanism is then of no value in solving the problem. However, the velocity pole P_{1X}, which we shall now denote simply as P, can be found without defining an equivalent mechanism. For each member construct a line from the center of rotation through the center of curvature. The intersection of these two lines defines the velocity pole P for the motion of the contact point B.

Let us now determine the velocities of the mechanism of Fig. 5-23 by using the direct approach. With ω_2 given, the velocity of B_2 on link 2 is computed and Eq. (4-24), for the velocity of the coincident point B_3, becomes

$$\mathbf{V}_{B_3} = \mathbf{V}_{B_2} + \mathbf{V}^R \tag{a}$$

with

$$\mathbf{V}^R = \mathbf{V}_{B_3/2} \tag{b}$$

We should note that $\mathbf{V}_{B_3/2}$ means the velocity of B_3, a point on body 3, relative to body 2. The vector $\mathbf{V}_{B_3/2}$ is the same vector as $\mathbf{V}_{B_3B_2}$. However, when we study the acceleration relations in direct-contact mechanisms we shall find, for example, that $\mathbf{A}_{B_3B_2}$ (the acceleration of B_3 relative to B_2) is *not* the same vector as $\mathbf{A}_{B_3/2}$ (the acceleration of B_3 relative to link 2). For this reason it is desirable to write $\mathbf{V}_{B_3/2}$ instead of $\mathbf{V}_{B_3B_2}$ even though they are the same for velocities. The shilling which has been added to the subscript is standard practice in many textbooks, but is generally unnecessary. It has been added here to avoid any confusion which might result in the interpretation of a vector like this:

$$\mathbf{V}_{D4}$$

Note that one might read this vector either as the absolute velocity of point D_4 or as the velocity of D relative to link 4. By rewriting it as

$$\mathbf{V}_{D/4}$$

we understand that it is the velocity of point D relative to link 4. We shall continue the use of subscripted subscripts, however. Thus the absolute velocity of point D_4 is written

$$\mathbf{V}_{D_4}$$

The velocity polygon of Fig. 5-23 shows the solution of Eq. (*a*). The velocity $\mathbf{V}_{B_3/2}$ is a vector from b_2 to b_3. Note that other velocity relations are also evident, such as

$$\mathbf{V}_{B_3} = \mathbf{V}_B + \mathbf{V}_{B_3B} \tag{c}$$

where $\mathbf{V}_{B_3B}$ is a vector from b to b_3.

It should be observed that both B and B_3 have motion along the common tangent relative to body 2. However, they are two distinct points. B is the point of contact, and its motion relative to link 2 is on the circle arc whose center is A. But B_3 is fixed to link 3 and hence it generates a different path on link 2 than does point B. The path generated by B_3 on body 2 will have its center of curvature somewhere on the common normal because $\mathbf{V}_{B_3/2}$ is along the common tangent. The path of B_3 on 2 can be found by using a tracing-paper overlay, holding

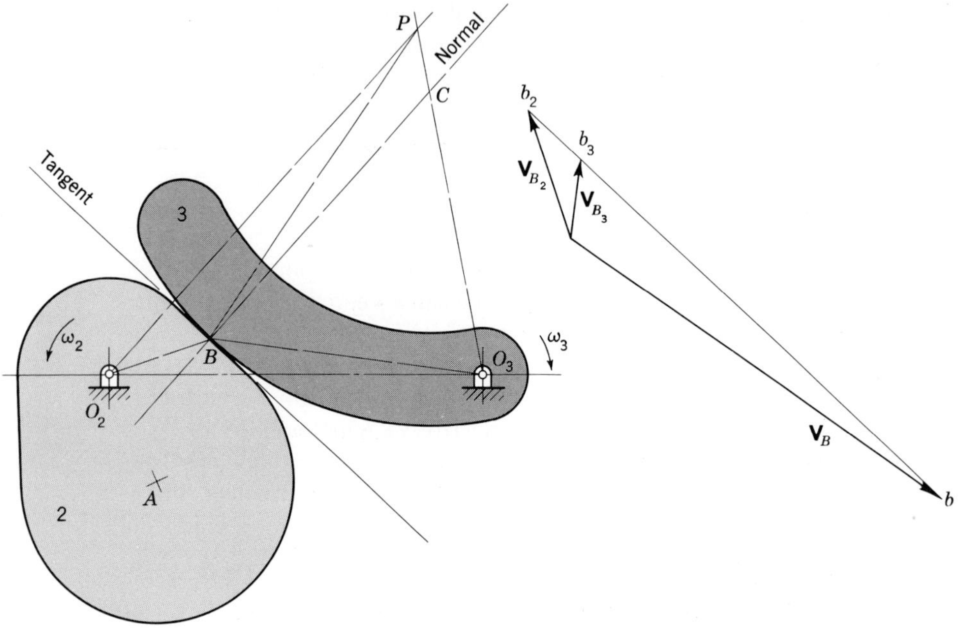

body 2 stationary, and rotating link 1 through a variety of positions.[1] The center of curvature of B_3 on body 2 is not, in general, located at the center of curvature A of the profile of body 2. We shall use these facts when we study the accelerations.

Let us now turn our attention to Fig. 5-24, which is the same mechanism, but the cam is turned to a new position such that the follower contacts a flat portion of the cam surface. The center of curvature of the cam profile is now on the common normal at infinity. Hence, we cannot use the equivalent mechanism in our analysis. The velocity pole P for the motion of the point of contact B is obtained as the intersection of a line from O_3 through the center of curvature C of body 3, and another line through O_2 parallel to the common normal. The velocity polygon, shown in the figure, is then obtained as the solution to Eqs. (a) and (c).

To solve for the accelerations of the cam mechanism of Fig. 5-23, we are tempted to write Eq. (5-14), for the acceleration of a point in a moving reference system, in the form

$$\mathbf{A}_{B_3} = \mathbf{A}_{B_2} + \mathbf{A}^R + \mathbf{A}^C \tag{d}$$

[1] See Charles R. Mischke, "Elements of Mechanical Analysis," p. 156, Addison-Wesley Publishing Company, Inc., Reading, Mass., 1963.

where

$$\mathbf{A}_{B_3} = \mathbf{A}_{B_3}{}^r + \mathbf{A}_{B_3}{}^t$$

with

$$A_{B_3}{}^r = O_3B_3\omega_3{}^2 \qquad A_{B_3}{}^t = O_3B_3\dot{\omega}_3$$

and with $\dot{\omega}_3$ as the unknown to be found. Since ω_2 is constant, the magnitude of the first term on the right of Eq. (*d*) is

$$A_{B_2} = A_{B_2}{}^r = O_2B_2\omega_2{}^2$$

The relative acceleration is $\mathbf{A}^R = \mathbf{A}_{B_3/2}$. Using Eqs. (5-3) and (5-4) for the acceleration of a point moving on a curved path, we see that

$$\mathbf{A}_{B_3/2} = \mathbf{A}^n_{B_3/2} + \mathbf{A}^t_{B_3/2}$$

with

$$\mathbf{A}^n_{B_3/2} = \frac{\dot{s}^2}{\rho}\,\hat{\mathbf{\mu}} \qquad (e)$$

$$\mathbf{A}^t_{B_3/2} = \ddot{s}\hat{\mathbf{\lambda}} \qquad (f)$$

where $\hat{\mathbf{\mu}}$ is the direction of the common normal, $\hat{\mathbf{\lambda}}$ the direction of the common tangent, $\dot{s}$ and $\ddot{s}$ the speed and acceleration of the point along the path, respectively, and ρ the instantaneous radius of curvature of the path. Therefore

$$\dot{s} = V_{B_3/2} \qquad \ddot{s} = A^t_{B_3/2}$$

But the radius of curvature ρ is unknown. As indicated previously, the path of B_3 on link 2 can be found by a graphical procedure, but the measurement of the radius of curvature by graphical means is not a very satisfactory solution.

Finally, the Coriolis acceleration is

$$\mathbf{A}^C = 2\boldsymbol{\omega}_2 \times \mathbf{V}_{B_3/2}$$

Both terms in this equation are known, and the direction is that of the common normal. Putting all this information back together in Eq. (*d*) yields

$$\overset{2}{\mathbf{A}_{B_3}{}^r} + \overset{1}{\mathbf{A}_{B_3}{}^t} = \overset{2}{\mathbf{A}_{B_2}{}^r} + \overset{1}{\mathbf{A}^n_{B_3/2}} + \overset{1}{\mathbf{A}^t_{B_3/2}} + \overset{2}{2\boldsymbol{\omega}_2 \times \mathbf{V}_{B_3/2}} \qquad (g)$$

All the directions are known, but the equation can be solved only if a solution to Eq. (*e*), for the magnitude of $\mathbf{A}^n_{B_3/2}$, is found.[1]

Professor Alexander Cowie[2] of the Illinois Institute of Technology has developed a very clever solution to this problem. Equation (5-14)

[1] Mischke, in *ibid.*, does this.
[2] *Op. cit.*, p. 170.

is written for the point of contact B, once using link 2 as the moving system and again using link 3 as the moving reference. For the mechanism of Fig. 5-23 Cowie's approach yields the two equations

$$\mathbf{A}_B = \mathbf{A}_{B_2}{}^r + \mathbf{A}^n_{B/2} + \mathbf{A}^t_{B/2} + 2\boldsymbol{\omega}_2 \times \mathbf{V}_{B/2} \tag{h}$$

$$\mathbf{A}_B = \mathbf{A}_{B_3}{}^r + \mathbf{A}_{B_3}{}^t + \mathbf{A}^n_{B/3} + \mathbf{A}^t_{B/3} + 2\boldsymbol{\omega}_3 \times \mathbf{V}_{B/3} \tag{i}$$

which are solved simultaneously to yield

$$\mathbf{A}_{B_3}{}^r + \mathbf{A}_{B_3}{}^t + \mathbf{A}^n_{B/3} + \mathbf{A}^t_{B/3} + 2\boldsymbol{\omega}_3 \times \mathbf{V}_{B/3} = \mathbf{A}_{B_2}{}^r + \mathbf{A}^n_{B/2} + \mathbf{A}^t_{B/2} + 2\boldsymbol{\omega}_2 \times \mathbf{V}_{B/2} \tag{5-19}$$

Let us examine these terms one by one. First, the magnitude of $\mathbf{A}_{B_3}{}^r$ is calculated from

$$A_{B_3}{}^r = O_3B_3\omega_3{}^2$$

since ω_3 is known from the velocity solution. The direction is from B_3 to O_3. The term

$$A_{B_3}{}^t = O_3B_3\dot{\omega}_3$$

is perpendicular to O_3B_3, but $\dot{\omega}_3$ is unknown. The magnitude of $\mathbf{A}^n_{B/3}$ is

$$A^n_{B/3} = \frac{(V_{B/3})^2}{BC} \tag{j}$$

because the motion of B relative to 3 is on a circle arc having its center at C. The direction is along the common normal.

The magnitude of $\mathbf{A}^t_{B/3}$ cannot be calculated, but its direction is along the common tangent. The next term is the Coriolis component referred to link 3. Its magnitude is

$$|2\boldsymbol{\omega}_3 \times \mathbf{V}_{B/3}| = 2\omega_3V_{B/3}$$

which can be computed because both ω_3 and $V_{B/3}$ are known. The direction is along the common normal.

Proceeding to the right side of Eq. (5-19), we compute the magnitude of the first term by using

$$A_{B_2}{}^r = O_2B_2\omega_2{}^2$$

and its direction is from B_2 to O_2. The magnitude of the next term is calculated in the same manner as in Eq. (j). Thus

$$A^n_{B/2} = \frac{(V_{B/2})^2}{AB} \tag{k}$$

and it is directed along the common normal. As before, we cannot find the magnitude of the tangential component of the relative acceleration

$A^t_{B/2}$, but its direction is along the common tangent. The last term in Eq. (5-19) is the Coriolis component referred to link 2 as the moving system. Its magnitude is

$$|2\omega_2 \times \mathbf{V}_{B/2}| = 2\omega_2 V_{B/2}$$

which can be computed; the direction is along the common normal.

Let us now rewrite Eq. (5-19), combine sets of terms, and summarize what we know about each term. Thus

$$\overset{2}{\mathbf{A}_{B_3}^r} + \overset{1}{\mathbf{A}_{B_3}^t} = \overset{2}{\mathbf{A}_{B_2}^r} + \overbrace{(\overset{2}{\mathbf{A}_{B/2}^n} + \overset{2}{2\omega_2 \times \mathbf{V}_{B/2}} - \overset{2}{\mathbf{A}_{B/3}^n} - \overset{2}{2\omega_3 \times \mathbf{V}_{B/3}})}^{2} + \overbrace{(\overset{1}{\mathbf{A}_{B/2}^t} - \overset{1}{\mathbf{A}_{B/3}^t})}^{1} \qquad (5\text{-}20)$$

The solution to this equation is shown in Fig. 5-25, and the numerical details in the example which follows. Note that it is necessary to treat $\mathbf{A}^t_{B/2} - \mathbf{A}^t_{B/3}$ as a single vector. For convenience, the solution to the equivalent four-bar mechanism is also illustrated.

Example 5-8 Analyze the cam mechanism of Fig. 5-25 by using the direct approach, and compare the results with a solution obtained by using an equivalent mechanism. The dimensions are shown in the figure.

Solution The velocity polygon is shown in Fig. 5-25. We omit the details of the velocity analysis, since it has been previously considered. The results are:

$V_{B_2} = 4.34$ fps perpendicular to O_2B_2

$V_{B_3} = 2.81$ fps perpendicular to O_3B_3

$V_A = 2.50$ fps perpendicular to O_2A

$V_B = 2.50$ fps perpendicular to PB

$V_C = 4.36$ fps perpendicular to O_3C

$V_{B/2} = 4.27$ fps along the common tangent

$V_{B/3} = 2.12$ fps along the common tangent

$V_{CA} = 4.27$ fps along the common tangent

$\omega_3 = 4.36$ rad/sec cw

$\omega_X = 2.56$ rad/sec cw

The acceleration components needed for Eq. (5-20) are determined next. Thus

$$A_{B_3}{}^r = O_3B_3\omega_3{}^2 = \frac{7.72}{12}(4.36)^2 = 12.3 \text{ fps}^2$$

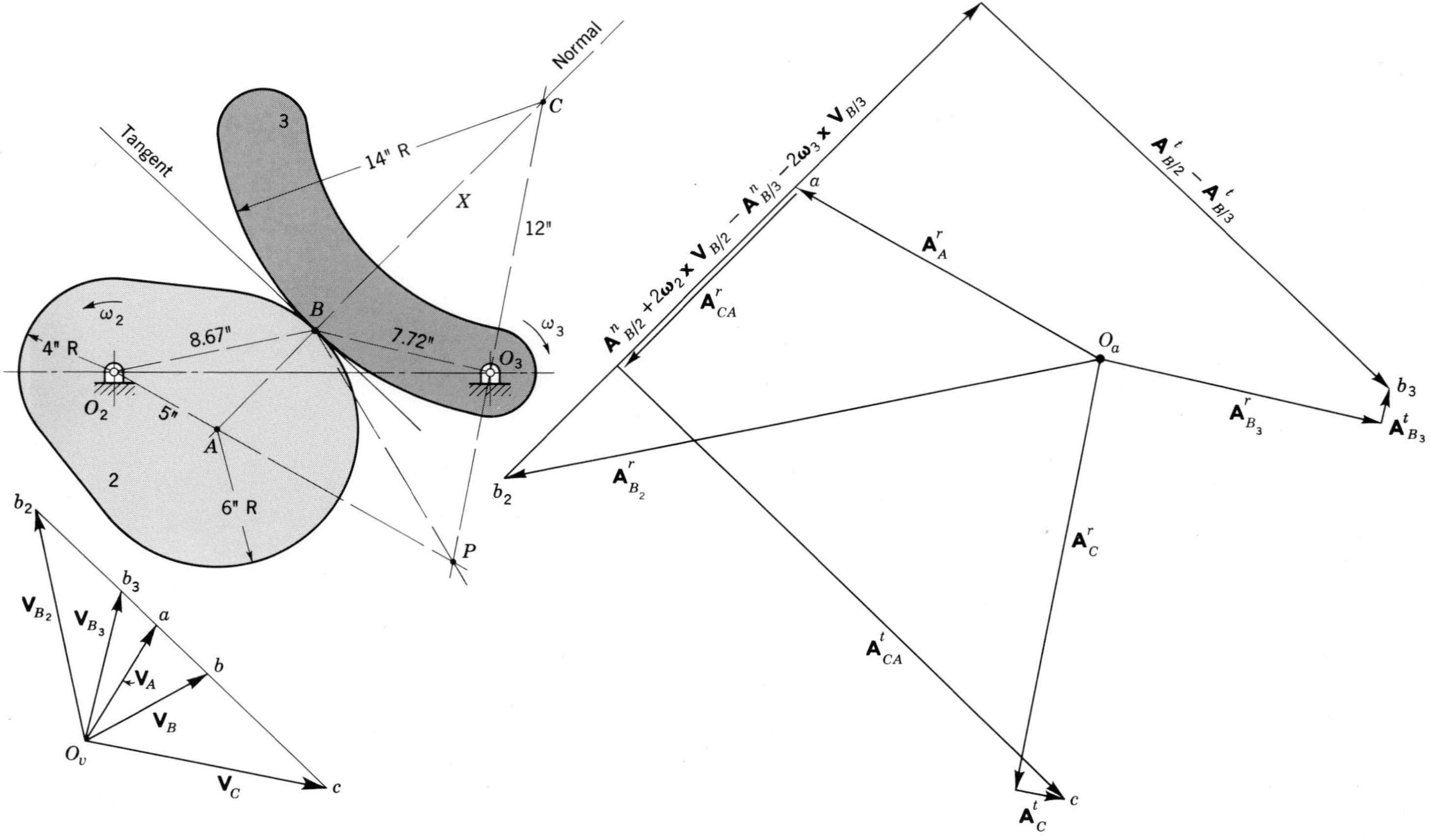

Fig. 5-25 $\omega_2 = 6$ rad/sec, $O_2O_3 = 16$ in.

and the direction is from B_3 to O_3. Next, $\mathbf{A}_{B_3}{}^t$ is along the common tangent, but the magnitude is unknown. Then,

$$A_{B_2}{}^r = O_2B_2\omega_2{}^2 = \frac{8.67}{12}(6)^2 = 26.0 \text{ fps}^2$$

with a direction from B_2 toward O_2. The group of components in the direction of the common normal have magnitudes of

$$A^n_{B/2} = \frac{(V_{B/2})^2}{AB} = \frac{(4.27)^2}{\frac{6}{12}} = 36.6 \text{ fps}^2$$

$$|2\omega_2 \times \mathbf{V}_{B/2}| = 2\omega_2 V_{B/2} = (2)(6)(4.27) = 51.2 \text{ fps}^2$$

$$A^n_{B/3} = \frac{(V_{B/3})^2}{BC} = \frac{(2.12)^2}{\frac{14}{12}} = 3.85 \text{ fps}^2$$

$$|2\omega_3 \times \mathbf{V}_{B/3}| = 2\omega_3 V_{B/3} = (2)(4.36)(2.12) = 18.1 \text{ fps}^2$$

If we accept the direction from B to A as positive, then the sum of this group of terms [see Eq. (5-20)] is

$$|\mathbf{A}^n_{B/2} + 2\omega_2 \times \mathbf{V}_{B/2} - \mathbf{A}^n_{B/3} - 2\omega_3 \times \mathbf{V}_{B/3}| = +36.6 + (-51.2) - (-3.85) - (+18.1) = 28.85 \text{ fps}^2$$

The last group of terms, $\mathbf{A}^t_{B/2} - \mathbf{A}^t_{B/3}$, is, of course, along the common tangent.

When all these vectors are plotted to scale, the acceleration polygon in Fig. 5-25 results. The component $A_{B_3}{}^t$ is scaled at 1.40 fps². Hence

$$\dot{\omega}_3 = \frac{A_{B_3}{}^t}{O_3B_3} = \frac{1.40}{7.72/12} = 2.17 \text{ rad/sec}^2 \text{ ccw}$$

The analysis using the equivalent linkage requires that we solve the equation

$$\mathbf{A}_{C^r} + \mathbf{A}_{C^t} = \mathbf{A}_{A^r} + \mathbf{A}_{CA^r} + \mathbf{A}_{CA^t} \qquad (1)$$

All the radial components for this equation can be calculated. They are as follows:

$$A_{C^r} = O_3C\omega_3{}^2 = \tfrac{12}{12}(4.36)^2 = 19.0 \text{ fps}^2 \qquad \text{from } C \text{ to } O_3$$

$$A_{A^r} = O_2A\omega_2{}^2 = \tfrac{5}{12}(6)^2 = 15.0 \text{ fps}^2 \qquad \text{from } A \text{ to } O_2$$

$$A_{CA^r} = AC\omega_X{}^2 = \tfrac{20}{12}(2.56)^2 = 11.0 \text{ fps}^2 \qquad \text{from } C \text{ to } A$$

$\mathbf{A}_{C^t}$ and $\mathbf{A}_{CA^t}$ are perpendicular to O_3C and AC, respectively. The polygon for the solution of Eq. (1) is superimposed on the polygon for the direct approach for purposes of comparison. A_{C^t} can be scaled from the polygon and used to calculate $\dot{\omega}_3$ as a check on the previous results. It is seen that the equivalent-mechanism approach is by far the easiest to use.

5-7 ACCELERATION POLES

While of little use in analysis, it is desirable to define the *instantaneous center of acceleration* or *acceleration pole* for plane-motion mechanisms here, at least to avoid the implication that the velocity pole is also the pole of acceleration. If we consider a fixed body and a moving body, then *the acceleration pole is the point on the moving body which has zero acceleration at the instant considered.*

In Fig. 5-26*a* let point J be the acceleration pole, a point of zero acceleration in the moving plane whose location is desired. Another point A in the moving plane has a known acceleration $\ddot{\mathbf{R}}$. We also specify $\dot{\boldsymbol{\omega}}$ and $\boldsymbol{\omega}$ as the angular acceleration and angular velocity of the moving member. The vector equation for the acceleration of J may then be written

$$\mathbf{A}_J = \ddot{\mathbf{R}} + \dot{\boldsymbol{\omega}} \times \mathbf{r} + \boldsymbol{\omega} \times (\boldsymbol{\omega} \times \mathbf{r}) = 0 \tag{5-21}$$

Since we are dealing with plane motion, $\hat{\boldsymbol{\omega}} = \hat{\mathbf{k}}$ and so Eq. (5-21) can be written

$$\ddot{\mathbf{R}} + r\dot{\omega}(\hat{\mathbf{k}} \times \hat{\mathbf{r}}) - r\omega^2\hat{\mathbf{r}} = 0 \tag{a}$$

Solving for $\ddot{\mathbf{R}}$ gives

$$\ddot{\mathbf{R}} = r\omega^2\hat{\mathbf{r}} - r\dot{\omega}(\hat{\mathbf{k}} \times \hat{\mathbf{r}}) \tag{b}$$

Now, $\hat{\mathbf{r}}$ and $(\hat{\mathbf{k}} \times \hat{\mathbf{r}})$ are at right angles to each other; hence the two terms on the right in Eq. (*b*) are the rectangular components of $\ddot{\mathbf{R}}$. We wish to know the distance and direction of point J from point A; hence we plot the graphical solution to Eq. (*b*) in Fig. 5-26*b*. Solving for the magnitude and direction of $\mathbf{r}$ gives

$$r = \frac{\ddot{R}}{\sqrt{\omega^4 + \dot{\omega}^2}} \tag{5-22}$$

$$\gamma = \tan^{-1} \frac{\dot{\omega}}{\omega^2} \tag{5-23}$$

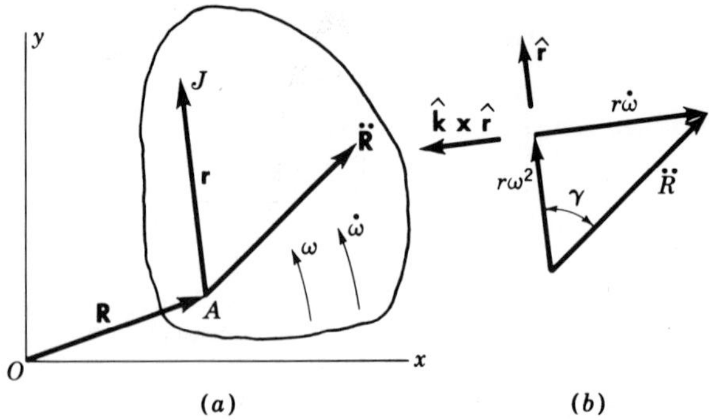

Fig. 5-26

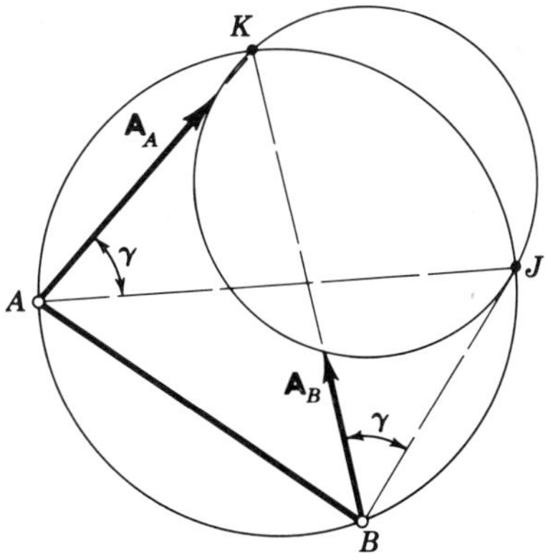

Fig. 5-27

Equation (5-22) states that the distance r to the acceleration pole depends upon the magnitude $\ddot{R}$ of the acceleration. Equation (5-23) gives the angle, and it applies for *any* point in the moving plane. Since the denominator ω^2 is always positive, the angle γ is always an acute angle.

There are many graphical methods of locating the acceleration center.[1,2] Here we shall present, without proof, the four-circle method as described by Rosenauer.[3] In Fig. 5-27 we have given points A and B on the moving link and their total accelerations $\mathbf{A}_A$ and $\mathbf{A}_B$. Extend $\mathbf{A}_A$ and $\mathbf{A}_B$ until they intersect at K. Construct a circle through points A, B, and K. Now draw another circle through the termini of the vectors and point K. The intersection of the two circles locates point J.

Rosenauer observes that only two circles are required to locate the pole but that it is called the four-circle method because the proof uses the four-circle theorem from geometry. This theorem states that the four circles which circumscribe each side of a quadrilateral and the intersection formed by the extension of the two adjacent sides intersect at one point.

5-8 ANALYTICAL METHODS

In this section we shall continue the analyses which were started in Sec. 4-16. The first mechanism studied was the slider-crank linkage. The acceleration is obtained by differentiating Eq. (4-31). The result, after

[1] N. Rosenauer and A. H. Willis, "Kinematics of Mechanisms," pp. 145–156, Associated General Publications, Sydney, Australia, 1953; republished by Dover Publications, Inc., New York, 1967.

[2] Kurt Hain (translated by T. P. Goodman, Douglas P. Adams, B. L. Harding, and D. R. Raichel), "Applied Kinematics," 2d ed., pp. 149–158, McGraw-Hill Book Company, New York, 1967.

[3] *Op. cit.*, p. 155.

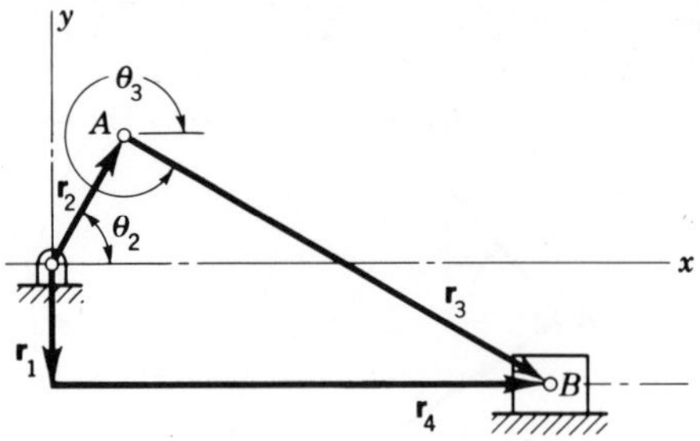

Fig. 5-28

considerable manipulation, is

$$\ddot{x} = -r\dot{\omega}\left(\sin\theta + \frac{r\sin 2\theta}{2l\cos\phi}\right) - r\omega^2\left(\cos\theta + \frac{r\cos 2\theta}{l\cos\phi} + \frac{r^3\sin^2 2\theta}{4l^3\cos^3\phi}\right) \tag{5-24}$$

This is an involved expression, and it becomes even more involved when the substitution

$$\cos\phi = \sqrt{1 - \left(\frac{r}{l}\sin\theta\right)^2} \tag{5-25}$$

is made. An approximate expression is sometimes used for the acceleration of the slider. The last term on the right in Eq. (5-24) can be neglected for large l/r ratios. Also, for large values of l/r, $\cos\phi$ is near unity. By using these approximations Eq. (5-24) can be written

$$\ddot{x} = -r\dot{\omega}\left(\sin\theta + \frac{r}{2l}\sin 2\theta\right) - r\omega^2\left(\cos\theta + \frac{r}{l}\cos 2\theta\right) \tag{5-26}$$

These expressions are quite complicated, and if the digital computer is to be used in the analysis, Raven's method is far the easiest to use. Let us use Raven's approach to analyze the accelerations for the offset slider-crank linkage shown in Fig. 5-28. The position equation is

$$r_2e^{j\theta_2} + r_3e^{j\theta_3} - r_1e^{-j\pi/2} - r_4e^{j0} = 0 \tag{a}$$

With r_1, r_2, r_3, and θ_2 given, this equation is solved for θ_3 and r_4. The results are

$$r_3^y = -r_2^y - r_1 \qquad r_3^x = \sqrt{r_3^2 - (r_3^y)^2} \tag{5-27}$$

$$\theta_3 = \tan^{-1}\frac{r_3^y}{r_3^x} \tag{5-28}$$

with

$$r_2^x = r_2\cos\theta_2 \qquad r_2^y = r_2\sin\theta_2 \tag{5-29}$$

where the superscripts designate the components along the x and y axes.

Differentiating Eq. (a) once with respect to time gives us the velocity equation, which is really the equation of the velocity polygon in exponential form. The result is

$$jr_2\omega_2 e^{j\theta_2} + jr_3\omega_3 e^{j\theta_3} - \dot{r}_4 e^{j0} = 0 \tag{b}$$

Raven's method, as we have seen, consists in making a trigonometric transformation, separating the result into real and imaginary terms, and solving for the two unknowns. When this procedure is carried out for Eq. (b), we find

$$\omega_3 = -\frac{r_2\omega_2 \cos \theta_2}{r_3} \tag{5-30}$$

$$\dot{r}_4 = -r_2\omega_2 \sin \theta_2 - r_3\omega_3 \sin \theta_3 \tag{5-31}$$

We use the same approach to obtain the acceleration equations by first differentiating Eq. (b). The final results are

$$\dot{\omega}_3 = \frac{r_2\omega_2^2 \sin \theta_2 + r_3\omega_3^2 \sin \theta_3}{r_3 \cos \theta_3} \tag{5-32}$$

$$\ddot{r}_4 = -r_2\omega_2^2 \cos \theta_2 - r_3\dot{\omega}_3 \sin \theta_3 - r_3\omega_3^2 \cos \theta_3 \tag{5-33}$$

Figure 5-29 shows the computer printout of the solution to these equations using data from Prob. 4-17. DTETA is the crank-angle increment, 15° in this example; it is an abbreviation of $\Delta\theta$.

The same procedure, when carried out for the four-bar linkage, gives results which can be used for computer solutions of both the crank-and-rocker linkage and the drag-link mechanism. They cannot be used for other four-bar linkages unless an arrangement is built into the computer program to cause the solution to stop whenever an extreme position is reached. If, for the four-bar linkage, we write

$$\mathbf{r}_1 + \mathbf{r}_2 + \mathbf{r}_3 - \mathbf{r}_4 = 0 \tag{c}$$

where the subscripts are the link numbers and where link 2 is the driver having a constant angular input velocity, then the acceleration relations which are obtained by Raven's method are

$$\dot{\omega}_3 = \frac{r_2\omega_2^2 \cos (\theta_2 - \theta_4) + r_3\omega_3^2 \cos (\theta_3 - \theta_4) - r_4\omega_4^2}{r_3 \sin (\theta_4 - \theta_3)} \tag{5-34}$$

$$\dot{\omega}_4 = \frac{r_2\omega_2^2 \cos (\theta_2 - \theta_3) - r_4\omega_4^2 \cos (\theta_3 - \theta_4) + r_3\omega_3^2}{r_4 \sin (\theta_4 - \theta_3)} \tag{5-35}$$

In using the Chace approach, we first differentiate Eq. (c) to get the velocities. Since r_2, r_3, and r_4 are constants for the four-bar linkage,

BEGIN RUN 1

R1	R2	R3	DTETA	OMEGA2
.150000E+01	.250000E+01	.100000E+02	.150000E+02	.100000E+02

THETA2 DEGS	THETA3 DEGS	OMEGA3 RPS	VELB IPS	ALPHA3 RPSS	ACCELB IPSS
.000000E+00	.351374E+03	-.250000E+01	-.375000E+01	-.948229E+00	-.313215E+03
.150000E+02	.347603E+03	-.241482E+01	-.115552E+02	.534304E+01	-.286963E+03
.300000E+02	.344039E+03	-.216507E+01	-.184539E+02	.116605E+02	-.229508E+03
.450000E+02	.340928E+03	-.176777E+01	-.234543E+02	.176240E+02	-.148721E+03
.600000E+02	.338501E+03	-.125001E+01	-.262319E+02	.226543E+02	-.565095E+02
.750000E+02	.336955E+03	-.647060E+00	-.266812E+02	.260645E+02	.334790E+02
.900000E+02	.336423E+03	-.150874E-04	-.250001E+02	.272772E+02	.109107E+03
.105000E+03	.336955E+03	.647031E+00	-.216152E+02	.260646E+02	.162889E+03
.120000E+03	.338501E+03	.124998E+01	-.170695E+02	.226545E+02	.193491E+03
.135000E+03	.340928E+03	.176775E+01	-.119012E+02	.176242E+02	.204833E+03
.150000E+03	.344039E+03	.216505E+01	-.654628E+01	.116608E+02	.203505E+03
.165000E+03	.347603E+03	.241481E+01	-.128597E+01	.534334E+01	.196000E+03
.180000E+03	.351374E+03	.250000E+01	.374978E+01	-.947943E+00	.186785E+03
.195000E+03	.355108E+03	.241482E+01	.852997E+01	-.699305E+01	.177416E+03
.210000E+03	.358569E+03	.216508E+01	.130410E+02	-.126208E+02	.166491E+03
.225000E+03	.153423E+01	.176779E+01	.172041E+02	-.176000E+02	.150252E+03
.240000E+03	.381325E+01	.125003E+01	.208191E+02	-.215943E+02	.123773E+03
.255000E+03	.524879E+01	.647089E+00	.235561E+02	-.242113E+02	.826877E+02
.270000E+03	.573919E+01	.452995E-04	.250000E+02	-.251259E+02	.251305E+02
.285000E+03	.524893E+01	-.647001E+00	.247402E+02	-.242115E+02	-.467194E+02
.300000E+03	.381351E+01	-.124996E+01	.224822E+02	-.215948E+02	-.126223E+03
.315000E+03	.153460E+01	-.176773E+01	.181514E+02	-.176007E+02	-.203297E+03
.330000E+03	.358569E+03	-.216504E+01	.119593E+02	-.126216E+02	-.266518E+03
.345000E+03	.355109E+03	-.241480E+01	.441146E+01	-.699386E+01	-.305545E+03
.360000E+03	.351375E+03	-.250000E+01	-.374925E+01	-.948800E+00	-.313216E+03

Fig. 5-29

this yields

$$\boldsymbol{\omega}_2 \times \mathbf{r}_2 + \boldsymbol{\omega}_3 \times \mathbf{r}_3 - \boldsymbol{\omega}_4 \times \mathbf{r}_4 = 0 \tag{d}$$

When this equation is solved for the velocities as in Sec. 4-16, we obtain

$$\omega_3 = -\frac{r_2\omega_2}{r_3}\frac{(\hat{\mathbf{k}} \times \hat{\mathbf{r}}_2) \cdot \hat{\mathbf{r}}_4}{(\hat{\mathbf{k}} \times \hat{\mathbf{r}}_3) \cdot \hat{\mathbf{r}}_4} \tag{5-36}$$

$$\omega_4 = \frac{r_2\omega_2}{r_4}\frac{(\hat{\mathbf{k}} \times \hat{\mathbf{r}}_2) \cdot \hat{\mathbf{r}}_3}{(\hat{\mathbf{k}} \times \hat{\mathbf{r}}_4) \cdot \hat{\mathbf{r}}_3} \tag{5-37}$$

If the reader will compare Eq. (5-37) with (4-38) he will note a sign change. This occurs because of the decision to reverse the direction of $\mathbf{r}_4$ in Eq. (*c*) as compared with the direction used in Sec. 4-16.

We now differentiate Eq. (*d*) to get the accelerations. The result, in vector form, is

$$\boldsymbol{\omega}_2 \times (\boldsymbol{\omega}_2 \times \mathbf{r}_2) + \dot{\boldsymbol{\omega}}_3 \times \mathbf{r}_3 + \boldsymbol{\omega}_3 \times (\boldsymbol{\omega}_3 \times \mathbf{r}_3) - \dot{\boldsymbol{\omega}}_4 \times \mathbf{r}_4 - \boldsymbol{\omega}_4 \times (\boldsymbol{\omega}_4 \times \mathbf{r}_4) = 0 \tag{e}$$

The unknowns are $\dot{\omega}_3$ and $\dot{\omega}_4$. For plane motion we can write Eq. (*e*) in the form

$$-r_2\omega_2^2\hat{\mathbf{r}}_2 + r_3\dot{\omega}_3(\hat{\mathbf{k}} \times \hat{\mathbf{r}}_3) - r_3\omega_3^2\hat{\mathbf{r}}_3 - r_4\dot{\omega}_4(\hat{\mathbf{k}} \times \hat{\mathbf{r}}_4) + r_4\omega_4^2\hat{\mathbf{r}}_4 = 0 \tag{f}$$

Let us now dot through this equation with the unit vector $\hat{\mathbf{r}}_3$. Then

$$r_3\dot{\omega}_3(\hat{\mathbf{k}} \times \hat{\mathbf{r}}_3) \cdot \hat{\mathbf{r}}_3 = 0$$

because $\hat{\mathbf{k}} \times \hat{\mathbf{r}}_3$ is at right angles to $\hat{\mathbf{r}}_3$. Also

$$r_3\omega_3^2\hat{\mathbf{r}}_3 \cdot \hat{\mathbf{r}}_3 = r_3\omega_3^2$$

And so Eq. (*f*) becomes

$$-r_2\omega_2^2\hat{\mathbf{r}}_2 \cdot \hat{\mathbf{r}}_3 - r_3\omega_3^2 - r_4\dot{\omega}_4(\hat{\mathbf{k}} \times \hat{\mathbf{r}}_4) \cdot \hat{\mathbf{r}}_3 + r_4\omega_4^2\hat{\mathbf{r}}_4 \cdot \hat{\mathbf{r}}_3 = 0 \tag{g}$$

The only unknown in this equation is $\dot{\omega}_4$. Since it is a scalar equation, we solve for it and get

$$\dot{\omega}_4 = \frac{r_2\omega_2^2\hat{\mathbf{r}}_2 \cdot \hat{\mathbf{r}}_3 + r_3\omega_3^2 - r_4\omega_4^2\hat{\mathbf{r}}_4 \cdot \hat{\mathbf{r}}_3}{-r_4(\hat{\mathbf{k}} \times \hat{\mathbf{r}}_4) \cdot \hat{\mathbf{r}}_3} \tag{5-38}$$

By using a similar procedure, the angular acceleration of link 3 is found to be

$$\dot{\omega}_3 = \frac{r_2\omega_2^2\hat{\mathbf{r}}_2 \cdot \hat{\mathbf{r}}_4 + r_3\omega_3^2\hat{\mathbf{r}}_3 \cdot \hat{\mathbf{r}}_4 - r_4\omega_4^2}{r_3(\hat{\mathbf{k}} \times \hat{\mathbf{r}}_3) \cdot \hat{\mathbf{r}}_4} \tag{5-39}$$

The reader should prove for himself that Eqs. (5-38) and (5-39) are indeed identical with (5-34) and (5-35).

PROBLEMS

5-1. The position vector of a point is defined by the equation $\mathbf{r} = [4t - (t^3/3)]\hat{\imath} + 10\hat{\jmath}$, where $\mathbf{r}$ is in inches and t is in seconds. Find the acceleration of the point after 2 sec has elapsed.

5-2. Find the acceleration after 3 sec of a point which moves according to the relation $\mathbf{r} = [t^2 - (t^3/6)]\hat{\imath} + (t^3/3)\hat{\jmath}$. The units are feet and seconds.

5-3. A point moves according to the equations $x = 8t^2 \cos \omega t$ and $y = -4t^3 \sin \omega t$, where x and y are in inches, t is in seconds, and ω is 10π rad/sec cw. Write the position vector for the point. Find the velocity and acceleration corresponding to $t = 0.5$ sec.

5-4. The position vector of a point, in complex polar form, is $\mathbf{r} = t^2 e^{j\pi t^3}$, where t is in seconds and r is in feet. Find the angular acceleration of the position vector and the linear acceleration of the point corresponding to $t = 0.60$ sec.

5-5. The path of a point is described by the equation $\mathbf{r} = (t^2 + 4)e^{-j\pi t/10}$, where r is in inches. For $t = 20$ sec find the tangent vector, the normal and tangential components of acceleration, and the radius of curvature.

5-6. The position of a moving particle is defined by the vector equation $\mathbf{r} = 2\hat{\imath} - 4t^2\hat{\jmath} + 3t\hat{\mathbf{k}}$ ft. Find the vectors for the velocity and acceleration, their magnitudes, the unit tangent vector, and the radius of curvature at $t = 0.30$ sec.

5-7. The position of a point moving on a helical path is given by the equation $\mathbf{r} = 5t\hat{\imath} + 3 \cos 2\pi t\hat{\jmath} + 3 \sin 2\pi t\hat{\mathbf{k}}$. For $t = 0.25$ sec find the normal and tangential acceleration vectors, their magnitudes, and the radius of curvature. What is the expression for the unit tangent vector?

5-8. The motion of a point in the xy plane is described by the equations $x = 4t \cos \pi t^3$ and $y = (t^3 \sin 2\pi t)/6$, where x and y are in inches. Find the acceleration of the point when $t = 1.40$ sec.

5-9. Link 2 in the figure has an angular velocity $\omega_2 = 120$ rad/sec and an angular acceleration $\dot{\omega}_2 = 4800$ rad/sec^2 at the instant shown. Determine the acceleration of point A.

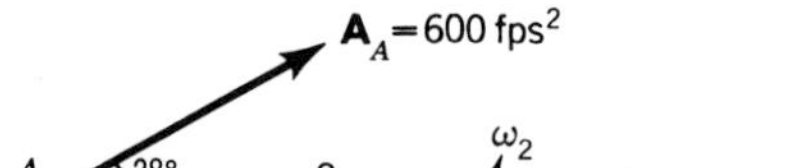

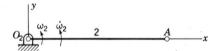

Prob. 5-9 $O_2A = 3$ in.

5-10. Link 2 is rotating clockwise, as shown in the figure. Find its linear and angular velocity and the angular acceleration.

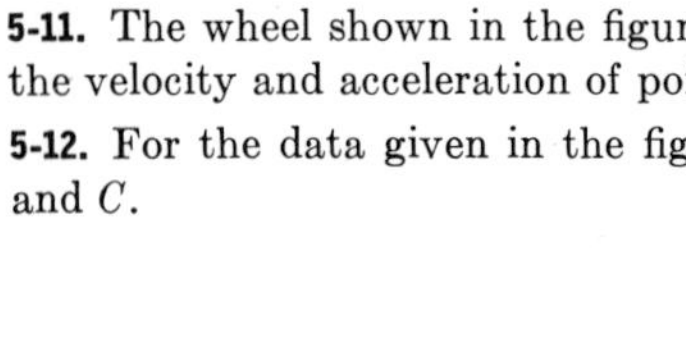

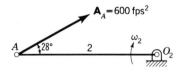

Prob. 5-10 $O_2A = 8$ in.

5-11. The wheel shown in the figure rolls on the flat surface without slipping. Find the velocity and acceleration of points A and B in terms of the given data.

5-12. For the data given in the figure find the velocity and acceleration of points B and C.

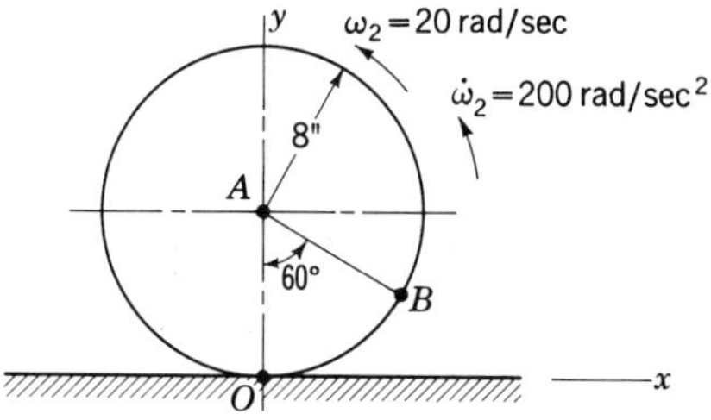

Prob. 5-11

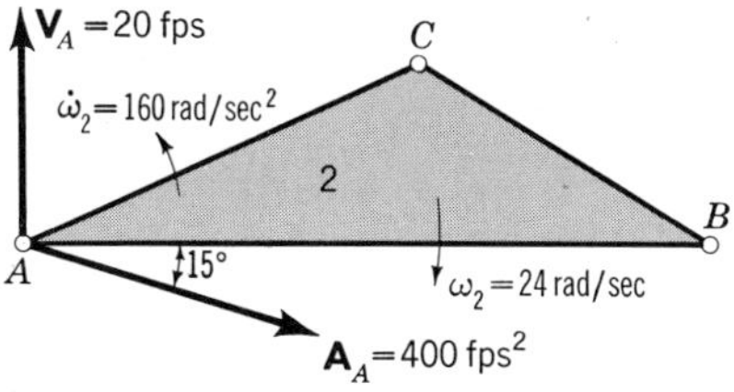

Prob. 5-12 $AB = 16$ in., $AC = 10$ in., $BC = 8$ in.

5-13. Wheels 3 and 4 in the figure roll without slipping. If $\omega_3 = 20$ rad/sec cw and $\dot{\omega}_2 = 64$ rad/sec² ccw, find $\mathbf{V}_A$, $\mathbf{A}_A$, and ω_4.

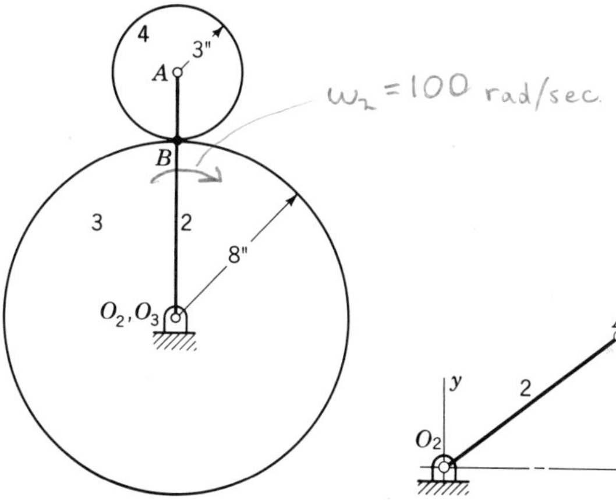

Prob. 5-13

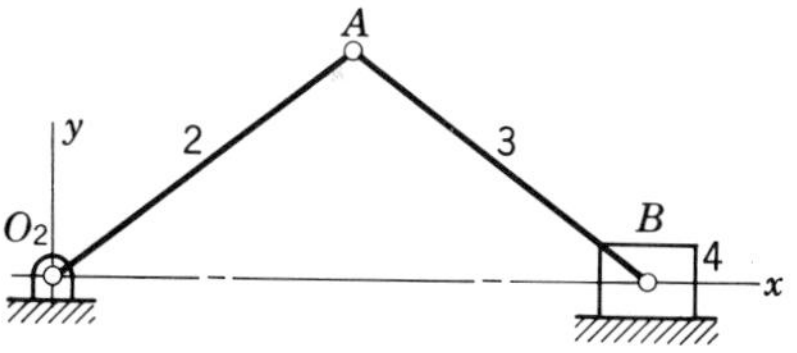

Prob. 5-14 $O_2A = AB = 5$ in., $O_2B = 8$ in.

5-14. In the figure $\mathbf{V}_B = -20\hat{\imath}$ fps. Find the angular velocity and angular acceleration of links 2 and 3.

5-15. For $\omega_2 = 24$ rad/sec and $\dot{\omega}_2 = 140$ rad/sec², find the velocity and acceleration of point B of the straight-line mechanism shown in the figure.

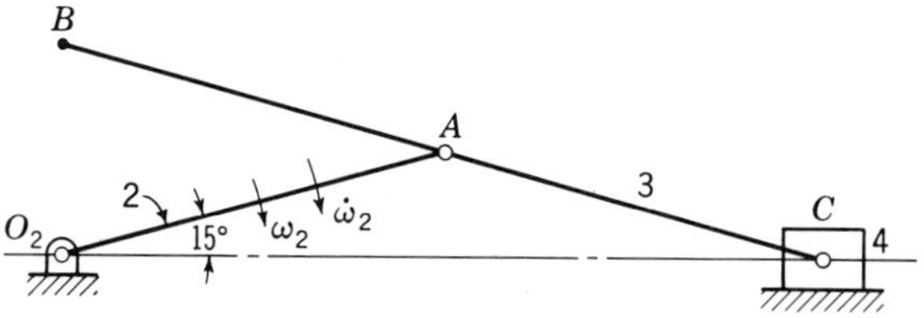

Prob. 5-15 $O_2A = 3$ in., $AB = AC = 3$ in.

5-16. If ω_2 is 32 rad/sec, find the acceleration of C and the angular acceleration of link 4.

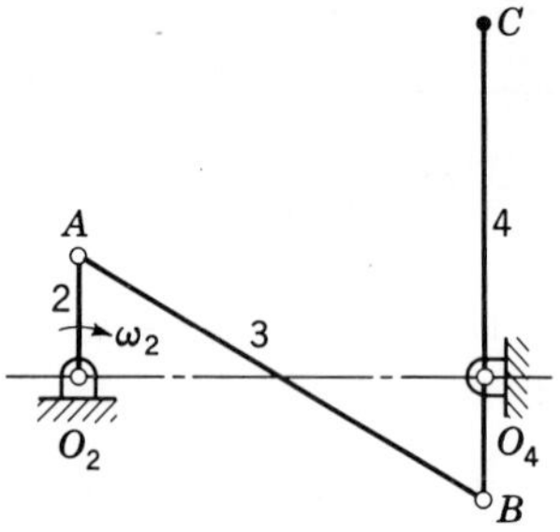

Prob. 5-16 $O_2A = O_4B = 2$ in., $AB = BC = 8$ in.

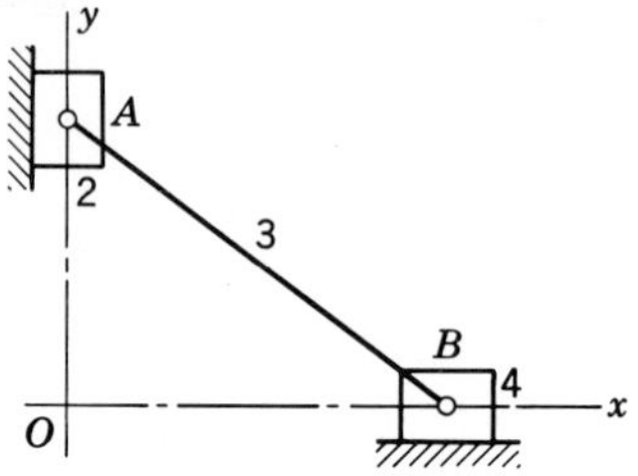

Prob. 5-17 $OA = 3$ in., $OB = 4$ in.

5-17. If $\mathbf{V}_A = -10\hat{\jmath}$ fps, find $\mathbf{A}_B$ and $\dot{\omega}_3$.

5-18. Point A in the figure has a velocity of 16 fps. Find A_C, ω_3, ω_4, $\dot{\omega}_3$, and $\dot{\omega}_4$.

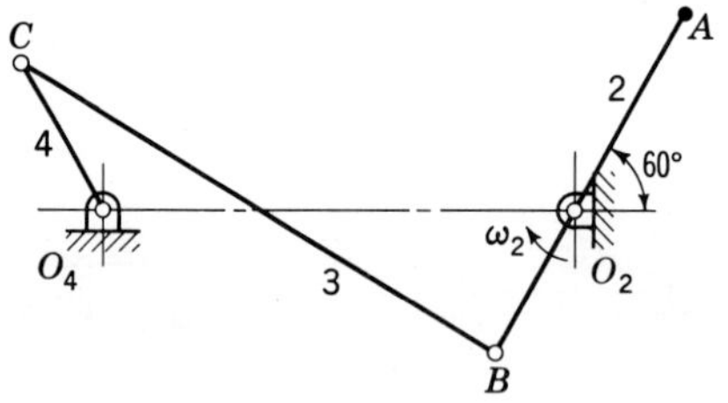

Prob. 5-18 $O_2A = 4$ in., $AB = 7$ in., $BC = 10$ in., $O_4C = 3$ in., $O_2O_4 = 8\frac{1}{2}$ in

5-19. Link 2 in the figure rotates at 60 rpm in a clockwise direction. What angular acceleration must link 2 have, for the position shown, to make the angular acceleration of link 4 zero?

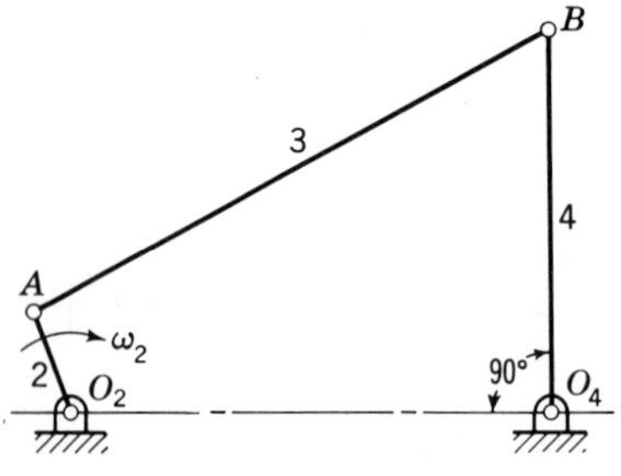

Prob. 5-19 $O_2A = 2$ in., $AB = 11$ in., $O_4B = 7$ in., $O_2O_4 = 9$ in.

5-20. Find V_B, A_B, ω_3, and $\dot{\omega}_3$. Link 3 is vertical for the phase shown.

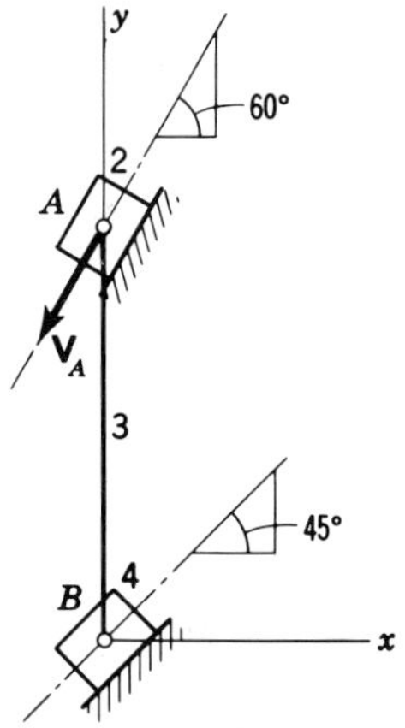

Prob. 5-20 $AB = 10$ in., $V_A = 24$ fps.

5-21. Using the data of Prob. 4-9, construct the acceleration polygon and find $\dot{\omega}_3$ and $\dot{\omega}_4$.

5-22. Solve Prob. 4-10 for the accelerations.

5-23. Solve Prob. 4-19 for the angular acceleration of links 3 and 4 for the position shown.

5-24 to **5-30.** The nomenclature for this group of problems is illustrated in the figure, and the dimensions and data for each problem are given in the accompanying table. For each problem determine θ_3, θ_4, ω_3, ω_4, $\dot{\omega}_3$, and $\dot{\omega}_4$. The input angular velocity is a constant in each problem, and a negative sign is used to indicate a clockwise angular velocity. All dimensions are in inches and ω_2 is in radians per second.

Problem number	r_1	r_2	r_3	r_4	θ_2, *deg*	ω_2
5-24	4	6	9	10	240	1
5-25	4	6	10	10	315	56
5-26	14	4	14	10	0	10
5-27	10	4	20	16	70	−6
5-28	8	2	10	6	40	12
5-29	16	5	12	12	210	−18
5-30	16	5	12	12	315	−18

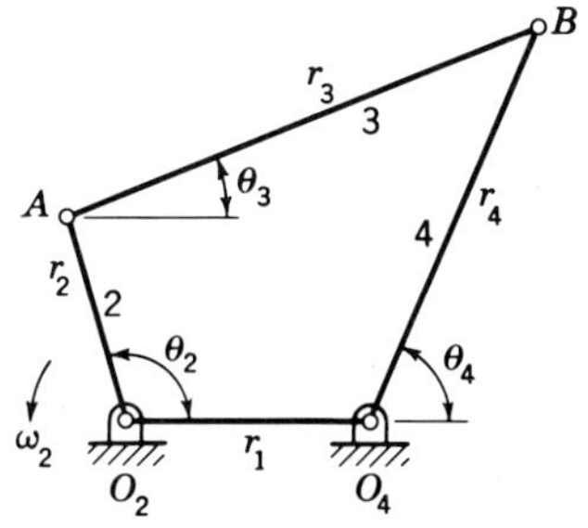

Probs. 5-24 to **5-30**

5-31. The mechanism shown in the figure is a marine steering gear called Rapson's slide. O_2B is the tiller, and AC the actuating rod. If the velocity of AC is $10\underline{/180°}$ in./min, find the angular acceleration of the tiller.

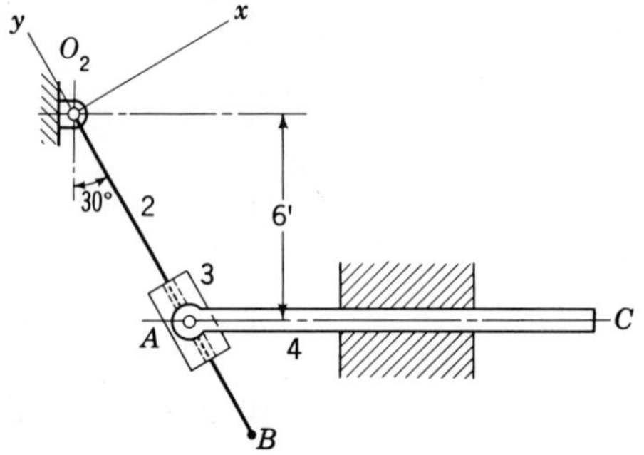

Prob. 5-31

5-32. The crank of the crank-and-lever system shown in the figure has a speed of 60 rpm ccw. Find the velocity and acceleration of point B.

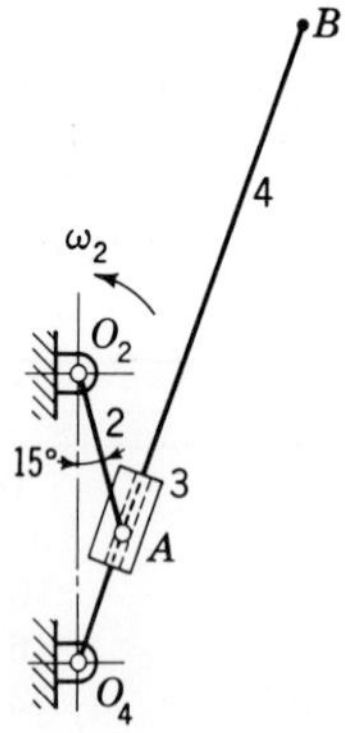

Prob. 5-32 $O_2O_4 = 12$ in., $O_2A = 7$ in., $O_4B = 28$ in.

5-33. In the hydraulic actuator shown in the figure the speed of A is to be maintained a constant 20 fps. Find the velocity and acceleration of the piston, and also the angular velocity and angular acceleration of the cylinder which is necessary to retain a constant V_A for the phase shown.

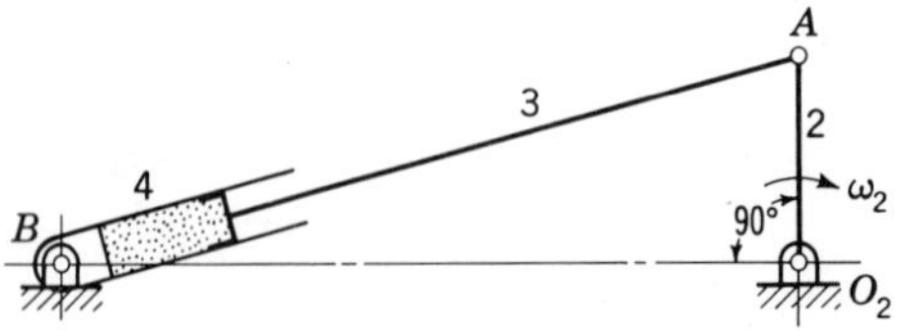

Prob. 5-33 $O_2B = 64$ in., $O_2A = 18$ in.

5-34. Draw the velocity and acceleration polygons for the linkage shown in the figure and find the image of C_3 and C_4 in both cases. Determine V^R, A^R, and A^C for the motion of C_4 relative to C_3 with a moving reference system fixed to link 3.

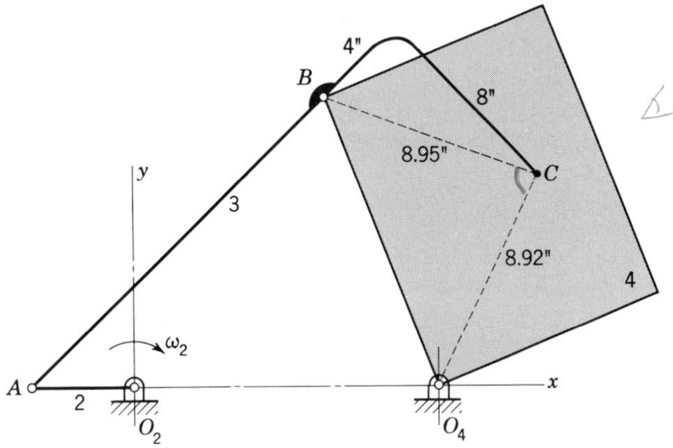

Prob. 5-34 $O_2A = 4$ in., $AB = 16$ in., $O_4B = O_2O_4 = 12$ in., $\omega_2 = 6$ rad/sec.

5-35. The figure shows a cam and oscillating roller follower. If the cam rotates counterclockwise at 400 rpm, find the angular velocity and angular acceleration of the follower.

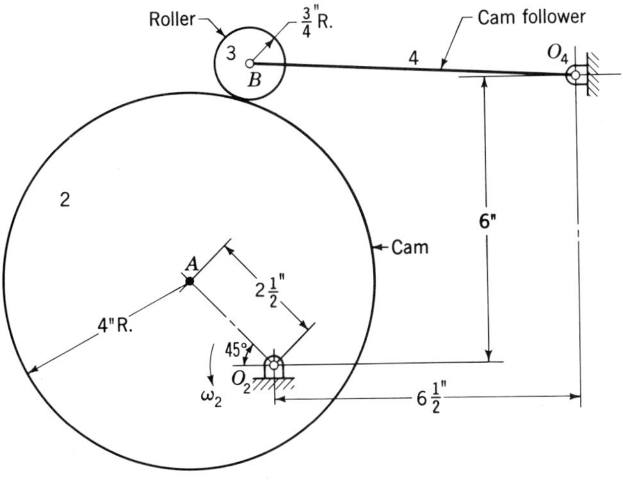

Prob. 5-35 $O_4B = 7$ in.

5-36. A cam and offset flat-face translating follower are shown in the figure. Find the velocity and acceleration of the follower for a cam speed of 240 rpm.

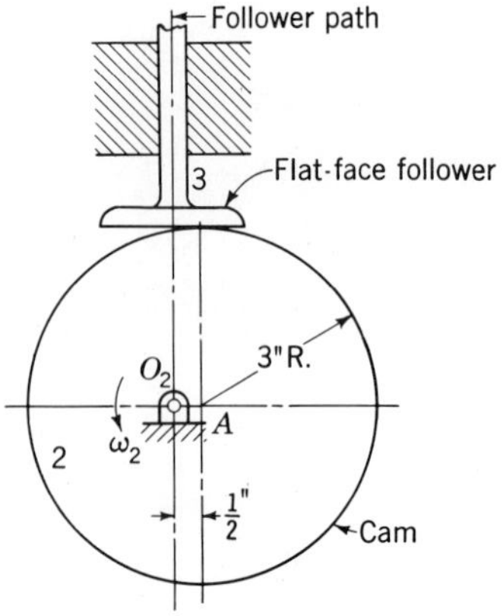

Prob. 5-36

5-37. Construct the acceleration image of the linkage of Prob. 4-17 and compute the acceleration of points C and D and the angular acceleration of link 6.

5-38. Make a complete graphical acceleration analysis of the linkage of Prob. 4-21. Shade the image of link 3.

5-39. Make a complete acceleration analysis of the linkage of Prob. 4-23.

5-40. Make a complete acceleration analysis of the V-engine mechanism of Prob. 4-24.

6
The Geometry of Motion

They aren't really "lower" pairs unless they have only one degree of freedom

Now that we have learned the use of some of the important tools of kinematic analysis it is time to turn to a study of the geometry of mechanisms. In this chapter we shall develop some important geometrical concepts useful in analysis and synthesis of mechanisms.

6-1 DEFINITIONS

Reuleaux defines a machine as follows:[1,2] "*A machine is a combination of resistant bodies so arranged that by their means the mechanical forces of nature can be compelled to do work accompanied by certain determinate motions.*"

[1] Franz Reuleaux (translated by Alexander B. W. Kennedy), "The Kinematics of Machinery," p. 35, Macmillan and Company, 1876; republished by Dover Publications, Inc., New York, 1963.
[2] There appears to be no agreement at all on the proper definition of a machine. In a footnote Reuleaux gives 17 definitions and his translator gives seven more and discusses the whole problem in detail.

Reuleaux continues and defines "*Kinematics, the Science of Pure Mechanism, as the study of those arrangements of the machine by which the mutual motions of its parts, considered as changes of position, are determined.*"

In summary, a machine has to do with the transmission of energy, whereas a mechanism is concerned with the transmission of motion.

We shall employ the word *link* to designate a component of a mechanism. Generally, a link is assumed to be rigid, though it may possess only one-way rigidity. A belt, or chain, for example, may be considered a rigid member in tension but not in compression.

The links of a mechanism must be connected together in some manner in order to transmit motion from the *driver* to the *follower* or output member. These connections are called *kinematic pairs* because a connection must consist of *two elements,* one from each link. Thus, we can also define a link as *the rigid connection between at least two elements which are members of different kinematic pairs.*

In our previous studies we have always had one fixed or grounded link in each mechanism which served as the machine frame. But it is often convenient to unfix this link to study the relative motions or the geometry of the entire linkage. Thus we shall employ the phrase *kinematic chain* for a group of links connected together, but having no link fixed to the frame.

A link composed of only two elements is called a *binary link.* Thus, the four-bar linkage of Fig. 6-2 has four binary links. But a link may be connected to more than two other links. In Fig. 6-9*a* links 2 and 6 each have three elements and they are called *ternary links.* In Fig. 6-9*b* link 1 has four elements and hence is called a *quaternary link.*

Hartenberg and Denavit[1] use the phrase *pair variable* for the coordinate or coordinates used to describe the relative motion of the elements of a pair.

There are six kinematic pairs which are identified as *lower pairs* by Reuleaux. Table 6-1 lists the names and symbols employed by Hartenberg and Denavit for these, together with the number of degrees of freedom and the pair variables for each of the six. These pairs are illustrated in Fig. 6-1.

The *turning pair* or *revolute,* shown in Fig. 6-1*a*, permits only relative rotation and hence has one freedom.

The *prismatic pair,* shown in Fig. 6-1*b*, permits only a relative

[1] R. S. Hartenberg and Jacques Denavit, "Kinematic Synthesis of Linkages," p. 35, McGraw-Hill Book Company, New York, 1964. This book is one of the few classics on kinematics available in the English language. The title is misleading; a considerable amount of material on kinematic theory and analysis is also included.

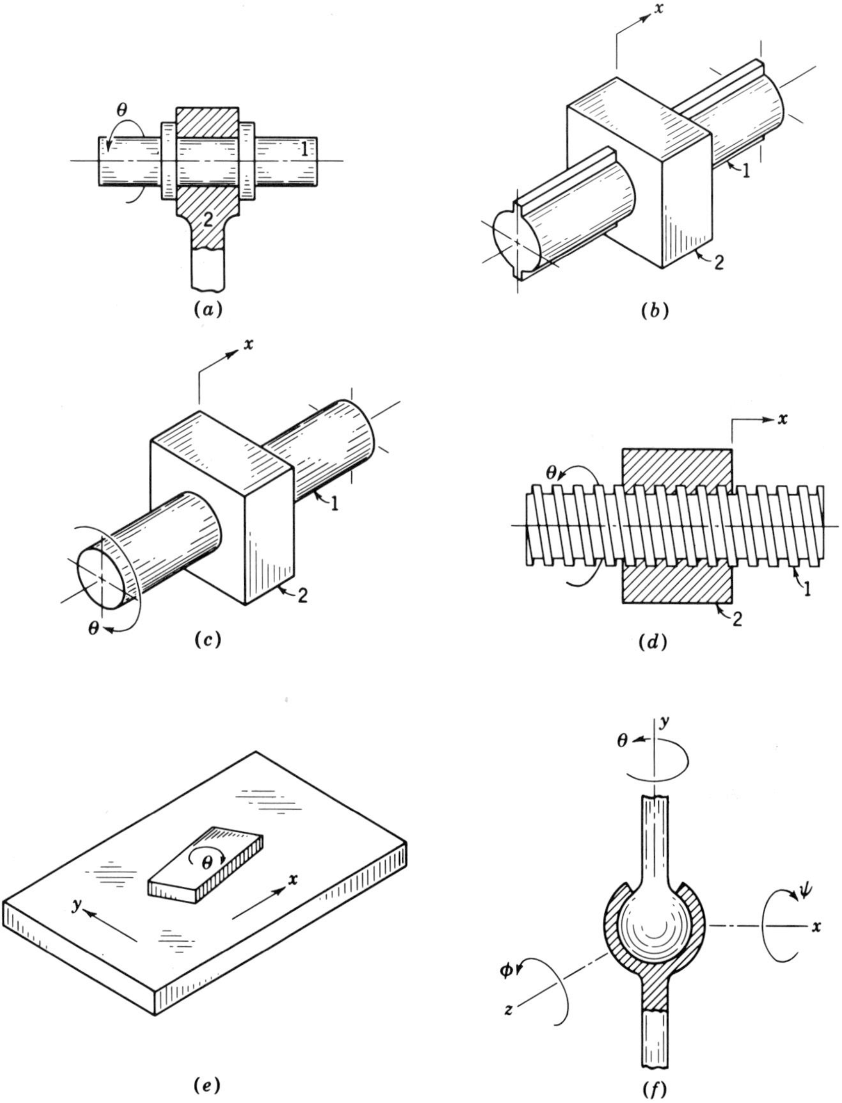

Fig. 6-1 Illustrations of all lower pairs. (*a*) Revolute or turning pair; (*b*) prismatic; (*c*) cylindric; (*d*) screw; (*e*) planar; (*f*) globular.

motion of sliding. The *cylindric pair* in Fig. 6-1*c* permits both an angular rotation and a sliding motion. Thus the degree of freedom of the prismatic pair is one, and of the cylindric pair is two.

The *screw pair* (Fig. 6-1*d*) has only one degree of freedom because the lead and the rotation are related by the helix angle of the thread. Thus the pair variable is either x or θ, but not both. Note that the screw pair reverts to a revolute if the helix angle is made zero and to a prismatic pair if it is made 90°.

Table 6-1 The six lower kinematic pairs

Name	*Symbol*	*Degree of freedom*	*Pair variable*
Revolute	R	1	θ
Prismatic	P	1	x
Cylindric	C	2	x, θ
Screw	S_L	1	x or θ
Planar	F	3	x, y, θ
Globular	G	3	θ, ϕ, ψ

The *planar pair* is seldom, if ever, found in mechanisms. The pair variables are x, y, and θ.

The *globular* or *spherical pair* of Fig. 6-1*f* is a ball-and-socket joint, and so three rotations, one about each coordinate axis, are possible.

A *lower pair* is usually defined as any pair of mating elements having surface contact. We note, too, that all pairs in Fig. 6-1 do indeed have surface contact. But a pivot bearing, for example, is a revolute with line contact. Thus we should not be too strict with this definition of lower pairs. The lower pairs are all said to have *form closure* because, in each case, one element surrounds the other. Again, the intent is that the pair have form closure. A partial journal bearing, for example, is a lower pair that does not have form closure.

The connection between a belt and its pulley, between a rope and a drum, or between a chain and a sprocket, is called a *wrapping pair*. In each case one of the links has one-way rigidity.

A connection between two links in which the contacting elements are lines or points, instead of surfaces, is described by Reuleaux as a *higher pair*. Examples of higher pairs are mating gear teeth, a wheel rolling on a rail, a ball rolling on a flat surface or in the raceway of a ball bearing, and a cam contacting its roller follower. There are an infinite number of higher pairs and so they cannot be easily classified. Thus, in this book, we shall treat each problem as a separate case when we encounter it.

6-2 GRASHOF'S LAW

It is customary to classify all planar four-link mechanisms having only revolute pairs, that is, all four-bar linkages, into two classes. *Class* 1 includes all those chains in which a link, the shortest, can make a complete revolution relative to each of the other three. Any four-bar linkage in which no link can make a complete revolution relative to any of the other three is a *class* 2 chain. In this section we shall find that Grashof's law provides an excellent means of classifying four-bar linkages in even greater detail.

Grashof's law, as given by Harding,[1] states that *the sum of the shortest and the longest links of a planar four-bar linkage cannot be greater than the sum of the remaining two links if there is to be continuous relative rotation between two members.* Harding's notation is easy to remember. Divide the linkage into two pairs such that the members of a pair are opposite to each other. In other words, the members of a pair cannot be adjacent to each other. Let the pair having the longest total length be called the *major pair* and designate its members a and b such that $a > b$. Call the other pair the *minor pair* and designate its members c and d such that $c > d$. Then the notation may be summarized by the following inequalities:

$$a + b > c + d \tag{6-1}$$

$$a > b \tag{6-2}$$

$$c > d \tag{6-3}$$

where a is opposite to b, and c is opposite to d.

With this notation Grashof's law is written

$$a - b < c - d \tag{6-4}$$

which defines a class 1 linkage. Therefore, the inequality

$$a - b > c - d \tag{6-5}$$

defines a class 2 linkage.

Attention is called to the fact that nothing in Grashof's law specifies anything about a fixed link in the chain. We are free to fix any one of the four links. When we do so, we find that by fixing different links we

[1] Bruce L. Harding, Proposed Standardized System for Notation and Classification of the Four-bar Linkage, *ASME Paper* 57-F-28.

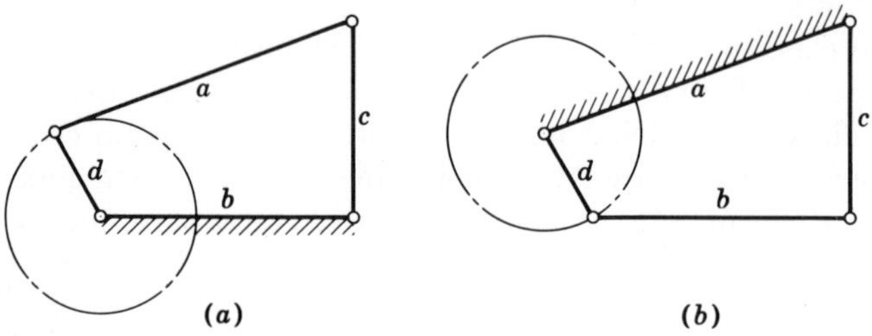

Fig. 6-2 The two crank-and-rocker mechanisms obtained from a class 1 chain.

can create a variety of mechanisms. These mechanisms are called the *inversions* of the chain.

One of the most useful inversions of the class 1 chain is the *crank-and-rocker* linkage shown in Fig. 6-2. Note that two inversions are obtained, depending upon whether link a or link b is fixed. Link d, of course, is the crank, and link c the rocker in each case.

The *drag-link mechanism*, also called the *double-crank linkage*, is obtained by fixing the shortest link (Fig. 6-3). With this inversion, link b is usually the driver. Since both links a and b rotate completely with respect to d, they are properly described as cranks. Note, too, that link c executes one complete revolution during a rotation of crank b. The reader will find an interesting challenge in devising a practical working model of this mechanism.

By fixing link c, the link opposite the shortest one, we obtain the *double-rocker mechanism* (Fig. 6-4). Note, in Fig. 6-4b, that the rockers

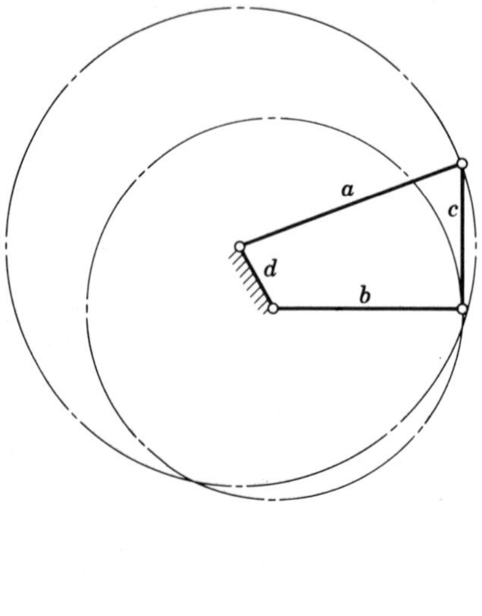

Fig. 6-3 Drag-link mechanism obtained from a class 1 chain by fixing the shortest link.

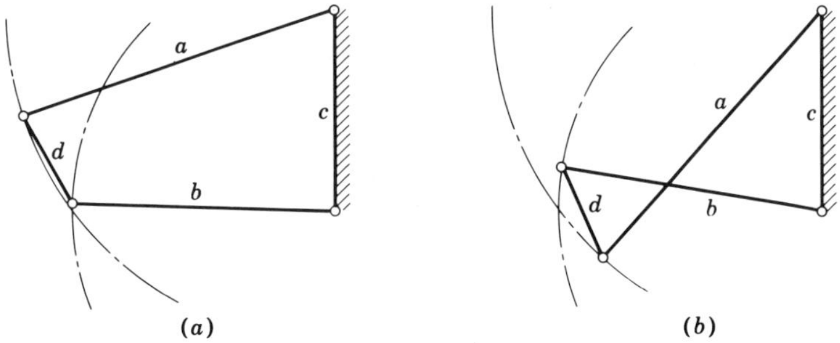

Fig. 6-4 Two versions of the double-rocker linkage obtained from a class 1 chain.

can also be crossed. This means that, theoretically, link d can make a complete rotation during operation of the mechanism.

All the inversions of the class 2 chains are double-rocker mechanisms.

There are many special cases that arise when equalities exist. The Watt's linkage of Fig. 6-5a has $c = d$; hence $a - b > c - d$, and so it is a class 2 linkage. The tracing point P generates an approximate straight line over a portion of its path.

The *parallelogram linkage* of Fig. 6-5b has $a = b$ and $c = d$; hence it is a borderline case between class 1 and class 2. Note that cranks c and d can rotate in a full circle. However, for the phase in which all four links lie on the same straight line, *change points* exist and, in passing through this phase, the mechanism may become crossed.

The *isosceles linkage* of Fig. 6-5c is a borderline case, too, because $a - b = c - d$.

The tracing point P of *Roberts' mechanism* in Fig. 6-5d generates an approximate straight line. This is a class 1 chain. The dashed lines in the figure indicate that the linkage is defined by forming three congruent isosceles triangles. Thus $d = c/2$.

Tracing point P of the *Chebyshev chain* in Fig. 6-5e also generates an approximate straight line. The linkage is formed by creating a 3-4-5 triangle with link a in the vertical position, as shown by the dashed lines in the figure. Thus $O_aB' = 3$, $O_aO_b = 4$, and $O_bB' = 5$. Since $a = b$, $O_aA' = 5$, and the tracing point P' is the midpoint of link d. Note that $O_aP'A$ also forms a 3-4-5 triangle and hence that P and P' are two points on a straight line parallel to link c.

Points P, Q, and R are on a line connected with the fixed pivot of the pantograph chain in Fig. 6-5f. The motions of these three points are directly proportional to their distances from the fixed pivot.

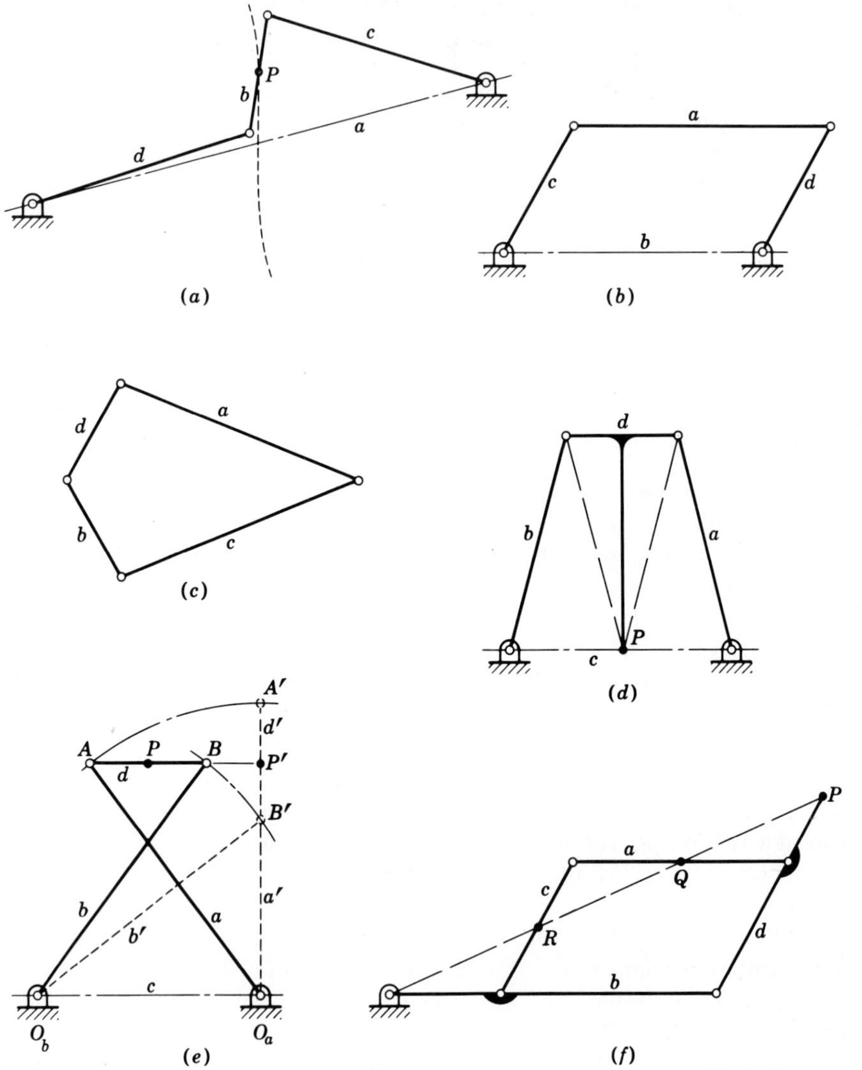

Fig. 6-5 Special cases of Grashof's law. (*a*) Watt's mechanism; $c = d$. (*b*) Parallelogram linkage; $a = b$, $c = d$. (*c*) Isosceles chain; $a = c$, $b = d$. (*d*) Roberts' mechanism; $a = b$. (*e*) Chebyshev linkage; $a = b$. (*f*) Pantograph chain.

6-3 GRÜBLER'S CRITERION FOR PLANAR MECHANISMS[1]

Unlike the Grashof criterion, which is concerned with the dimensions of the four-bar mechanism, Grübler's criterion is concerned only with the number of links in the mechanism and with the number and kinds of pairs.

The fixed link in a planar mechanism has zero degrees of freedom. The remaining links have three degrees of freedom for each, two translations and a rotation, if they are not connected to other links. When a link is joined to another link by a lower pair, then its degrees of freedom are reduced by two. But if the links are joined by a higher pair, then only one degree of freedom is lost. Thus, to obtain the degrees of freedom of a kinematic chain we multiply the number of links by the degrees of freedom of each link and subtract those which are lost because of the constraints produced by each lower or higher pair. So Grübler's criterion for plane-motion mechanisms is written

$$f = 3(n - 1) - h - 2l \tag{6-6}$$

where f = number of degrees of freedom of chain
n = number of links
h = number of higher pairs
l = number of lower pairs

pure rolling is a lower pair; like a screw pair

Equation (6-6) must be employed with some care. A rolling-contact higher pair, for example, destroys two freedoms instead of one.

If the solution to Eq. (6-6) is $f = 0$, then motion is impossible and the linkage forms a *structure;* in particular, a statically determinate structure.

If the solution is $f = -1$, then there is one redundant member and the chain is a *statically indeterminate structure.*

If $f = 1$, then the chain is a mechanism and constrained motion is possible. But if $f = 2$, then constrained motion is possible only if there are two separate input motions to the chain.

In Fig. 6-6*a* and *b* are statically determinate structures; note that two of the connections in *b* are double revolute pairs. A statically indeterminate structure is shown in Fig. 6-6*c*. The four-bar mechanism and the slider-crank linkage in Fig. 6-6*d* and *e*, respectively, both have one degree of freedom. The five-bar linkage of *f*, however, does not have constrained motion, since $f = 2$.

In Fig. 6-7*a* we illustrate a six-link mechanism. It has $f = 1$, however, because links 3, 4, and 5 form a structure. Note, in Fig. 6-7*b*, that this structure can also be treated as a single link, making a four-bar

[1] Here we follow the approach suggested by Thomas P. Goodman in his report, Four Cornerstones of Kinematic Design, *General Electric Company Rept.* 60GL90, Schenectady, N.Y., Aug. 15, 1960.

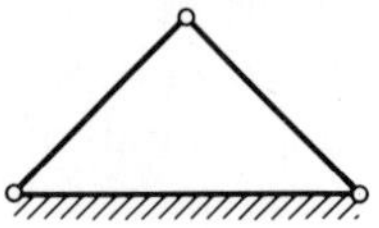

$n = 3,\ h = 0$
$l = 3,\ f = 0$
(a)

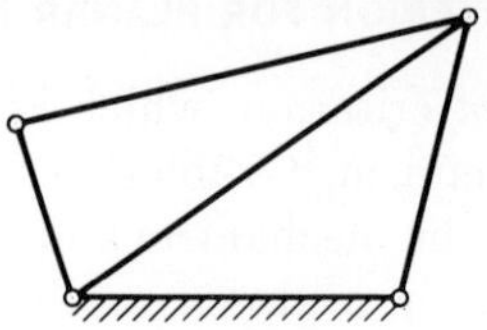

$n = 5,\ h = 0,$
$l = 6,\ f = 0$
(b)

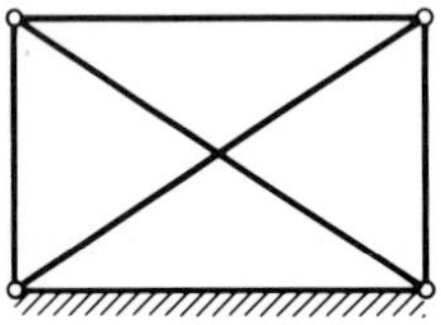

$n = 6,\ h = 0,$
$l = 8,\ f = -1$
(c)

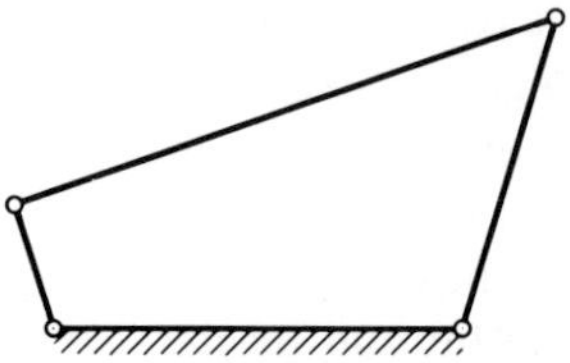

$n = 4,\ h = 0,$
$l = 4,\ f = 1$
(d)

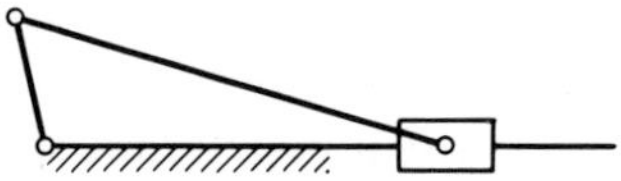

$n = 4,\ h = 0,$
$l = 4,\ f = 1$
(e)

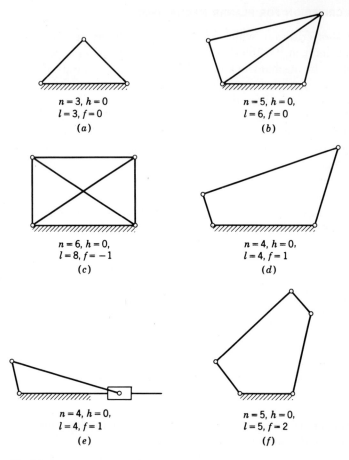

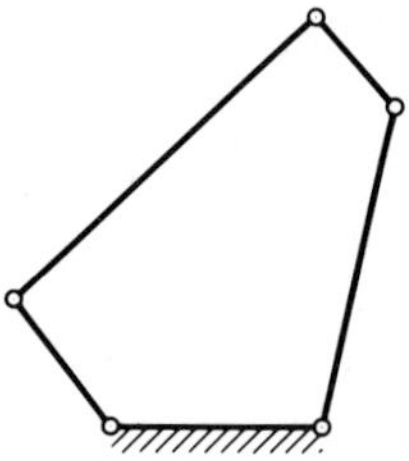

$n = 5,\ h = 0,$
$l = 5,\ f = 2$
(f)

Fig. 6-6

mechanism with a triangular coupler. In Fig. 6-7*a* there are two revolutes at *A* because link 2 is paired with link 3 and also with link 4. For the same reason there are two revolutes at *B*. In Fig. 6-7*b*, however, only a single revolute exists at *A* and at *B*.

Sometimes Grübler's criterion will give incorrect results. Notice that Fig. 6-7*c* is a structure. But if link 5 is arranged as in Fig. 6-7*d*, then the result is a double-parallelogram linkage with one degree of freedom even though Eq. (6-6) indicates it to be a structure.

Figure 6-8*a* illustrates a three-link mechanism with one higher pair having constrained motion. In Fig. 6-8*b* the higher pair has been replaced by link 4 and two lower pairs.

Of course we could have determined the number of degrees of free-

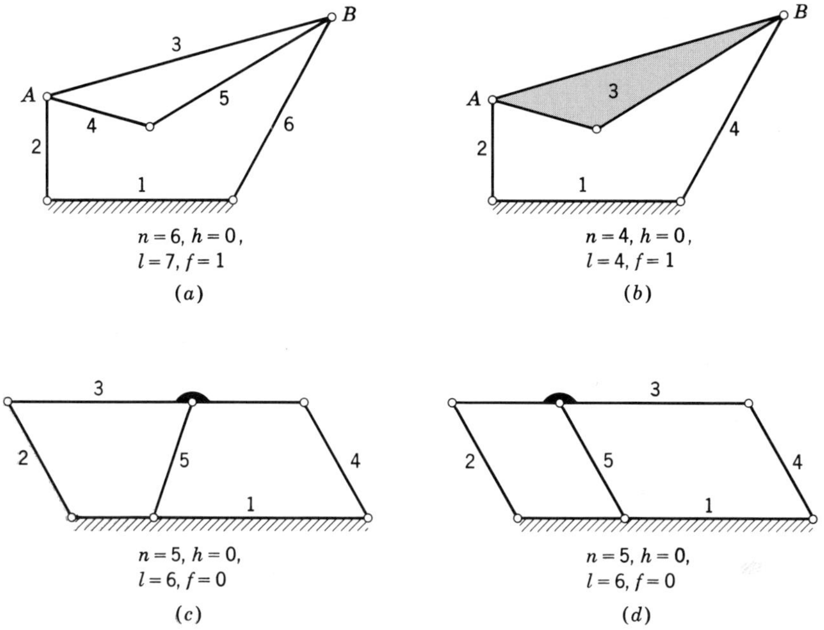
B
3
A
5
4
6
2
1
n = 6, h = 0,
l = 7, f = 1
(a)
B
A
3
2
4
1
n = 4, h = 0,
l = 4, f = 1
(b)
3
2
5
4
1
n = 5, h = 0,
l = 6, f = 0
(c)
3
2
5
4
1
n = 5, h = 0,
l = 6, f = 0
(d)

Fig. 6-7

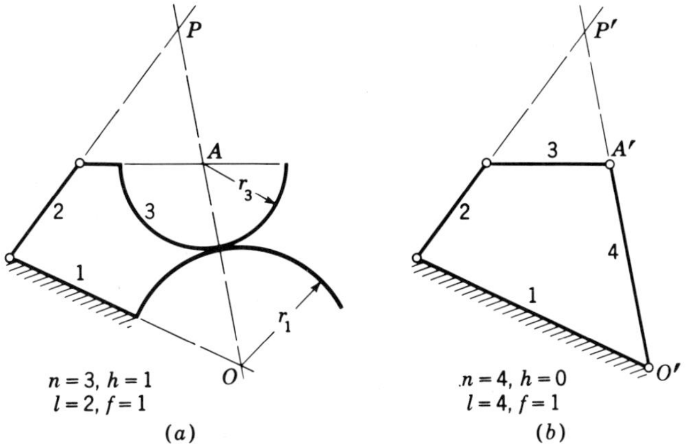
P
A
2
3
1
O
n = 3, h = 1
l = 2, f = 1
(a)
P′
3
A′
2
4
1
O′
n = 4, h = 0
l = 4, f = 1
(b)

Fig. 6-8

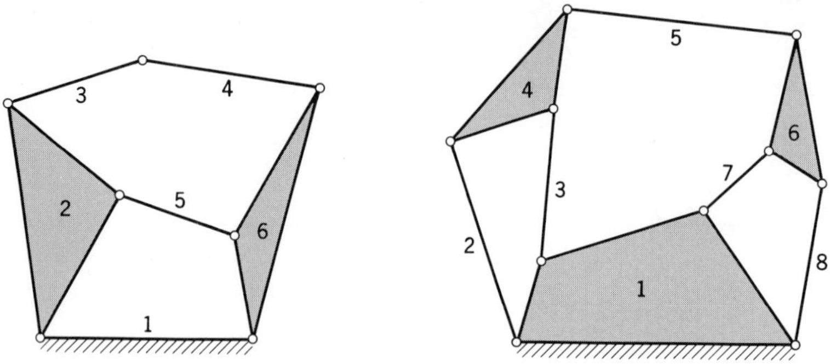

$n = 6$, $h = 0$, $l = 7$, $f = 1$

(a)

$n = 8$, $h = 0$, $l = 10$, $f = 1$

(b)

Fig. 6-9

dom of the mechanisms analyzed thus far by inspection. So Grübler's criterion is most useful for linkages such as those in Fig. 6-9 for which the number of degrees of freedom of each is not quite so evident.

Grübler's criterion is also useful in synthesizing complex linkages by the method of permutations. One simply finds all possible solutions to Eq. (6-6) for various n's, h's, and l's and tabulates them.[1]

6-4 INDEXES OF MERIT

In this section we shall study some of the various ratios, parameters, and angles which occur in mechanisms and which tell us when a mechanism is a good one and when it is a poor one. In the analysis and synthesis of mechanisms it is often desirable to plot these indexes of merit for a revolution of the driving crank to obtain a graph which is then used to evaluate the mechanism.

In Sec. 4-13 we learned that the ratio of the angular velocity of the output link to the input link is inversely proportional to the segments into which the common pole cuts the line of centers. Thus, in Fig. 6-10, if links 2 and 4 are the input and output links, respectively, then

$$\frac{\omega_4}{\omega_2} = \frac{p_2}{p_4} \tag{6-7}$$

is the equation for the velocity ratio. We also learned in Sec. 4-15 that

[1] For example, see Kurt Hain (translated by T. P. Goodman, Douglas P. Adams, B. L. Harding, and D. R. Raichel), "Applied Kinematics," 2d ed., chap. 3, McGraw-Hill Book Company, New York, 1967.

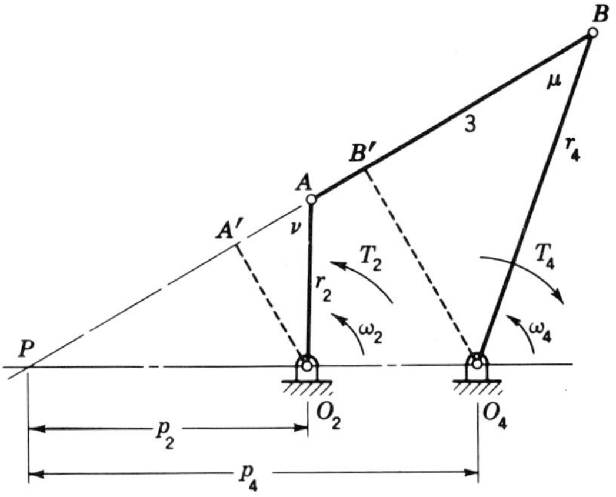

Fig. 6-10

an extreme value of the velocity ratio occurs when the collineation axis is perpendicular to the connecting link (link 3).

If we now assume that the linkage of Fig. 6-10 can have neither friction forces nor inertia forces during its operation, then we can derive a relation between the input torque T_2 applied to link 2, which is required to drive link 4, and the resisting load torque T_4. With the assumption of no friction, the power applied to link 2 is the same as the power applied by link 4 to the load; hence

$$T_2\omega_2 = T_4\omega_4$$

or

$$\frac{\omega_4}{\omega_2} = \frac{T_2}{T_4} \tag{6-8}$$

The *mechanical advantage* of a mechanism is the instantaneous ratio of the output force (torque) to the input force (torque). Therefore, from Eq. (6-8), we see that mechanical advantage is the reciprocal of the velocity ratio.

The mechanism of Fig. 6-10 has been redrawn in Fig. 6-11 so that

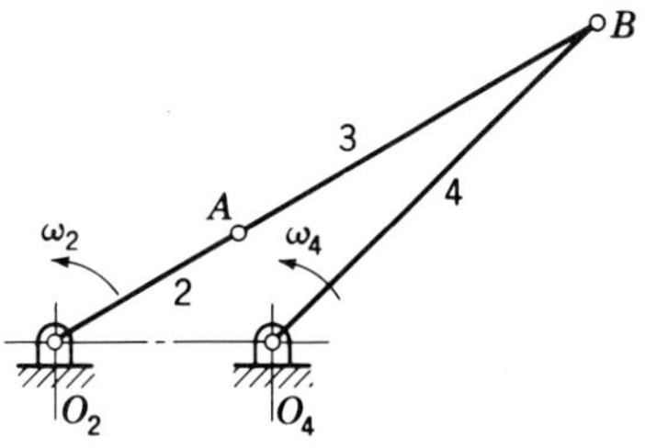

Fig. 6-11

links 2 and 3 are on the same straight line. At this position ω_4 is passing through zero; hence an extreme value of the mechanical advantage (infinity) is obtained. A mechanism in this phase is said to be *in toggle*. For a rotating-crank or class 1 mechanism there will be two such positions.

Proceeding further, construct O_2A' and O_4B' perpendicular to the line PAB in Fig. 6-10. Then, by similar triangles,

$$\frac{p_4}{O_4B'} = \frac{p_2}{O_2A'} \tag{a}$$

As shown in Fig. 6-10, let μ and ν be the smallest of the two angles made by the coupler, or its extension, with the output and input links, respectively. Then $O_4B' = r_4 \sin \mu$ and $O_2A' = r_2 \sin \nu$. Therefore, from Eq. (*a*),

$$\frac{p_4}{p_2} = \frac{r_4 \sin \mu}{r_2 \sin \nu} \tag{b}$$

Then, using Eqs. (6-7) and (6-8), we see that another expression for mechanical advantage is

$$\frac{T_4}{T_2} = \frac{r_4 \sin \mu}{r_2 \sin \nu} \tag{6-9}$$

Equation (6-9) shows that the mechanical advantage is infinite whenever the angle ν is 0 or 180°, that is, whenever the mechanism is in toggle.

The angle μ between the coupler and the output link is called the *transmission angle*. Equation (6-9) shows that the mechanical advantage is largest when the transmission angle is 90° for a fixed value of ν. If links 3 and 4 were to be coincident with each other, the transmission angle would become zero and the mechanism would lock or jam. Actually even a very small amount of friction will cause this to happen before $\mu = 0$; hence μ should always be greater than about 45 or 50°. The best mechanism, based on the quality of its force transmission, will therefore have a transmission angle which deviates from 90° by the smallest amount.

In higher-pair contact, such as the contact of meshing gear teeth, the concept of pressure angle is used. The *pressure angle* is the complement of the transmission angle.

6-5 COUPLER CURVES

The connecting rod or *coupler* of a planar four-bar linkage may be imagined as an infinite plane. During motion of the linkage, points fixed to this plane generate paths with respect to the fixed plane which are called *coupler curves.*

One of the best sources of coupler curves of the four-bar linkage is the Hrones-Nelson compilation.[1] This consists of a set of 11- by 17-in. charts containing 7000 coupler curves of the crank-and-rocker linkage. Figure 6-12 is a reproduction of one of the pages. In each case the length of the driving crank is unity; the lengths of the remaining links are varied to produce the different combinations. This atlas of coupler curves is invaluable to the engineer who needs a linkage to generate curves of specified characteristics.

The equations of the coupler curves are of sixth order; hence they are usually solved by machine computation. If many solutions are desired, digital computation can be quite expensive. The speed and the graphical output capabilities of the analog computer make it the best choice.[2]

One of the unusual properties of the planar four-bar linkage is that there are not one but three four-bar linkages which will generate the same coupler curve. It was discovered by Roberts[3] in 1875 and by Chebyshev in 1878 and hence is known as the Roberts-Chebyshev theorem. Though mentioned in an English publication in 1954,[4] it did not appear in the American literature, interestingly enough, until it was presented independently and almost simultaneously by Richard S. Hartenberg and Jacques Denavit of Northwestern University and by Rolland T. Hinkle of Michigan State University.[5,6]

In Fig. 6-13 let O_1ABO_2 be the original four-bar linkage with a coupler point P attached to AB. The remaining two linkages defined by the Roberts-Chebyshev theorem were termed *cognate linkages* by Hartenberg and Denavit. Each of the cognate linkages is shown in Fig. 6-13, one using short dashes for the links and the other using long dashes. The construction is evident by observing that there are four similar triangles, each containing the angles α, β, and γ, and three different parallelograms.

[1] John A. Hrones and George L. Nelson, "Analysis of the Four-bar Linkage," published jointly by The Technology Press of the Massachusetts Institute of Technology, Cambridge, Mass., and John Wiley & Sons, Inc., New York, 1951.

[2] See Joseph E. Shigley, "Simulation of Mechanical Systems," p. 304, McGraw-Hill Book Company, New York, 1967, for methods of solving this problem.

[3] By S. Roberts, a mathematician; he was not the same Roberts of the approximate-straight-line generator (Fig. 6-5*d*).

[3] P. Grodzinski and E. M'Ewan, Link Mechanisms in Modern Kinematics, *Proc. Inst. Mech. Engrs.*, vol. 168, no. 37, pp. 877–896, 1954.

[5] R. S. Hartenberg and Jacques Denavit, The Fecund Four-bar, *Trans. 5th Conf. Mech.*, Purdue University, Lafayette, Ind., p. 194, 1958.

[6] R. T. Hinkle, Alternate Four-bar Linkages, *Prod. Eng.*, vol. 29, p. 54, October, 1958.

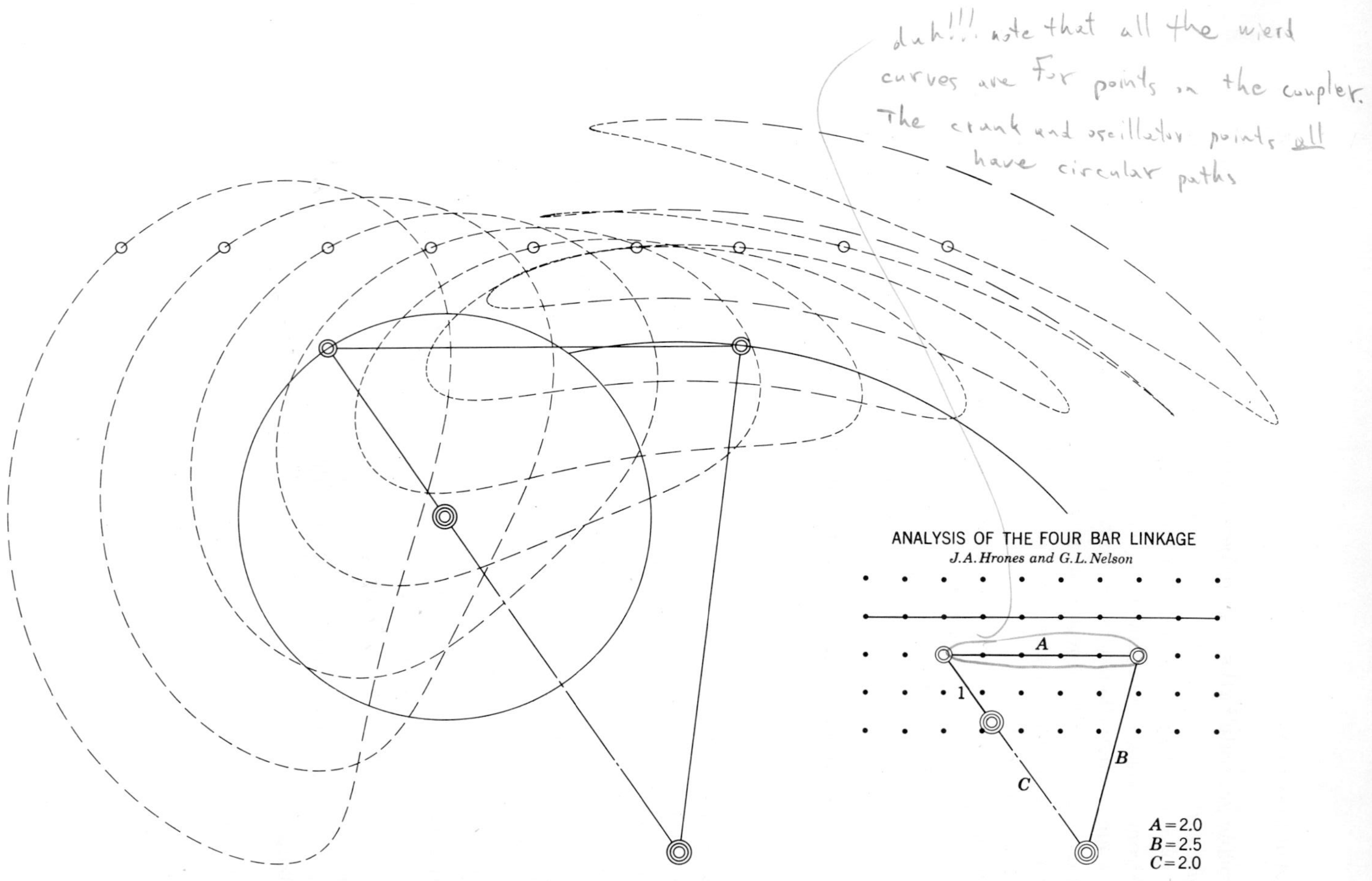

Fig. 6-12 A reproduction of one of the Hrones-Nelson pages. (*Reproduced by permission of the publishers,* The *Technology Press, M.I.T., Cambridge, Mass., and John Wiley & Sons, Inc., New York.*)

A good way to obtain the dimensions of the two cognate linkages is to imagine that the frame connections O_1, O_2, and O_3 can be unfastened. Then "pull" O_1, O_2, and O_3 away from each other until a straight line is formed by the crank, coupler, and follower of each linkage. We get Fig. 6-14 if we do this to Fig. 6-13. Note that the frame distances are incorrect, but all the movable links are of the correct length and all the angles are correct. Given any four-bar linkage and its coupler point, one can create a drawing like Fig. 6-14 and obtain the dimensions of the other two cognate linkages. This approach was discovered by A. Cayley and is called the *Cayley diagram*.[1]

The advantage of the Roberts-Chebyshev theorem is that one of the other two cognates may have better motion characteristics or a better transmission angle, or it may fit into a smaller space.

If the tracing point P is on the straight line AB or its extension, then a figure like Fig. 6-14 is of little help because all three linkages are compressed into a single straight line. An example is shown in Fig. 6-15, where O_1ABO_2 is the original linkage having a coupler point P on an extension of AB. To find the cognate linkages, locate O_3 on an extension of O_1O_2 in the same ratio as AB is to BP. Then construct, in order, the parallelograms O_1A_1PA, O_2B_2PB, and $O_3C_1PC_2$.

[1] A. Cayley, On Three-bar Motion, *Proc. London Math. Soc.*, vol. 7, pp. 136–166, 1876. *Note:* In Cayley's time a four-bar linkage was described as a three-bar mechanism because the idea of a kinematic chain had not yet been conceived.

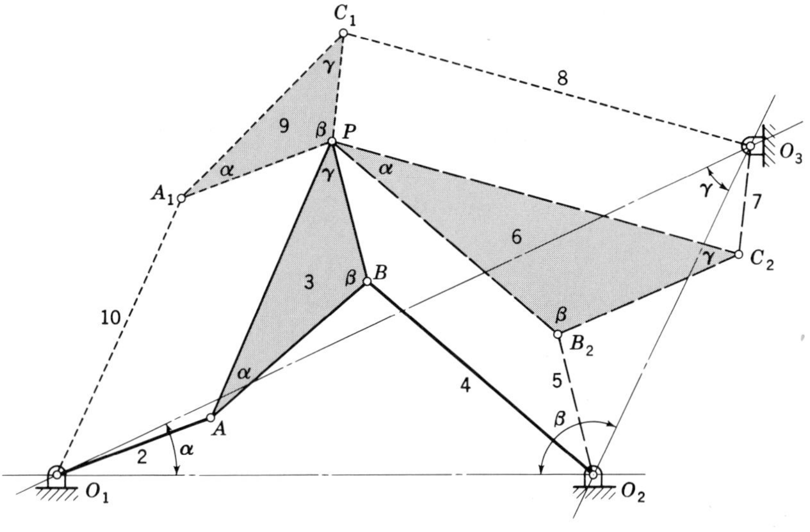

Fig. 6-13

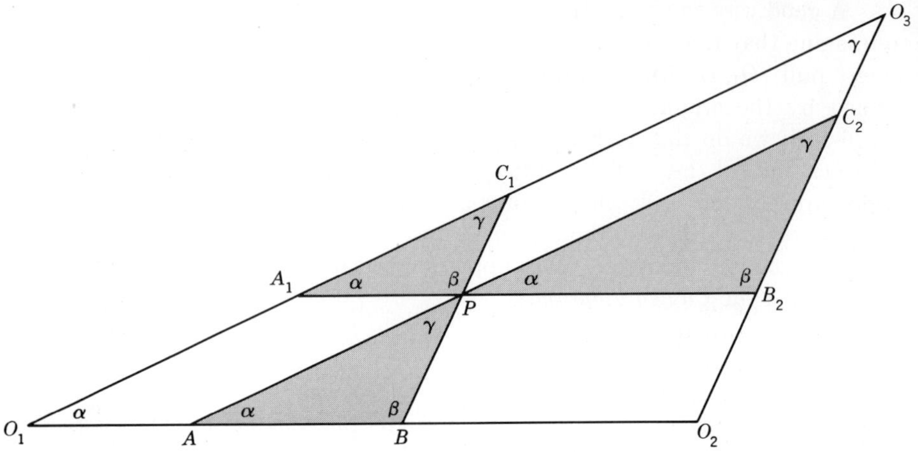

Fig. 6-14 The Cayley diagram.

Hartenberg and Denavit show that the angular-velocity relations between the links in Fig. 6-13 are

$$\begin{aligned} \omega_9 &= \omega_2 = \omega_7 \\ \omega_{10} &= \omega_3 = \omega_5 \qquad\qquad (6\text{-}10) \\ \omega_8 &= \omega_4 = \omega_6 \end{aligned}$$

They also observe that if crank 2 is driven at a constant angular velocity and if the velocity relationships are to be preserved during generation of the coupler curve, then the cognate mechanisms will have to be driven at variable angular velocities.

* Note that A_1 and C_1 are both end points for the coupler on O_1O_3 and O_2O_3

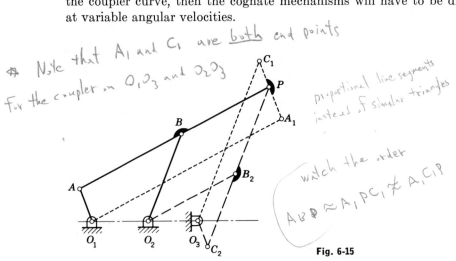

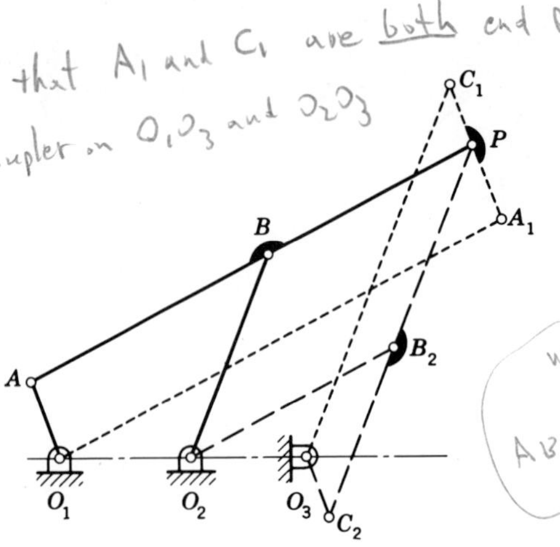

proportional line segments instead of similar triangles

watch the order $ABP \approx A_1PC_1 \not\approx A_1C_1P$

Fig. 6-15

6-6 POLODES

The paths generated by the motion of a velocity pole are called *polodes.*[1] In Fig. 6-16 pole 13 is located at the intersections of the extensions of links 2 and 4. As the linkage is moved through all possible positions, pole 13 traces out the path called the *fixed polode.*

Figure 6-17 shows an inversion of the linkage in which link 3 is *fixed* and link 1 is movable. When this inversion is moved through all possible positions, pole 13 traces out a different path called the *movable polode.*

Only portions of the polodes are shown in Figs. 6-16 and 6-17.

In Fig. 6-18 the movable polode has been attached to link 3, and links 2 and 4 are imagined to be unfastened. If, now, the movable polode is permitted to roll on the fixed polode with no sliding, then link 3 will have exactly the same motion as it had before. This remarkable property of the two polodes turns out to be quite useful in the synthesis of linkages.

We can restate this property as follows: *The plane motion of one rigid body relative to another can be replaced by the rolling motion of one polode on another.* The instantaneous point of contact is the velocity pole, as shown in Fig. 6-18. Also shown are the *pole tangent,* the common tangent to the two polodes, and the *pole normal.*

The polodes of Fig. 6-18 were generated by pole 13, the instantaneous center common to the coupler and the frame. Another set of

[1] Opinion seems to be about equally divided on whether these paths should be termed *polodes* or *centrodes.* Generally, writers employing the words *instantaneous center, instant center,* or *centro* call them *centrodes.* And those who prefer to use *pole* instead of *instantaneous center* call them *polodes.*

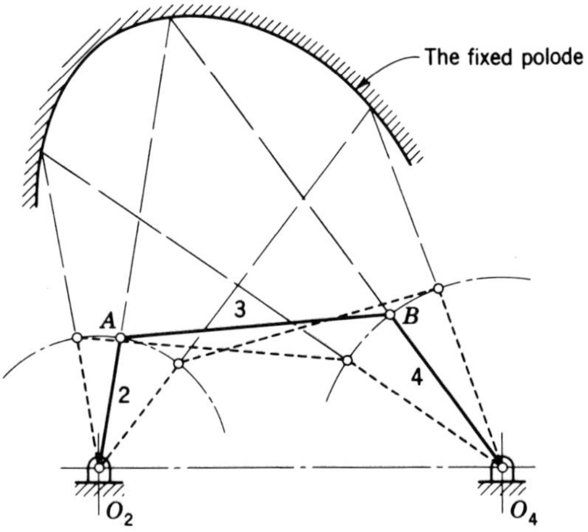

Fig. 6-16

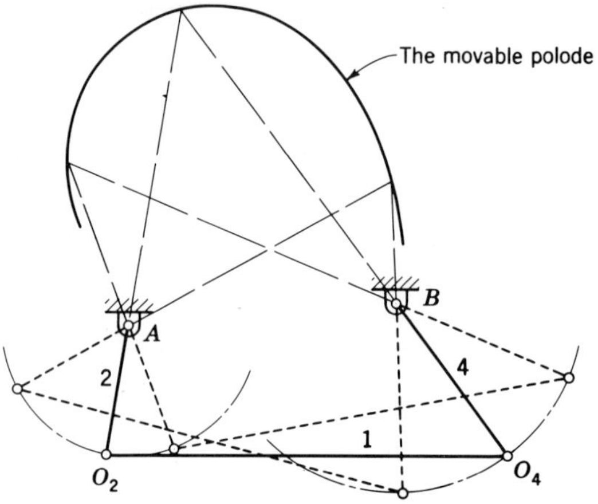

Fig. 6-17

polodes, both movable, are generated when pole 24 is used. Figure 6-19 shows that these are both ellipses for the special situation of a crossed double-crank linkage. To obtain the polode for link 4, for example, fix link 4 and rotate O_2 about O_4 in a full circle while plotting the path of pole 24. Having obtained the two polodes in this manner, one can roll one polode upon the other and get the identical motion that would result from operation of the linkage system itself.

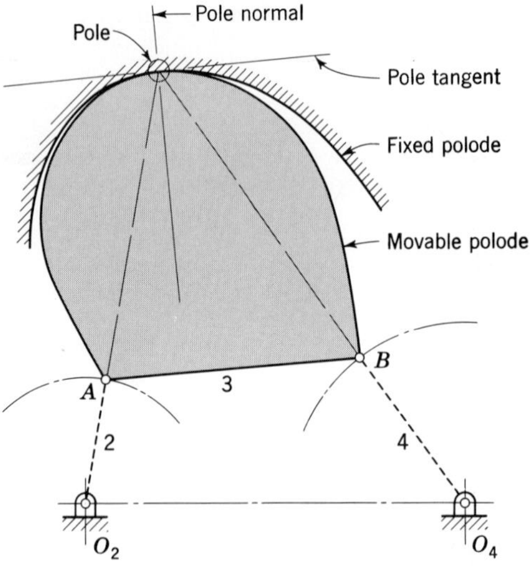

Fig. 6-18

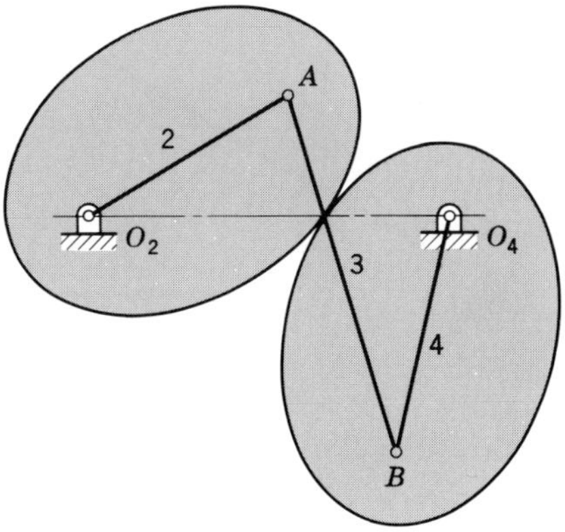

Fig. 6-19

6-7 THE EULER-SAVARY EQUATION[1]

When two rigid bodies move relative to one another with plane motion, a point in the first body will describe a path relative to the second. The Euler-Savary equation defines the relationship between two points, one in each body. These two points have the property that either one is the instantaneous center of curvature of the path traced out by the other.

In Fig. 6-20 define two circles. Let us call the circle with center at O_2 the fixed polode, though it need not be fixed. The circle with center at O_3 is called the movable polode. In actuality the polodes are only rarely circles, but we are interested only in instantaneous values and so, for convenience, we show them as circles. The movable polode is to move around the fixed polode by pure rolling. The instantaneous point of contact is the pole O_{32}. If we give the movable polode an angular velocity ω, then the instantaneous velocity of O_3 is

$$v = r_3\omega \qquad (a)$$

Defining the pole velocity as u and noting the similar triangles, we write

$$\frac{r_2 + r_3}{v} = \frac{r_2}{u} \qquad (b)$$

[1] The most important and most useful references on this subject are as follows:

N. Rosenauer and A. H. Willis, "Kinematics of Mechanisms," chap. 4, Associated General Publications, Sydney, Australia, 1953; republished by Dover Publications, Inc., New York, 1967.

A. E. R. de Jonge, A Brief Account of Modern Kinematics, *Trans. ASME,* vol. 65, pp. 663–683, 1943.

Hartenberg and Denavit, *op. cit.,* chap. 7.

Allen S. Hall, Jr., "Kinematics and Linkage Design," chap. 5, Prentice-Hall, Inc., Englewood Cliffs, N.J., 1961. This book is a real classic on the theory of mechanisms and contains many useful examples.

Hain, *op. cit.,* chap. 4.

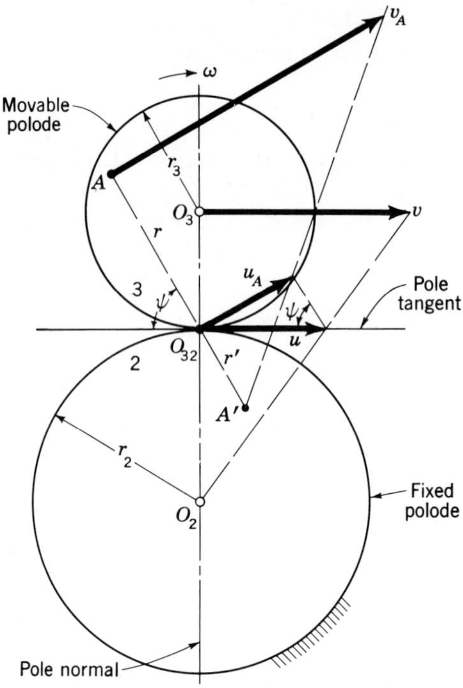

Fig. 6-20

Then, using Eq. (*a*), we find

$$\frac{r_2 + r_3}{r_2 r_3} = \frac{\omega}{u}$$

Therefore

$$\frac{1}{r_2} + \frac{1}{r_3} = \frac{\omega}{u} = \text{const} \tag{c}$$

In using Eq. (*c*) a sign convention for r_2 and r_3 is required. Let us agree on the following:

1. If the *outside* of r_2 rolls on the *outside* of r_3, both r_2 and r_3 are positive.
2. If the *outside* of r_2 rolls on the *inside* of r_3, r_2 is positive and r_3 is negative.
3. If the *inside* of r_2 rolls on the *outside* of r_3, r_2 is negative and r_3 is positive.

Now, in Fig. 6-20, choose any point A on the movable polode at a distance r from the pole. The velocity of A is

$$v_A = r\omega \tag{d}$$

as shown in the figure. Next, find the component of u in the direction of v_A graphically; or find it analytically, using the equation

$$u_A = u \sin \psi \tag{e}$$

where ψ is the angle between the pole tangent and $O_{32}A$. The intersection of AO_{32} extended and a line drawn through the heads of u_A and v_A locates A', the *conjugate* of A. This is called the *Hartmann construction*. Denoting r' as the distance $O_{32}A'$ and noting the similar triangles, we have

$$\frac{r + r'}{v_A} = \frac{r'}{u_A}$$

or

$$\frac{r + r'}{r'} = \frac{r\omega}{u \sin \psi}$$

Therefore

$$\left(\frac{1}{r} + \frac{1}{r'}\right) \sin \psi = \frac{\omega}{u} = \text{const} \tag{6-11}$$

This is the Euler-Savary equation. The rules for the signs of r and r' are the same as those for r_2 and r_3.

In some books Eq. (6-11) is written in the form

$$\left(\pm \frac{1}{r} \pm \frac{1}{r'}\right) \sin \psi = \text{const} \tag{6-12}$$

where, now, r and r' are always treated as positive and the rules given above are applied to the reciprocals.

If we wish to deal only with points on the pole normal, then $\psi = 90°$, $\sin \psi = 1$, and Eq. (6-11) becomes

$$\frac{1}{r} + \frac{1}{r'} = \text{const} \tag{f}$$

The limiting value of either r or r' is infinity. Specifying r_i as the conjugate of r when $r = \infty$ gives

$$\frac{1}{r_i} + \frac{1}{\infty} = \text{const}$$

or

$$\frac{1}{r_i} = \text{const}$$

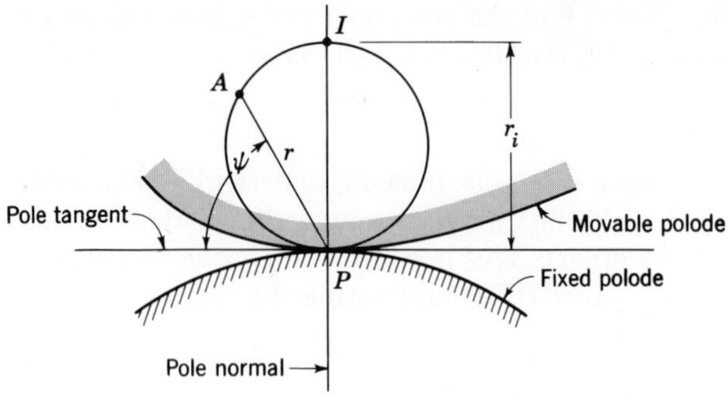

Fig. 6-21

Thus Eq. (6-11) can be written

$$\left(\frac{1}{r} + \frac{1}{r'}\right) \sin \psi = \frac{1}{r_i} \tag{6-13}$$

If we make $r' = \infty$ in Eq. (6-13), then it can be solved for r. The result is

$$r = r_i \sin \psi \tag{6-14}$$

which is the equation of a circle of diameter r_i, as shown in Fig. 6-21. Any point A on this circle has $r' = \infty$ and hence has an infinite radius

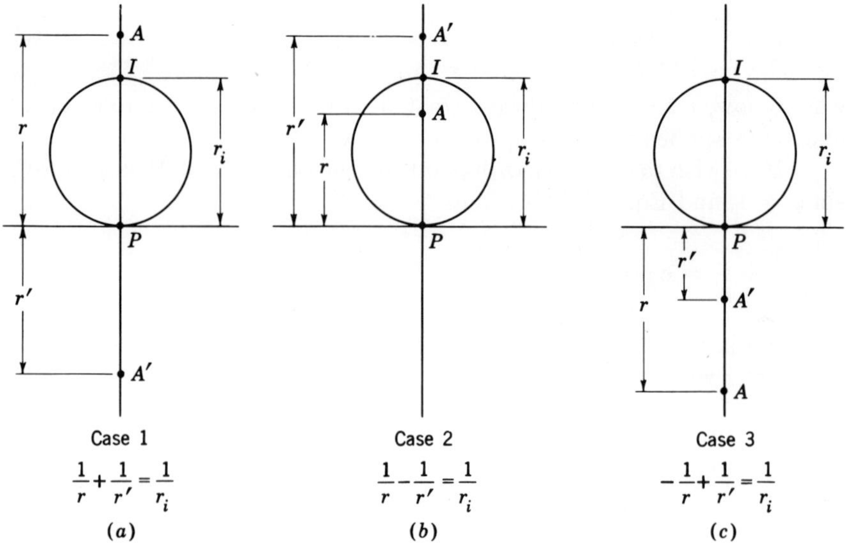

Fig. 6-22

of curvature. For this reason, the circle is called the *inflection circle.* Point I, on the pole normal, is called the *inflection point*, or the *inflection pole*. According to Hall[1] the inflection pole has the same velocity u as the velocity pole, and zero acceleration if $\dot{\omega} = 0$. Equation (5-3) shows that the normal acceleration of every point on the inflection circle is zero because the radius of curvature is infinite.

Figure 6-22 illustrates the sign convention for points on the common normal and relates their location to the inflection circle.

Example 6-1 Find the inflection circle for the motion of the coupler of the slider-crank linkage in Fig. 6-23 and determine the instantaneous radius of curvature of the path of coupler of point C.

Solution We begin in Fig. 6-24 by locating the pole P at the intersection of link 2 extended and a line through B perpendicular to the direction of sliding. Points B and P must both lie on the inflection circle by definition; hence we need only one additional point to construct this circle.

The center of curvature of A is, of course, at O_2, which we now call A'. We measure $PA = r_A = 2.64$ in. and $PA' = r'_A = 4.64$ in. This is identified as case 2, and so the Euler-Savary equation is written

$$\frac{1}{r_A} - \frac{1}{r'_A} = \frac{1}{r_i \sin \psi}$$

Substituting r_A and r'_A and solving yields

$$r_i \sin \psi = 6.15 \text{ in.}$$

Now lay off 6.15 in. from P to locate I_A, the third point on the inflection circle. The circle can now be constructed through these three points, the pole normal and the pole tangent drawn in; we then find $PI = r_i = 6.28$ in.

Next draw the ray PCI_C and measure $r_C = 3.1$ in., $r_i \sin \varphi = PI_C = 4.85$ in., and substitute these in the equation

$$\frac{1}{r_C} - \frac{1}{r'_C} = \frac{1}{r_i \sin \psi}$$

We then find $r'_C = 8.6$ in. and lay this off in Fig. 6-24, as shown. The instantaneous radius of curvature of the path generated by C is

$$\rho_C = r'_C - r_C = 5.5 \text{ in.}$$

[1] *Op. cit.*, p. 71.

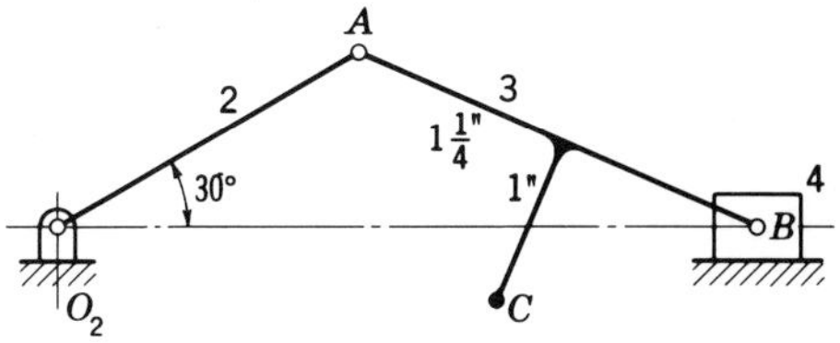

Fig. 6-23 $O_2A = 2$ in., $AB = 2.5$ in.

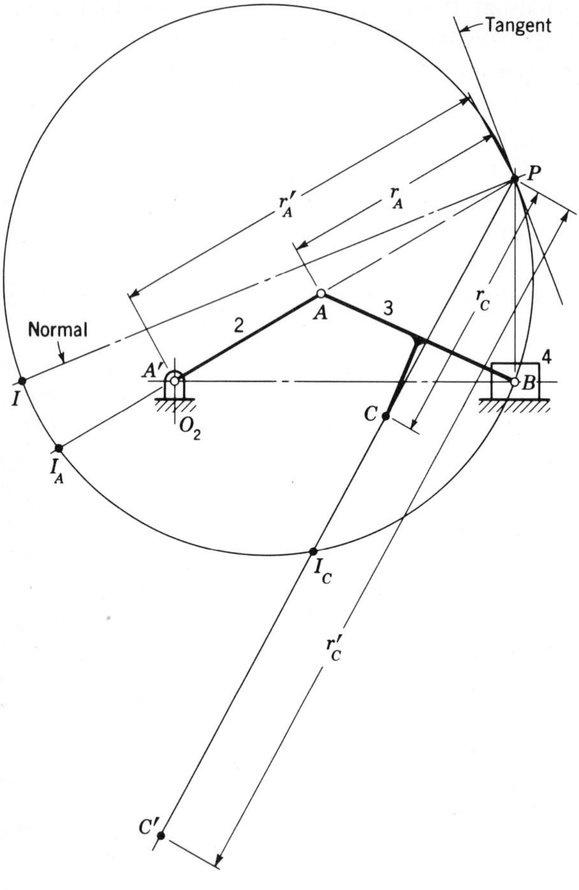

Fig. 6-24

6-8 THE BOBILLIER CONSTRUCTIONS

The Euler-Savary equation can be solved graphically to obtain the inflection circle if two points and their respective centers of curvature are given. Alternatively, if the inflection circle is given, the Euler-Savary equation can be solved graphically to obtain the center of curvature of any specified point in the moving plane. The graphical solutions to be presented in this section are called the *Bobillier constructions*.

Having given an inflection circle together with the pole normal and pole tangent, let us select any two points A and B (Fig. 6-25) not on a single straight line through P. Now, we use the Euler-Savary equation and find the conjugates A' and B' of these points. Draw a line $A'B'$ and let this intersect a second line AB at Q. An axis drawn from Q through

P is called the *collineation axis.* This axis applies only to the two rays APA' and PBB', and so it is said to *belong* to these two rays; point Q, however, will move to a different place on the axis if another set of points is chosen on the same rays. Bobillier's theorem states that *the angle between the pole tangent and the first ray is the same as the angle between the second ray and the collineation axis.* The angles may be measured in either direction as long as both of them are measured in the *same* direction.

In applying the Euler-Savary equation to a mechanism two pairs of conjugate points are usually known, and we desire, first, to find the inflection circle. For example, a four-bar linkage with a crank O_2A and a follower O_4B has A and O_2 as one set of points and B and O_4 as the other when we are interested in the motion of the coupler relative to the frame. Thus, given two pairs of conjugate points, how do we use the Bobillier theorem to find the inflection circle?

In Fig. 6-26*a* let A and A' and B and B' be the given pairs of conjugate points. Rays constructed through each pair intersect at P, the velocity pole, giving one point on the inflection circle. Point Q is located next by the intersection of a ray through A and B with a ray through A' and B'. Then the collineation axis is drawn as the line PQ.

The next step is shown in Fig. 6-26*b*. Draw a line through P parallel to $A'B'$. Point W is the intersection of this line with AB. Now, through W, draw a second line parallel to the collineation axis. This line intersects APA' in I_A and PBB' in I_B, the two points on the inflection circle for which we were searching.

We could construct the circle through the three points I_A, I_B, and

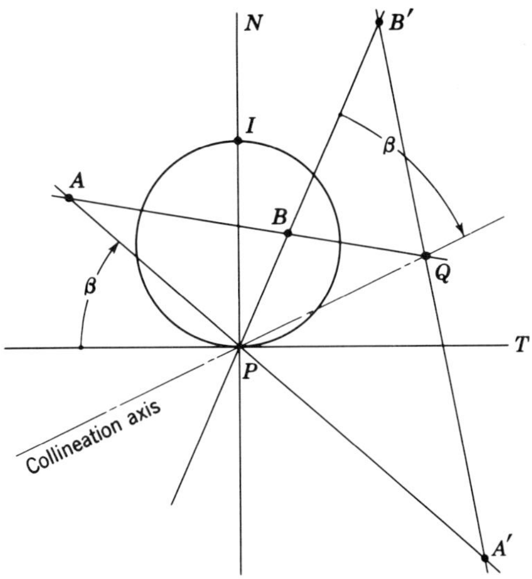

Fig. 6-25

P, but there is an easier way. Remember that triangles inscribed in semicircles are always right triangles having the diameter as the hypotenuse. Therefore, erect a perpendicular to AP at I_A and another to PB extended at I_B. The intersection of these perpendiculars gives point I, the inflection pole, as shown in Fig. 6-26c. Since PI is a diameter, the inflection circle, the pole normal N, and the pole tangent T can all be constructed.

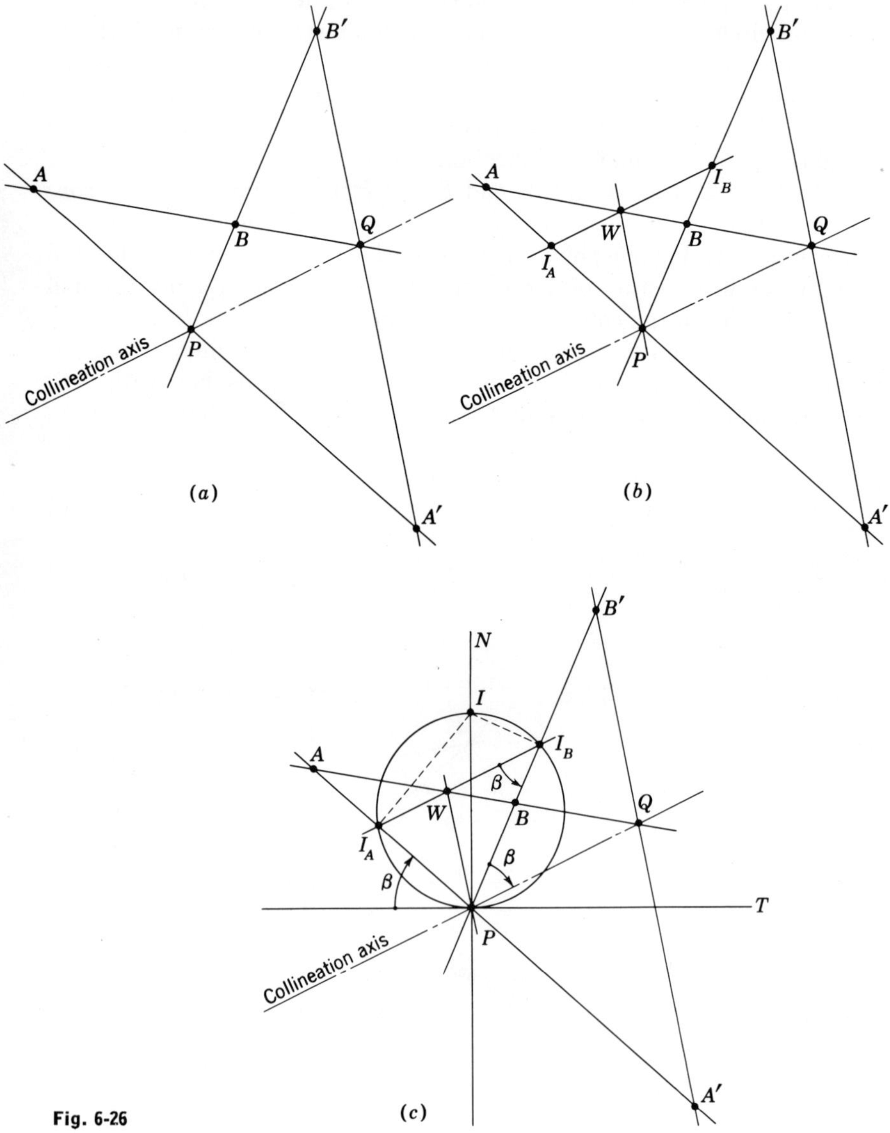

Fig. 6-26

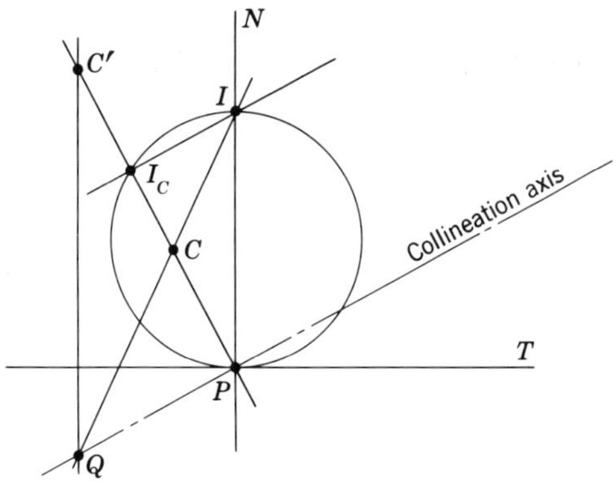

Fig. 6-27

To show that this construction satisfies the Bobillier theorem, note that the circle arc from I_A to P is inscribed by the angle that I_AP makes with the pole tangent. But this same arc is also inscribed by the angle I_AI_BP. Therefore these two angles are equal. But the line I_AWI_B was constructed parallel to the collineation axis. Therefore the line PI_B also makes the same angle β with the collineation axis.

Our final problem is to learn how to use the Bobillier theorem to find the conjugate of, say, a point C when the inflection circle is given. In Fig. 6-27 join C with pole P and locate point I_C on the inflection circle. This ray will serve as one of the two necessary to locate the collineation axis. For the other we may as well use the pole normal because I and its conjugate I', at infinity, are both known. For these two rays the collineation axis is a line through P parallel to the line I_CI, as we learned in Fig. 6-26*c*. The balance of the construction is similar to that of Fig. 6-25. Q is located by the intersection of a line through I and C with the collineation axis. Then a line through Q and I', at infinity, intersects the ray PC, extended, at C', the conjugate of C.

Example 6-2 Find the center of curvature of point C on the coupler of the four-bar linkage shown in Fig. 6-28, using the Bobillier theorem.

Solution Locate P at the intersection of BB' and AA'. Locate Q_1 at the intersection of $A'B'$ and AB. PQ_1 is the first collineation axis. Through P draw a line parallel to $A'B'$ to locate W on AB. Through W draw a line parallel to PQ_1 to locate I_A on AA' and I_B on BB'. Through I_A drop a perpendicular to AA' and through I_B a perpendicular to BB'. These perpendiculars intersect at I and define the inflection circle, the pole normal N, and the pole tangent T.

To obtain the conjugate of C draw the ray PC, and locate I_C on the inflection circle. The collineation axis PQ_2 for the pair of rays PC and PI is a line through P parallel to a line, not shown, from I to I_C. Q_2 is obtained as the extension of a line from I to C. Through Q_2 draw a line parallel to the pole normal. Its intersection with the ray PC yields C', the center of curvature of C.

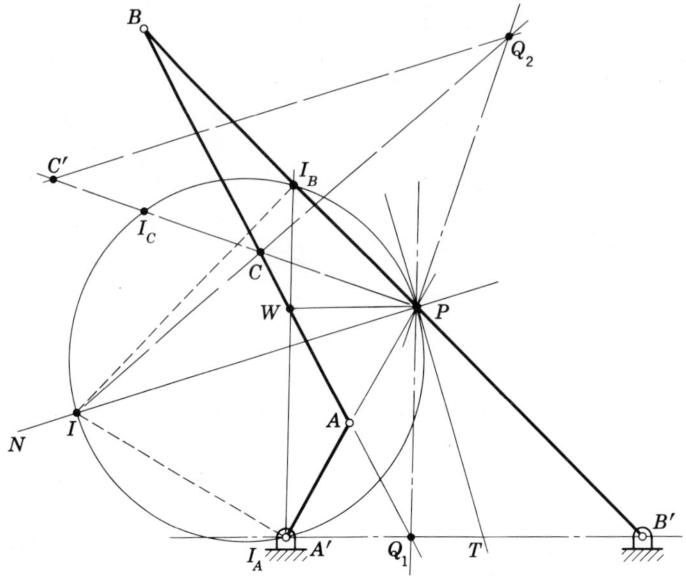

Fig. 6-28

6-9 THE CUBIC OF STATIONARY CURVATURE

Consider a point on the coupler of a four-bar linkage which generates a path relative to the frame, at the instant considered, whose radius of curvature is ρ. Then, by *stationary curvature,* we mean

$$\frac{d\rho}{ds} = 0 \qquad (a)$$

where s is the distance along the path. The *cubic of stationary curvature,* also called the *circling-point curve,* is the locus of all points on the coupler or moving plane having stationary curvature at the instant. It should be noted that stationary curvature is not, necessarily, constant curvature but, rather, that the continually varying radius of curvature of the path of the point on the moving plane has reached a maximum or a minimum value.

Here we shall present a fast and simple graphical method of obtaining the cubic of stationary curvature, as described by Hain.[1] In Fig. 6-29 we have the four-bar linkage $A'ABB'$, with A' and B' the frame connections. Then A and B have stationary curvature because they

[1] *Op. cit.,* pp. 498–502.

describe circles about A' and B', respectively, and hence lie on the cubic.

The first step is to obtain the pole normal and pole tangent. Since the inflection circle is not required, find the collineation axis PQ and draw the pole tangent T at the angle ψ from the line PB' opposite in direction to the angle ψ made by the collineation axis with the line PA'. This construction follows from Bobillier's theorem.

Next, construct the pole normal N. At this point it is convenient to loosen the drawing from the board and reorient it so that the pole normal is coincident with the T-square or horizontal edge of the drafting machine.

Now construct a line through A perpendicular to PA and another line through B perpendicular to PB. These lines intersect the pole normal and pole tangent, respectively, in A_N, A_T and B_N, B_T, as shown in Fig. 6-29. Now draw the two rectangles $PA_NA_GA_T$ and $PB_NB_GB_T$. The points A_G and B_G define the auxiliary line G which is used to obtain the cubic.

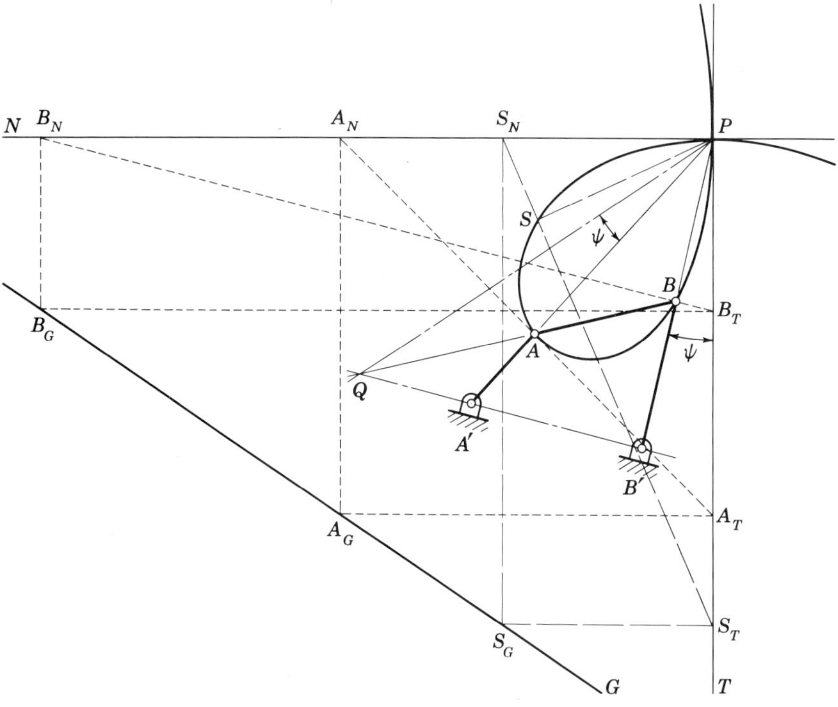

Fig. 6-29

Choose any point S_G on G. A ray parallel to N locates S_T and another, parallel to T, locates S_N. Connect S_T with S_N and drop a perpendicular to the resulting line through P. This locates point S, a point on the cubic of stationary curvature. Repeat this process as often as desired by choosing other points on G, and draw a smooth curve through all the points.

We note that there are two tangents at P, the *pole-normal tangent* and the *pole-tangent tangent*. The radius of curvature of the cubic at these tangents is obtained as follows: Extend G to intersect T in G_T and N in G_N (not shown). Then half the distance PG_T is the radius of curvature of the cubic at the pole-normal tangent, and half the distance PG_N is the radius of curvature at the pole-tangent tangent.

A point with very interesting properties occurs at the intersection of the cubic with the inflection circle. It is called *Ball's point.* A point on the coupler coincident with Ball's point will describe a path which is approximately a straight line because it has stationary curvature and is also located at an inflection point in its path.

The equation of the cubic of stationary curvature[1] is

$$\frac{1}{M \sin \psi} + \frac{1}{N \cos \psi} - \frac{1}{r} = 0 \qquad (6\text{-}15)$$

where r is the distance from the pole to a point on the cubic measured at the angle ψ from the pole tangent. The constants M and N are obtained by using any two points, such as A and B of Fig. 6-29, known to lie on the cubic. It turns out[2] that M and N are, respectively, the diameters PG_T and PG_N of the circles on the pole tangent and pole normal whose radii represent the curvatures of the cubic at the pole.

PROBLEMS

6-1. The link lengths of a planar four-bar linkage are 1, 2, 3, and 4 in. Assemble these in all possible combinations and classify each, using Grashof's law. Find the four inversions of each linkage and describe them by name, for example, a crank-and-rocker mechanism or a drag-link mechanism.

6-2. The same as Prob. 6-1 except that the link lengths are 1, 3, 5, and 5 in.

6-3. The crank and coupler of a slider-crank mechanism are 1 in. and 4 in., respectively. Make a drawing of the mechanism and then draw the three inversions. Choose one of the movable links as a driving member and describe the resulting motion of the output member. Can you think of an application of each inversion?

6-4. Find the number of freedoms of each mechanism shown in the figure.

[1] For a derivation of this equation see Hall, *op. cit.*, p. 98, or Hartenberg and Denavit, *op. cit.*, p. 206.

[2] See D. C. Tao, "Applied Linkage Synthesis," p. 111, Addison-Wesley Publishing Company, Inc., Reading, Mass., 1964.

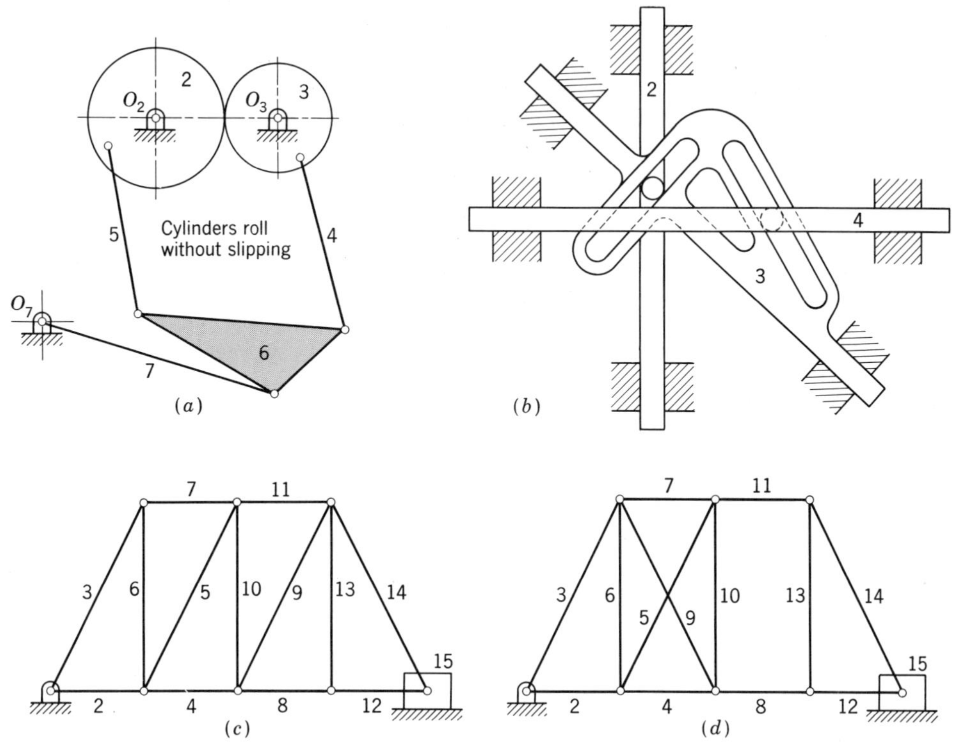

Prob. 6-4

6-5. A crank-and-rocker linkage has a 4-in. frame, a 1-in. crank, a $3\frac{1}{2}$-in. coupler, and a 3-in. rocker. Plot curves of velocity ratio, mechanical advantage, and transmission angle versus crank angle for one complete revolution of the crank, using 15° intervals of crank angle. Now find the crank angles corresponding to the extreme values of the velocity ratio. Find both toggle positions and record the corresponding crank angles and transmission angles. Does the transmission angle pass through 90°? Locate these positions exactly on the drawing of the linkage and record the corresponding values of the crank angle.

6-6. In the figure, point C is a point on the coupler; plot its complete path.

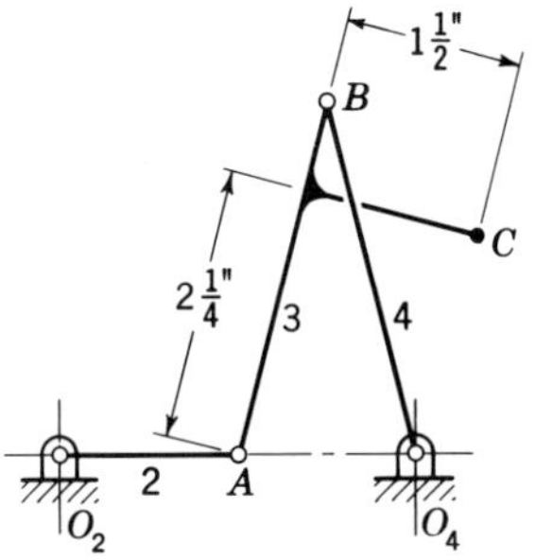

Prob. 6-6

6-7. Find the two cognate linkages for the four-bar linkage of Prob. 6-6.

6-8. Use Grübler's criterion and find a mechanism containing a floating or moving quaternary link. How many inversions of this mechanism can you find?

6-9. Plot the complete coupler curve for point P of Roberts' mechanism of Fig. 6-5d. Use $a = b = c = 2\frac{1}{2}$ in. and $d = 1\frac{1}{4}$ in. Find the two remaining cognate linkages.

6-10. Find the cognate linkages for the mechanism of Prob. 6-15.

6-11. Find the cognate linkages for Watt's straight-line mechanism (Fig. 6-5a). Use $a = 4$ in., $b = 1\frac{1}{2}$ in., and $c = d = 2$ in.

6-12. Find the fixed and movable polodes for the frame and coupler of a drag-link mechanism. By constructing simple paper models, show that the coupler can be disconnected from the cranks and fastened to the movable polode and that, when rolled on the fixed polode, it will have the same motion as if it were connected to the linkage system.

6-13. Devise a practical working model of the drag-link mechanism.

6-14. Find the inflection circle for motion of the coupler of the double-slider mechanism shown in the figure. Select several points on the pole normal and find their conjugates. Plot portions of the paths of these points to demonstrate for yourself that the conjugates are indeed the centers of curvature.

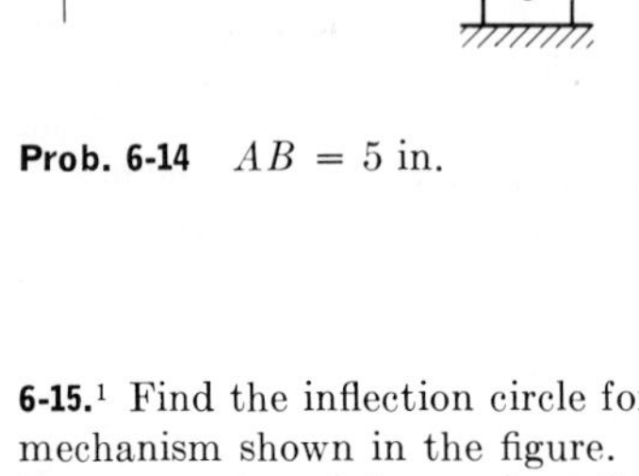

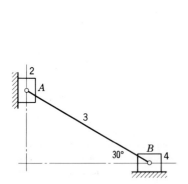

Prob. 6-14 $AB = 5$ in.

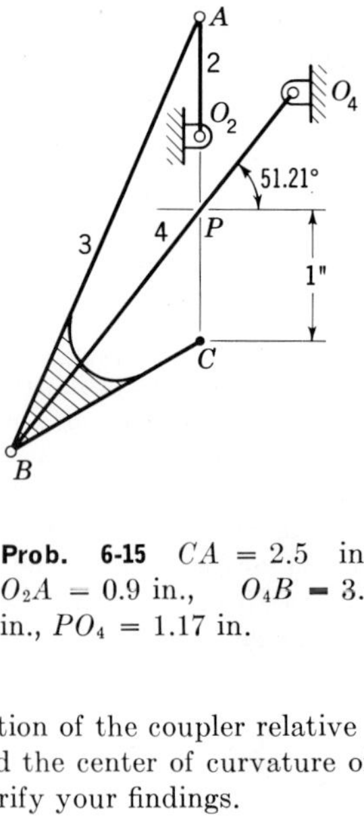

Prob. 6-15 $CA = 2.5$ in., $O_2A = 0.9$ in., $O_4B = 3.5$ in., $PO_4 = 1.17$ in.

6-15.[1] Find the inflection circle for motion of the coupler relative to the frame of the mechanism shown in the figure. Find the center of curvature of point C, and generate a portion of the path of C to verify your findings.

6-16. For motion of the coupler relative to the frame, find the inflection circle, the pole normal, the pole tangent, and the center of curvature of C for the mechanism illustrated.

[1] This is one of the solutions to a kinematic synthesis problem by Delbert Tesar and James C. Wolford appearing in the paper, Five Point Exact Four-bar Straight-line Mechanisms, *Trans. 7th Conf. Mechanisms,* Penton Publishing Co., Cleveland, Ohio, 1962.

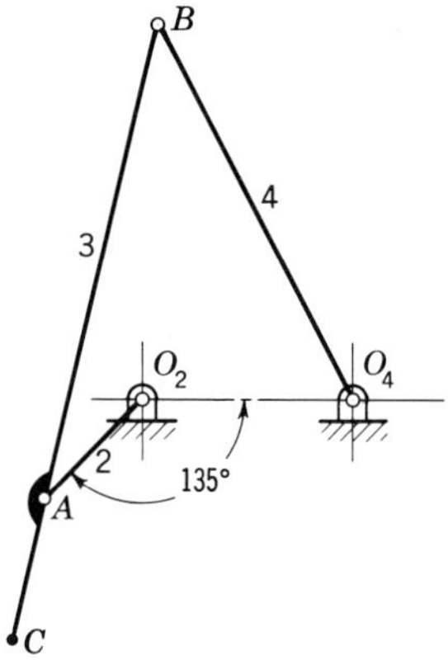

Prob. 6-16 $O_2A = 1$ in., $AB = 3\frac{1}{2}$ in., $O_4B = 3$ in., $O_2O_4 = 1\frac{1}{2}$ in., $BC = 4\frac{1}{2}$ in.

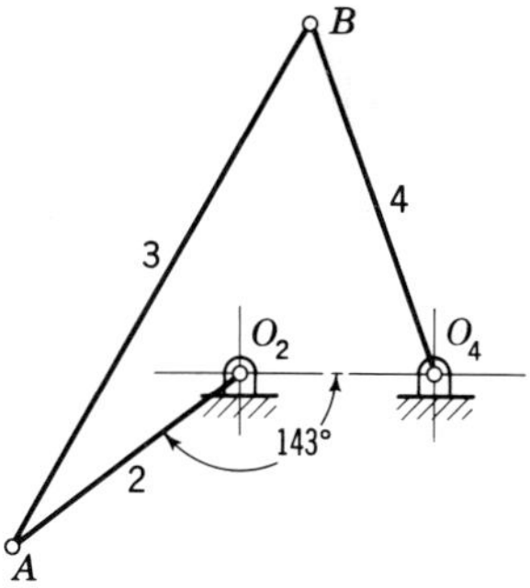

Prob. 6-17 $O_2A = 1\frac{3}{4}$ in., $AB = 3\frac{3}{4}$ in., $O_4B = 2\frac{3}{8}$ in., $O_2O_4 = 1\frac{1}{4}$ in.

6-17. Find the inflection circle for the motion of the coupler, using the linkage shown in the figure. Choose points on the coupler coincident with the velocity and inflection poles and plot their paths.

6-18. Draw the linkage shown in the figure full size on 18- by 24-in. paper, and place A' 6 in. from the lower edge and 7 in. from the right edge. Better utilization of the paper will be obtained by tilting the frame through about 15°, as shown.

(*a*) Find the inflection circle.

(*b*) Draw the cubic of stationary curvature.

(*c*) Choose a coupler point C coincident with the cubic, and plot a portion of the coupler path in the vicinity of the cubic. Find its conjugate C'. Draw a circle through C with center at C' and compare this circle with the actual path of C.

(*d*) Find Ball's point. Locate a point D of the coupler there, plot a portion of its path, and compare the result with a straight line.

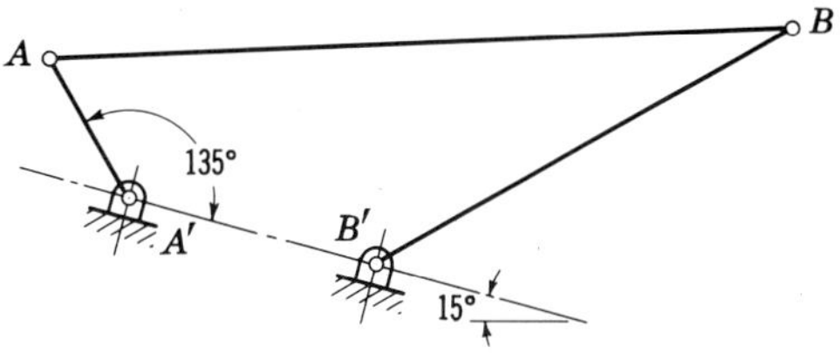

Prob. 6-18 $AA' = 1$ in., $BB' = 3\frac{1}{4}$ in., $AB = 5$ in., $A'B' = 1\frac{3}{4}$ in.

7 Cams

A *cam* is a mechanical element of a machine which is used to drive another element, called the *follower*, through a specified movement by direct contact. Cam mechanisms are simple, are easy to design, and occupy a very small space. Furthermore, follower motions having almost any desired characteristics are not difficult to obtain. For these reasons cam mechanisms are used extensively in modern machinery.

7-1 CLASSIFICATION OF CAMS AND FOLLOWERS

Cam followers are classified according to whether their motion is *translation* or *oscillation* and whether the displacement path is *radial* or *offset* from the cam center. The follower surface which contacts the cam is also used in the classification. Thus, we may speak of a *flat-face*, *spherical-face*, or *roller-face* follower, since these are the usual shapes of the contacting follower surfaces.

Cams are classified according to their shape. The *plate cam*, also called a *disk cam* or a *radial cam*, is used most frequently. Figure 7-1

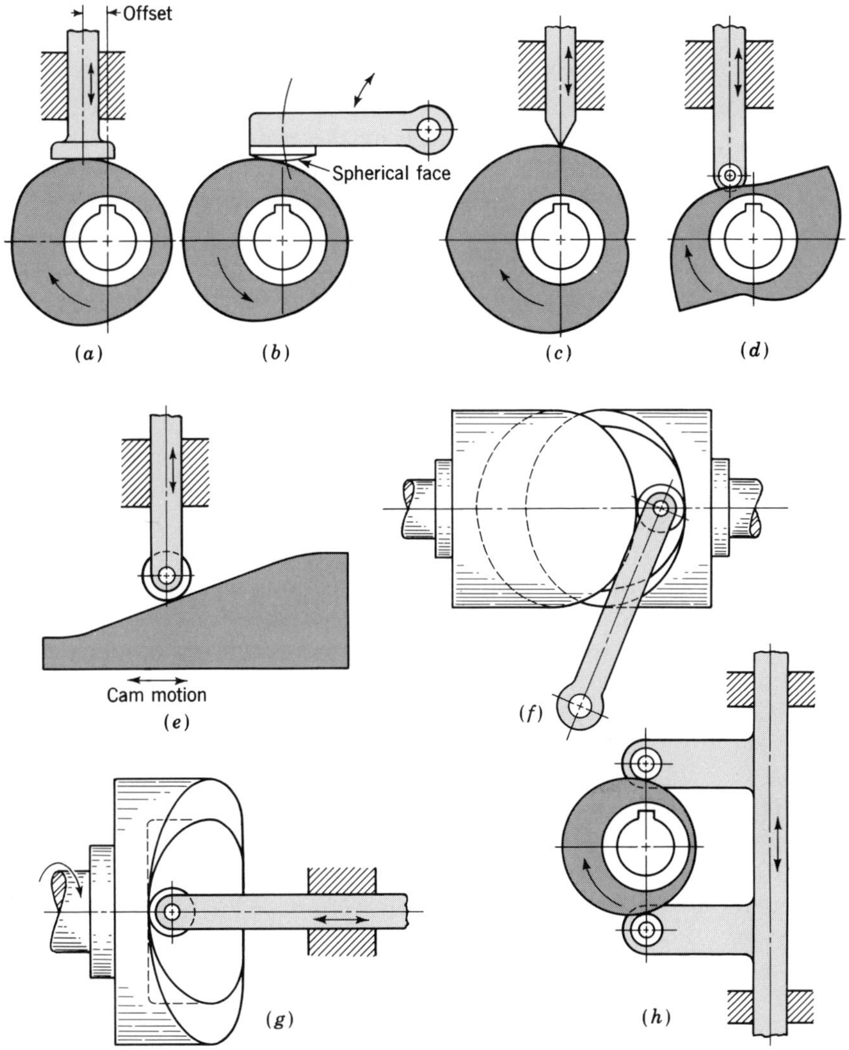

Fig. 7-1 Types of cams. (*a*) Radial cam and flat-face offset translating follower; (*b*) radial cam and spherical-face oscillating follower; (*c*) radial (heart) cam and translating knife-edge follower; (*d*) radial two-lobe frog cam and translating offset roller follower; (*e*) wedge cam and translating roller follower; (*f*) cylindrical cam and oscillating roller follower; (*g*) end or face cam and translating roller follower; (*h*) yoke cam and translating roller follower.

illustrates four different plate cams with different followers together with a *wedge cam*, a *cylinder cam*, an *end* or *face cam*, and a *yoke cam*.

In all cases the follower must be constrained to follow the cam. This may be done by a spring, by gravity, or by a mechanical constraint. In Fig. 7-1*f* the follower is constrained by the groove. In Fig. 7-1*h* the follower has two rollers at a fixed distance apart which acts as the constraint; the mating cam, in such an arrangement, is often called a *constant-diameter* cam. The mechanical constraint can also be introduced by employing *dual* or *conjugate* cams in an arrangement similar to Fig. 7-1*h*. Each cam then has its own roller, but the rollers are mounted on the same reciprocating or oscillating follower.

7-2 GEOMETRY OF THE RADIAL CAM

Let us now examine the problem of determining the exact shape of a cam surface required to deliver a specified follower motion. We choose a simple plate cam, to rotate at a constant angular velocity, and a radial translating follower with roller contact. Our problem is to find the shape of the cam, in numerical or graphical form, such that when it is rotated the follower will move as specified.

Since we wish to find the cam profile, we employ the principle of inversion. The rule is: *To develop the cam surface, hold the cam stationary and rotate the follower opposite to the cam rotation.* An application of this rule is shown in Fig. 7-2, but first let us note the nomenclature of the figure.

The *base circle* is the smallest circle that can be drawn tangent to the cam surface.

The *trace point* is a theoretical point on the follower; it corresponds to the point of a fictitious knife-edge follower and is used to generate the *pitch curve*. For a knife-edge follower the pitch curve and the cam surface are identical.

The *pressure angle*, which is the complement of the transmission angle for linkages, is the angle between the direction of follower motion and a normal to the pitch curve. If the pressure angle is too large, a translating follower will jam in its bearings. The pressure angle varies during a complete cycle of motion and reaches both minimum and maximum values, as we shall learn soon. Its existence, however, implies that the forces between the cam and the follower are not in the same direction as the follower motion.

The *pitch point* designates the location of the maximum pressure angle. The *pitch circle* is a circle from the cam center through the pitch point.

The *prime circle* is the smallest circle from the cam center through the pitch curve.

Turning now to the construction of the cam profile we begin, usually, by plotting a *displacement diagram* (Fig. 7-2*b*). The abscissa represents one revolution of the cam and has a length equal to the circumference of the prime circle. The ordinate is the follower travel. This diagram identifies the *rise*, which is the motion of the follower away from the cam center; the *dwells*, or those periods during which the follower is at rest; and the *return*, which is the motion of the follower toward the cam center. Sometimes the rise is called the *lift*.

The *transition* or *inflection points* correspond to the pitch points

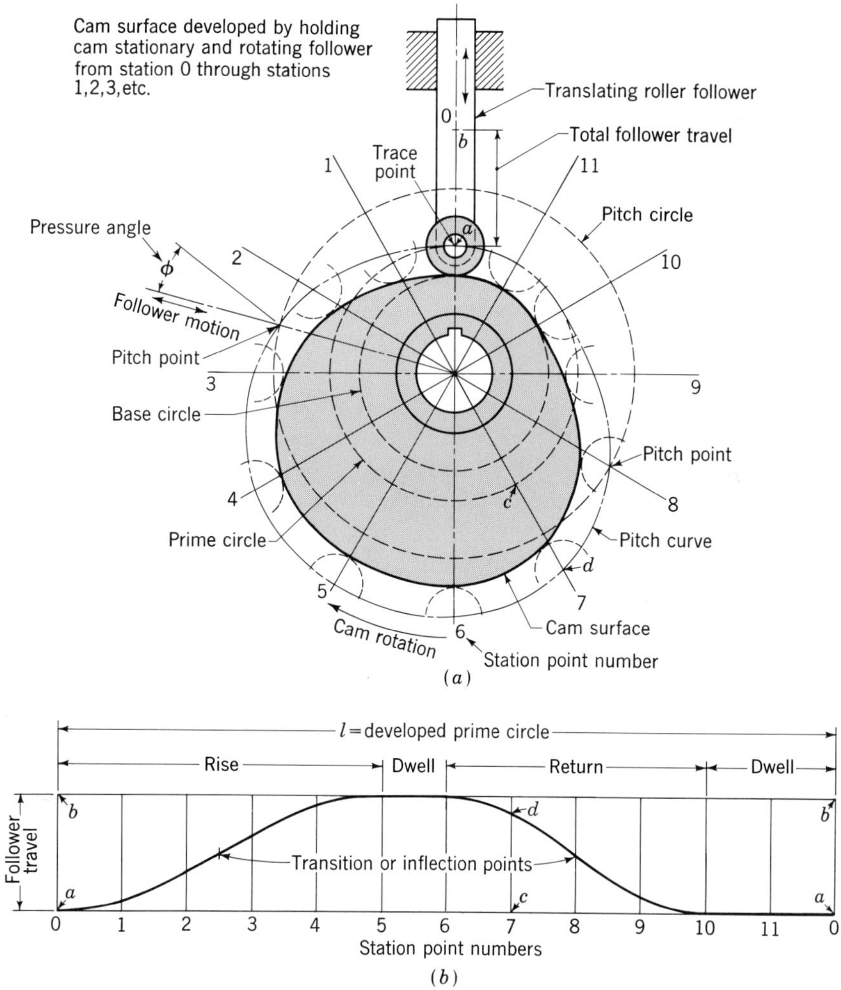

Fig. 7-2 Layout of a plate cam from the displacement diagram. (*a*) Cam nomenclature; (*b*) displacement diagram.

on the cam layout. These points identify the maximum steepness of the pitch curve, which is the same as the maximum pressure angle. Though shown at the midpoint of follower travel in Fig. 7-2*b*, they need not occur at midtravel.

In constructing the cam profile, we divide both the displacement diagram and the cam pitch circle into an equal number of segments. We can then assign station numbers to the boundaries of these segments and transfer, with dividers, distances from the displacement diagram to the cam layout. A smooth curve constructed through these points is the pitch curve and is the path generated by the trace point. If the follower is a roller follower, as it is in this example, we simply draw the roller in its proper position at each station and then construct the cam surface as a smooth curve tangent to all these roller positions.

7-3 DISPLACEMENT DIAGRAMS

During rotation of the cam through one turn the follower executes a series of events consisting of rises, dwells, and returns. There are many possible follower motions which can be used for the rises and the returns. In this section we shall display graphical methods of constructing the displacement diagram for rises with *modified uniform*, *simple harmonic*, *parabolic*, and *cycloidal motions*.

The displacement diagram for *uniform motion* is a straight line with a constant slope. Thus the velocity of the follower during the motion is a constant. This motion is seldom used because of the bumps at the beginning and end of rise. By smoothing these bumps, as shown in Fig. 7-3, we obtain modified uniform motion. In this example the modification was made by using a circle arc tangent to the dwell which preceded the rise, but any suitable modification can be used.

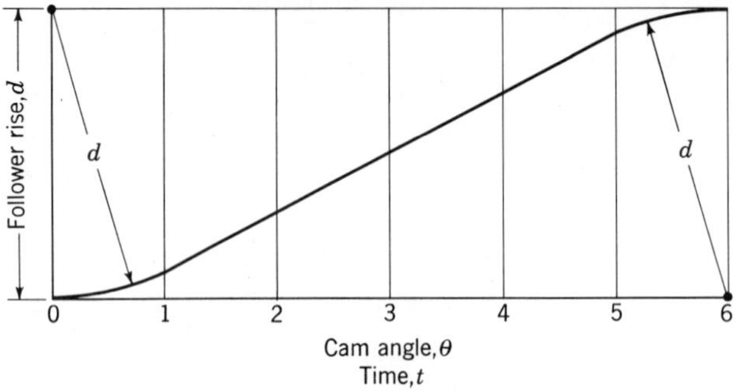

Fig. 7-3 Uniform motion modified by a circular arc.

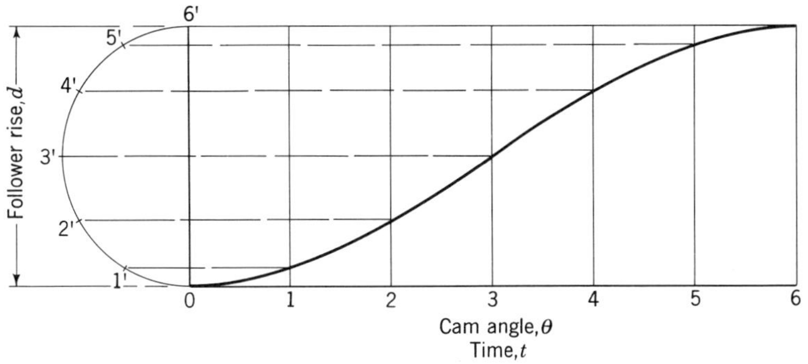

Fig. 7-4 Simple harmonic motion.

The displacement diagram for simple harmonic motion is shown in Fig. 7-4. A semicircle having a diameter equal to the rise d is divided equally into the same number of parts as are used for the abscissa or time axis.

Sometimes the semicircle of Fig. 7-4 is drawn as an ellipse in order to modify the motion. The ellipse is then divided equally in the same manner. If the ellipse is constructed with its major axis parallel to the abscissa, the beginning and ending follower velocities are slower than for harmonic motion.

Parabolic motion, as we shall soon learn, is a constant acceleration motion. The graphical construction is shown in Fig. 7-5. Use an even number of time divisions and use at least six divisions.[1] Through the origin of the displacement diagram construct any line at any convenient angle to the ordinate (Fig. 7-5). If six divisions are used, divide this line into parts proportional to 1, 3, 5, 5, 3, 1; or 1, 3, 5, 7, 7, 5, 3, 1, if eight divisions are used. Connect the end of the last division with the end of the ordinate. Draw the remaining lines through each point parallel to this line.

Cycloidal motion is obtained by rolling a circle of radius $d/2\pi$, where d is the rise (return) in inches, on the ordinate. We find the construction shown in Fig. 7-6 to be more convenient than rolling, for graphical purposes. Use point B as a center and draw a circle of radius $d/2\pi$. Divide this circle into the same number of parts as are to be used for the time axis. Project these points horizontally until they intersect

[1] In the graphical layout of a cam for actual construction purposes, a great many divisions must be employed to obtain good accuracy. At the same time the drawing is made to a very large scale. But the use of so many in an explanation would make the figures very difficult to read. For this reason only the minimum number of divisions required to define the curve have been employed here.

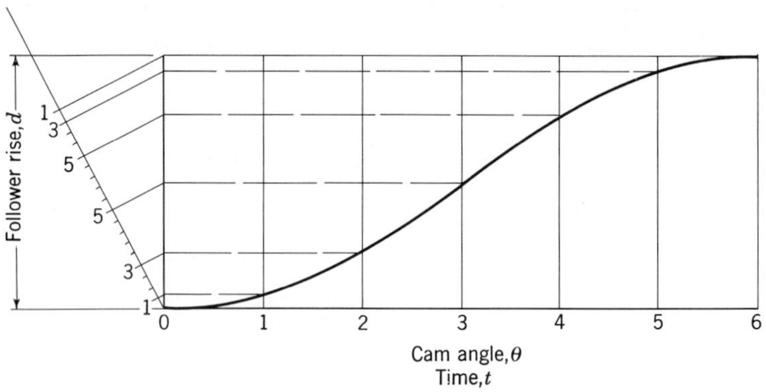

Fig. 7-5 Parabolic motion.

the ordinate. Now, from the ordinate, project each point parallel to the diagonal OB to obtain the points on the displacement diagram corresponding to each follower position.

7-4 GRAPHICAL LAYOUT OF CAM PROFILES

Figure 7-7 shows the construction of the profile of a plate cam which drives a reciprocating offset roller follower. As shown by the displacement diagram, there is a rise in 180° and a return in 150° of cam rotation, both with cycloidal motion. The diagram also shows a 30° dwell following the return motion. Note that the abscissa is 360° long, that the rise is in six 30° divisions, and the return in six 25° divisions.

Begin by constructing the *offset circle*, using a radius equal to the amount of the offset. The radial lines which extend to the circumfer-

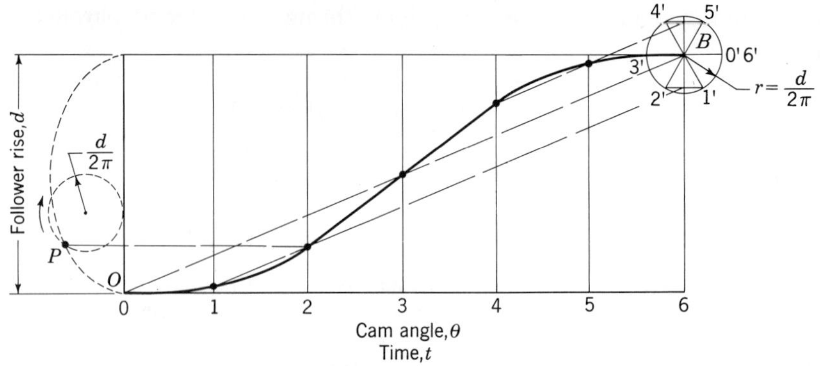

Fig. 7-6 Cycloidal motion.

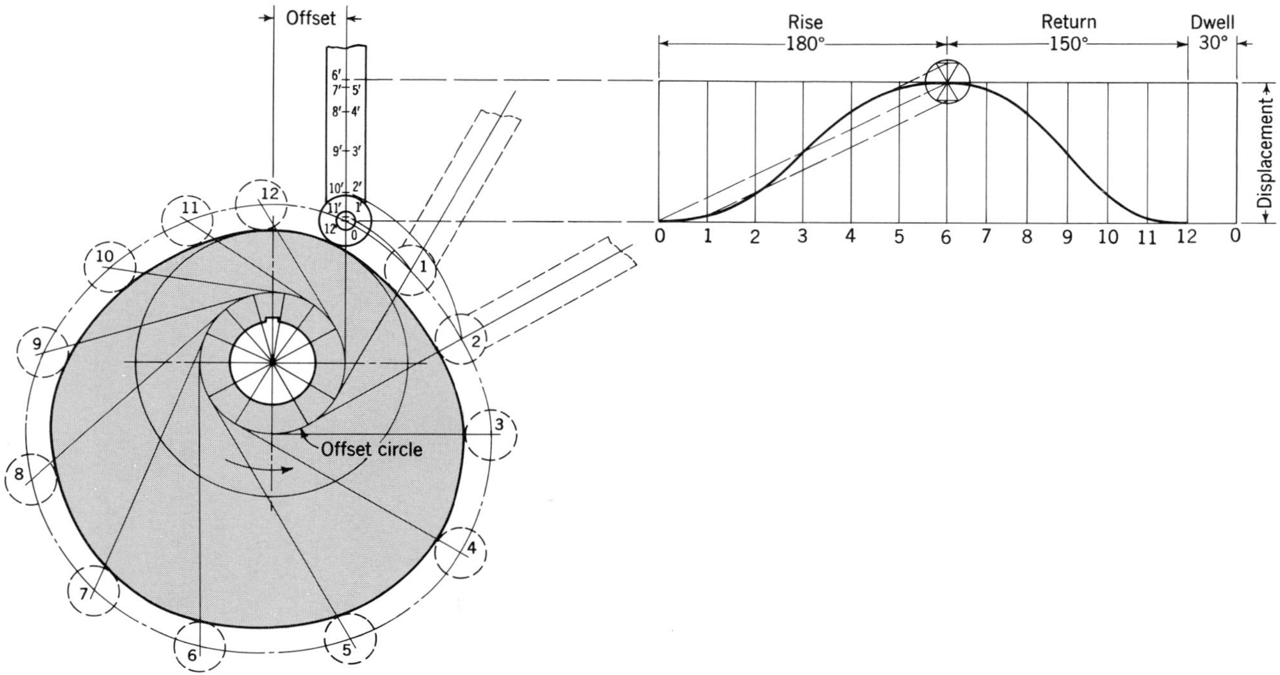

Fig. 7-7 Layout of a cam curve for an offset translating roller follower with cycloidal motion.

ence of this circle are perpendicular to the follower centerlines, and the follower centerlines are all tangent to the offset circle. So begin with the radial line corresponding with the bottom or zero-rise position of the follower and mark off six 30° segments and six 25° segments, leaving a 30° segment to complete the circumference.

The displacement diagram represents the pitch curve or path of the center of the roller. Project or transfer these positions to the cam layout, obtaining points 0′, 1′, 2′, etc. Now, remember to rotate the follower *opposite* to cam rotation. In this problem the cam is to rotate counterclockwise, as shown by the arrow on the figure; so an arc from 1′ to the center of the roller at 1 is clockwise. The roller centers or stations 2, 3, 4, etc., are established in a similar manner. A smooth curve tangent to all the roller circles is the required cam profile.

Figure 7-8 shows the construction for a plate cam with a flat-face translating follower with modified uniform motion for the rise and return strokes. The pitch curve is constructed by using a method quite similar to the one previously described. A replica of the flat face of the follower is then constructed in each position. This produces a series of straight lines, and the cam profile is a smooth curve tangent to all of them.

It may be helpful to extend each straight line representing a position of the follower face so as to form a series of triangles. If these triangles are lightly shaded, as suggested in the illustration, it will be easier to draw the cam profile. The length of the flat face can also be obtained from this layout as indicated.

Figure 7-9 shows the layout of the cam profile of a plate cam with an oscillating roller follower. In this example the follower motion, as indicated by the construction of the displacement diagram, is parabolic. In this case we must rotate the pivot point or center of rotation of the follower opposite to cam rotation to develop the cam profile. At each station around the cam the follower must be correctly positioned. First draw a circle from the camshaft center through the pivot point of the follower. Divide this circle to correspond with the divisions on the displacement diagram. In this case the ordinate of the displacement diagram is the rectified arc of the path of motion of the roller center. Thus corresponding points may be transferred to the arc traveled by the roller center by using dividers.

Next, construct all the roller-center arcs around the cam, using as centers the points previously located on the circle through the center of oscillation. Then the pitch curve is obtained by rotating each point into its corresponding position about the camshaft center. Finally, circles representing the roller diameter are drawn with centers on the pitch curve, and the cam profile is the curve which is tangent to all these roller circles.

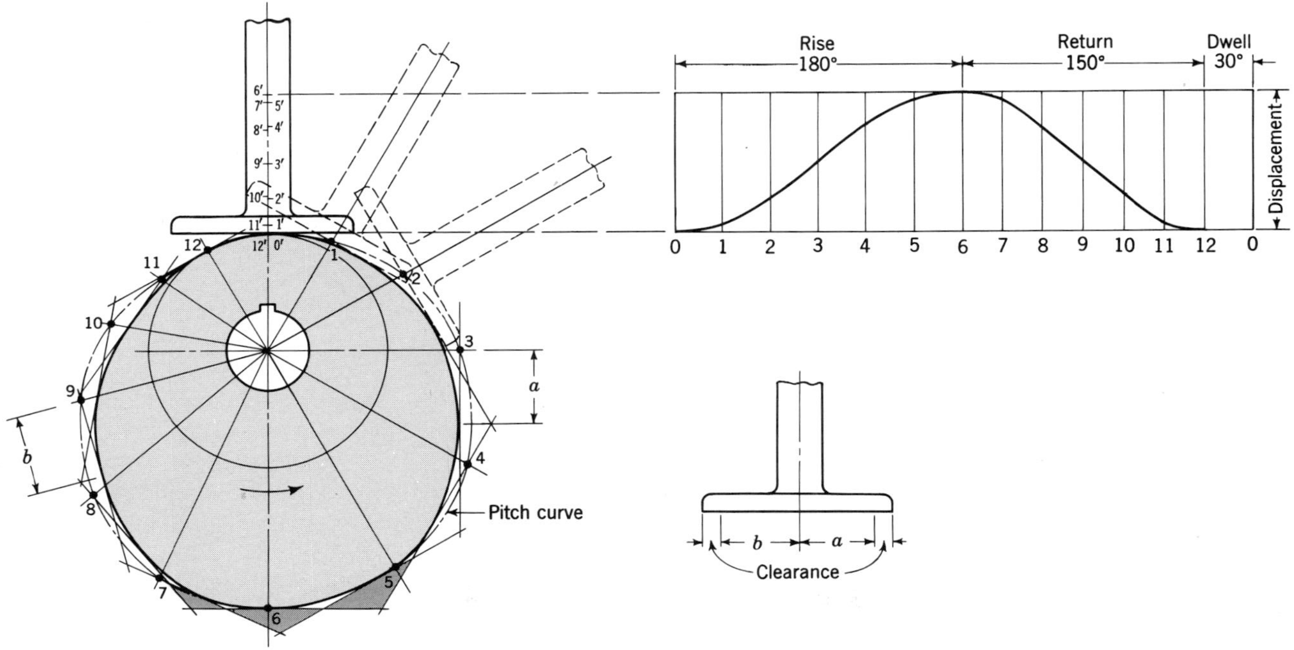

Fig. 7-8 Layout of a cam curve for a flat-face translating follower with modified uniform motion.

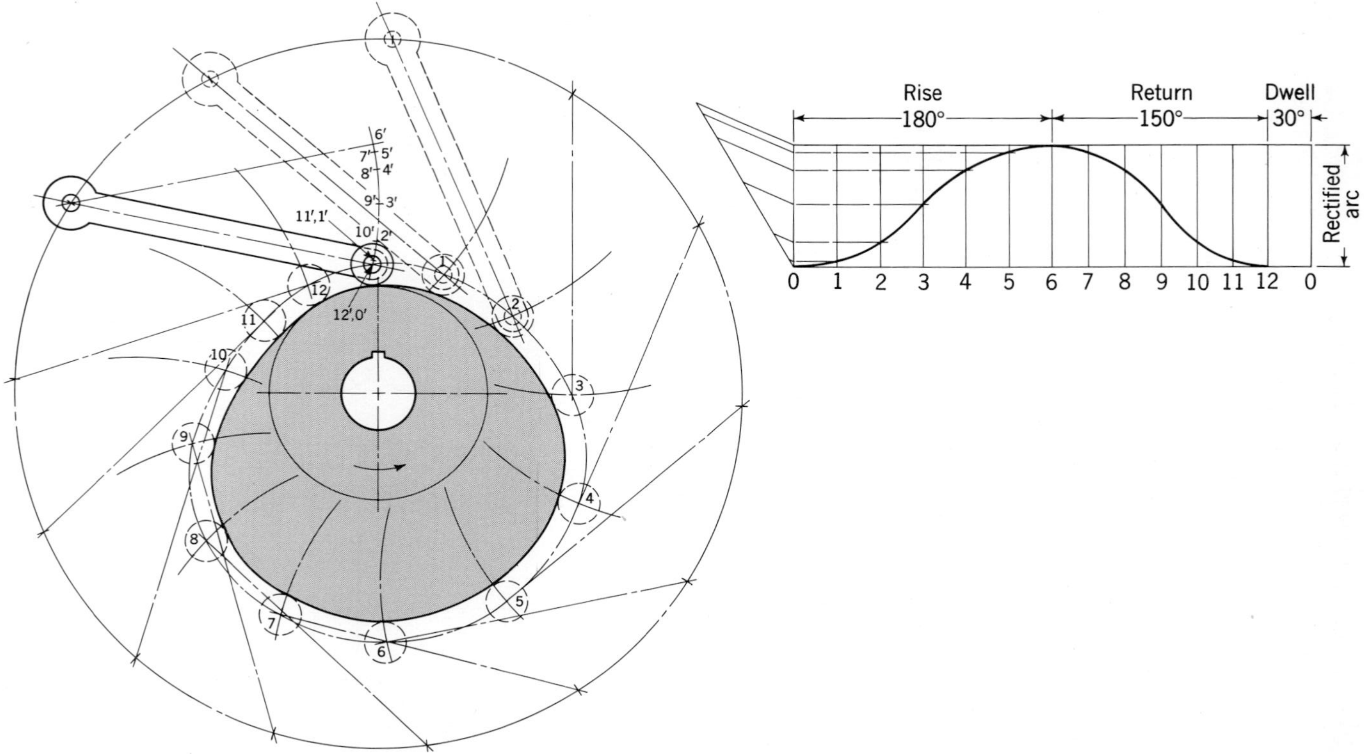

Fig. 7-9 Layout of a cam curve for an oscillating roller follower with parabolic motion.

Note, in the examples above, that there is no particular reason, except convenience, for using the same motions for return as for rise. In fact, there usually exist sound reasons for using different motions for these two parts of the cycle.

7-5 BASIC FOLLOWER MOTIONS

We have seen that the displacement diagram is plotted with the follower motion as the ordinate and the cam rotation angle as the abscissa. It is possible to use the diagram as a rectangular-coordinate system and obtain the displacement as a function of the cam angle. Then by relating the cam dimensions to the displacement diagram we may create a cam surface corresponding to any curve which can be described by an algebraic equation in rectangular coordinates.

Let us begin with the simple equation

$$y = C\theta \tag{7-1}$$

where y is the follower displacement, C is a constant, and θ is the cam angle. Designating the total rise as d to occur in a cam angle of β rad gives

$$d = C\beta \qquad \text{or} \qquad C = \frac{d}{\beta}$$

Substituting this value of C in Eq. (7-1) gives

$$y = \frac{d}{\beta}\theta \tag{7-2}$$

This is the equation for straight-line, or uniform, motion.

The velocity and acceleration of the follower are, of course, the first and second time derivatives of Eq. (7-2). Thus

$$\dot{y} = \frac{d}{\beta}\frac{d\theta}{dt} = \frac{d}{\beta}\omega \tag{7-3}$$

$$\ddot{y} = \frac{d}{\beta}\frac{d\omega}{dt} = 0 \tag{7-4}$$

where ω is the angular velocity of the cam and in our development is assumed to be constant. Figure 7-10 is a plot of the displacement, velocity, and acceleration of the uniform-motion cam follower. A dwell at the beginning and at the end of rise has been assumed. Note that the velocity is constant and the acceleration is zero except at the beginning and end of rise, where it reaches infinity instantaneously.

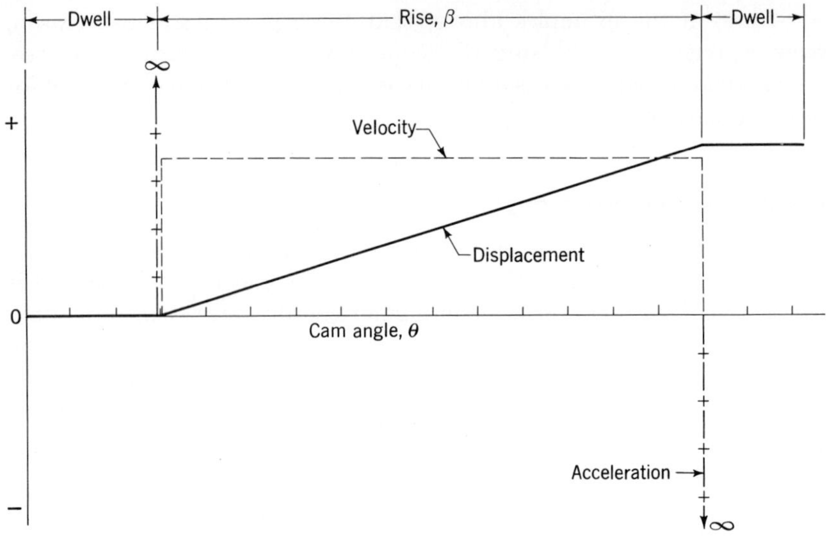

Fig. 7-10 Displacement, velocity, and acceleration relations for uniform motion.

Now let us repeat the preceding analysis using the equation of a parabola. For the first half of the motion the equation is

$$y = C\theta^2 \tag{7-5}$$

This equation is valid only to the inflection point. If we choose this to be at $y = d/2$, then $\theta = \beta/2$ and substitution into Eq. (7-5) yields

$$C = \frac{2d}{\beta^2}$$

so that

$$y = 2d\left(\frac{\theta}{\beta}\right)^2 \tag{7-6}$$

The velocity and acceleration are

$$\dot{y} = \frac{4d\omega}{\beta^2}\,\theta \tag{7-7}$$

$$\ddot{y} = \frac{4d\omega^2}{\beta^2} \tag{7-8}$$

The acceleration is constant. The maximum velocity occurs at the inflection point where $\theta = \beta/2$. Its value is

$$\dot{y}_{\text{max}} = \frac{2d\omega}{\beta} \tag{7-9}$$

For the second half of the displacement the equation

$$y = C_1 + C_2\theta + C_3\theta^2 \tag{7-10}$$

is used. Substituting the condition at $\theta = \beta$ where $y = d$, we have

$$d = C_1 + C_2\beta + C_3\beta^2 \tag{7-11}$$

Also, when $\theta = \beta$ the velocity is zero. So, from Eq. (7-10),

$$\dot{y} = C_2\omega + 2C_3\omega\theta$$

$$0 = C_2\omega + 2C_3\omega\beta \tag{7-12}$$

Since the maximum velocity occurs when $\theta = \beta/2$, we also have, from Eq. (7-9),

$$\frac{2d\omega}{\beta} = C_2\omega + 2C_3\omega\frac{\beta}{2} \tag{7-13}$$

Solving Eqs. (7-11) to (7-13) simultaneously gives

$$C_1 = -d \qquad C_2 = \frac{4d}{\beta} \qquad C_3 = -\frac{2d}{\beta^2}$$

When these constants are substituted into Eq. (7-10), the displacement equation for the second half of the rise is found to be

$$y = d\left[1 - 2\left(1 - \frac{\theta}{\beta}\right)^2\right] \tag{7-14}$$

Differentiating again to obtain the velocity and acceleration,

$$\dot{y} = \frac{4d\omega}{\beta}\left(1 - \frac{\theta}{\beta}\right) \tag{7-15}$$

$$\ddot{y} = -\frac{4d\omega^2}{\beta^2} \tag{7-16}$$

The term *jerk* or *second acceleration* is often used to describe the third time derivative. Jerk is a useful index of the quality of various cam-follower motions. Displacement, velocity, acceleration, and jerk diagrams for parabolic motion are shown in Fig. 7-11. Point B is the inflection point; to the left of this point Eqs. (7-5) to (7-9) apply, and to

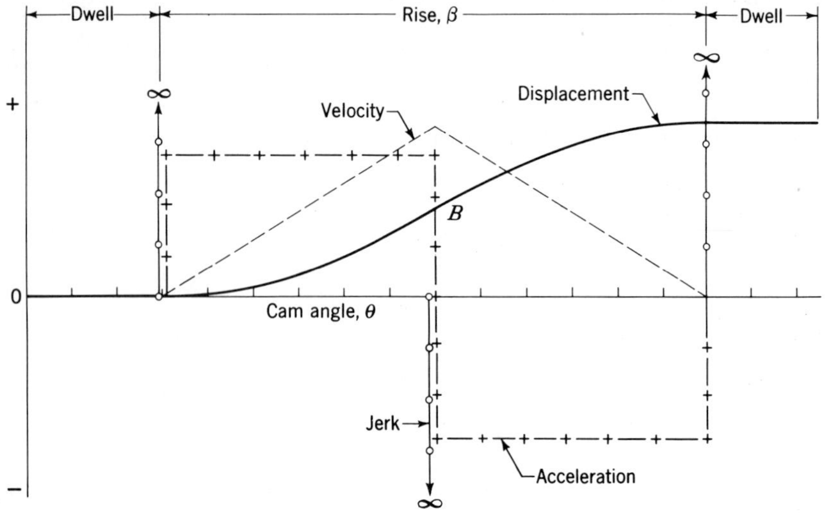

Fig. 7-11 Displacement, velocity, acceleration, and jerk relations for parabolic motion.

the right Eqs. (7-10) to (7-16) apply. Although parabolic motion produces constant acceleration, it is noted that the jerk, which is proportional to the change in force, is infinite at the beginning and end of the rise and at the inflection point.

The above discussion has been somewhat detailed in order to illustrate thoroughly the process of obtaining the equations of displacement, velocity, and acceleration for parabolic motion. We present the equations for simple harmonic and cycloidal motion without development. The procedure is the same.

The equations for simple harmonic motion are

$$y = \frac{d}{2}\left(1 - \cos\frac{\pi\theta}{\beta}\right) \tag{7-17}$$

$$\dot{y} = \frac{\pi d\omega}{2\beta}\sin\frac{\pi\theta}{\beta} \tag{7-18}$$

$$\ddot{y} = \frac{d}{2}\left(\frac{\pi\omega}{\beta}\right)^2\cos\frac{\pi\theta}{\beta} \tag{7-19}$$

As shown in Fig. 7-12, the displacement is a versed sine, the velocity a sine, and the acceleration a cosine curve. Unlike parabolic motion, there is no discontinuity at the inflection point.

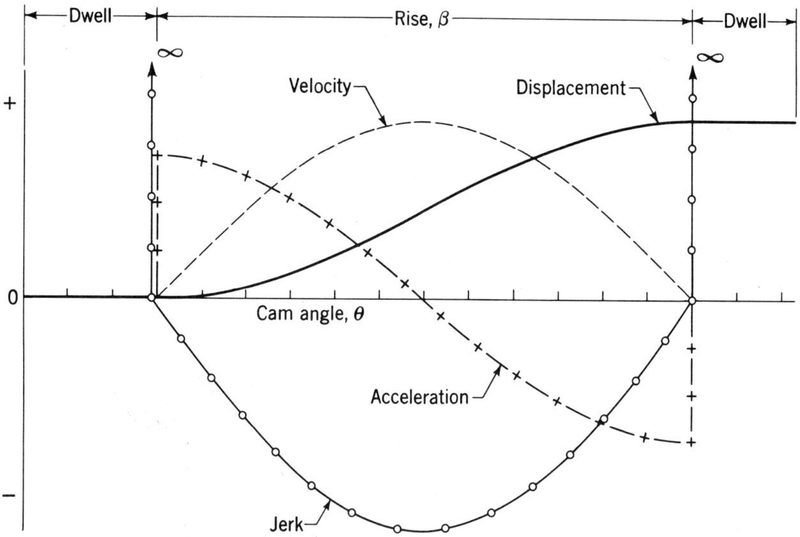

Fig. 7-12 Displacement, velocity, acceleration, and jerk relations for simple harmonic motion.

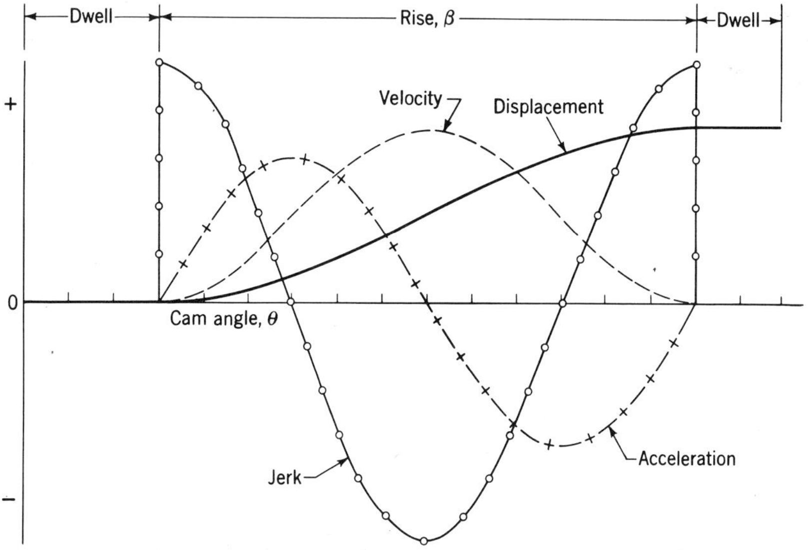

Fig. 7-13 Displacement, velocity, acceleration, and jerk relations for cycloidal motion.

Cycloidal motion has a sine acceleration curve; the equations are

$$y = d\left(\frac{\theta}{\beta} - \frac{1}{2\pi}\sin\frac{2\pi\theta}{\beta}\right) \tag{7-20}$$

$$\dot{y} = \frac{d\omega}{\beta}\left(1 - \cos\frac{2\pi\theta}{\beta}\right) \tag{7-21}$$

$$\ddot{y} = 2d\pi\left(\frac{\omega}{\beta}\right)^2 \sin\frac{2\pi\theta}{\beta} \tag{7-22}$$

The diagrams for these equations are plotted in Fig. 7-13.

7-6 COMPARISON OF FOLLOWER MOTIONS

If they are poorly designed, cam mechanisms used in high-speed machinery may be noisy, have frequent fatigue failures, require considerable maintenance, and generally perform less than satisfactorily. In these cases we must study the acceleration and jerk diagrams rather carefully to make a good selection of the cam motion. In slow-speed applications, however, often only the displacement and perhaps the velocity are significant.

In order to provide a basis of comparison for the motions we have already studied, let us use unity for d, ω, and β and then compute the maximum values of each kinematic quantity. If we do not consider the discontinuities at the beginning or end of the motion, nor at the inflection point for parabolic motion, then the results of such an investigation can be summarized as shown in Table 7-1.

The data presented in Table 7-1 might lead a person to rate uniform motion as the best and cycloidal motion as the worst. Since this is exactly opposite to the facts, we must examine the problem in a different manner.

Figure 7-14 shows a plot of the velocity, acceleration, and jerk for

Table 7-1 Maximum velocity, acceleration, and jerk for basic motions

	Type of motion			
	Uniform	*Parabolic*	*Simple harmonic*	*Cycloidal*
$\dot{y}_{max}$	1	2	$\pi/2$	2
$\ddot{y}_{max}$	0	4	$\pi^2/2$	2π
$\dddot{y}_{max}$	0	0	$\pi^3/2$	$4\pi^2$

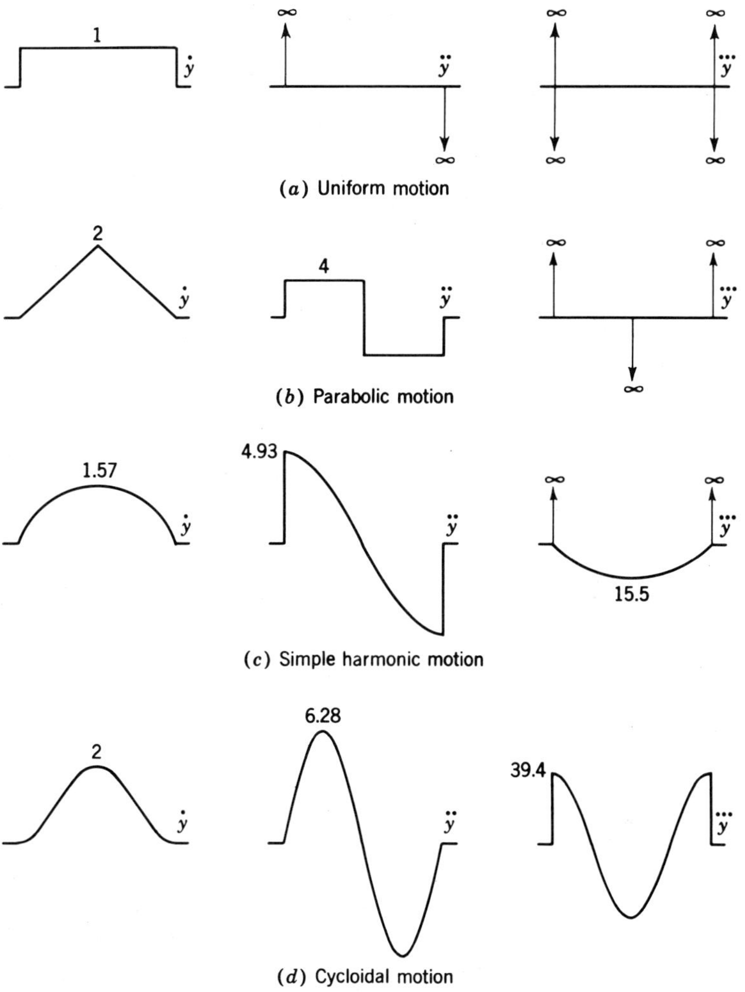

Fig. 7-14 Comparison of the kinematic characteristics of four basic motions having dwells at the beginning and the end of rise.

the four motions under consideration, drawn to scale.[1] Uniform motion is presented only for its theoretical interest. The infinite acceleration and jerk will cause serious difficulties in cam mechanisms even if they are operated at slow speeds.

We note, in the figure, that the acceleration is positive when the

[1] The book by Harold A. Rothbart, "Cams," John Wiley & Sons, Inc., New York, 1956, is a real classic on cams. It contains a similar comparison of 11 different motions. See p. 184.

velocity is increasing and is negative when the velocity is decreasing. The velocity is zero at both the beginning and the end of rise. Since velocity is the integral of the acceleration, the total area under the acceleration curve is zero. In other words, the areas of the positive and negative halves are equal to each other. Using this line of reasoning, we are tempted to conclude that a rectangle gives an area with the smallest height and hence is the most economical shape to use for an acceleration diagram. This would lead at once to the conclusion that parabolic motion is the best of all. But the infinite jerk which occurs in three places for parabolic motion really makes it one of the worst selections if high speeds are involved.

The figure shows that simple harmonic motion also has an infinite jerk, and so we conclude that it is not much better.

Though the higher acceleration in cycloidal motion may cause some problems in high-speed operation, the jerk is finite and so we conclude that this is the best of the basic motions presented.

7-7 PRESSURE ANGLE

The pressure angle is the angle between the direction in which the cam is pushing and the direction in which the follower is traveling. The size of a cam is dependent on the size of the camshaft, the curvature of the cam profile, and the pressure angle. By making the cam larger we can reduce the pressure angle, but a large cam produces more camshaft unbalance at high speeds, uses more space, and results in larger dimensions for other parts of the design. Thus, there is a basic conflict between designing a cam that occupies a small space and, at the same time, obtaining a small pressure angle. A high pressure angle or a slight increase in friction may cause a translating roller follower to chatter and perhaps even to jam in its guides.

If we design a group of cams all having the same pressure angle but each with a different motion, then the size of the cam depends upon the motion used. The displacement diagram of Fig. 7-15 is a graph of all the basic motions we have studied, plotted at the same pressure angle. Since the length of the diagram is the length of the pitch-circle arc during rise, it is evident that some motions require larger cams than others if the maximum pressure angle is not to exceed a specified value.

It is not difficult to express the pressure angle in terms of a mathematical equation. Depending upon the cam motion, however, the equation may be difficult to solve except by graphical or computer methods. In Fig. 7-16 a cam drives an offset translating roller follower. Let r_A be the radius of the prime circle, e be the offset, and ϕ the pressure angle measured between a normal to the pitch curve through the trace point A

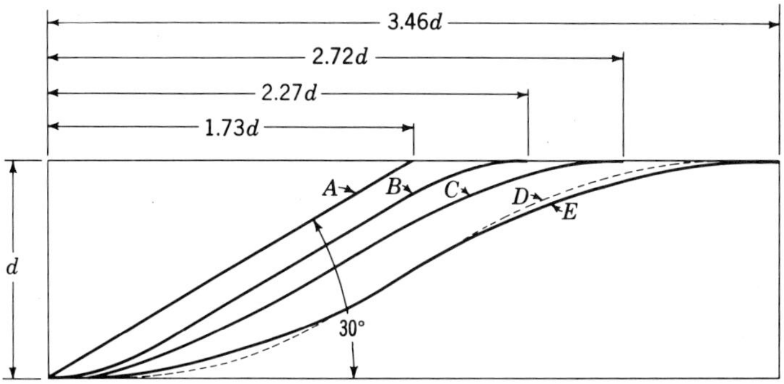

Fig. 7-15 Length of pitch-circle arc required to give the same pressure angle for different basic motions. A, uniform motion; B, modified uniform motion; C, simple harmonic motion; D, cycloidal motion; E, parabolic motion. In all cases shown the maximum pressure angle is 30°. Note that the length of the pitch-circle arc is given in terms of the rise d, which is the same for each motion.

and the path of follower travel. Then point P is the velocity pole 24 common to cam 2 and follower 4. The velocity of the follower $\dot{y}$ is therefore the distance O_2P multiplied by the cam angular velocity ω. To express this in equation form, note that

$$a = \sqrt{r_A^2 + e^2} \tag{a}$$

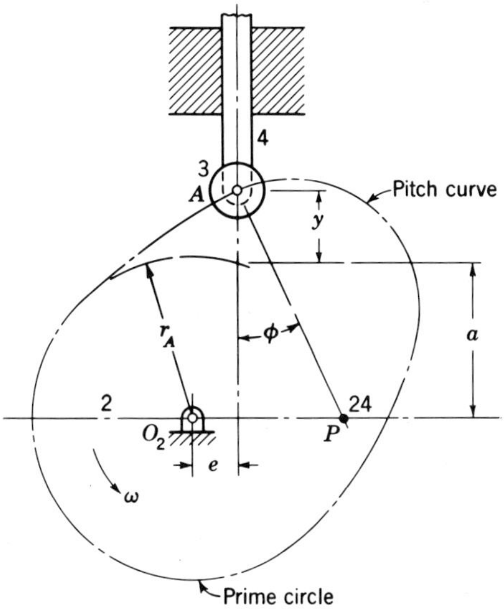

Fig. 7-16

Then

$$\dot{y} = \omega[e + (a + y) \tan \phi] \tag{b}$$

or

$$\tan \phi = \frac{(\dot{y}/\omega) - e}{a + y} \tag{7-23}$$

Though both $\dot{y}$ and ω appear in this equation, the pressure angle *does not* depend upon the velocity. To solve this equation for, say, cycloidal motion, Eq. (7-21) for $\dot{y}$ would be substituted in the numerator and Eq. (7-20) for y in the denominator. The angular velocity ω would cancel, leaving the result dependent only on the geometry.

For a radial follower $e = 0$, and Eq. (7-23) reduces to

$$\tan \phi = \frac{\dot{y}/\omega}{r_A + y} \tag{7-24}$$

Cam pressure angles of about 30 to 35° are about the largest that one can use without running into difficulties. Equation (7-23) shows that the offset e makes the numerator smaller and the denominator larger and hence is very effective in reducing the pressure angle. It is for this reason that pressure angle need not be an important factor in the design of cam mechanisms using an offset roller follower. This is also true, incidentally, for flat-face and oscillating followers.

For translating roller followers an approximate approach to design works very satisfactorily. We begin by defining a quantity called *cam factor* as

$$f = \frac{l}{d} \tag{7-25}$$

where f = cam factor
l = arc of pitch circle, in.
d = follower rise, in.

Examining Fig. 7-15 and letting $l = 1.73$ in Eq. (7-25), we see that $f = 1.73$ for uniform motion and a 30° pressure angle. Thus the cam factor is the length of the arc of the pitch circle, in terms of the rise, which subtends the rise (or return) motion for a given pressure angle and type of motion. Values of the cam factor for design purposes are listed in Table 7-2 for various pressure angles and basic motions.

The length of the arc of the pitch circle and its radius are related by the equation

$$l = r_P \beta$$

Table 7-2 Cam factors for basic motions

Pressure angle ϕ, deg	Type of motion: Uniform	Modified uniform	Simple harmonic	Parabolic and cycloidal
10	5.67	5.84	8.91	11.34
15	3.73	3.99	5.85	7.46
20	2.75	3.10	4.32	5.50
25	2.14	2.58	3.36	4.28
30	1.73	2.27	2.72	3.46
35	1.43	2.06	2.24	2.86
40	1.19	1.92	1.87	2.38
45	1.00	1.83	1.57	2.00

Therefore the pitch-circle radius is

$$r_P = \frac{l}{\beta} = \frac{fd}{\beta} \tag{7-26}$$

Example 7-1 A plate cam is to drive a radial translating roller follower through a rise of 2 in. with cycloidal motion and a pressure angle not greater than 25° in 180° of cam rotation. The follower is to return with simple harmonic motion with a pressure angle not greater than 35° and dwell as long as possible. Corresponding to a roller diameter of $\frac{1}{2}$ in., compute the radii of the pitch, base, and prime circles and the cam angles corresponding to the return and dwell events.

Solution The size of the cam is uniquely determined by the rise motion. Entering Table 7-2 with $\phi = 25°$, we find $f = 4.28$ for cycloidal motion. Since β is 180° or π rad, Eq. (7-26) yields

$$r_P = \frac{fd}{\beta} = \frac{(4.28)(2)}{\pi} = 2.72 \text{ in.}$$

Since the pitch circle passes through the center of the roller at the midpoint of rise,[1] the radius of the prime circle is

$$r_A = r_P - \frac{d}{2} = 2.72 - 1 = 1.72 \text{ in.}$$

The base circle is the radius of the prime circle less the roller radius. Hence

$$r_B = 1.72 - 0.25 = 1.47 \text{ in.}$$

Next, using Table 7-2 again for the return motion, we find $f = 2.24$ for simple harmonic motion and a 35° pressure angle. Since the pitch-circle radius is known, we solve Eq. (7-26) for the cam angle β corresponding to return. This

[1] This is a convenience, not a restriction.

yields

$$\beta = \frac{fd}{r_P} = \frac{(2.24)(2)}{2.72} = 1.65 \text{ rad}$$

or

$$\beta = \frac{1.65}{\pi}(180) \approx 95°$$

for the return. Thus we find 75° remaining for the dwell.

7-8 RADIUS OF CURVATURE

Even though a cam has been proportioned to give a satisfactory pressure angle, the follower still may not complete the desired motion if the curvature of the pitch curve is too sharp. Figure 7-17 shows two cam curves generated by different-sized rollers moving on the same pitch curve. The cam curve generated by the large roller loops over itself, which is an impossibility. The result will be a pointed cam, causing the roller to deviate from the pitch curve. It is also clear from the figure that a smaller roller moving on the same pitch curve will generate a satisfactory cam curve. Similarly, if the cam size is increased enough, the large roller will operate satisfactorily.

In Fig. 7-18 we see that the cam will be pointed when

$$r_R = \rho_K$$

where r_R is the radius of the roller, and ρ_K is the radius of curvature of the pitch curve. Therefore, for roller followers, the radius of curvature

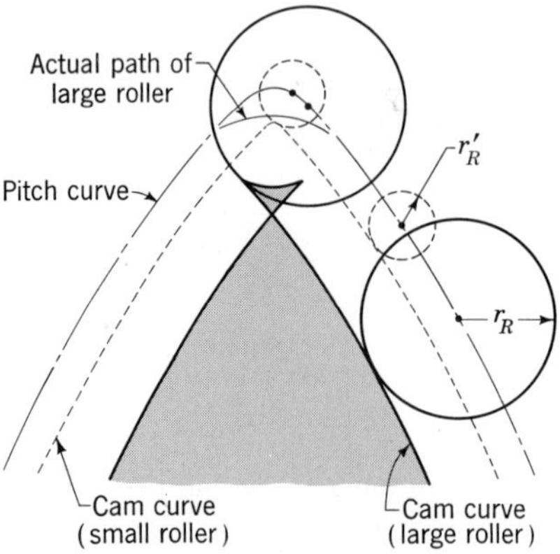

Fig. 7-17 The required follower motion cannot be obtained with the large roller.

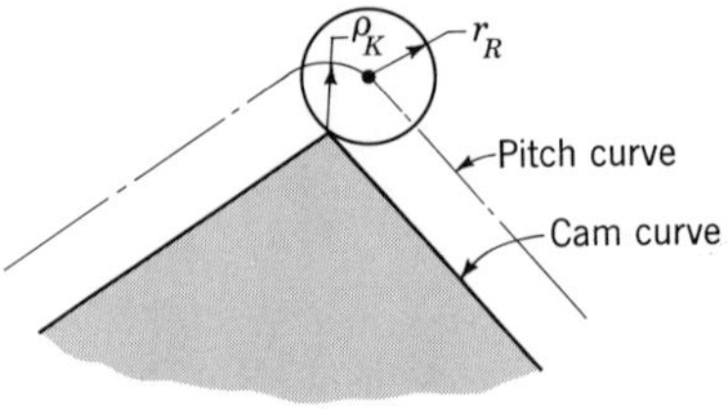

Fig. 7-18

of the pitch curve must always be greater than the roller radius; that is,

$$\rho_K > r_R \tag{7-27}$$

Analytical expressions for the minimum radius of curvature may easily be obtained. In Fig. 7-19 let r_A be the radius of the prime circle and y the displacement of the follower. Then the radius to any point on the pitch curve depends upon the cam angle θ and is

$$r = r_A + y \tag{a}$$

The general equation for radius of curvature, from differential calculus, is

$$\rho = -\frac{[r^2 + (dr/d\theta)^2]^{\frac{3}{2}}}{r^2 + 2(dr/d\theta)^2 - r(d^2r/d\theta^2)} \tag{b}$$

From Eq. (a) we have $dr/d\theta = dy/d\theta$, so that Eq. (b) becomes

$$\rho_K = -\frac{[(r_A + y)^2 + (dy/d\theta)^2]^{\frac{3}{2}}}{(r_A + y)^2 + 2(dy/d\theta)^2 - (r_A + y)(d^2y/d\theta^2)} \tag{c}$$

where the negative sign indicates that the curvature is concave. In cams ρ_K usually has a minimum positive value when $dy/d\theta = 0$; this occurs at $\theta = 0$ and $y = 0$. Now $y = f(t)$ and $\theta = \omega t$. Therefore

$$\frac{dy}{d\theta} = \frac{dy}{dt}\frac{dt}{d\theta} = \frac{dy/dt}{d\theta/dt} \tag{d}$$

Also

$$\frac{d^2y}{d\theta^2} = \frac{d}{d\theta}\frac{dy}{d\theta} = \frac{d}{dt}\frac{dy}{d\theta}\frac{dt}{d\theta} = \frac{d(dy/d\theta)/dt}{d\theta/dt} \tag{e}$$

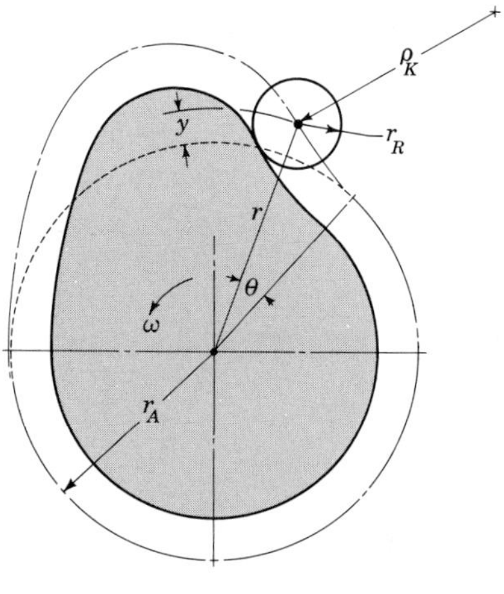

Fig. 7-19

From Eq. (*d*) we have for the numerator of Eq. (*e*)

$$\frac{d}{dt}\frac{dy}{d\theta} = \frac{d}{dt}\frac{dy/dt}{d\theta/dt} = \frac{(d\theta/dt)(d^2y/dt^2) - (dy/dt)(d^2\theta/dt^2)}{(d\theta/dt)^2} \tag{f}$$

Consequently,

$$\frac{d^2y}{d\theta^2} = \frac{(d\theta/dt)(d^2y/dt^2) - (dy/dt)(d^2\theta/dt^2)}{(d\theta/dt)^3} \tag{g}$$

Now $d\theta/dt = \omega$, $dy/dt = \dot{y}$, $d^2\theta/dt^2 = 0$ because ω is a constant, and $d^2y/dt^2 = \ddot{y}$. Equation (*g*) therefore may be written

$$\frac{d^2y}{d\theta^2} = \frac{\omega\ddot{y}}{\omega^3} = \frac{\ddot{y}}{\omega^2} \tag{h}$$

Using $dy/d\theta = 0$, $\theta = 0$, $y = 0$, $d^2y/dt^2 = \ddot{y}_0$ at $\theta = 0$, and Eq. (*h*), we find from Eq. (*c*)

$$\rho_{K,\min} = -\frac{r_A{}^2}{r_A - (\ddot{y}_0/\omega^2)} \tag{7-28}$$

Equation (7-28) gives the *minimum* radius of curvature of the *pitch curve* and shows that it depends upon the follower acceleration and upon the roller radius. Equation (*h*), however, shows that the units of time cancel, and so it is strictly a geometric property, as might be expected. From Fig. 7-19, the minimum radius of curvature of the cam curve ρ_C must be

$$\rho_{C,\min} = \rho_{K,\min} + r_R \tag{7-29}$$

The convex pitch curve is like Fig. 7-18, and Eq. (7-27) rules. Equation (*c*) must have a positive sign, however, and may be written

$$\rho_K = \frac{[(r_A + y)^2 + (\dot{y}/\omega)^2]^{3/2}}{(r_A + y)^2 + 2(\dot{y}/\omega)^2 - (r_A + y)(\ddot{y}/\omega^2)} \tag{7-30}$$

In this case the minimum radius of curvature occurs at the point of maximum negative acceleration. To use this equation, compute the numerical values of y, $\dot{y}$, and $\ddot{y}$ corresponding to the location of $\ddot{y}_{\max}$ when it is negative.

In the case of a flat-face follower the cam must be large enough to prevent the radius of curvature of the profile from becoming zero during the negative acceleration periods. In Fig. 7-20*a* let point C be the instantaneous center of curvature of the cam profile at the point of contact and ρ_C the instantaneous radius of curvature. By using the Aronhold-Kennedy theorem, three velocity poles can be located. Point 12 is located at O_2; point 13 is at infinity because follower 3 is a slider; and point 23, the common pole, is located at P on the line of centers PO_2.

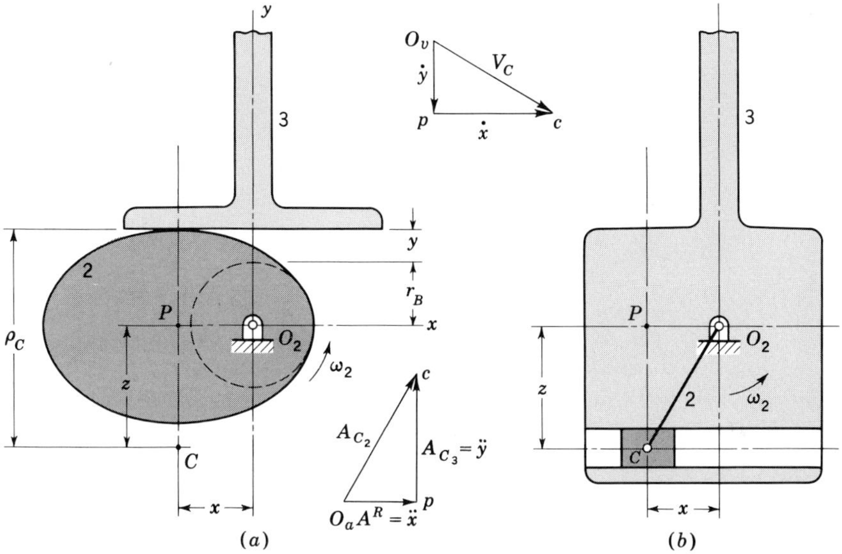

Fig. 7-20

The velocity and acceleration relations may be found rather quickly by constructing the equivalent mechanism of Fig. 7-20*b*, which turns out to be the Scotch-yoke mechanism. The velocity of C is computed by using the equation

$$V_C = O_2C\omega_2$$

and is shown in the accompanying velocity polygon. This vector has two components; the vertical component

$$\dot{y} = -x\omega_2 \tag{i}$$

is the velocity of the follower, while the horizontal component

$$\dot{x} = z\omega_2 \tag{j}$$

is the sliding component. The polygon O_vpc is the velocity image of triangle O_2PC; the two triangles are congruent for $\omega_2 = 1$ rad/sec.

The acceleration relations are obtained from the equation

$$\mathbf{A}_{C_2}{}^t + \mathbf{A}_{C_2}{}^r = \mathbf{A}_{C_3} + \mathbf{A}^R + \mathbf{A}^C \tag{k}$$

But $A_{C_2}{}^t = O_2C\dot{\omega}_2 = 0$ because ω_2 is a constant. Also $A^C = 2\omega_3 V^R = 0$ because follower 3 is a slider and hence $\omega_3 = 0$. We therefore have

$$\mathbf{A}_{C_2} = \mathbf{A}_{C_3} + \mathbf{A}^R \tag{l}$$

as the equation of the acceleration polygon. The result, shown in Fig. 7-20, is also the image of O_2PC. Since $A_{C_2} = O_2C\omega_2^2$, we have

$$A^R = \ddot{x} = x\omega_2^2 \tag{m}$$

and

$$A_{C_3} = \ddot{y} = -z\omega_2^2 \tag{n}$$

Though the center of curvature is a geometric property, Eq. (n) enables us to locate it when the follower acceleration $\ddot{y}$ is known. Thus

$$z = -\frac{\ddot{y}}{\omega_2^2} \tag{o}$$

Equation (o) shows that the center of curvature is located above the line of centers when the acceleration $\ddot{y}$ is negative. Thus the possibility exists during negative acceleration periods that the radius of curvature of the profile will become zero and result in undercutting or sharp corners on the profile. To avoid this we establish the criterion for minimum radius of curvature as

$$\rho_{C,\min} = r_B + y + \frac{\ddot{y}}{\omega^2} > 0 \tag{7-31}$$

Note that r_B and y are always positive in Eq. (7-31). Thus the curvature is always sufficient during the periods of positive acceleration.

7-9 ACCELERATION RATIO

During a complete rise motion or a complete return motion, we have seen that the acceleration has both a positive and a negative maximum. The *acceleration ratio* is the ratio of the absolute values of these two maximums. Sometimes, in the design of cam mechanisms, it is necessary to limit one of these maximums but not the other. In the case of a follower retained against the cam by gravity, for example, contact will be lost if the negative acceleration is greater than that due to gravity. In such a case the cam speed would be limited by the acceleration due to gravity. A solution to this problem, besides the use of retaining springs, is to use an acceleration ratio different from unity.

If we employ the subscripts 1 and 2, respectively, to define the first and second parts of any rise, or return, motion, then the acceleration ratio is

$$K = \frac{|\ddot{y}_{1,\max}|}{|\ddot{y}_{2,\max}|} \tag{7-32}$$

Thus, if $K > 1$ the positive acceleration is the larger, and for $K < 1$ the reverse holds. As an example, let us develop the required relations for parabolic motion. We first divide the total cam angle β into two parts, β_1 and β_2. For the first part of the rise, the equation of motion is

$$y_1 = C\theta \tag{a}$$

Using the boundary conditions that $y = d/2$ at $\theta = \beta_1$, we have $C = d/2\beta_1^2$; hence the equation of motion for the first part is

$$y_1 = \frac{d\theta^2}{2\beta_1^2} \tag{7-33}$$

The velocity and acceleration then become

$$\dot{y}_1 = \frac{d\omega}{\beta_1^2}\theta \qquad \ddot{y}_1 = \frac{d\omega^2}{\beta_1^2} \tag{b}$$

The maximum velocity occurs at $\theta = \beta_1$ and is

$$\dot{y}_{1,\max} = \frac{d\omega}{\beta_1} \tag{c}$$

For the second part of the motion the equation to be evaluated is

$$y_2 = C_1 + C_2\theta + C_3\theta^2 \tag{d}$$

with the boundary conditions

$$\text{At } \theta = \beta_1 + \beta_2: \qquad y_2 = d \qquad \dot{y}_2 = 0$$

$$\text{At } \theta = \beta_1: \qquad \dot{y}_2 = \frac{d\omega}{\beta_1}$$

When C_1, C_2, and C_3 are evaluated by using these conditions and the results are substituted back into Eq. (*d*), we obtain

$$y_2 = -\frac{d}{2\beta_1\beta_2}[\beta_1^2 + \beta_2^2 - 2(\beta_1 + \beta_2)\theta + \theta^2] \tag{7-34}$$

The second derivative of Eq. (7-34) yields the constant acceleration

$$\ddot{y}_2 = -\frac{d\omega^2}{\beta_1\beta_2} \tag{e}$$

Thus, the acceleration ratio, from Eqs. (*b*) and (*e*), is

$$K = \frac{|\ddot{y}_1|}{|\ddot{y}_2|} = \frac{d\omega^2/\beta_1^2}{d\omega^2/\beta_1\beta_2} = \frac{\beta_2}{\beta_1} \tag{7-35}$$

And so we see that the total cam angle β is to be divided into two parts in proportion to the desired acceleration ratio. We also note that

$$|\beta_1\ddot{y}_1| = |\beta_2\ddot{y}_2| \tag{f}$$

and hence that the area enclosed by the positive acceleration curve is the same as the area enclosed by the negative part of the curve.

7-10 CIRCLE-ARC AND TANGENT CAMS

Sometimes the problems involved in the manufacture of cams in large quantities are so important that compromises must be made with the velocity and acceleration relationships. In these cases it may be simpler to employ a reverse procedure and design the cam first, with the displacement diagram obtained as the second step. Such cams are usually composed of some combination of curves such as straight lines, circle arcs, involutes, and the like, which are readily produced by machine tools. The design approach is by iteration. A trial cam is designed and the kinematic characteristics which determine its performance computed. This process is then repeated a number of times until a cam having the desired performance is obtained.

The *circle-arc cam*, shown in Fig. 7-21*a*, is made of three arcs of differing radii. Points A, B, C, and D are the tangent or blending points. The equivalent linkage, shown in Fig. 7-21*b*, may be used to analyze the performance of the mechanism. It is worth noting that the acceleration may change abruptly at the blending points because of the instantaneous change in the radius of curvature.

The *tangent cam*, shown in Fig. 7-22, is composed of straight lines tangent to circle arcs. A roller follower is shown since there would be no reason for employing such cams with flat-face followers. The velocity and acceleration analysis may be carried out by defining the equiva-

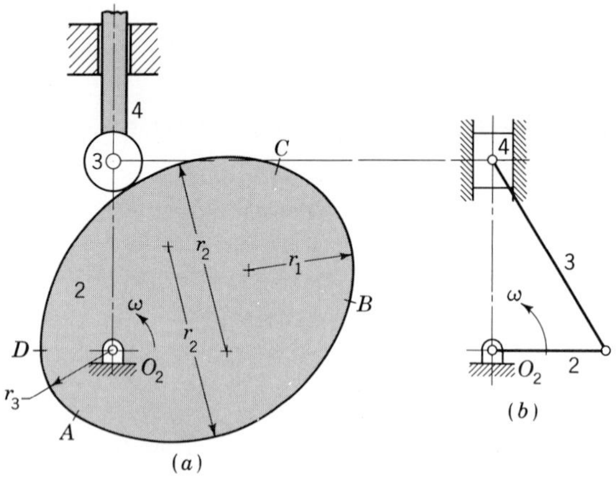

Fig. 7-21 (*a*) Circle-arc cam. (*b*) Equivalent linkage.

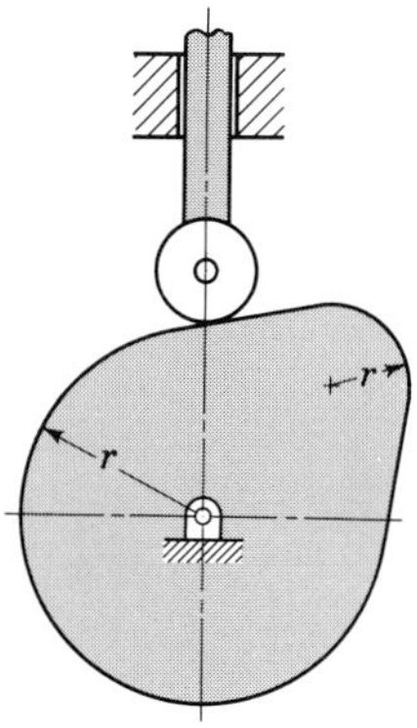

Fig. 7-22 A tangent cam.

lent linkage. It is noted, however, that the crank length of the equivalent mechanism changes continuously during contact of the roller with the straight-line portion of the profile.

7-11 ADVANCED CAM CURVES

Especially in the case of high-speed machinery the basic curves which we have studied in the previous sections may be inadequate in regard to displacement, velocity, or acceleration. Sometimes the deficiencies may be corrected by combining portions of several basic curves. When this is done the displacement curves should be tangent at the junction and the accelerations should be the same.

Another method of eliminating some of the deficiencies of the basic curves is to use a polynomial curve. This type of curve can be used to approximate almost any cam curve, although in some instances the calculations may become quite involved. The polynomial equation is

$$y = C_0 + C_1\theta + C_2\theta^2 + C_3\theta^3 + \cdots \qquad (7\text{-}36)$$

where y and θ are the rise and cam angle as before. The constant C depends upon the boundary conditions. Equation (7-36) is used by defining as many boundary conditions as are desired. As many terms are used as there are conditions to be fulfilled.

As an example of the use of Eq. (7-36) let us select the following conditions to be satisfied:

$$\begin{array}{ll} \theta = 0 & \theta = \beta \\ y = 0 & y = d \\ \dot{y} = 0 & \dot{y} = 0 \\ \ddot{y} = 0 & \ddot{y} = 0 \end{array}$$

There are six conditions, and so Eq. (7-36) is written

$$y = C_0 + C_1\theta + C_2\theta^2 + C_3\theta^3 + C_4\theta^4 + C_5\theta^5 \tag{a}$$

The first and second derivatives with respect to time of (*a*) are

$$\dot{y} = \omega C_1 + 2\omega C_2\theta + 3\omega C_3\theta^2 + 4\omega C_4\theta^3 + 5\omega C_5\theta^4 \tag{b}$$

$$\ddot{y} = 2\omega^2 C_2 + 6\omega^2 C_3\theta + 12\omega^2 C_4\theta^2 + 20\omega^2 C_5\theta^3 \tag{c}$$

The following six equations are obtained when the boundary conditions are substituted in Eqs. (*a*), (*b*), and (*c*):

$$0 = C_0 \tag{d}$$

$$d = C_0 + C_1\beta + C_2\beta^2 + C_3\beta^3 + C_4\beta^4 + C_5\beta^5 \tag{e}$$

$$0 = \omega C_1 \tag{f}$$

$$0 = \omega C_1 + 2\omega C_2\beta + 3\omega C_3\beta^2 + 4\omega C_4\beta^3 + 5\omega C_5\beta^4 \tag{g}$$

$$0 = 2\omega^2 C_2 \tag{h}$$

$$0 = 2\omega^2 C_2 + 6\omega^2 C_3\beta + 12\omega^2 C_4\beta^2 + 20\omega^2 C_5\beta^3 \tag{i}$$

When these equations are solved simultaneously, there results

$$C_0 = 0,\ C_1 = 0,\ C_2 = 0,\ C_3 = \frac{10d}{\beta^3},\ C_4 = -\frac{15d}{\beta^4},\ C_5 = \frac{6d}{\beta^5}$$

The displacement equation is obtained by substituting these constants into Eq. (*a*).

$$y = \frac{10d}{\beta^3}\theta^3 - \frac{15d}{\beta^4}\theta^4 + \frac{6d}{\beta^5}\theta^5 \tag{7-37}$$

This is called the 3-4-5 *polynomial* because of the terms remaining. The velocity, acceleration, and jerk are:

$$\dot{y} = \frac{30d\omega}{\beta^3}\theta^2 - \frac{60d\omega}{\beta^4}\theta^3 + \frac{30d\omega}{\beta^5}\theta^4 \tag{7-38}$$

$$\ddot{y} = \frac{60d\omega^2}{\beta^3}\theta - \frac{180d\omega^2}{\beta^4}\theta^2 + \frac{120d\omega^2}{\beta^5}\theta^3 \tag{7-39}$$

$$\dddot{y} = \frac{60d\omega^3}{\beta^3} - \frac{360d\omega^3}{\beta^4}\theta + \frac{360d\omega^3}{\beta^5}\theta^2 \tag{7-40}$$

These relationships are plotted in Fig. 7-23. The results are similar to the cycloidal curve, yet distinctly different.

In general, this type of cam will begin and end its motion slower than other types. In order to satisfactorily produce such a cam, extreme

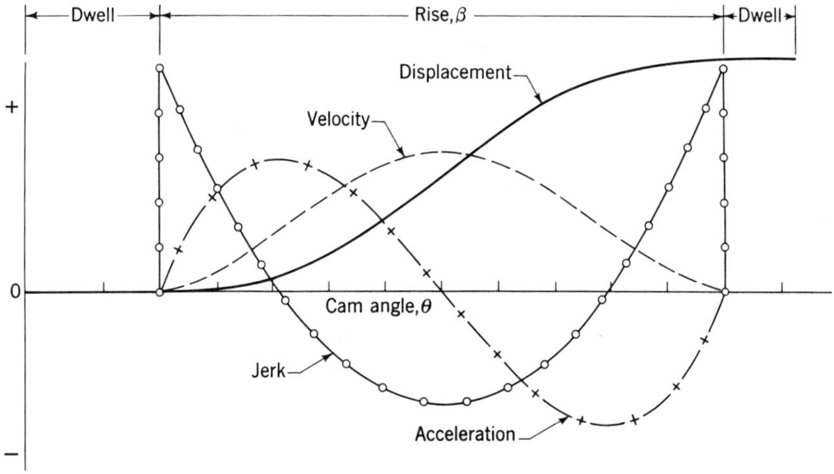

Fig. 7-23 Displacement, velocity, acceleration, and jerk relations for the 3-4-5 polynomial motion.

accuracy of manufacture is required at the beginning and the end of the motion.

Rothbart[1] describes many cam motions which have not been investigated here. One of the most important is the *modified trapezoidal acceleration motion*. This curve is obtained as a combination of cycloidal and parabolic motion, as shown in Fig. 7-24. Note that only the positive portion of the acceleration curve is shown. In synthesizing the motion, mathematical continuity for all the kinematic quantities up to and including the jerk should be preserved.

[1] *Op. cit.*

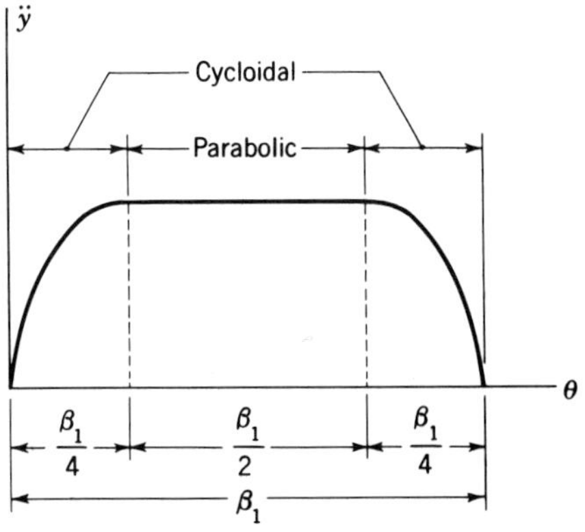

Fig. 7-24

7-12 THE GENEVA MECHANISM

The *Geneva wheel*, or *Maltese cross*, is a camlike mechanism which provides intermittent rotary motion and is widely used in both low- and high-speed machinery. Although originally developed as a stop to prevent overwinding of watches, it is now extensively used in automatic machinery, as, for example, where a spindle, turret, or worktable must be indexed. It is also used in motion-picture projectors to provide the intermittent advance of the film.

A drawing of a six-slot Geneva mechanism is shown in Fig. 7-25. Notice that the centerlines of the slot and crank are mutually perpendicular at engagement and at disengagement. The crank, which usually rotates at a uniform angular velocity, carries a roller to engage with the slots. During one revolution of the crank the Geneva wheel rotates a fractional part of a revolution, the amount of which is dependent upon the number of slots. The circular segment attached to the crank effectively locks the wheel against rotation when the roller is not in engagement and also positions the wheel for correct engagement of the roller with the next slot.

The design of a Geneva mechanism is initiated by specifying the crank radius, the roller diameter, and the number of slots. At least 3 slots are necessary, but most problems can be solved with wheels having from 4 to 12 slots. The design procedure is shown in Fig. 7-26.

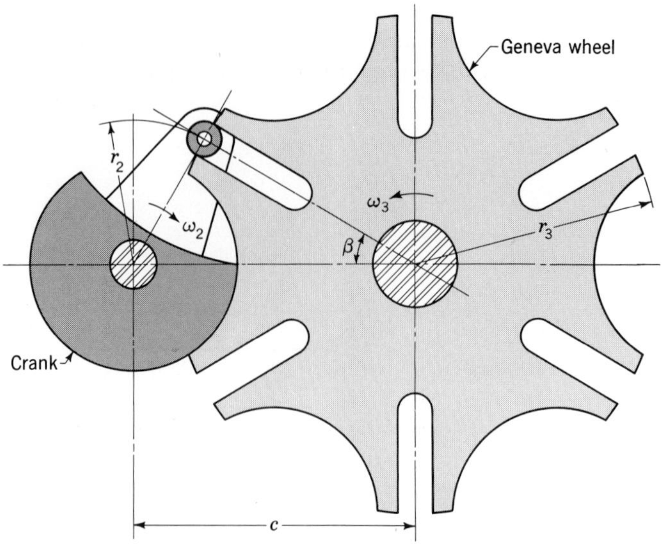

Fig. 7-25 The Geneva mechanism.

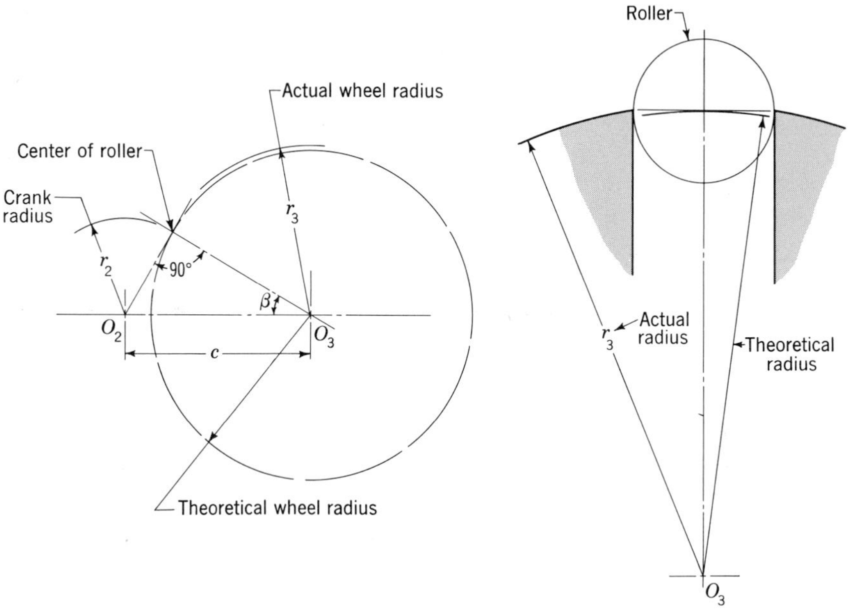

Fig. 7-26 Design of a Geneva wheel.

The angle β is half the angle subtended by adjacent slots; that is,

$$\beta = \frac{360}{2n} \tag{a}$$

where n is the number of slots in the wheel. Then, defining r_2 as the crank radius,

$$c = \frac{r_2}{\sin \beta} \tag{b}$$

where c is the center distance. Note, too, from Fig. 7-26, that the actual Geneva-wheel radius is more than that which would be obtained by a zero-diameter roller. This is due to the difference between the sine and the tangent of the angle subtended by the roller, measured from the wheel center.

After the roller has entered the slot and is driving the wheel, the geometry is that of Fig. 7-27. Here θ_2 is the crank angle and θ_3 the wheel angle. They are related trigonometrically by the equation

$$\tan \theta_3 = \frac{\sin \theta_2}{(c/r_2) - \cos \theta_2} \tag{c}$$

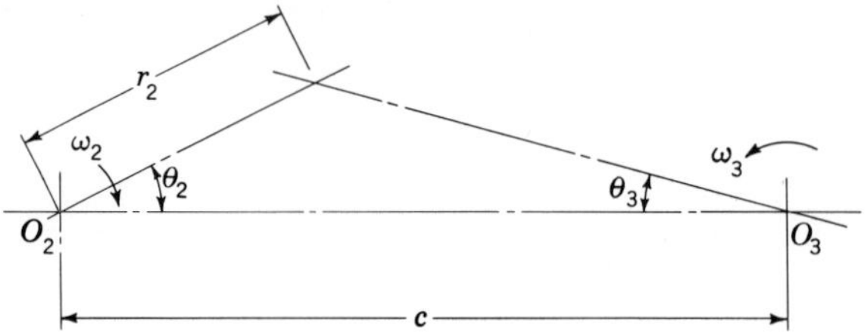

Fig. 7-27

We may determine the angular velocity of the wheel for any value of θ_2 by differentiating Eq. (*c*) with respect to time. This produces

$$\omega_3 = \omega_2 \frac{(c/r_2) \cos \theta_2 - 1}{1 + (c^2/r_2^2) - 2(c/r_2) \cos \theta_2} \tag{7-41}$$

The maximum wheel velocity occurs when the crank angle is zero. So, substituting $\theta_2 = 0$ gives

$$\omega_3 = \omega_2 \frac{r_2}{c - r_2} \tag{7-42}$$

The angular acceleration is obtained by differentiating Eq. (7-41) with respect to time. It is

$$\dot{\omega}_3 = \omega_2{}^2 \frac{(c/r_2) \sin \theta_2 (1 - c^2/r_2^2)}{[1 + (c/r_2)^2 - 2(c/r_2) \cos \theta_2]^2} \tag{7-43}$$

The angular acceleration reaches a maximum when

$$\theta_2 = \cos^{-1} \left\{ \pm \sqrt{\left[\frac{1 + (c^2/r_2^2)}{4(c/r_2)}\right]^2 + 2} - \frac{1 + (c/r_2)^2}{4(c/r_2)} \right\} \tag{7-44}$$

This occurs when the roller has advanced about 30 percent into the slot.

Several methods have been employed to reduce the wheel acceleration in order to reduce inertia forces and the consequent wear on the sides of the slot. Among these is the idea of using a curved slot. This does reduce the acceleration, but it increases the deceleration and consequently the wear on the other side of the slot.

Another method utilizes the Hrones-Nelson synthesis. The idea is to place the roller on the connecting link of a four-bar linkage. The path of the roller should be curved during the period in which it drives the wheel. Also, during this period, it should have a low value of acceleration. Figure 7-28 shows one solution and includes the path taken by

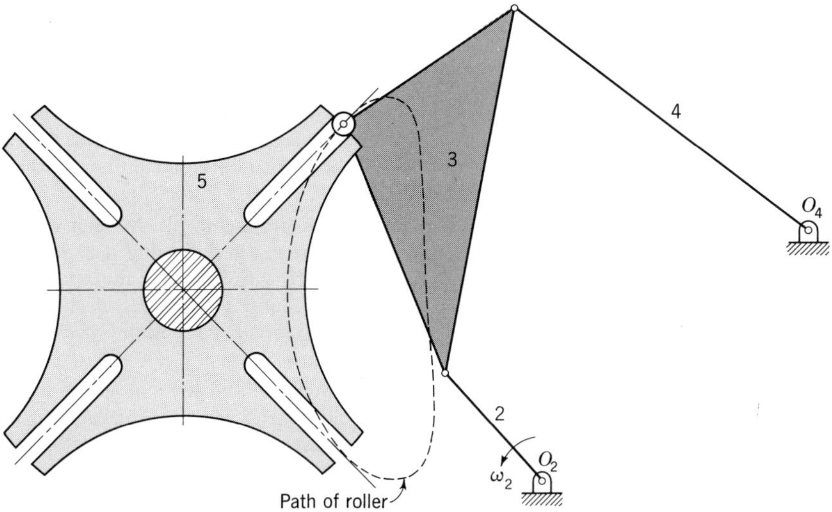

Fig. 7-28 Geneva wheel driven by a four-bar linkage synthesized by the Hrones-Nelson method. Link 2 is the driving crank.

the roller. This is the path which is sought for in leafing through the book.

The inverse Geneva mechanism of Fig. 7-29 enables the wheel to rotate in the same direction as the crank and requires less radial space. The locking device is not shown, but this can be a circular segment attached to the crank, as before, which locks by wiping against a built-up rim on the periphery of the wheel.

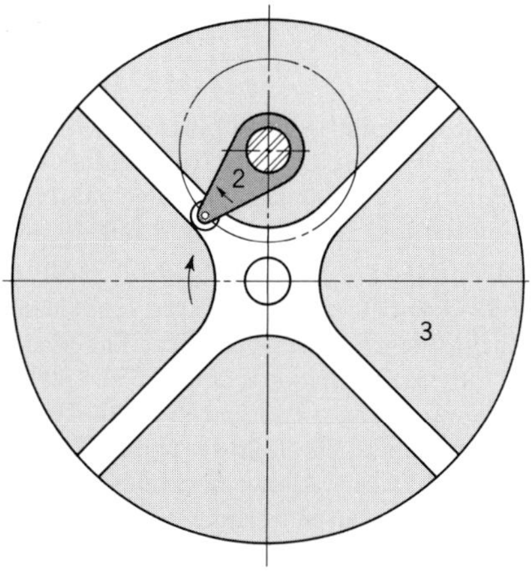

Fig. 7-29 The inverse Geneva mechanism.

PROBLEMS

(Problems 7-1 to 7-15 require drawing equipment for their solution; 11 by 17 paper is recommended.)

7-1. Construct the rise portion of the displacement diagram and all the cam curve for a disk cam with translating radial roller follower using the following data: roller diameter $\frac{5}{8}$ in.; rise 2 in. in 180° and return in 180°, both with harmonic motion; pitch-circle diameter 6 in.; clockwise cam rotation.

7-2. Construct the rise portion of the displacement diagram and all the cam curve for a disk cam with translating offset roller follower using the following data: roller diameter $\frac{3}{4}$ in.; rise $1\frac{1}{2}$ in. in 180° and return in 150°, both with parabolic motion, then dwell 30°; pitch-circle diameter 5 in.; cam rotation counterclockwise; offset $\frac{3}{4}$ in. Offset the follower in a direction such that the bending forces are a minimum during the rise.

7-3. Construct the rise portion of the displacement diagram and the cam curve for a disk cam to drive the oscillating roller follower shown in the figure. The follower is to rise through an arc of 30° with modified uniform motion in 150° of cam rotation, dwell for 30°, return in 150° with the same motion, and dwell for 30°. The abscissa of the rise portion of the displacement diagram should be equal in length to the corresponding arc of the pitch circle. The ordinate should be the rectified arc of the path of the roller center.

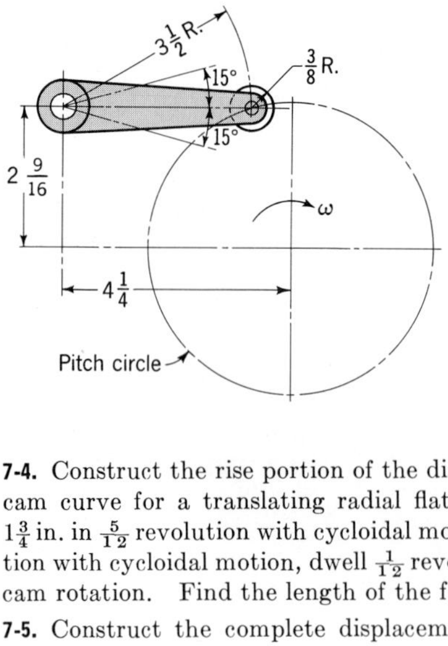

Prob. 7-3

7-4. Construct the rise portion of the displacement diagram, the pitch curve, and the cam curve for a translating radial flat-face follower using the following data: rise $1\frac{3}{4}$ in. in $\frac{5}{12}$ revolution with cycloidal motion, dwell $\frac{1}{12}$ revolution, return in $\frac{5}{12}$ revolution with cycloidal motion, dwell $\frac{1}{12}$ revolution; pitch-circle diameter $5\frac{1}{2}$ in.; clockwise cam rotation. Find the length of the follower face allowing $\frac{3}{16}$ in. clearance.

7-5. Construct the complete displacement diagram, the pitch curve, and the cam curve for a translating offset flat-face follower according to the following data: rise $1\frac{1}{4}$ in. with harmonic motion in 180° of cam rotation, dwell 30°, and return in the remaining angle with parabolic motion; pitch-circle diameter 3 in.; rotation clockwise; offset $\frac{1}{2}$ in. The follower should be offset in the direction that reduces bending of the follower face. Allow $\frac{3}{16}$ in. clearance and find the length of face required on each side of the follower centerline.

7-6. The oscillating flat-face follower shown in the figure is to be driven by a disk cam and is to rise through an arc of 24° with harmonic motion in 180° of cam rotation, dwell for 15°, and return in the remaining angle with the same motion. Construct the cam curve, and determine the dimensions of the flat face. Allow $\frac{3}{16}$ in. clearance at each end. Indicate the correct direction of rotation.

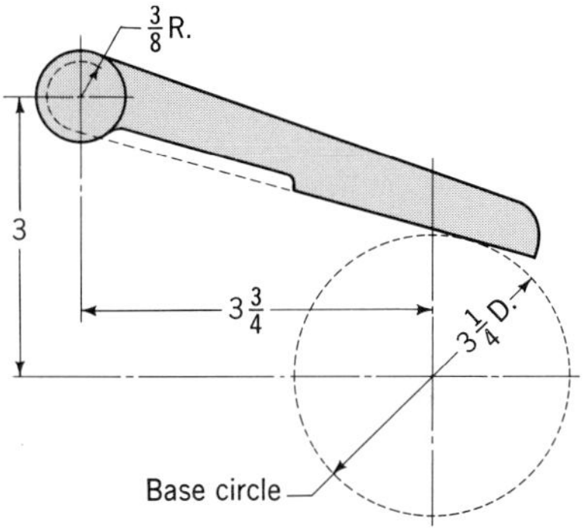

Prob. 7-6

7-7. Construct the rise portion of the displacement diagram and all the cam curve for a disk cam with translating offset roller follower according to the following data: roller diameter $\frac{5}{8}$ in.; rise $1\frac{3}{8}$ in., with 3-4-5 polynomial motion in 180° of cam rotation, and return with the same motion; pitch-circle diameter 6 in.; offset $\frac{3}{4}$ in.; clockwise cam rotation. Offset the follower in a direction such that bending is reduced.

7-8. A disk cam rotates at 400 rpm and is to lift a translating radial roller follower $1\frac{1}{2}$ in. with cycloidal motion and return the follower with parabolic motion. There are no dwell periods. Determine the size of the pitch circle for a maximum pressure angle during rise of 30°. The roller diameter is $\frac{5}{8}$ in. Draw the cam curve and check for satisfactory curvature. Calculate the maximum velocity during rise and return. Find the maximum acceleration during each motion. The cam rotation is counterclockwise.

7-9. A disk cam rotates at 600 rpm and is to lift a flat-face translating follower 2 in. with parabolic motion in 180° of cam rotation. The follower motion is radial. The follower then dwells for 30° of rotation and returns in the remaining time with harmonic motion. Using a trial-and-error method, determine a satisfactory base-circle radius. It is suggested that the base-circle radius be increased (or decreased) in $\frac{1}{4}$-in. increments until a satisfactory value is obtained. Complete the cam curve and determine a suitable length for the follower face. The cam rotation is clockwise. What is the maximum velocity of the follower during rise and return?

7-10. Draw the rise portion of a cam displacement diagram having $d = 1$ in., $\beta = 1$ rad, and $\omega = 1$ rad/sec. Use a modified harmonic motion obtained by constructing an ellipse on the rise axis such that its major axis is horizontal, parallel to the time axis, and $1\frac{1}{2}$ times as long as the minor axis. Differentiate graphically to obtain the velocity and acceleration diagrams and record the maximum values of each.

7-11. The same as Prob. 7-10 except that the rise axis is to be the major axis of the ellipse.

7-12. Determine the displacement diagram and the maximum pressure angle for the circle-arc cam and roller follower shown in the figure. Use $\omega = 1$ rad/sec.

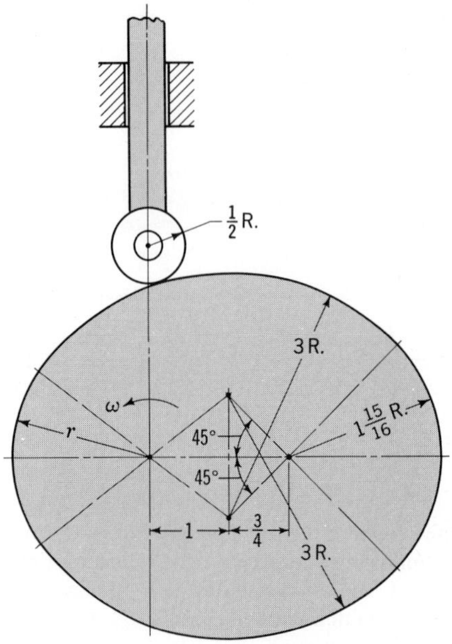

Prob. 7-12

7-13. Using a graphical method, determine the velocity and acceleration diagrams for the follower of Prob. 7-12.

7-14. Find the displacement diagram and the value of the maximum pressure angle for the tangent cam and roller follower shown in the figure.

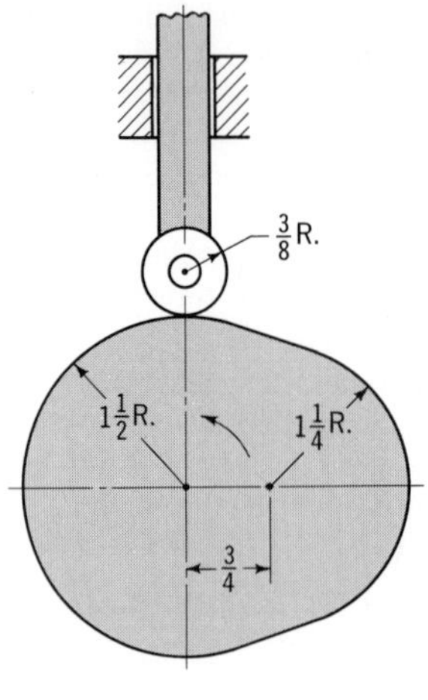

Prob. 7-14

7-15. Determine the velocity and acceleration diagrams for the cam and follower of Prob. 7-14.

7-16. For simple harmonic motion write the equations for the velocity and jerk at the midpoint of the motion. Also, determine the acceleration at the beginning of the motion.

7-17. For cycloidal motion determine the values of θ for which the acceleration is maximum. What is the formula for the acceleration at these points? Find the equation for the velocity and jerk at the midpoint of the motion.

7-18. A disk cam rotates at 300 rpm and drives a translating radial roller follower through a rise of 3 in. in 180° of cam rotation. Find the diameter of the pitch circle if harmonic motion and a 25° pressure angle are to be used. Find the maximum velocity and acceleration of the follower which correspond to this pitch circle.

7-19. The same as Prob. 7-18, except use cycloidal motion.

7-20. Use a roller diameter of $\frac{3}{4}$ in. for Prob. 7-18. Is the radius of curvature satisfactory?

7-21. A translating radial roller follower rises 2 in. in 120° of cam rotation with cycloidal motion, returns in 90° with parabolic motion, and dwells for the remaining period of time. The cam is a disk cam and rotates at 750 rpm. Find the diameter of the pitch circle if the pressure angle during rise is not to exceed 30°. Calculate the resulting maximum velocity and acceleration during rise and also during the return stroke.

7-22. A flat-face radial translating follower is to rise 4 in. with cycloidal motion in 180° of cam rotation, return in 20° with the same motion, and dwell for the remaining angle. Find the minimum radius of the base circle if the cam rotates at 150 rpm.

7-23. The boundary conditions for a polynomial cam motion are: when $\theta = 0$, $y = 0$, and $\dot{y} = 0$; when $\theta = \beta$, $y = d$, and $\dot{y} = 0$. Find the displacement, velocity, and acceleration equations, and sketch the corresponding diagrams.

8
Spur Gears

8-1 TERMINOLOGY

Spur gears are used to transmit rotary motion between parallel shafts; they are usually cylindrical in shape, and the teeth are straight and parallel to the axis of rotation.

The terminology of gear teeth is illustrated in Fig. 8-1, and most of the following definitions are shown:

The *pitch circle* is a theoretical circle upon which all calculations are usually based. The pitch circles of a pair of mating gears are tangent to each other.

A *pinion* is the smaller of two mating gears. The larger is often called the *gear*.

The *circular pitch* p_c is the distance, in inches, measured on the pitch circle, from a point on one tooth to a corresponding point on an adjacent tooth.

The *diametral pitch* P is the number of teeth on the gear per inch of pitch diameter. The units of diametral pitch are the reciprocal of inches. Note that the diametral pitch cannot actually be measured on the gear itself.

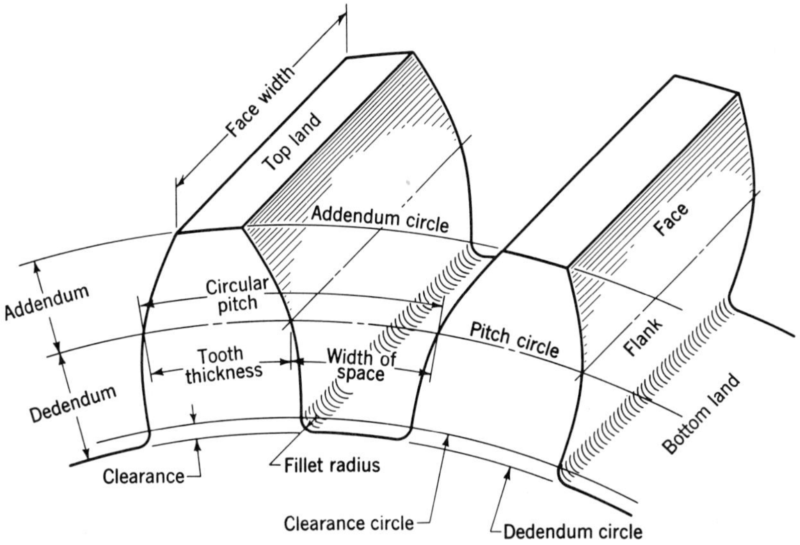

Fig. 8-1 Terminology.

The *addendum* a is the radial distance between the top land and the pitch circle.

The *dedendum* b is the radial distance from the bottom land to the pitch circle.

The *whole depth* h_t is the sum of the addendum and dedendum.

The *clearance circle* is a circle that is tangent to the addendum circle of the mating gear.

The *clearance* c is the amount by which the dedendum in a given gear exceeds the addendum of its mating gear.

The *backlash* is the amount by which the width of a tooth space exceeds the thickness of the engaging tooth on the pitch circles.

The reader should satisfy himself as to the validity of the following useful relations:

$$P = \frac{N}{d} \tag{8-1}$$

where P = diametral pitch, teeth per inch
N = number of teeth
d = pitch diameter, in.

$$p_c = \frac{\pi d}{N} \tag{8-2}$$

where p_c = circular pitch, in.

$$p_c P = \pi \tag{8-3}$$

The numerical value of the pitch describes, roughly, the size of the tooth. If the teeth on a gear are 10 diametral pitch, then they have a circular pitch, according to Eq. (8-3), of 0.314 in. The pitch is the same whether the teeth are on a 2-in. gear or a 10-in. gear. A large value of the circular pitch indicates a large tooth and a small value represents a small tooth. The reverse is true for diametral pitch: Teeth whose diametral pitch is 10, for example, are smaller than those whose diametral pitch is 6.

8-2 CONJUGATE ACTION

Mating gear teeth acting against each other to produce rotary motion may be likened to a cam and follower. When the tooth profiles (or cam and follower profiles) are shaped so as to produce a constant angular-velocity ratio during meshing, then the surfaces are said to be conjugate. It is possible to specify any profile for one tooth and then to find a profile for the mating tooth such that the surfaces are conjugate. One of these solutions is the *involute* profile, which, with few exceptions, is in universal use for gear teeth.

The action of a single pair of mating teeth as they pass through an entire phase of action must be such that the ratio of the angular velocity of the driving gear to that of the driven gear remains constant. This is the fundamental criterion which governs the choice of the tooth profiles. If it were not true of gearing, then very serious vibration and impact problems would exist, even at low speeds.

In Sec. 4-13 we learned that the angular-velocity-ratio theorem states: The angular-velocity ratio of any mechanism is inversely proportional to the segments into which the common pole cuts the line of centers. In Fig. 8-2 two profiles are in contact at A; let profile 2 be the driver and 3 the driven. A normal to the profiles at the point of contact A intersects the line of centers O_2O_3 at the common pole P.

In gearing, P is called the *pitch point,* and BC the *line of action.* Designating the pitch-point radii of the two profiles as r_2 and r_3, then, from Eq. (4-28),

$$\frac{\omega_2}{\omega_3} = \frac{r_3}{r_2} \tag{8-4}$$

This equation is frequently used to define the *law of gearing,* which states that *the pitch point must remain fixed on the line of centers.* This means that all the lines of action for every instantaneous point of contact must

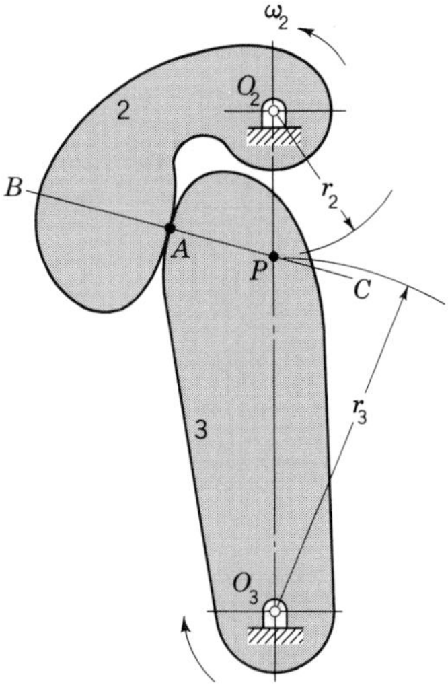

Fig. 8-2

pass through the pitch point. Our problem now is to determine the shape of the mating surfaces to satisfy the law of gearing.

It should not be assumed that just any shape or profile for which a conjugate can be found will be satisfactory. Even though conjugate curves are found, the practical problems of reproducing these curves in great quantities on steel gear blanks, as well as other materials, and using existing machinery still exist. In addition, the changes in the shaft centers due to misalignment and large forces must be considered. And finally, the tooth profile selected must be one which can be reproduced economically. A major portion of this chapter is devoted to illustrating how the involute profile fulfills these requirements.

8-3 INVOLUTE PROPERTIES

If mating tooth profiles have the shape of involute curves, then the condition that the common normal at all points of contact is to pass through the pitch point is satisfied. An involute curve is the path generated by a tracing point on a cord as the cord is unwrapped from a cylinder called the *base cylinder*. This is shown in Fig. 8-3, where T is the tracing point. Note that the cord AT is normal to the involute at T and that the distance AT is the instantaneous value of the radius of curvature. As the involute is generated from the origin T_0 to T_1, the radius of curvature varies continuously; it is zero at T_0 and greatest at T_1. Thus the cord is the generating line, and it is always normal to the involute.

Let us now examine the involute profile to see how it satisfies the requirement for the transmission of uniform motion. In Fig. 8-4 two

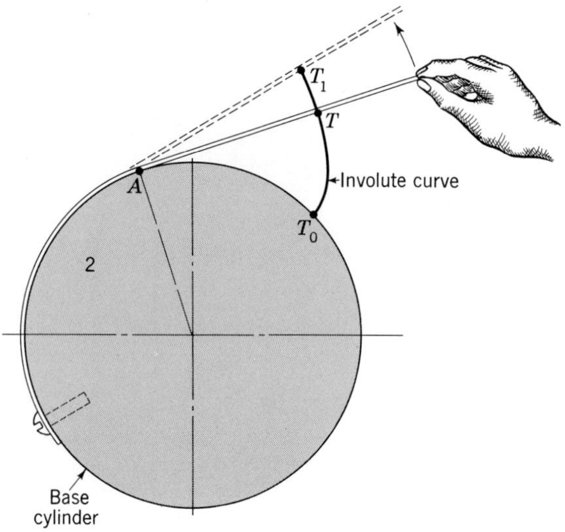

Fig. 8-3

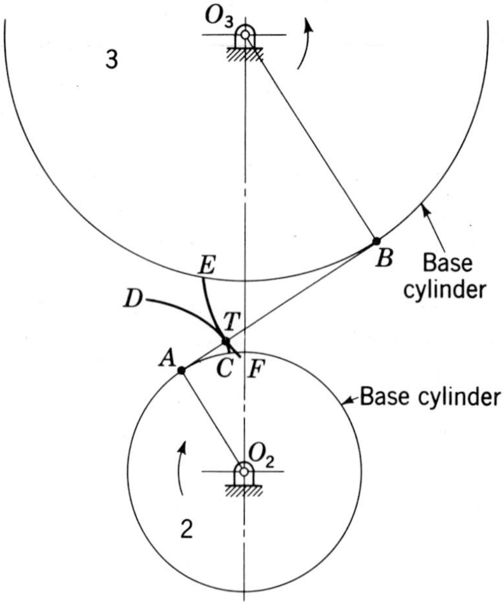

Fig. 8-4 Involute action.

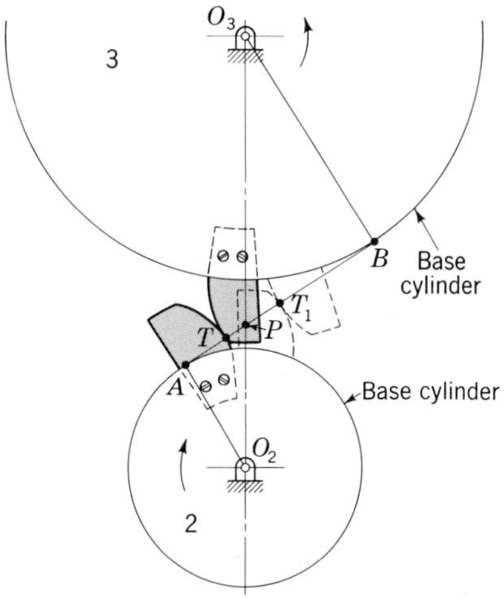

Fig. 8-5

gear blanks with fixed centers O_2 and O_3 are shown having base cylinders whose respective radii are O_2A and O_3B. We now imagine that a cord is wound clockwise around the base cylinder of gear 2, pulled tightly between points A and B, and wound counterclockwise around the base cylinder of gear 3. If now the base cylinders are rotated in different directions so as to keep the cord tight, a point T will trace out the involutes CD on gear 2 and EF on gear 3. The involutes thus generated simultaneously by the single tracing point are conjugate profiles.

Next imagine that the involutes of Fig. 8-4 are scribed on plates and the plates cut along the scribed curves and then bolted to the respective cylinders in the same positions. The result is shown in Fig. 8-5. The cord can now be removed, and if gear 2 is moved clockwise, gear 3 is caused to move counterclockwise by the camlike action of the two curved plates. The path of contact will be the line AB formerly occupied by the cord. Since the line AB is the generating line for each involute, it is normal to both profiles at all points of contact. Also, it always occupies the same position because it is tangent to both base cylinders. Therefore point P is the pitch point; it does not move; and so the involute curve satisfies the law of gearing.

Before concluding this section the reader should observe that a change in center distance, which might be caused by incorrect mounting, will have no effect upon the shape of the involute. In addition, the pitch point is still fixed, and so the law of gearing is satisfied.

8-4 FUNDAMENTALS

In order to illustrate the fundamentals of spur gears, we shall proceed, step by step, through the actual layout of a pair of spur gears. The dimensions used will be obtained from Sec. 8-13, which lists standard tooth forms. New terms will be introduced and explained as we progress through the layout.

The purpose of a layout of gear teeth is *not* to use it in the shop, but rather to use it for analysis. In the production of large quantities of gears the shop requires only the drawings of the gear blanks together with a specification (not a drawing) of the tooth form and size. On the other hand, if tools are to be manufactured for cutting gear teeth, then drawings of the tooth form and shape must be made. Sometimes these drawings are made at a scale many times larger than the tooth itself in order that accurate dimensions may be obtained.

For given information we shall select a 2-in.-diameter, 10-diametral-pitch pinion to drive a 50-tooth gear. The tooth form selected is the 20° full depth. The various steps in correct order are illustrated in Figs. 8-6 and 8-7 and are as follows:

Step 1. Calculate the pitch diameters and draw the pitch circles tangent to each other (Fig. 8-6). The numbers 2 and 3 will be employed as subscripts to designate the pinion and gear, respectively. From Eq. (8-1) the pitch diameter of the gear is

$$d_3 = \frac{N_3}{P_d} = \frac{50}{10} = 5 \text{ in.}$$

Step 2. Draw a line perpendicular to the line of centers through the pitch point (Fig. 8-6). The pitch point is the point of tangency of the pitch circles. Draw the pressure line at an angle equal to the pressure angle from the perpendicular. The *pressure line* corresponds to the generating line, or the line of action, defined in the previous sections. As shown, it is always normal to the involutes at the point of contact and passes through the pitch point. It is called the pressure line because the resultant tooth force during action is along this line. The *pressure angle* is the angle that the pressure line makes with a perpendicular to the line of centers through the pitch point. In this example the pressure angle is 20°.

Step 3. Through the centers of each gear construct perpendiculars O_2A and O_3B to the pressure line (Fig. 8-6). These radial distances, from the centers to the pressure line, are the radii of the two *base circles*. The base circles correspond to the base cylinders of the previous section. The involute curve originates on these base circles. Draw each base circle.

Step 4. Generate an involute curve on each base circle (Fig. 8-6). This is illustrated on gear 3. First, divide the base circle into equal parts A_0, A_1, A_2, etc. Now construct the radial lines O_3A_0, O_3A_1, O_3A_2, etc. Next, construct perpendiculars to these radial lines. The involute begins at A_0.

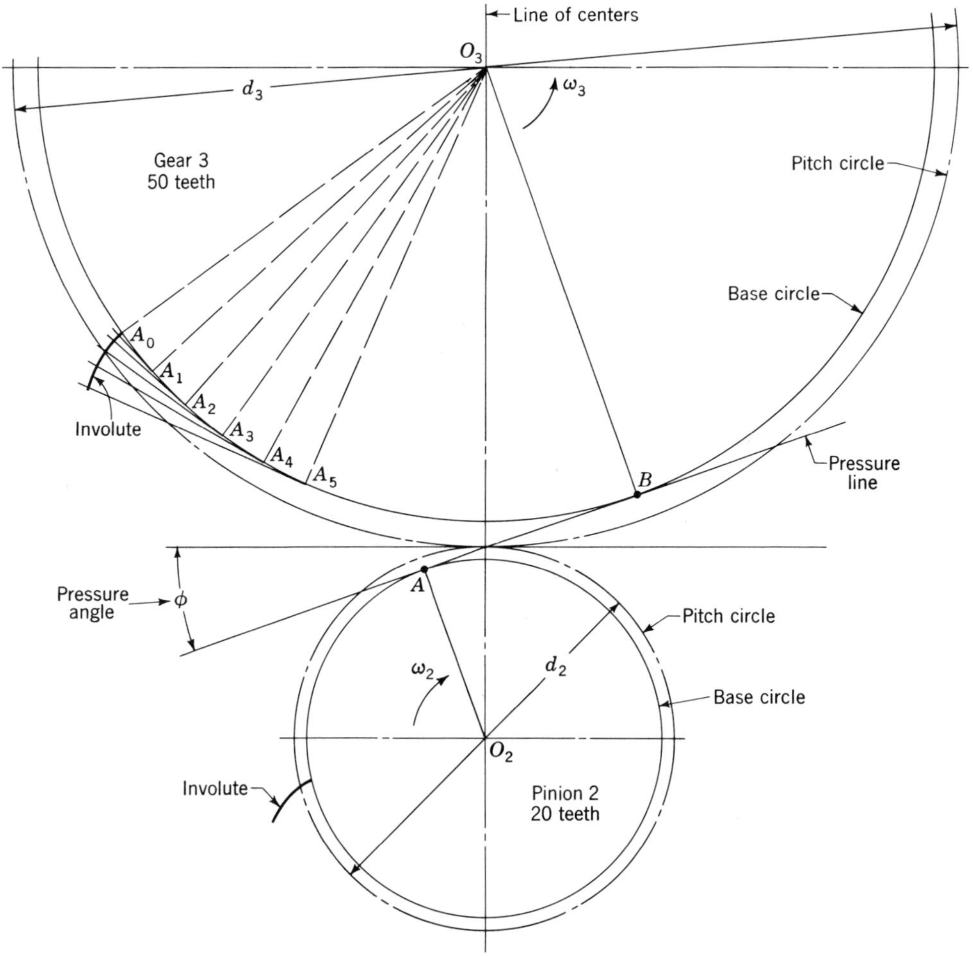

Fig. 8-6 Layout of a pair of spur gears.

The second point is obtained by laying the distance A_0A_1 on the perpendicular through A_1. The next point is found by laying off twice A_0A_1 on the perpendicular through A_2, and so on. The curve constructed through these points is the involute. The involute for the pinion is constructed in the same manner on the pinion base circle.

Step 5. Using cardboard or, preferably, a sheet of clear plastic, cut a template for each involute and mark on it the corresponding center of each gear. These templates are then used to draw the involute portion of each tooth. They can be turned over to draw the opposite side of the tooth. In some cases it may be desirable to construct a template for the entire tooth.

Step 6. Calculate the circular pitch. The width of tooth and width of space are constructed equal to half the circular pitch. Mark these distances off on the pitch circles. From Eq. (8-3),

$$p_c = \frac{\pi}{P} = \frac{\pi}{10} = 0.31412 \text{ in.}$$

so the width of tooth and space is 0.31412/2, or 0.15706 in. These points are marked off on the pitch circles of Fig. 8-7.

Step 7. Draw the addendum and dedendum circles for the pinion and gear (Fig. 8-7). From Table 8-1, the addendum is

$$a = \frac{1}{P} = \frac{1}{10} = 0.10 \text{ in.}$$

The dedendum is

$$b = \frac{1.25}{P} = \frac{1.25}{10} = 0.125 \text{ in.}$$

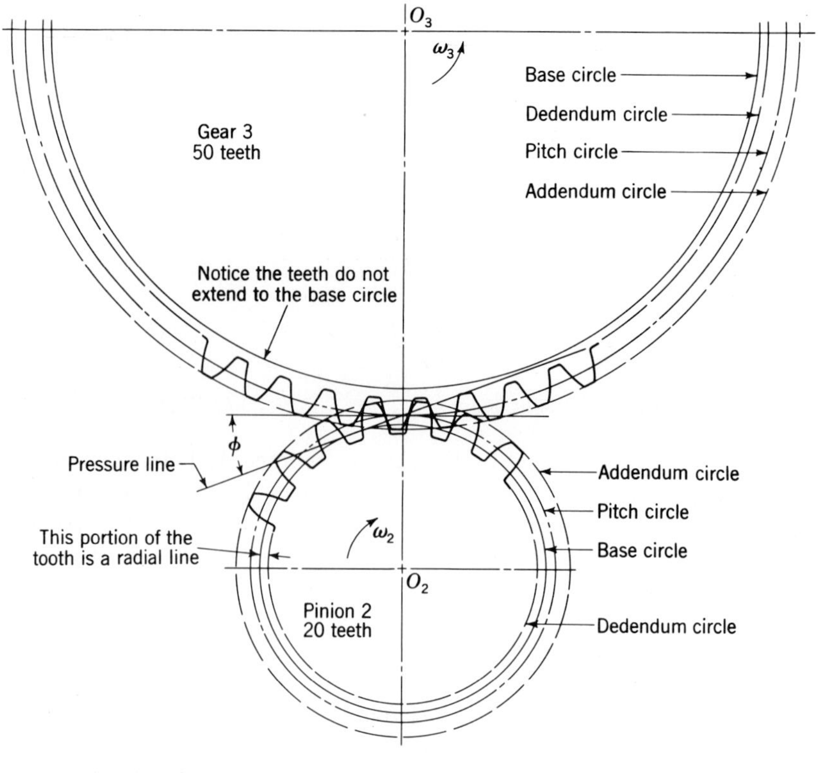

Fig. 8-7 Layout of a pair of spur gears (*continued*).

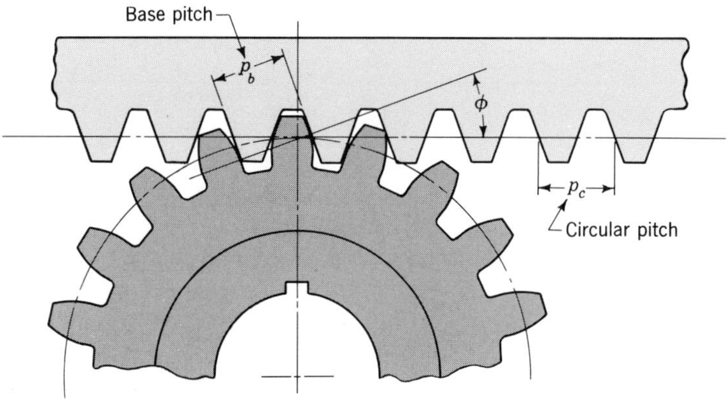

Fig. 8-8 Involute pinion and rack.

Step 8. Now draw the involute portion of the profiles of the teeth on the pinion and gear (Fig. 8-7). The portion of the tooth between the clearance and dedendum circles may be used for a fillet. Notice that the base circle of the gear is smaller than the dedendum circle, and so, except for the fillet, the profile of the tooth is all involute. On the other hand, the pinion base-circle radius is larger than the radius of the dedendum circle. This means that the portion of the tooth below the base circle is not involute. For the present we shall construct this portion as a radial line except for the fillet. This completes the construction.

Involute rack We may imagine a *rack* as a spur gear having an infinitely large pitch diameter. Therefore the rack has an infinite number of teeth, and the base circle, too, is located an infinite distance from the pitch point. For involute teeth the sides become straight lines making an angle to the line of centers equal to the pressure angle. Figure 8-8 shows an involute rack in mesh with the pinion of the previous example.

Base pitch Corresponding sides of involute teeth are parallel curves; the *base pitch* is the constant and fundamental distance between them along a common normal (Fig. 8-8). The base pitch and the circular pitch are related as follows:

$$\frac{p_b}{p_c} = \cos \phi \qquad (8\text{-}5)$$

where p_b is the base pitch in inches.

Internal gear Figure 8-9 depicts the pinion of the preceding example mating with an *internal*, or *annular*, gear. With internal contact both centers are on the same side of the pitch point. Thus the positions

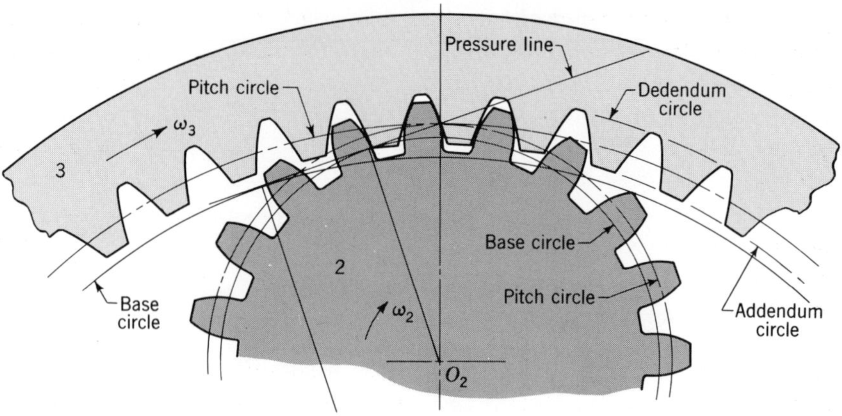

Fig. 8-9 Internal gear and pinion.

of the addendum and dedendum circles with respect to the pitch circle are reversed. As shown in Fig. 8-9, the addendum circle of the internal gear lies *inside* the pitch circle; similarly, the dedendum circle lies outside the pitch circle.

It is also seen from Fig. 8-9 that the base circle lies inside the pitch circle near the addendum circle.

8-5 ARC OF ACTION

It will be profitable, at this stage, to trace the action of a pair of teeth from the time they begin contact until they leave contact. In Fig. 8-10 we have reproduced the pitch circles of the gears of the preceding section. Let the pinion be the driver, and let it be rotating clockwise. Our problem is to locate the initial and final points of contact as a pair of mating teeth go through the meshing cycle.

To solve the problem we construct the pressure line and the addendum and dedendum circles of both gears. We have seen, for involute teeth, that contact must take place along the pressure line. This explains why the pressure line is also called the line of action. As shown in the figure, contact begins where the addendum circle of the driven gear crosses the line of action. Thus initial contact is on the tip of the gear tooth and on the flank of the pinion tooth.

As the pinion tooth drives the gear tooth, both approach the pitch point. Near the pitch point, contact *slides up* the flank of the pinion tooth and *down* the face of the gear tooth. At the pitch point, contact exists at the pitch circles. Note that the motion is pure rolling only at the pitch point.

As the teeth recede from the pitch point, the point of contact is traveling in the same direction as before. Contact *slides up* the face of the pinion tooth and *down* the flank of the gear tooth. The last point of contact occurs at the tip of the pinion tooth and the flank of the gear tooth. This is located at the intersection of the line of action and the addendum circle of the pinion.

The *approach* phase of the action is the period between initial contact and the pitch point. During the approach phase, contact is sliding down the face of the gear tooth toward the pitch circle. This kind of action may be likened to *pushing* a stick over a surface.

At the pitch point there is no sliding. The action is pure rolling.

The *recess* phase of the action is the period between contact at the pitch point and final contact. During the recess phase, contact is sliding down the flank of the gear tooth, away from the pitch circle. This kind of action may be likened to *pulling* a stick over a surface.

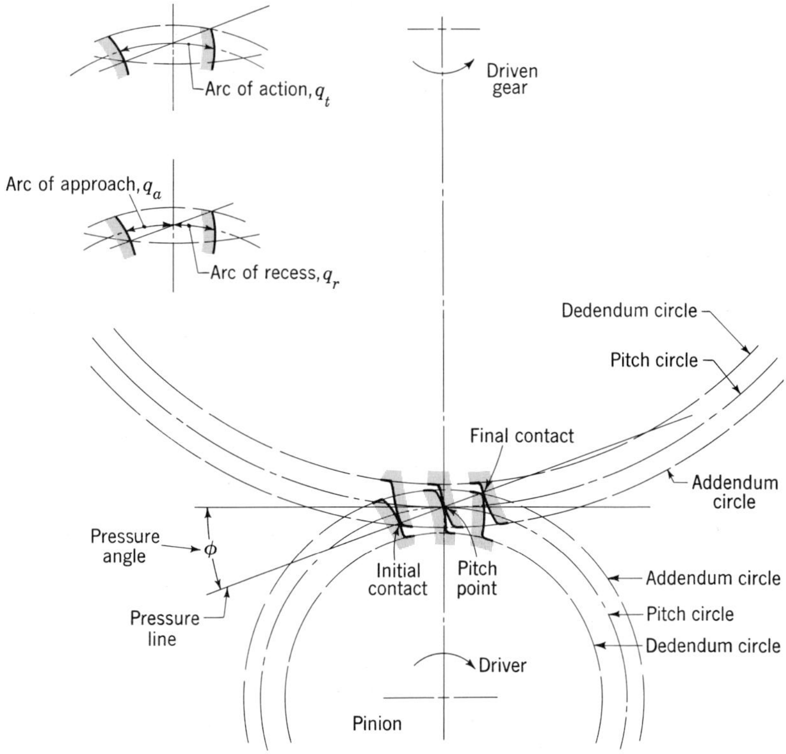

Fig. 8-10 Approach and recess phases of gear-tooth action.

Pinion and gear-tooth profiles are now constructed through the initial and final points of contact in Fig. 8-10. The intersection of these profiles with the pitch circles defines the arcs of action, approach, and recess.

The *arc of action* q_t is the arc of the pitch circle through which a tooth profile moves from the beginning to the end of contact with a mating profile.

The *arc of approach* q_a is the arc of the pitch circle through which a tooth profile moves from its beginning of contact until the point of contact arrives at the pitch point.

The *arc of recess* q_r is the arc of the pitch circle through which a tooth profile moves from contact at the pitch point until contact ends.

A definition useful in the evaluation of different gearsets is the contact ratio. The *contact ratio* m_c is defined as the ratio of the arc of action to the circular pitch. In equation form it may be written

$$m_c = \frac{q_t}{p_c} \tag{8-6}$$

The contact ratio may be considered as giving the average number of teeth in contact. Thus the opportunity exists, for a gearset with a high contact ratio, for transmitting more power because of a greater division of load between teeth. It is also probable that high contact ratios result in less noise when the gears are operated at high speeds.

8-6 THE FORMING OF GEAR TEETH

There are a very large number of ways of forming the teeth of gears, such as *sand casting, shell molding, investment casting, permanent-mold casting, die casting,* or *centrifugal casting.* They can be formed by using the *powder-metallurgy process;* or, by using *extrusion,* a single bar of aluminum may be formed and then sliced into gears. Gears which carry large loads in comparison with their size are usually made of steel and are cut with either *form cutters* or *generating cutters.* In form cutting, the tooth space takes the exact shape of the cutter. In generating, a tool having a shape different from the tooth profile is moved relative to the gear blank so as to obtain the proper tooth shape.

Probably the oldest method of cutting gear teeth is *milling.* A form milling cutter corresponding to the shape of the tooth space is used to cut one tooth space at a time, after which the gear is indexed through one circular pitch to the next position. Theoretically, with this method, a different cutter is required for each gear to be cut because, for example, the shape of the space in a 25-tooth gear is different from the space in, say, a 24-tooth gear. Actually, the change in space is not too great, and eight cutters can be used to cut any gear in the range of 12 teeth to a rack with reasonable accuracy. Of course, a separate set of cutters is required for each pitch.

Shaping is a highly favored method of generating gear teeth. The cutting tool may be either a rack cutter or a pinion cutter. The operation may best be explained by reference to Fig. 8-11. Here the reciprocating rack cutter is first fed into the blank until the pitch circles are tangent. Then, after each cutting stroke, the gear blank and cutter roll slightly on their pitch circles. When the blank and cutter have rolled a distance equal to the circular pitch, the cutter is returned to the starting point, and the process is continued until all the teeth have been cut. Figure 8-12 shows the shaping of a helical gear using a pinion shaper.

Hobbing is a method of generating gear teeth which is quite similar to generating them with a rack cutter. The hob is a cylindrical cutter with one or more helical threads quite like a screw-thread tap, and has straight sides as in a rack. The hob and the blank are rotated continuously at the proper angular-velocity ratio, and the hob is then fed slowly across the face of the blank from one end of the teeth to the other.

Following the cutting process, *grinding*, *lapping*, *shaving*, and *burnishing* are often used as final finishing processes when tooth profiles of very good accuracy and surface finish are desired.

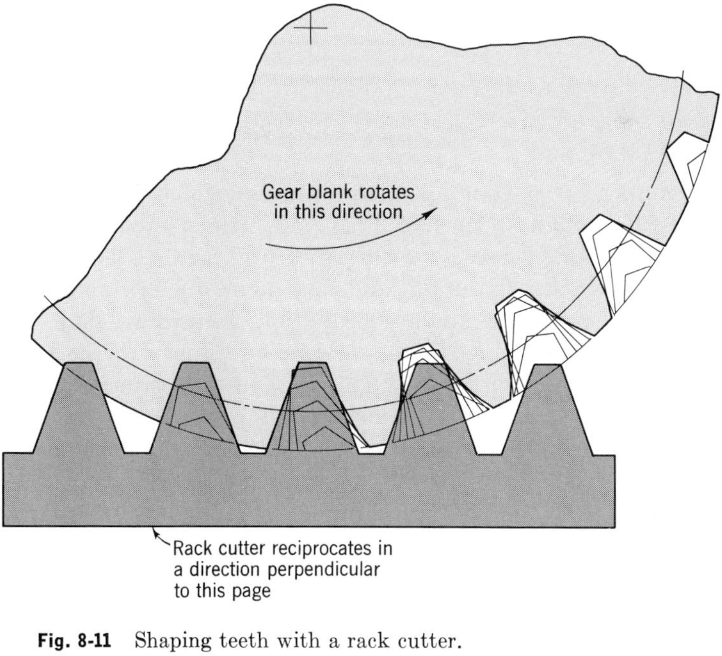

Fig. 8-11 Shaping teeth with a rack cutter.

Fig. 8-12 Shaping a helical gear. (*Fellows Gear Shaper Co., Springfield, Vt.*)

8-7 INTERFERENCE

The contact of portions of tooth profiles which are not conjugate is called *interference*. Consider Fig. 8-13. Illustrated are two 16-tooth, $14\frac{1}{2}°$-pressure-angle gears, with full-depth teeth. The driver, gear 2, turns clockwise. The initial and final points of contact are designated A and B, respectively, and are located on the pressure line. Now notice that the points of tangency of the pressure line with the base circles C and D are located *inside* of points A and B. Interference is present.

The interference is explained as follows. Contact begins when the tip of the driven tooth contacts the flank of the driving tooth. In this case the flank of the driving tooth first makes contact with the driven tooth at point A, and this occurs *before* the involute portion of the driving tooth comes within range. In other words, contact is occurring below the base circle of gear 2 on the *noninvolute* portion of the flank. The actual effect is that the involute tip or face of the driven gear tends to dig out the noninvolute flank of the driver.

In this example the same effect occurs as the teeth leave contact. Contact should end at point D or before. Since it does not end until point B, the effect is for the tip of the driving tooth to dig out, or interfere with, the flank of the driven tooth.

When gear teeth are produced by a generation process, interference is automatically eliminated because the cutting tool removes the interfering portion of the flank. This effect is called *undercutting;* if undercutting is at all pronounced, then the undercut tooth is considerably weakened. Thus the effect of eliminating interference by a generation process is merely to substitute another problem for the original one.

The importance of the problem of teeth which have been weakened by undercutting cannot be overemphasized. Of course, interference can be eliminated by using more teeth on the gears. However, if the gears are to transmit a given amount of power, then more teeth can be used

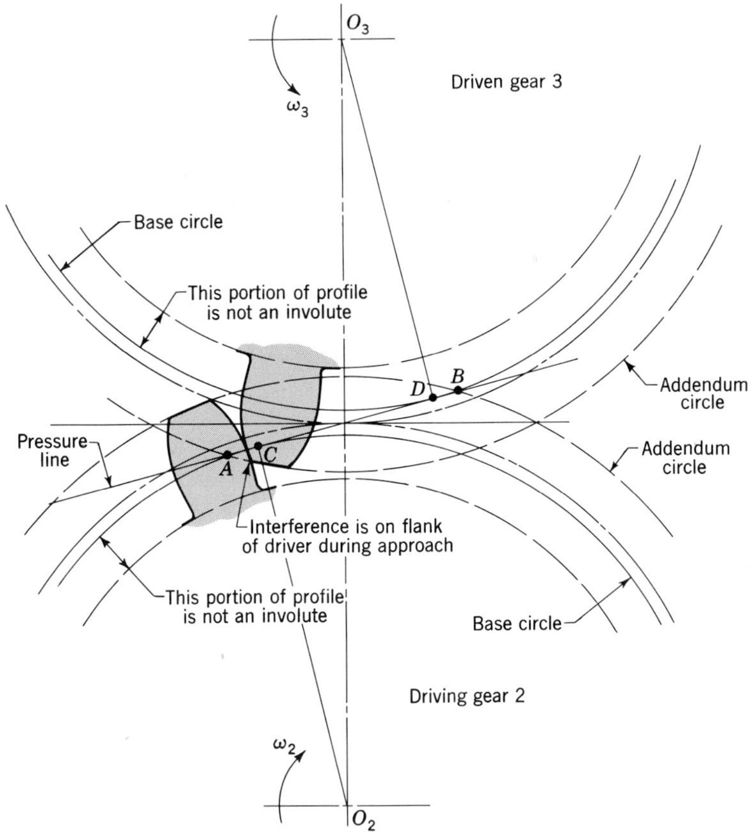

Fig. 8-13 Interference in the action of gear teeth.

only by increasing the pitch diameter. This makes the gears larger, which is seldom desirable, and it also increases the pitch-line velocity. This increased pitch-line velocity makes the gears noisier and reduces the power transmission somewhat, although not in direct ratio. In general, however, the use of more teeth to eliminate interference or undercutting is seldom an acceptable solution.

Another method of reducing the interference and the resulting degree of undercutting is to employ a larger pressure angle. The larger pressure angle creates a smaller base circle so that a greater portion of the tooth profile has an involute shape. In effect this means that fewer teeth may be employed, and as a result, gears with a large pressure angle are smaller.

8-8 SYNTHESIS OF SPUR-GEAR TEETH

In order to appreciate the full meaning of the effect of the uniform-velocity-ratio requirement upon the shape of the tooth profile, it will be interesting to specify a tooth profile for the driving gear and then to find the shape of the correct mating profile. This is called *synthesis*. The problem is, given a tooth shape, what is the shape of the mating tooth?

The solution is initiated by establishing the moving coordinate system xy on gear 2 with origin at O_2. The system is to rotate with gear 2 at an angular velocity represented by ω_2 (Fig. 8-14). It will be convenient to fix the x axis through the centerline of one of the teeth, as shown in the figure. Let the equation of the tooth profile on this gear be

$$y = F(x) \qquad (a)$$

Now, the law of gearing states that the normal to the tooth profile, at the point of contact, must always pass through the pitch point. The pitch point is identified as P in Fig. 8-14, and the normal is the line AB,

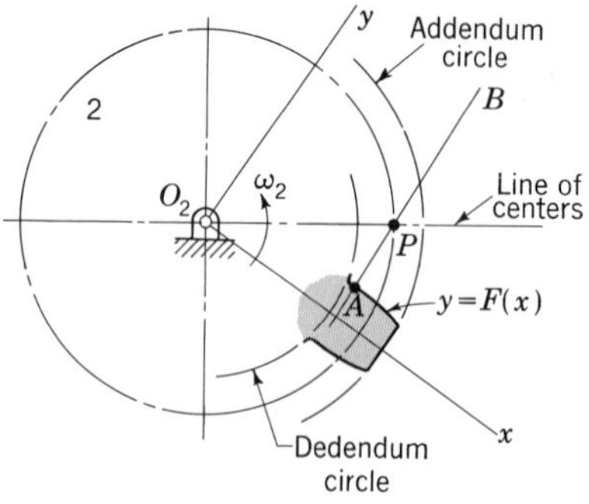

Fig. 8-14

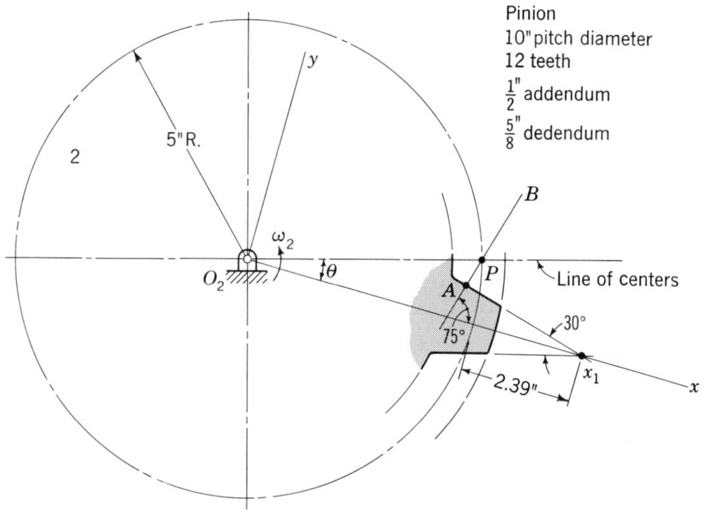

Fig. 8-15

normal to the tooth profile at A. Thus the mating tooth must be in contact at point A. So the simultaneous solution of the equation of line AB with Eq. (a) should give the coordinates in the xy system of all the points of contact as the x axis rotates.

Let us see how this works in an actual example. We shall choose a 10-in., 12-tooth pinion to drive a 20-in., 24-tooth gear. A trapezoidal tooth having a 30° included angle is selected for the pinion, as shown in Fig. 8-15. The circular pitch is determined first. From Eq. (8-2),

$$p_c = \frac{\pi d}{N} = \frac{\pi(2)(5)}{12} = 2.62 \text{ in.} \tag{b}$$

Now, from the geometry of the figure, the intercept of the tooth profile on the x axis is

$$x_1 = 7.39 \text{ in.} \tag{c}$$

The slope of the tooth profile is $-\tan 15°$, so the equation of the tooth becomes

$$y = m(x - x_1)$$

which is the equation of a straight line intersecting the x axis at x_1 and at a slope of m. This becomes

$$\begin{aligned} y &= -\tan 15°(x - 7.39) \\ &= 1.98 - 0.268x \end{aligned} \tag{d}$$

which is the equation of a straight line intersecting the x axis at $x_1 = 7.39$ and with a slope of $-\tan 15°$.

The angular location of the x axis is designated as θ with respect to the fixed line of centers. The pitch point P, too, is fixed, and so its coordinates, with respect to the moving xy system, are

$$x_P = 5 \cos \theta \qquad y_P = 5 \sin \theta \tag{e}$$

The slope of the normal to the tooth profile through P is tan 75°. Therefore the equation of line AB is

$$y - y_P = m(x - x_P) \tag{f}$$

where m is the slope. Substituting Eqs. (e) and the slope, and rearranging, produces

$$y = x \tan 75 - 5 \tan 75 \cos \theta + 5 \sin \theta$$

or

$$y = 3.73x - 18.66 \cos \theta + 5 \sin \theta \tag{g}$$

Thus Eq. (d) is the equation of the tooth profile, and Eq. (g) is that of the normal to the profile through the pitch point. As indicated above, a simultaneous solution of these two equations gives the coordinates of the point of contact. Solving Eqs. (d) and (g), then, gives

$$x = 0.495 + 4.66 \cos \theta - 1.25 \sin \theta \tag{h}$$

$$y = 1.85 - 1.25 \cos \theta + 0.335 \sin \theta \tag{i}$$

The coordinates of the point of contact A in the xy system may be easily obtained by substituting various values of θ in Eqs. (h) and (i). These points must be common to both profiles, and so they define the shape of the mating tooth. However, they have no value in defining the unknown profile unless they are referred to a coordinate system which moves with the tooth whose profile is to be found. Thus it is necessary to define a second coordinate system containing the unknown tooth and moving with that tooth. This is done in Fig. 8-16.

In Fig. 8-16 the coordinate system containing gear 3 is designated as the $x'y'$ system, and it is positioned so that the x' axis coincides with the centerline of the tooth space. As shown, the x axis intersects its pitch circle at C, the x' axis intersects its pitch circle at D, and the arc distances PC and PD are equal. The pitch circles have pure rolling contact so that the angular position of the $x'y'$ system is dependent upon the position of the xy system.

Let us designate the radius of each gear by r_2 and r_3; then the center

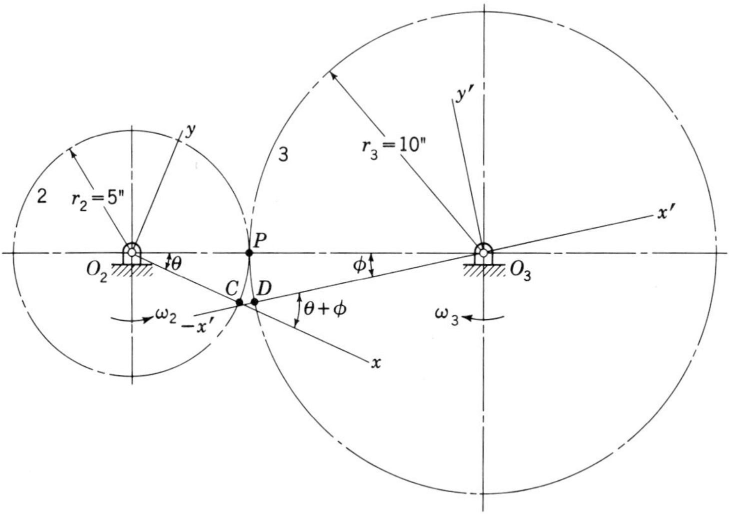

Fig. 8-16

distance is $r_2 + r_3$. Since the pitch circles roll without slipping,

$$r_2\theta = r_3\phi$$

or

$$\phi = \frac{r_2}{r_3}\theta \tag{j}$$

so that

$$\theta + \phi = \frac{r_2 + r_3}{r_3}\theta \tag{k}$$

And for this example,

$$\theta + \phi = \frac{5 + 10}{10}\theta = 1.5\theta \tag{l}$$

From Fig. 8-16, the coordinates of O_3 in the xy system are

$$x = (r_2 + r_3)\cos\theta \qquad y = (r_2 + r_3)\sin\theta \tag{m}$$

or, substituting values for r_2 and r_3,

$$x_{O_3} = 15\cos\theta \tag{n}$$

$$y_{O_3} = 15\sin\theta \tag{o}$$

Equations (n) and (o) are the coordinates of the origin of the $x'y'$ system relative to the xy system, and the angle $(\theta + \phi)$ is the direction, also relative to the xy system.

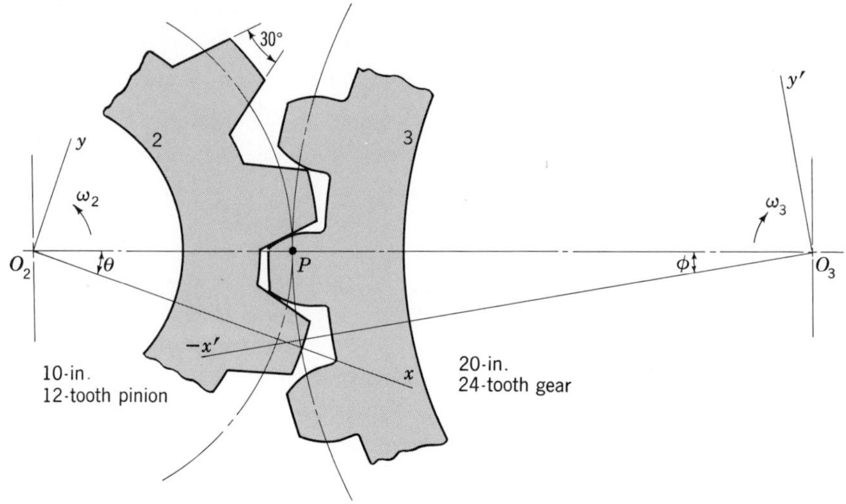

Fig. 8-17 Synthesis of gear teeth.

To obtain the coordinates of the mating-tooth profile we must transform the coordinates of point A [given by Eqs. (h) and (i)] to the new $x'y'$ axes. This constitutes a translation of the origin from O_2 to O_3 and a rotation through the angle $(\theta + \phi)$. The transformation equations may be found in any mathematics reference book and are:

$$x' = (y - y_{O_3}) \sin (\theta + \phi) + (x - x_{O_3}) \cos (\theta + \phi) \tag{p}$$

$$y' = (y - y_{O_3}) \cos (\theta + \phi) - (x - x_{O_3}) \sin (\theta + \phi) \tag{q}$$

Substituting values from Eqs. (l), (n), and (o),

$$x' = (y - 15 \sin \theta) \sin 1.5\theta + (x - 15 \cos \theta) \cos 1.5\theta \tag{r}$$

$$y' = (y - 15 \sin \theta) \cos 1.5\theta - (x - 15 \cos \theta) \sin 1.5\theta \tag{s}$$

Then, replacing x and y by Eqs. (h) and (i) and simplifying,

$$x' = (1.85 - 1.25 \cos \theta - 14.66 \sin \theta) \sin 1.5\theta + (0.495 - 10.34 \cos \theta - 1.25 \sin \theta) \cos 1.5\theta \tag{t}$$

$$y' = (1.85 - 1.25 \cos \theta - 14.66 \sin \theta) \cos 1.5\theta - (0.495 - 10.34 \cos \theta - 1.25 \sin \theta) \sin 1.5\theta \tag{u}$$

The coordinates x' and y' of the mating tooth are found by substituting values of θ in Eqs. (t) and (u); the resulting tooth form is illustrated in Fig. 8-17.

Referring again to Fig. 8-16, note that the inclination of the pressure line AB depends upon θ. Since θ varies as the teeth pass through contact, the inclination of the pressure line with the line of centers also varies. This means that, in this example, the pressure angle is *not* constant. In fact, the involute profile is the only one which gives a constant pressure angle.

The method of tooth synthesis illustrated above is quite general and may be used to find the shape of any tooth when the shape of its mate is specified. For example, Beggs[1] uses it to find the mating profile when the given tooth is a pin of circular cross section.

8-9 CYCLOIDAL PROPERTIES

The cycloidal tooth profile was extensively used for gear manufacture about a century ago because it is easy to form by casting. It is seldom used today, for reasons which we shall discover in this section.

The construction of a cycloidal tooth profile is shown in Fig. 8-18. Two generating circles, shown in dashed lines, roll on the inside and outside, respectively, of the pitch circle and generate the hypocycloidal flank and the epicycloidal face of the gear tooth. The same two circles are also used to generate the profile of the mating pinion teeth, but now

[1] Joseph Stiles Beggs, "Mechanism," p. 72, McGraw-Hill Book Company, New York, 1955.

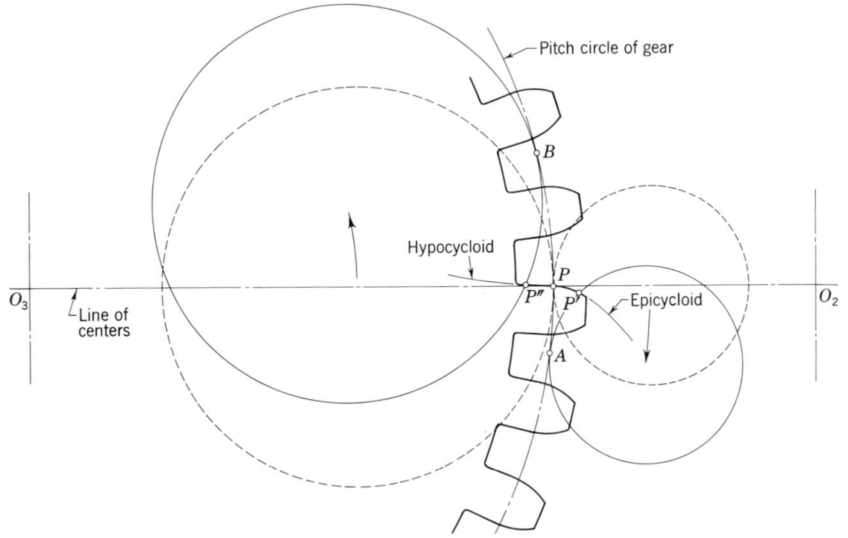

Fig. 8-18 Generating cycloidal teeth on a gear.

the role of the generating circles is reversed. The circle which generated the flank of the gear tooth now generates the epicycloidal face of the pinion tooth. And, similarly, the circle which generated the face of the gear tooth now generates the flank of the pinion tooth.

Notice that in generating one side of a tooth, the two generating circles roll in opposite directions.

In Fig. 8-19 the pinion and gear, produced by this method, have been placed in mesh. Consider the pinion as the driver and let it rotate counterclockwise. The two pitch circles are tangent at the pitch point P, and they roll upon each other without slipping. The two generating circles have stationary centers at A and B and also roll with the moving pitch circles. A point of contact C exists at the intersection of the generating circle with center at A and the two contacting profiles. Let C_2 be a point on the flank of the pinion tooth and C_3 a point on the face of the gear tooth. As the two pitch circles and the generating circle roll upon one another, a point on the generating circle simultaneously traces out the face of the tooth upon the moving gear and the flank of the tooth on the moving pinion. So point C is an instantaneous position of this moving point, and the arc CP, of the generating circle, is its path. Initial contact will occur at D where the addendum circle of the

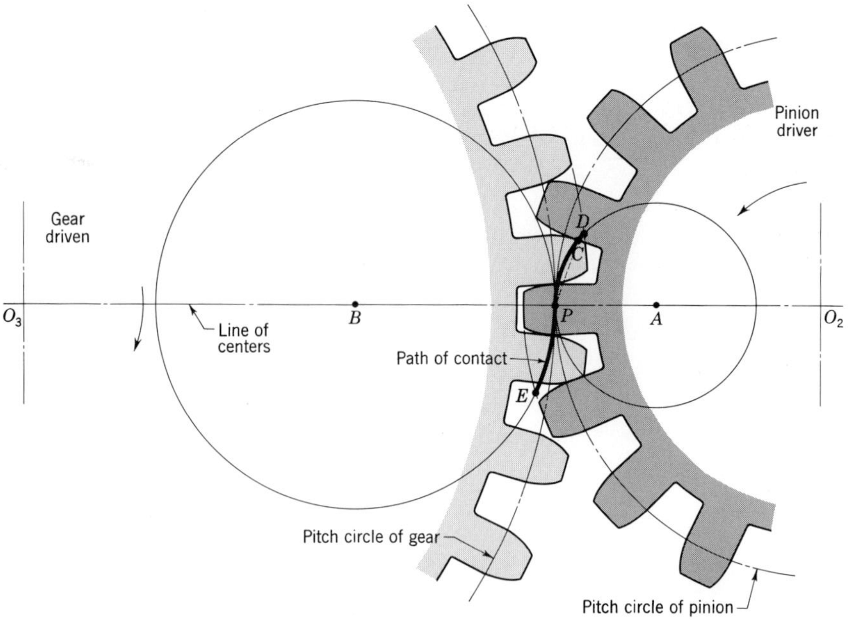

Fig. 8-19

driven gear cuts the generating circle. Thus the complete approach path is the arc DP. During approach, only the portions of the tooth profiles generated by the circle with center at A have been used.

Returning to Fig. 8-19, notice that the pitch point P is the instantaneous center of rotation of the generating circle no matter which of the two pitch circles it is considered to be rolling upon. Thus P is the instantaneous center of rotation of point C, on the generating circle, and consequently the line PC is normal to both tooth profiles. It is clear that as both gears rotate this will always be so. Thus cycloidal gearing satisfies the law of gearing in that the normal to the tooth profile always passes through the pitch point.

Notice, however, that line PC, which is the pressure line, will not have a constant inclination. As the point of contact nears the pitch point, the pressure line approaches a perpendicularity with the line of centers.

During recess action the generating circle with center at B takes over. Contact occurs on the face of the pinion tooth and the flank of the gear tooth. Notice that each of these profiles is generated by the circle with center at B. During recess the pressure line swings back toward an inclination similar to that during approach. The final point of contact is at E, where the addendum circle of the pinion intersects the generating circle. Thus the path of contact during recess is the arc distance PE.

The varying pressure angle of cycloidal teeth results in additional noise and wear, and also in changing bearing reactions at the shaft supports. Also, the double curvature which frequently occurs introduces problems in the cutting of teeth which are not present with the involute form. In order to run properly, cycloidal gears must be operated at exactly the correct center distance, otherwise the contacting portions of the profiles will not be conjugate. Since deflections due to the transmission of load are bound to occur, it would be virtually impossible to maintain correct center distance under all loading conditions. Thus for most existing applications it appears that the cycloidal tooth form has little to offer over the involute profile.

8-10 VARYING THE CENTER DISTANCE

Figure 8-20*a* illustrates a pair of meshing gears having involute teeth at a 20° pressure angle. Both sides of the teeth are in contact and so the center distance O_2O_3 cannot be shortened without jamming or deforming the teeth.

In Fig. 8-20*b* the same pair of gears have been separated by increasing the center distance slightly. Clearance or *backlash* now exists

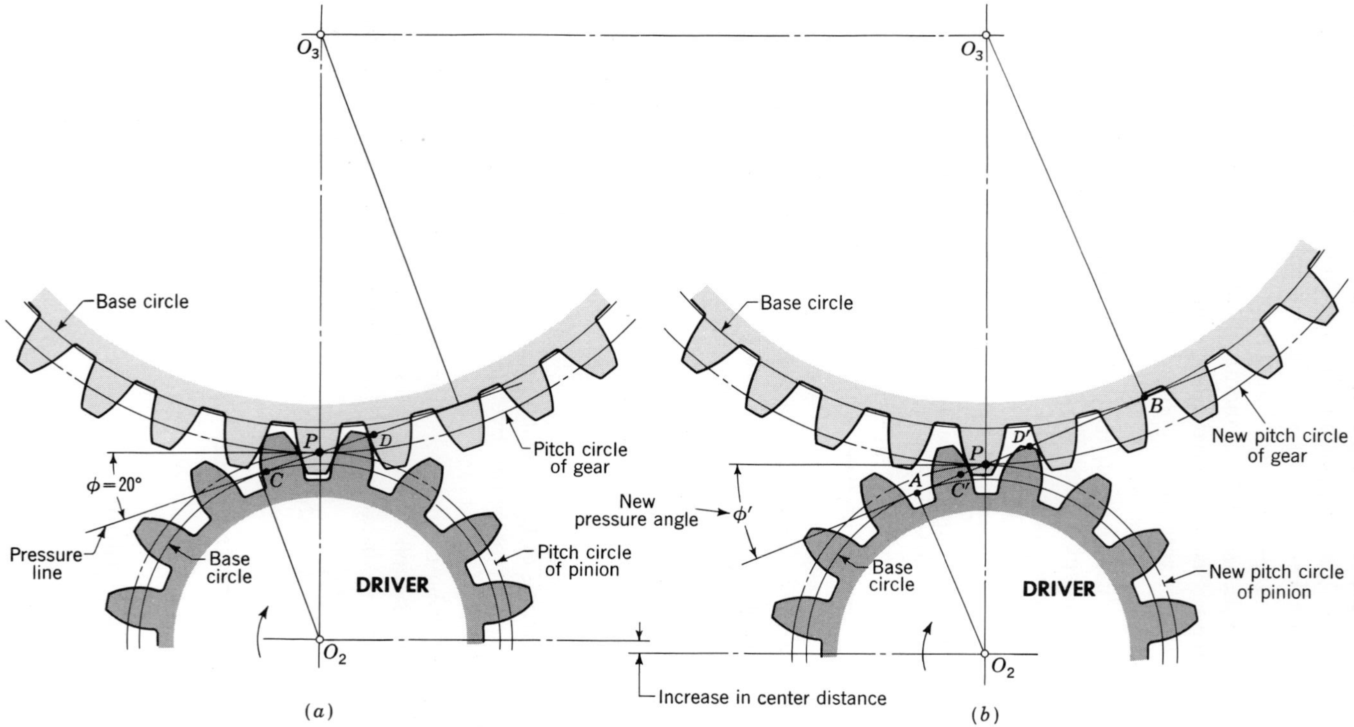

Fig. 8-20 Effect of increased center distance upon the action of involute gearing. (*a*) Mounting at normal center distance; (*b*) mounting at increased center distance.

between the teeth, as shown. When the center distance is increased, new pitch circles having larger radii are created because the pitch circles are always tangent to each other. However, the base circles are a constant and fundamental characteristic of the gears. This means that an increase in center distance changes the inclination of the line of action and results in a larger pressure angle. Notice, too, that a tracing point on the new pressure line will still generate the same involutes as in Fig. 8-20*a*, the normal to the tooth profiles still passes through the same pitch point, and hence the law of gearing is satisfied for any center distance.

To see that the velocity ratio has not changed in magnitude, we observe that the triangles O_2AP and O_3BP are similar. Also, since O_2A and O_3B are fixed distances and do not change with varying center distances, the ratio of the pitch radii, O_2P and O_3P, will remain fixed too.

A second effect of increasing the center distance observable in Fig. 8-20 is the shortening of the path of contact. The original path of contact CD has been shortened to $C'D'$. The contact ratio [Eq. (8-6)] can be defined as the ratio of the length of the path of contact to the base pitch. The limiting value of this ratio is unity; otherwise, periods would occur in which there would be no contact at all. Thus the center distance cannot be larger than that corresponding to a contact ratio of unity.

It is interesting to conclude, from the preceding discussion, that two gears of slightly different tooth numbers may be mounted upon the same axis, though not fixed to each other nor to the shaft, and mated with the same pinion or rack, provided the limitations discussed above are not exceeded.

8-11 INVOLUTOMETRY

The study of the geometry of the involute is called *involutometry*.[1] In Fig. 8-21 a base circle with center at O is used to generate the involute BC. AT is the generating line, ρ the instantaneous radius of curvature of the involute, and r the radius to any point T on the curve. If we designate the radius of the base circle as r_b, then the generating line AT has the same length as the arc AB and so

$$\rho = r_b(\alpha + \varphi) \tag{a}$$

where α is the angle between radius vectors defining the origin of the involute and any point, such as T, on the involute; and φ is the angle between radius vectors defining any point T on the involute and the origin A at the base circle of the corresponding generating line. Since OTA

[1] Pronounced ĭn′vōl ū tŏ′m e trē.

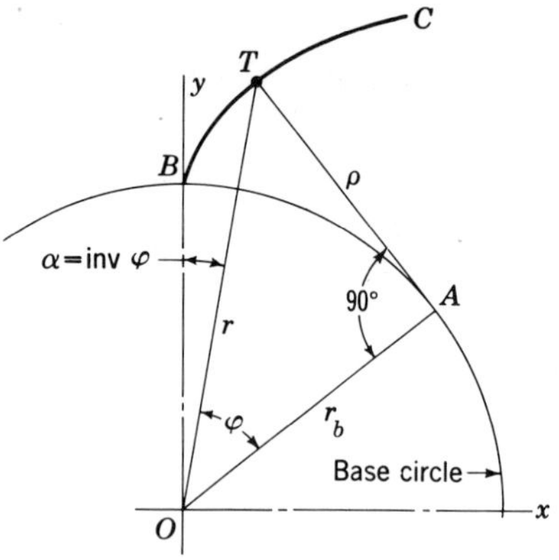

Fig. 8-21

is a right triangle,

$$\rho = r_b \tan \varphi \tag{8-7}$$

Solving Eqs. (*a*) and (8-7) simultaneously to eliminate ρ gives

$$\alpha = \tan \varphi - \varphi$$

which may be written

$$\text{inv } \varphi = \tan \varphi - \varphi \tag{8-8}$$

and defines the involute function. The angle φ in this equation is the variable involute pressure angle, and it must be specified in radians. If φ is known, then inv φ may be readily determined; but tables must be used to find the pressure angle when inv φ is given and φ is to be found (see Appendix).

Referring again to Fig. 8-21, we see that

$$r = \frac{r_b}{\cos \varphi} \tag{8-9}$$

To illustrate the use of the relations obtained above, the tooth dimensions of Fig. 8-22 will be determined. Here the portion of the tooth profile extending above the base circle has been drawn, and the arc thickness of the tooth, t_p, at the pitch circle (point A) is given. The problem is to determine the tooth thickness at any other point, say T. The various quantities shown in Fig. 8-22 are identified as follows:

r_b = radius of base circle, in.
r_p = radius of pitch circle, in.
r = radius at which tooth thickness is to be determined, in.
t_p = arc tooth thickness at pitch circle, in.

t = arc thickness to be determined, in.
ϕ = pressure angle corresponding to pitch radius r_p
φ = pressure angle corresponding to any point T
β_p = angular half-tooth thickness at pitch circle
β = angular half-tooth thickness at any point T

The half-tooth thicknesses at points A and T are

$$\frac{t_p}{2} = \beta_p r_p \qquad \frac{t}{2} = \beta r \tag{b}$$

so that

$$\beta_p = \frac{t_p}{2r_p} \qquad \beta = \frac{t}{2r} \tag{c}$$

Now we may write

$$\begin{aligned} \text{inv}\ \varphi - \text{inv}\ \phi &= \beta_p - \beta \\ &= \frac{t_p}{2r_p} - \frac{t}{2r} \end{aligned} \tag{d}$$

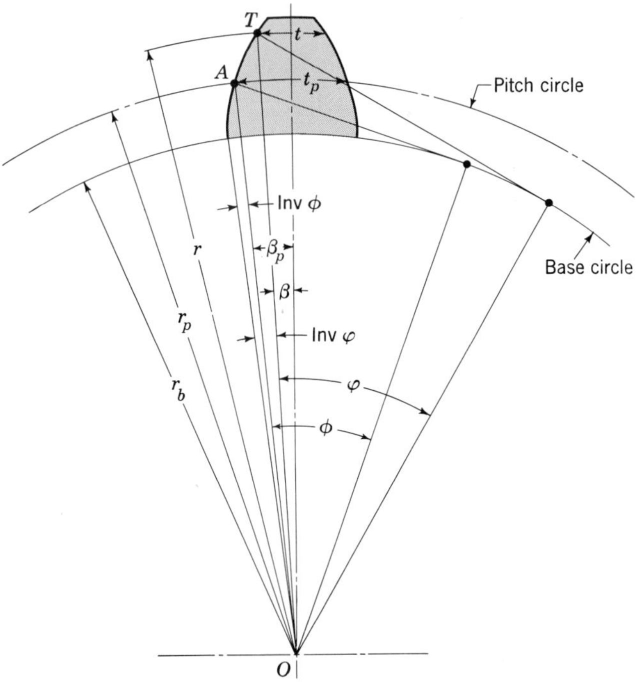

Fig. 8-22

The tooth thickness corresponding to any point T is obtained by solving Eq. (d) for t:

$$t = 2r\left(\frac{t_p}{2r_p} + \text{inv}\,\phi - \text{inv}\,\varphi\right) \tag{8-10}$$

Example 8-1 A gear has 20° teeth cut full depth, a diametral pitch of 2 teeth per inch, and 22 teeth. (a) Calculate the radius of the base circle. (b) Find the thickness of the tooth at the base circle, and also at the addendum circle.

Solution By employing Sec. 8-13 and the equations of Sec. 8-1, the following quantities are determined: addendum $a = 0.500$ in., dedendum $b = 0.5785$ in., pitch radius $r_p = 5.500$ in., circular pitch $p_c = 1.571$ in. The radius of the base circle is obtained from Eq. (8-9).

$$r_b = r_p \cos\phi = (5.500)\cos 20° = 5.168 \text{ in.}$$

The thickness of the tooth at the pitch circle is

$$t_p = \frac{p_c}{2} = \frac{1.571}{2} = 0.7854 \text{ in.}$$

Converting the tooth pressure angle 20° to radians gives $\phi = 0.349$ rad. Then

$$\text{inv}\,\phi = \tan 0.349 - 0.349 - 0.015 \text{ rad}$$

At the base circle $\varphi_b = 0$, so inv $\varphi_b = 0$. By utilizing Eq. (8-10), the tooth thickness at the base circle is

$$\begin{aligned} t_b &= 2r_b\left(\frac{t_p}{2r_p} + \text{inv}\,\phi - \text{inv}\,\varphi_b\right) \\ &= (2)(5.168)\left[\frac{0.7854}{(2)(5.500)} + 0.015 - 0\right] \\ &= 0.886 \text{ in.} \end{aligned}$$

The radius of the addendum circle is $r_a = 6.000$ in. The involute pressure angle corresponding to this radius is, from Eq. (8-9),

$$\varphi_a = \cos^{-1}\frac{r_b}{r_a} = \cos^{-1}\frac{5.168}{6.000} = 0.532 \text{ rad}$$

Thus

$$\text{inv}\,\varphi_a = \tan 0.532 - 0.532 = 0.058 \text{ rad}$$

and Eq. (8-10) gives the tooth thickness at the addendum circle as

$$\begin{aligned} t_a &= 2r_a\left(\frac{t_p}{2r_p} + \text{inv}\,\phi - \text{inv}\,\varphi_a\right) \\ &= (2)(6.000)\left[\frac{0.7854}{(2)(5.500)} + 0.015 - 0.058\right] \\ &= 0.341 \text{ in.} \end{aligned}$$

8-12 CONTACT RATIO

Equation (8-6) defines the contact ratio m_c as the ratio of the arc of action, measured on the pitch circle, to the circular pitch. But we have seen that the diameters of pitch circles may vary since they depend upon the center distance. Therefore, a more fundamental definition can be obtained by use of the base circle because its diameter is fixed.

In Fig. 8-23, with gear 2 as the driver, contact begins at point B, where the addendum circle of the driven gear crosses the line of action, and ends at C, where the addendum circle of the driver crosses the line of action. The length of the path of action or contact is

$$u = u_a + u_r \tag{a}$$

where the subscripts a and r designate the approach and recess phases, respectively. During approach, contact occurs along the line BP and the gear rotates through the angle α, called the *approach angle. This angle subtends an arc of the base circle obtained by constructing tooth profiles through B and P to intersect the base circle.*

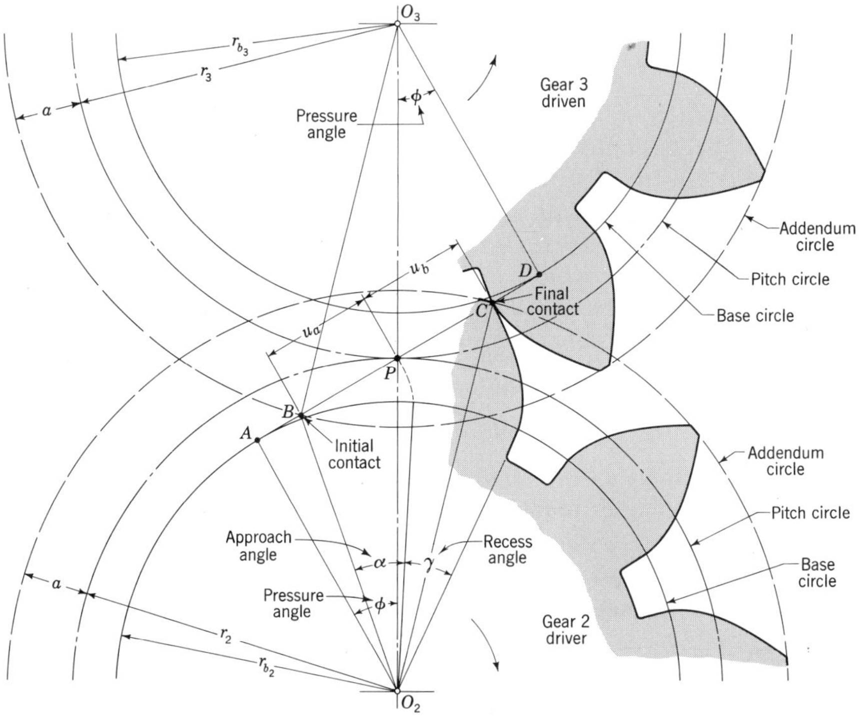

Fig. 8-23

During recess, contact occurs along PC while the gear rotates through the angle γ, called the *recess angle*. Note that the recess angle also subtends an arc of the base circle obtained by finding the intersection of tooth profiles through P and C with the base circle.

The base pitch is the distance between corresponding tooth profiles measured on the line of action. Therefore the contact ratio is

$$m_c = \frac{u_a + u_r}{p_b} \tag{8-11}$$

The values of u_a and u_b can be obtained analytically by observing the two right triangles O_2AC and O_3DB in Fig. 8-23. From triangle O_2AC we can write

$$u_a = [(r_3 + a)^2 - r_{b_3}{}^2]^{\frac{1}{2}} - r_3 \sin \phi \tag{8-12}$$

Similarly, from triangle O_2AC we have

$$u_r = [(r_2 + a)^2 - r_{b_2}{}^2]^{\frac{1}{2}} - r_2 \sin \phi \tag{8-13}$$

The contact ratio is then obtained by substituting Eqs. (8-12) and (8-13) into (8-11). We might note that Eqs. (8-12) and (8-13), however, are valid only for the conditions

$$u_a \leq r_2 \sin \phi \qquad u_r \leq r_3 \sin \phi \tag{8-14}$$

because contact cannot begin before point A (Fig. 8-23) nor end after point D. Thus if the value of u_a or u_r as calculated by Eq. (8-12) or Eq. (8-13) fail to satisfy the inequalities of Eq. (8-14), then Eq. (8-14) should be used to calculate u_a or u_r, whichever the case may be, using the equality sign.

The largest possible contact ratio is obtained by adjusting the addendums of each gear so as to utilize the entire distance AD (Fig. 8-23). The action is then defined by triangles O_2AD and O_3AD. Therefore

$$a_2 = [r_{b_2}{}^2 + (r_2 + r_3)^2 \sin^2 \phi]^{\frac{1}{2}} - r_2 \tag{8-15}$$

$$a_3 = [r_{b_3}{}^2 + (r_2 + r_3)^2 \sin^2 \phi]^{\frac{1}{2}} - r_3 \tag{8-16}$$

as the addendums a_2 and a_3, respectively, for gears 2 and 3. If either or both of these addendums are exceeded, then undercutting will occur during generation of the profiles.

8-13 INTERCHANGEABLE GEARS

The use of gears which are completely interchangeable is probably more of a convenience than a necessity today because, in the manufacture of most machines, pairs or groups of gears are made to mate with each other and no others. The following conditions must exist for gears to be interchangeable:

1. All gears must have the same diametral pitch.
2. All gears must have the same pressure angle.
3. The addendums and dedendums must be equal.
4. The thickness of the teeth must be the same and equal to half the circular pitch.

A *tooth system* is a *standard*[1] which specifies the relationships involving addendum, dedendum, working depth, tooth thickness, and pressure angle, to attain interchangeability of gears of all tooth numbers but of the same pressure angle and pitch.

Table 8-1 lists the tooth proportions for completely interchangeable gears and for operation on standard center distances.

[1] Standardized by the American Gear Manufacturers' Association (AGMA) and the United States of America Standards Institute. The spur-gear standards are the AGMA publications 201.02 and 201.02A, "Tooth Proportions for Coarse-pitch Involute Spur Gears," and 207.04, "20-degree Involute Fine-pitch Systems for Spur and Helical Gears," and are available from the AGMA.

Table 8-1 Standard AGMA and USASI tooth systems for spur gears

Quantity	*Coarse pitch* (up to 20P) full depth*		*Fine pitch (20P and up) full depth*
Pressure angle ϕ, deg	20°	25°	20°
Addendum a	$\frac{1.000}{P}$	$\frac{1.000}{P}$	$\frac{1.000}{P}$
Dedendum b	$\frac{1.250}{P}$	$\frac{1.250}{P}$	$\frac{1.200}{P} + 0.002$ in.
Working depth h_k	$\frac{2.000}{P}$	$\frac{2.000}{P}$	$\frac{2.000}{P}$
Whole depth h_t, min	$\frac{2.250}{P}$	$\frac{2.250}{P}$	$\frac{2.200}{P} + 0.002$ in.
Circular tooth thickness t_p	$\frac{\pi}{2P}$	$\frac{\pi}{2P}$	$\frac{1.5708}{P}$
Fillet radius of basic rack r_f	$\frac{0.300}{P}$	$\frac{0.300}{P}$	Not standardized
Basic clearance c, min	$\frac{0.250}{P}$	$\frac{0.250}{P}$	$\frac{0.200}{P} + 0.002$ in.
Clearance c (shaved or ground teeth)	$\frac{0.350}{P}$	$\frac{0.350}{P}$	$\frac{0.3500}{P} + 0.002$ in.
Minimum number of pinion teeth	18	12	18
Minimum width of top land t_0	$\frac{0.25}{P}$	$\frac{0.25}{P}$	Not standardized

* But not including 20*P*.

Table 8-2 Preferred diametral pitches

Coarse pitch	2, $2\frac{1}{4}$, $2\frac{1}{2}$, 3, 4, 6, 8, 10, 12, 16
Fine pitch	20, 24, 32, 40, 48, 64, 80, 96, 120, 150, 200

The 0.002-in. additional dedendum shown in Table 8-1 for fine-pitch gears provides space for the accumulation of dirt at the roots of the teeth.

The working depths shown in Table 8-1 are for, and define, *full-depth* teeth; for *stub* teeth, use

$$h_k = \frac{1.600}{P} \tag{a}$$

It should be noted rather particularly that the standards shown in Table 8-1 are *not* intended to restrict the freedom of the designer. Standard tooth proportions lead to interchangeability and standard cutters which are economical to purchase, but the need for high-performance gears may well dictate considerable deviation from these systems.

Some of the tooth systems which are now obsolete are the two AGMA $14\frac{1}{2}°$ systems, the Fellows 20° stub-tooth system, and the Brown & Sharpe system.[1] The obsolete systems should not be used for new designs, but it is sometimes necessary to refer to them when redesigning existing machinery which utilizes these older systems.

The diametral pitches listed in Table 8-2 should be used whenever possible in order to keep to a minimum the inventory of gear-cutting tools.

8-14 NONSTANDARD GEARS

In this section, we shall investigate the effects obtained by modifying such things as pressure angle, tooth depth, addendum, or center distance. Some of these modifications do not eliminate interchangeability; all of them are made with the intent of obtaining improved performance or more economical production.

There are three principal reasons for the use of nonstandard gears. The designer is often under great pressure to produce gear designs which are small and yet which will transmit large amounts of power. Consider, for example, a gearset which must have a 4:1 velocity ratio. If the smallest pinion that will carry the load has a pitch diameter of 2 in., then the gear will have a pitch diameter of 8 in., making the overall space required for the two gears slightly more than 10 in. On the other

[1] For details of these systems, see Darle W. Dudley (ed.), "Gear Handbook," pp. 5-28, 5-40, McGraw-Hill Book Company, New York, 1962.

hand, if the pitch diameter of the pinion can be reduced by only $\frac{1}{4}$ in., then the pitch diameter of the gear is reduced a full 1 in. and the overall size of the gearset reduced by $1\frac{1}{4}$ in. This reduction assumes considerable importance when it is realized that the sizes of associated machine elements, such as shafts, bearings, and enclosures, are also reduced. If a tooth of a certain pitch is required to carry the load, then the only method of decreasing the pinion diameter is to use a smaller number of teeth. We have already seen that problems involving interference, undercutting, and contact ratio are encountered when the tooth numbers are made less than prescribed minima. Thus the principal reasons for employing nonstandard gears are to eliminate undercutting, to prevent interference, and to maintain a reasonable contact ratio. It should be noted, too, that if a pair of gears is manufactured of the same material, then the pinion is the weaker and is subject to greater wear because its teeth are in contact a greater portion of the time. Thus undercutting weakens the tooth, which is already the weaker of the two. So another advantage of nonstandard gears is the tendency toward a better balance of strength between the pinion and gear.

As an involute curve is generated from the base circle, its radius of curvature becomes larger and larger. Near the base circle the radius of curvature is quite small, being exactly zero at the base circle. Contact near this region of sharp curvature should be avoided if possible because of the difficulty of obtaining good cutting accuracy in areas of small curvature and, too, because the contact stresses are likely to be very high. Nonstandard gears present the opportunity of designing to avoid these sensitive areas.

Clearance modifications A larger fillet at the root of the tooth increases the fatigue strength of the tooth and provides extra depth for shaving the tooth profile. Since interchangeability is not lost, the clearance is sometimes increased to $0.400/P$ to obtain this larger fillet.

In some applications a pressure angle of $17\frac{1}{2}°$ has been used with a clearance of $0.300/P$ to produce a contact ratio of 2.

Center-distance modifications When gears of low tooth numbers are to be paired with each other or when they are to be mated with larger gears, reduction in interference and improvement in the contact ratio can be obtained by increasing the center distance. Although this system changes the tooth proportions and the pressure angle of the gears, the resulting teeth can be generated with rack cutters (or hobs) of standard pressure angles or with standard pinion shapers. Before introducing this system it will be of value to develop additional relations in the geometry of gears.

The first relation to be obtained is that of finding the thickness of

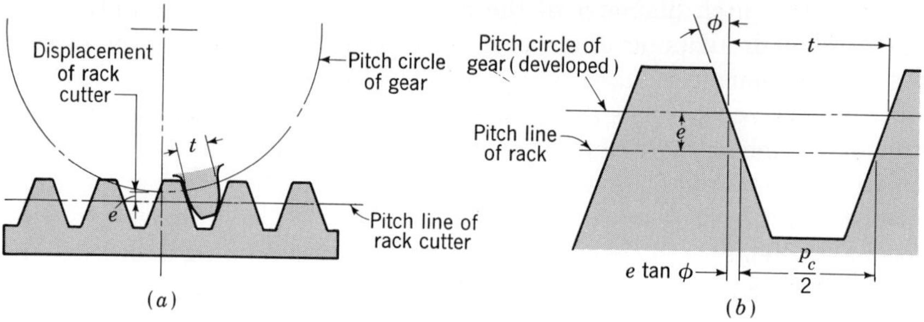

Fig. 8-24

a tooth which is cut by a rack cutter (or hob) when the pitch line of the rack has been displaced or offset a distance e from the pitch circle of the gear. What we are doing here is moving the rack cutter farther away from the center of the gear which is being cut. This will produce teeth which are thicker than before, and this thickness must be found. Figure 8-24*a* shows the problem, and Fig. 8-24*b* the solution. The increase over the standard amount is $2e \tan \phi$, so

$$t = 2e \tan \phi + \frac{p_c}{2} \tag{8-17}$$

where ϕ is the pressure angle of the rack cutter, and t is the thickness of the gear tooth on its own pitch circle.

Now suppose two gears of different tooth numbers have been cut with the cutter offset from the pitch circles as in the previous paragraph. Since the teeth have been cut with an offset cutter, they will mate at a new pressure angle and with new pitch circles and, consequently, new center distances. The word *new* is used in the sense, here, of not being standard. Our problem is to determine the radius of these new pitch circles and the value of this new pressure angle.

In the following notation the word *standard* refers to values which would have been obtained had the usual, or standard, systems been employed to obtain the dimensions.

ϕ = pressure angle of rack generating cutter
ϕ' = new pressure angle at which gears will mate
r_2 = standard pitch radius of pinion
r_2' = new pitch radius of pinion when meshing with given gear
r_3 = standard pitch radius of gear
r_3' = new pitch radius of gear when meshing with given pinion
t_2 = actual thickness of pinion tooth at standard pitch radius

t_3 = actual thickness of gear tooth at standard pitch radius
t_2' = thickness of pinion tooth at new pitch radius r_2'
t_3' = thickness of gear tooth at new pitch radius r_3'
N_2 = number of teeth on pinion
N_3 = number of teeth on gear

From Eq. (8-10),

$$t_2' = 2r_2'\left(\frac{t_2}{2r_2} + \text{inv}\,\phi - \text{inv}\,\phi'\right) \tag{a}$$

$$t_3' = 2r_3'\left(\frac{t_3}{2r_3} + \text{inv}\,\phi - \text{inv}\,\phi'\right) \tag{b}$$

The sum of these two thicknesses must be the same as the circular pitch, or from Eq. (8-2),

$$t_2' + t_3' = p_c = \frac{2\pi r_2'}{N_2} \tag{c}$$

The pitch diameters of a pair of mating gears are proportional to their tooth numbers, so

$$r_3 = \frac{N_3}{N_2} r_2 \qquad \text{and} \qquad r_3' = \frac{N_3}{N_2} r_2' \tag{d}$$

Substituting Eqs. (*a*), (*b*), and (*d*) in (*c*) and rearranging,

$$\text{inv}\,\phi' = \frac{N_2(t_2 + t_3) - 2\pi r_2}{2r_2(N_2 + N_3)} + \text{inv}\,\phi \tag{8-18}$$

Equation (8-18) gives the pressure angle ϕ' at which a pair of gears will operate when the tooth thicknesses on their standard pitch circles have been modified to t_2 and t_3.

It has been demonstrated that gears have no pitch circles until a pair of them are brought into contact. Bringing a pair of gears into contact creates a pair of pitch circles which are tangent to each other at the pitch point. Throughout this discussion the idea of a pair of so-called standard pitch circles has been used in order to define a certain point on the involute curves. These standard pitch circles, we have seen, are the ones which would have come into existence, when the gears were paired, *if the gears had not been modified from the standard dimensions.* On the other hand, the base circles are fixed circles which are not changed by tooth modifications. The base circle remains the same whether the tooth dimensions are changed or not; so we can determine the base-circle radius, using either the standard pitch circle or the new pitch circle. Equation (8-9) may therefore be written in either of the following ways:

$$r_b = r_2 \cos \phi \qquad \text{or} \qquad r_b = r_2' \cos \phi'$$

Thus

$$r_2' \cos \phi' = r_2 \cos \phi$$

or

$$r_2' = \frac{r_2 \cos \phi}{\cos \phi'} \tag{8-19}$$

Similarly, for the gear,

$$r_3' = \frac{r_3 \cos \phi}{\cos \phi'} \tag{8-20}$$

These equations give the values of the actual pitch radii when the two gears with modified teeth are *brought* into mesh without backlash. The new center distance is, of course, the sum of these radii.

All the necessary relations have now been developed to create nonstandard gears with changes in the center distance. The usefulness of these relations is best illustrated by an example.

Figure 8-25 is a drawing of a 20°, 1-pitch, 12-tooth pinion generated with a rack cutter with a standard clearance of $0.250/P$. In the 20°-full-depth system, interference is severe whenever the number of teeth is less than 14. The resulting undercutting is evident from the drawing.

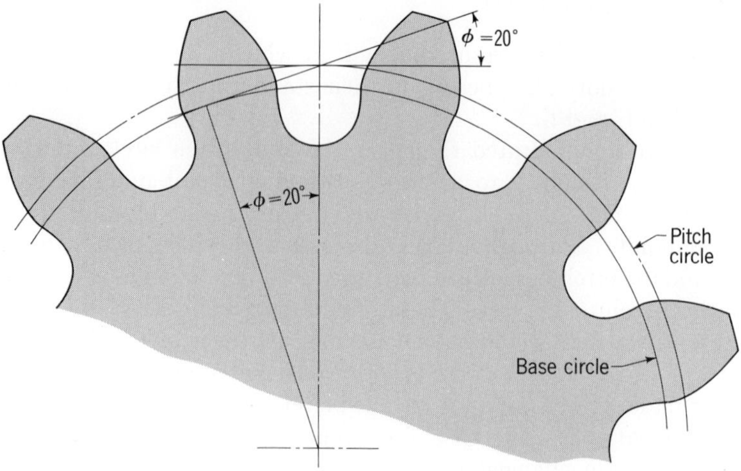

Fig. 8-25 Standard 12-tooth, 20°-full-depth gear showing undercut.

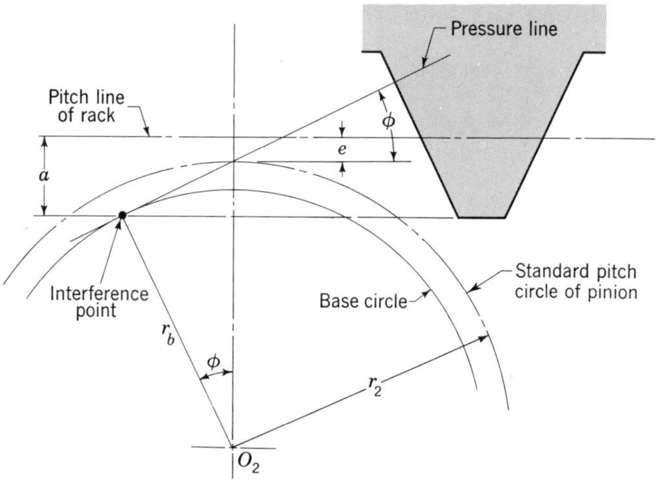

Fig. 8-26 Offsetting a rack to cause its addendum line to pass through the interference point.

If this pinion were mated with a standard 40-tooth gear, the contact ratio would be 1.41, which may easily be verified by Eq. (8-11).

In an attempt to eliminate the undercutting, improve the tooth action, and increase the contact ratio, let the 12-tooth pinion be cut from a larger blank. Then the resulting pinion will be paired again with the 40-tooth standard gear to determine the degree of improvement. If we designate the pinion as subscript 2 and the gear as 3, the following values may be found:

$\phi = 20°$ $\quad p_c = 3.1416$ in.

$r_2 = 6$ in. $\quad t_3 = 1.5708$ in.

$r_3 = 20$ in. $\quad N_2 = 12$ teeth

$P = 1$ $\quad N_3 = 40$ teeth

We shall offset the rack cutter so that its addendum line passes through the interference point of the pinion, that is, the point of tangency of the 20° pressure line and the base circle, as shown in Fig. 8-26. From Eq. (8-9) we have

$$r_b = r_2 \cos \phi \tag{e}$$

Then, from Fig. 8-26,

$$e = a + r_b \cos \phi - r_2 \tag{f}$$

Substituting Eq. (*e*) in (*f*),

$$e = a + r_2 \cos^2 \phi - r_2 = a - r_2 \sin^2 \phi \tag{8-21}$$

For a standard rack the addendum is $a = 1/P$; so $a = 1$ in. for this problem. The offset to be used is

$$e = 1 - 6 \sin^2 20° = 0.2981 \text{ in.}$$

Then, solving Eq. (8-17) for the thickness of the pinion tooth at its 6-in. pitch circle,

$$t_2 = 2e \tan \phi + \frac{p_c}{2} = (2)(0.2981) \tan 20° + \frac{3.1416}{2}$$
$$= 1.7878 \text{ in.}$$

The pressure angle at which these gears (and only these gears) will operate is found from Eq. (8-18).

$$\text{inv } \phi' = \frac{N_2(t_2 + t_3) - 2\pi r_2}{2r_2(N_2 + N_3)} + \text{inv } \phi$$
$$= \frac{12(1.7878 + 1.5708) - 2\pi 6}{(2)(6)(12 + 40)} + \text{inv } 20°$$
$$= 0.019077 \text{ rad}$$

From the Appendix,

$$\phi' = 21°39'4''$$

By using Eqs. (8-19) and (8-20), the new pitch radii are found to be:

$$r_2' = \frac{r_2 \cos \phi}{\cos \phi'} = \frac{6 \cos 20°}{\cos (21°39'4'')} = 6.0662 \text{ in.}$$

$$r_3' = \frac{r_3 \cos \phi}{\cos \phi'} = \frac{20 \cos 20°}{\cos (21°39'4'')} = 20.220 \text{ in.}$$

So the new center distance is

$$r_2' + r_3' = 6.0662 + 20.220 = 26.286 \text{ in.}$$

Notice that the center distance has not increased so much as the offset of the rack cutter.

In the beginning, a clearance of $0.25/P$ was specified, making the standard dedendums equal to $1.25/P$. So the root radii of the two gears are:

$$\begin{aligned} \text{Root radius of pinion} &= 6.2981 - 1.25 = 5.0481 \text{ in.} \\ \text{Root radius of gear} &= 20.0000 - 1.25 = \underline{18.7500 \text{ in.}} \\ \text{Sum of root radii} &= 23.7981 \text{ in.} \end{aligned}$$

The difference between this sum and the center distance is the working depth plus twice the clearance. Since the clearance is 0.25 in. for each gear, the working depth is

$$26.286 - 23.7981 - (2)(0.25) = 1.9879 \text{ in.}$$

The outside radius of each gear is the sum of the root radius, the clearance, and the working depth.

$$\begin{aligned} \text{Outside radius of pinion} &= 5.0481 + 0.25 + 1.9879 \\ &= 7.2860 \text{ in.} \\ \text{Outside radius of gear} &= 18.75 + 0.25 + 1.9879 \\ &= 20.9879 \text{ in.} \end{aligned}$$

The result is shown in Fig. 8-27, and the pinion is seen to have a stronger-looking form than the one of Fig. 8-25. Undercutting has been

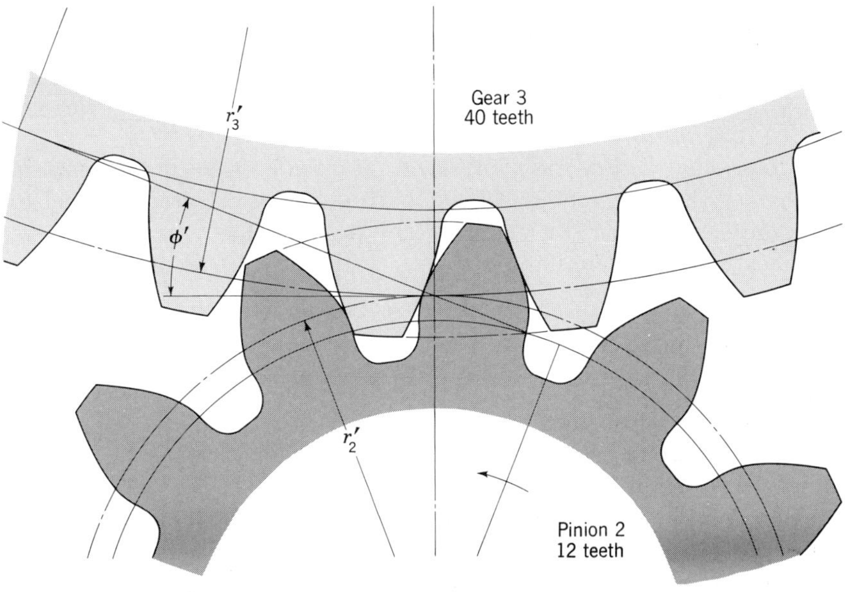

Fig. 8-27

completely eliminated. The contact ratio may be obtained from Eqs. (8-11) to (8-13). The following quantities are needed:

$$r_2' + a = 7.2860 \text{ in.} \qquad \text{outside radius of pinion}$$

$$r_3' + a = 20.9879 \text{ in.} \qquad \text{outside radius of gear}$$

$$r_{b_2} = r_2 \cos \phi = 6 \cos 20° = 5.6381 \text{ in.}$$

$$r_{b_3} = r_3 \cos \phi = 20 \cos 20° = 18.7938 \text{ in.}$$

$$p_b = p_c \cos \phi = 3.1416 \cos 20° = 2.9521 \text{ in.}$$

We then have

$$\begin{aligned} u_a &= [(r_3' + a)^2 - r_{b_3}{}^2]^{\frac{1}{2}} - r_3' \sin \phi' \\ &= [(20.9879)^2 - (18.7938)^2]^{\frac{1}{2}} - (20.220) \sin (21°39'4'') \\ &= 1.8826 \text{ in.} \end{aligned}$$

$$\begin{aligned} u_r &= [(r_2' + a)^2 - r_{b_2}{}^2]^{\frac{1}{2}} - r_2' \sin \phi' \\ &= [(7.2860)^2 - (5.6381)^2]^{\frac{1}{2}} - (6.0662) \sin (21°39'4'') \\ &= 2.3247 \text{ in.} \end{aligned}$$

Finally, from Eq. (8-11), the contact ratio is

$$m_c = \frac{u_a + u_r}{p_b} = \frac{1.8826 + 2.3247}{2.9521} = 1.425$$

Thus, the contact ratio has increased only slightly. The modification, however, is justified because of the elimination of undercutting and hence results in a very substantial improvement in the strength of the tooth.

Long-and-short-addendum systems It often happens in the design of machinery that the center distance between a pair of gears is fixed by some other characteristic or feature of the machine. In such a case, modifications to obtain improved performance cannot be made by varying the center distance.

We have seen, in the previous section, that improved action and tooth shape can be obtained by backing out the rack cutter from the pinion blank. The effect of this withdrawal is to create active tooth profile farther away from the base circle. An examination of Fig. 8-27 will reveal that more dedendum on the gear (not the pinion) could be utilized before the interference point is reached. If the rack cutter is advanced into the gear blank a distance equal to the offset from the pinion blank, then more of the gear dedendum will be utilized and, at the same time, the center distance will not have changed. This is called *the long-and-short-addendum system.*

In the long-and-short-addendum system there is no change in the pitch circles and, consequently, none in the pressure angle. The effect is to move the contact region away from the pinion center toward the gear center, thus shortening the approach action and lengthening the recess action.

The characteristics of the long-and-short-addendum system can be explained by reference to Fig. 8-28. Figure 8-28*a* illustrates a conventional (standard) set of gears having a dedendum equal to the addendum plus the clearance. Interference exists, and the tip of the gear tooth will have to be relieved as shown or the pinion will be undercut. This is so because the addendum circle of the gear crosses the pressure line at D, outside the tangency or interference point C; hence the distance CD is a measure of the degree of interference.

Now, to eliminate the undercutting or interference, the pinion addendum has been enlarged in Fig. 8-28*b* until the addendum circle of the pinion passes through the interference point (point A) of the gear. In this manner we shall be utilizing all the gear-tooth profile. The same whole depth is retained; hence the dedendum of the pinion is reduced by the same amount that the addendum is increased. This means that we must now lengthen the gear dedendum and shorten the addendum. With these changes the path of contact is the line BD in Fig. 8-28*b*. It is longer than the path BC in Fig. 8-28*a*, and so the contact ratio is higher. Notice, too, that the base circles, the pitch circles, the pressure angle, and the center distance have not changed. Both gears can be cut with standard cutters by advancing the cutter into the gear blank a distance equal to the amount of withdrawal, for this modification, from the pinion blank. Finally, note that the blanks, from which the gears are cut, are now of different diameters than standard blanks.

The tooth dimensions for the long-and-short-addendum system can be determined by using the equations developed in the previous sections.

A less obvious advantage of the long-and-short-addendum system is that more recess action than approach action is obtained. The approach action of gear teeth is analogous to pushing a piece of chalk across the blackboard; the chalk screeches. But when the chalk is pulled across the blackboard, it glides smoothly; this is analogous to recess action. Thus recess action is always preferred because of the smoothness and the lower frictional forces.

The long-and-short-addendum system has no advantage if the mating gears are of the same size. In this case, increasing the addendum of one gear would simply produce more undercutting of the mate. It is also apparent that the smaller gear of the pair should be the driver if the advantages of recess action are to be obtained.

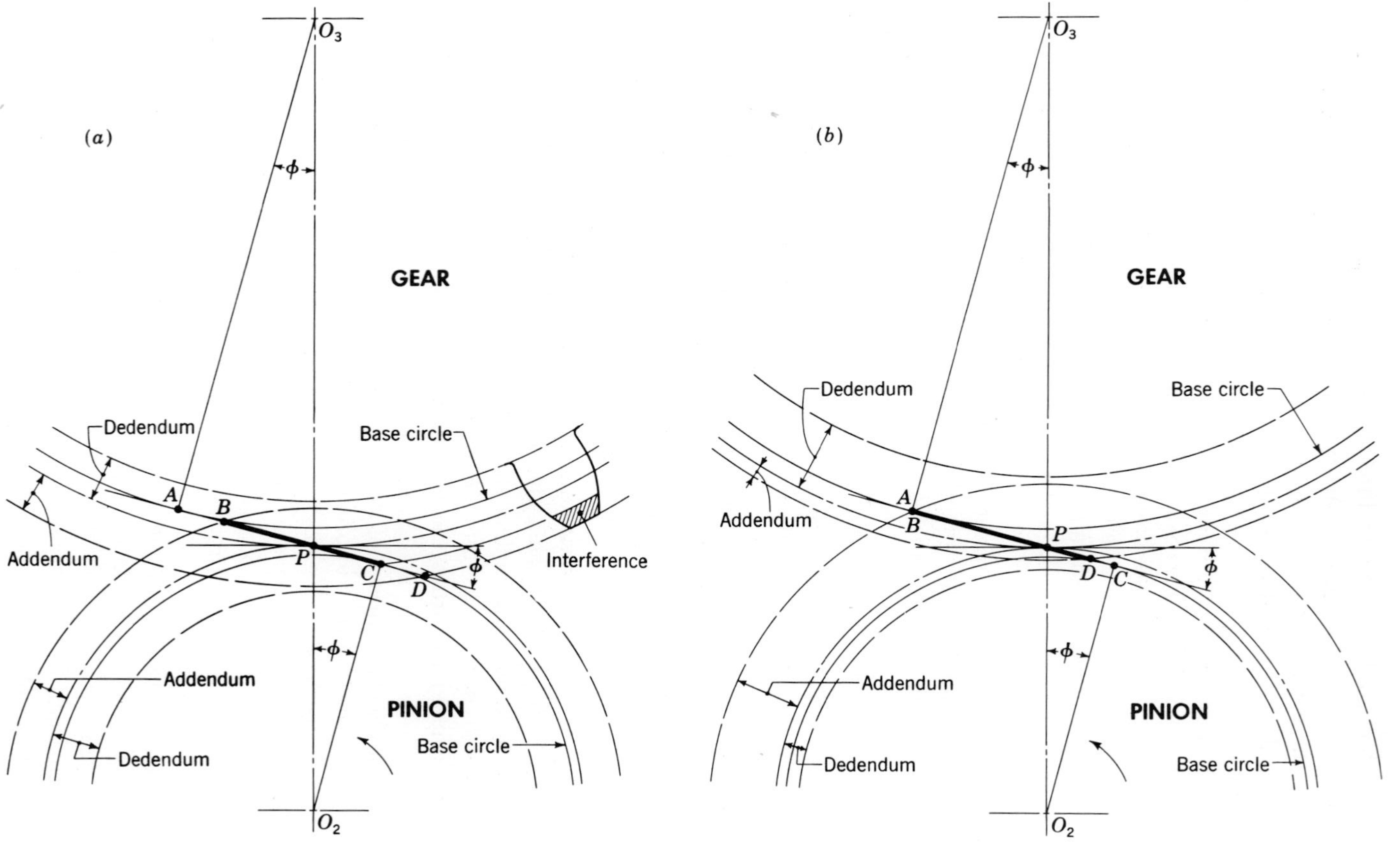

Fig. 8-28 Comparison of standard gears and gears cut by the long-and-short-addendum system. (*a*) Gear and pinion with standard addendum and dedendum; (*b*) gear and pinion with long-and-short addendum.

PROBLEMS

8-1. Find the diametral pitch of a pair of gears whose center distance is 0.3625 in. The gears have 32 and 84 teeth, respectively.

8-2. Find the number of teeth and the circular pitch of a 6-in.-pitch-diameter gear whose diametral pitch is 9.

8-3. What are the diametral pitch and the pitch diameter of a 40-tooth gear whose circular pitch is 3.50 in.?

8-4. The pitch diameters of a pair of mating gears are $3\frac{1}{2}$ and $8\frac{1}{4}$ in., respectively. If the diametral pitch is 16, how many teeth are there on each gear?

8-5. What is the diameter of a 33-tooth gear if the circular pitch is 0.875 in.?

8-6. A shaft carries a 30-tooth, 3-diametral-pitch gear which drives another gear at a speed of 480 rpm. How fast is the 30-tooth gear rotating if the shaft center distance is 9 in.?

8-7. Two gears having an angular-velocity ratio of 1.6:1 are mounted on shafts whose centers are 6.50 in. apart. If the diametral pitch of the gears is 3, how many teeth are there on each gear?

8-8. Calculate the center distance of a pair of gears if the tooth numbers are 17 and 41. The diametral pitch is 6.

8-9. A 4-diametral-pitch, 24-tooth pinion is to drive a 36-tooth gear. The gears are cut on the 20°-full-depth-involute system. Make a drawing of the gears showing one tooth on each gear. Find and tabulate the following results: the addendum, dedendum, clearance, circular pitch, tooth thickness, and base-circle diameters; the arcs of approach, recess, and action; and the contact ratio and base pitch.

8-10. A 5-diametral-pitch, 15-tooth pinion is to mate with a 30-tooth internal gear. The gears are 20° full-depth involute. Make a drawing of the gears showing several teeth on each gear. Can these gears be assembled in a radial direction? If not, what remedy should be used?

8-11. A $2\frac{1}{2}$-diametral-pitch, 17-tooth pinion and a 50-tooth gear are paired. The gears are cut on the 20°-full-depth-involute system. Make a drawing of the gears showing one tooth on each gear. Find the arcs of approach, recess, and action and the contact ratio, obtaining the data directly from the drawing.

8-12. Repeat Prob. 8-11, using the 25°-full-depth system.

8-13. Draw a 2-diametral-pitch, 26-tooth gear in mesh with a rack. The gears are 20° full-depth involute.

(*a*) Find the arcs of approach, recess, and action and the contact ratio.

(*b*) Draw a second rack in mesh with the same gear but offset $\frac{1}{8}$ in. away from the gear center. Determine the new contact ratio. Has the pressure angle changed?

8-14 to **8-18.** Shaper gear cutters have the advantage that they may be used for either external or internal gears, and also that only a small amount of "runout" is necessary at the end of the stroke. The generating action of a pinion shaper cutter may easily be simulated by employing a sheet of clear plastic. The figure illustrates one tooth of a 16-tooth-pinion cutter with 20° pressure angle as it may be cut from a plastic sheet. To construct the cutter, lay out the tooth on a sheet of drawing paper. Be sure to include the clearance at the top of the tooth. Draw radial lines through the pitch circle spaced at distances equal to one-fourth of the tooth thickness as shown in the figure. Now fasten the sheet plastic to the drawing and scribe the cutout, the pitch circle, and the radial lines onto the sheet. The sheet may then be removed, and the tooth outline trimmed with a razor blade. A small piece of fine sandpaper should then be used to remove any burrs.

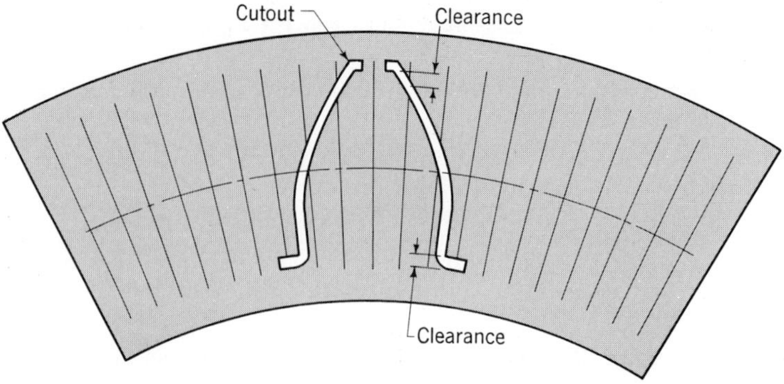

Probs. 8-14 to **8-18**

To generate a gear with the cutter, only the pitch circle and the addendum circle need be drawn. Divide the pitch circle into spaces equal to those used on the template, and construct radial lines through them. The tooth outlines are then obtained by rolling the template pitch circle upon that of the gear and drawing the cutter tooth lightly for each position. The resulting generated tooth upon the gear will be evident. The following problems all employ a standard 1-diametral-pitch, full-depth template constructed in the manner described above. In each case generate a few teeth and estimate the amount of undercutting.

Problem number	*Tooth number*
8-14	10
8-15	12
8-16	14
8-17	20
8-18	36

8-19. Using the template of the previous problems, generate a 10-tooth pinion on an 11-tooth blank and, again, estimate the amount of undercutting.

8-20. A 15-tooth pinion has 1.5-diametral-pitch, 20°-full-depth teeth. Calculate the thickness of the teeth at the base circle. What are the thickness and pressure angle at the addendum circle?

8-21. A tooth is 0.785 in. thick at a radius of 8 in. and a pressure angle of 25°. What is the thickness at the base circle?

8-22. A tooth is 1.57 in. thick at the pitch radius of 16 in. and a pressure angle of 20°. At what radius does the tooth become pointed?

8-23. A 25°-involute, 12-diametral-pitch pinion has 18 teeth. Calculate the tooth thickness at the base circle. What are the thickness and pressure angle at the addendum circle? Does the tooth actually exist at the base circle?

8-24. A special 10-tooth, 8-diametral-pitch pinion is to be cut with a $22\frac{1}{2}°$ pressure angle. What maximum addendum can be used before the teeth become pointed?

8-25. The accuracy of cutting gear teeth may be measured by fitting hardened and ground pins in diametrically opposite tooth spaces and measuring the distance over these pins. A gear has 96 teeth, is 10 diametral pitch, and is cut with the 20°-full-depth-involute system.

(*a*) Calculate the pin diameter which will contact the teeth at the pitch lines if there is to be no backlash.

(*b*) If the gear is accurately cut, what should be the distance measured over the pins?

8-26. A set of interchangeable gears is cut on the 20°-full-depth-involute system, having a diametral pitch of 4. The gears have tooth numbers of 24, 32, 48, and 96. For each gear, calculate the radius of curvature of the tooth profile at the pitch circle and at the addendum circle.

8-27. Calculate the contact ratio of a 17-tooth pinion which drives a 73-tooth gear. The gears are 96 diametral pitch and cut on the 20°-fine-pitch system.

8-28. A special 25°-pressure-angle, 11-tooth pinion is to drive a 23-tooth gear. The gears have a diametral pitch of 8 and are stub teeth. What is the contact ratio?

8-29. A 22-tooth pinion mates with a 42-tooth gear. The gears are full depth, have a diametral pitch of 16, and are cut with a $17\frac{1}{2}°$ pressure angle. Find the contact ratio.

8-30. A pair of mating gears are 24 diametral pitch and are generated on the 20° system. If the tooth numbers are 15 and 50, what maximum addendums may they have if interference is not to occur?

8-31. A set of gears is cast with a $17\frac{1}{2}°$ pressure angle and a circular pitch of $4\frac{1}{2}$ in. The pinion has 20 full-depth teeth. If the gear has 240 teeth, what must its addendum be in order to avoid interference?

8-32. Using the method described in Prob. 8-14, cut a 1-diametral-pitch, 20°-pressure-angle, full-depth rack tooth from a sheet of clear plastic. Use a modified clearance of $0.35/P$ in order to obtain a stronger fillet. This template may be employed to simulate the generating action of a hob. Now, using the variable-center-distance system, generate an 11-tooth pinion to mesh with a 25-tooth gear without interference. Record the values found for center distance, pitch radii, pressure angle, gear-blank diameters, cutter offset, and contact ratio. Note that more than one satisfactory solution exists.

8-33. Using the template constructed in Prob. 8-32, generate an 11-tooth pinion to mesh with a 44-tooth gear with the long-and-short-addendum system. Determine and record suitable values for gear and pinion addendum and dedendum and for the cutter offset and contact ratio. Compare the contact ratio with that which would have been obtained with standard gears.

8-34. A standard 20-tooth, 20°-pressure-angle, full-depth, 1-diametral-pitch pinion drives a 48-tooth gear. The speed of the pinion is 500 rpm. Using the length of the path of contact as the abscissa, plot a curve showing the sliding velocity at all points of contact. Notice that the sliding velocity changes sign when the point of contact passes through the pitch point.

9

Helical, Worm, and Bevel Gears

Most engineers prefer to use spur gears when power is to be transferred between parallel shafts, because they are easier to design and often more economical to manufacture. However, sometimes the design requirements are such that helical gears are a better choice. This is especially true when the loads are heavy, the speeds are high, or the noise level must be kept low.

When motion is to be transmitted between shafts which are not parallel, the spur gear cannot be used; the designer must then choose between crossed-helical, worm, bevel, or hypoid gears. Bevel gears have straight teeth, line contact, and high efficiencies. Crossed-helical and worm gears have a much lower efficiency because of the increased sliding action; however, if good engineering is used, crossed-helical and worm gears may be designed with quite acceptable values of efficiency. Hypoid and bevel gears are used for similar applications, and although hypoid gears have inherently stronger teeth, the efficiency is often much less. Worm gears are used when very high velocity ratios are required.

9-1 PARALLEL HELICAL GEARS

Helical gears are used for the transmission of motion between nonparallel and parallel shafts. When they are used with nonparallel shafts they are called *crossed-helical gears;* these are considered in Sec. 9-5.

The shape of the tooth of a helical gear is an involute helicoid, as illustrated in Fig. 9-1. If a piece of paper is cut into the shape of a parallelogram and wrapped around a cylinder, the angular edge of the paper becomes a helix. If the paper is then unwound, each point on the angular edge generates an involute curve. The surface obtained when every point on the edge generates an involute is called an *involute helicoid.*

The initial contact of spur-gear teeth is a line extending all the way across the face of the tooth. The initial contact of helical-gear teeth is a point which changes into a line as the teeth come into more engagement; in helical gears the line is diagonal across the face of the tooth. It is this gradual engagement of the teeth and the smooth transfer of load from one tooth to another which give helical gears the ability to transmit heavy loads at high speeds.

Double-helical (herringbone) gears are obtained when each gear has both right-hand and left-hand teeth cut on the same blank, and they operate on parallel axes. The thrust forces of the right- and left-hand halves are equal and opposite and cancel each other.

9-2 HELICAL-GEAR-TOOTH RELATIONS

Figure 9-2 represents a portion of the top view of a helical rack. Lines AB and CD are the centerlines of two adjacent helical teeth taken on the

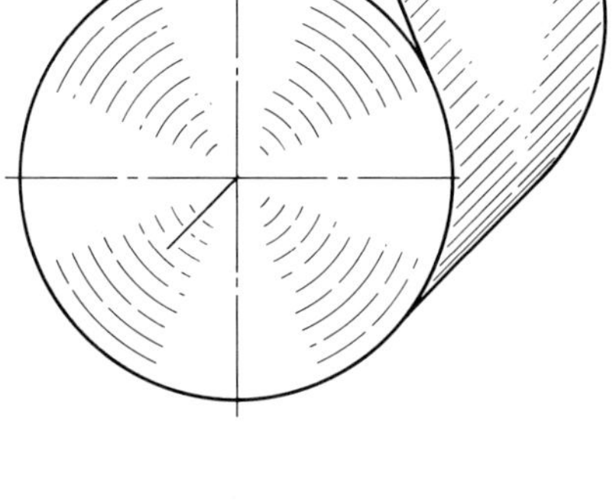

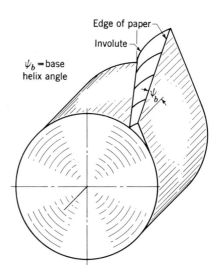

Fig. 9-1. An involute helicoid.

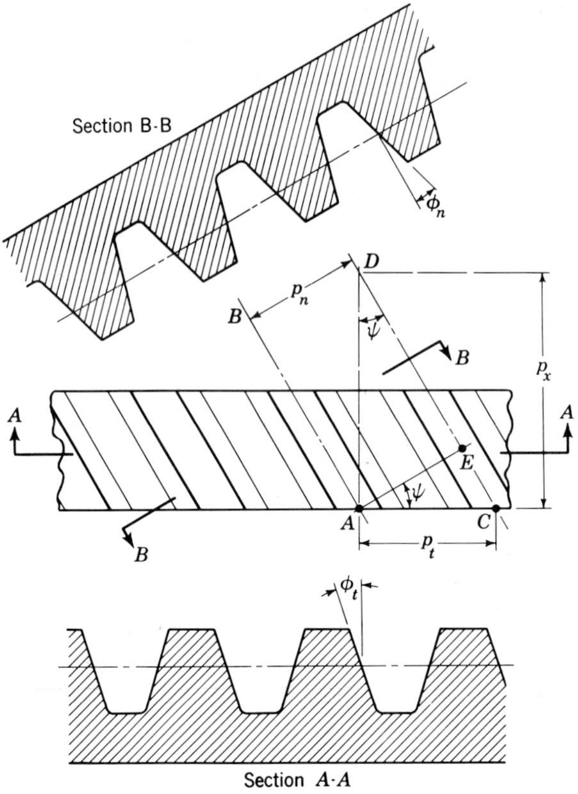

Fig. 9-2 Helical-gear-tooth relations.

pitch plane. The angle ψ is the *helix angle* and is to be measured at the pitch diameter unless otherwise specified. The distance AC is the *transverse circular pitch* p_t* in the plane of rotation. The distance AE is the *normal circular pitch* p_n and is related to the transverse circular pitch as follows:

$$p_n = p \cos \psi \tag{9-1}$$

The distance AD is called the *axial pitch* p_x and is

$$p_x = \frac{p}{\tan \psi} \tag{9-2}$$

*The subscript t is frequently omitted for the transverse pitch.

Since $p_n P_n = \pi$, the *normal diametral pitch* is

$$P_n = \frac{P}{\cos \psi} \tag{9-3}$$

where P is understood to be the *transverse* diametral pitch taken in the plane of motion.

Because of the angularity of the teeth we must define two pressure angles. These are the *transverse pressure angle* ϕ_t and the *normal pressure angle* ϕ_n, as shown in Fig. 9-2. These angles are related by the equation

$$\cos \psi = \frac{\tan \phi_n}{\tan \phi_t} \tag{9-4}$$

In applying these equations it is convenient to remember that all the equations and relations which are valid for spur gears apply equally for the transverse plane of a helical gear.

A better picture of the tooth relations may be obtained by an examination of Fig. 9-3. In order to obtain the geometric relations a helical gear has been cut by the oblique plane AA at an angle ψ to a right section. For convenience, only the pitch cylinder of radius r is shown.

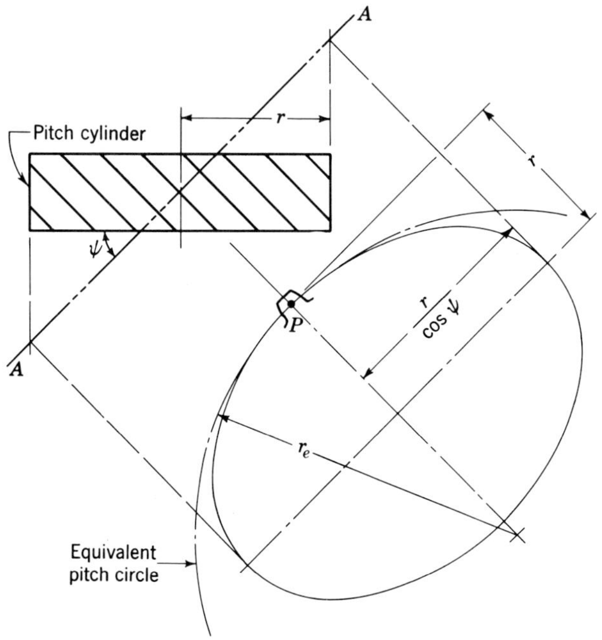

Fig. 9-3

The figure shows that the intersection of the plane and the pitch cylinder produces an ellipse whose radius at the pitch point P is r_e. This is called the *equivalent pitch radius*, and it is the radius of curvature of the pitch surface in the normal cross section. For the condition that $\psi = 0$, this radius of curvature is r. If we imagine the angle ψ to be slowly increased from zero to 90°, we see that r_e begins at a value of r and increases until, when $\psi = 90°$, $r_e = \infty$.

It can be shown[1] that

$$r_e = \frac{r}{\cos^2 \psi} \tag{9-5}$$

where r is the pitch radius of the helical gear, and r_e is the pitch radius of an equivalent spur gear. This equivalent gear is taken on the normal section of the helical gear. Let us define the number of teeth on the helical gear as N and on the equivalent spur gear as N_e. Then

$$N_e = d_e P_n \tag{d}$$

where $d_e = 2r_e$ is the pitch diameter of the equivalent spur gear. We can also write Eq. (d) as

$$N_e = \frac{d}{\cos^2 \psi} \frac{P}{\cos \psi} = \frac{N}{\cos^3 \psi} \tag{9-6}$$

9-3 HELICAL-GEAR-TOOTH PROPORTIONS[2]

Except for 20-diametral-pitch gears and finer, there is no standard for the proportions of helical-gear teeth. One reason for this is that it is

[1] The equation of an ellipse with its center at the origin of an xy system with a and b as the semimajor and semiminor axes, respectively, is

$$\frac{x^2}{a^2} + \frac{y^2}{b^2} = 1 \tag{a}$$

Also, the formula for radius of curvature is

$$\rho = \frac{[1 + (dy/dx)^2]^{\frac{3}{2}}}{d^2y/dx^2} \tag{b}$$

By using these two equations, it is not difficult to find the radius of curvature corresponding to $x = 0$ and $y = b$. The result is

$$\rho = \frac{a^2}{b} \tag{c}$$

Now referring to Fig. 9-3, we substitute $a = r/\cos \psi$ and $b = r$ into Eq. (c) and obtain Eq. (9-5).

[2] See also Darle W. Dudley (ed.), "Gear Handbook," pp. 5-41 to 5-48, McGraw-Hill Book Company, New York, 1962.

more economical to change the tooth design slightly than it is to purchase special tooling. Since helical gears are rarely used interchangeably anyway, and since many different designs will work well together, there is really little advantage in having them interchangeable.

As a general guide we present two sets of basic proportions in this section. Even these may be modified slightly to fit the dimensions of cutting tools which may be on hand. The dimensions of the first set are shown in Table 9-1. These are for a normal diametral pitch of 1 and for helix angles between 0 and 30°. For pitches different from unity, the values given in the table should be divided by the actual value of the normal diametral pitch. The values listed in Table 9-1 are for a normal pressure angle of 20°, and so all the various helix angles may be cut with the same hob. Of course, the normal diametral pitch of the hob and the gear to be cut must be the same.

The second set of proportions is based on a transverse diametral

Table 9-1 Helical-gear-tooth proportions for $\phi_n = 20°$*
Normal diametral pitch $P_n = 1$; normal circular pitch $p_n = 3.14159$ in.

Helix angle ψ, *deg*	*Diametral pitch* P	*Circular pitch* p	*Axial pitch* p_x	*Pressure angle* ϕ	*Working depth, in.*	*Whole depth, in.*
0	1.000	3.14159	. . .	20°	2.000	2.250
5	0.996195	3.15359	36.04560	20°4′ 13.1″	2.000	2.250
8	0.990268	3.17247	22.57327	20°10′50.6″	2.000	2.250
10	0.984808	3.19006	18.09171	20°17′ 0.7″	2.000	2.250
12	0.978148	3.21178	15.11019	20°24′37.1″	2.000	2.250
15	0.965926	3.25242	12.13817	20°38′48.8″	2.000	2.250
18	0.951057	3.30326	10.16640	20°56′30.7″	2.000	2.250
20	0.939693	3.34321	9.18540	21°10′22.0″	2.000	2.250
21	0.933580	3.36510	8.76638	21°17′56.4″	2.000	2.250
22	0.927184	3.38832	8.38636	21°25′57.7″	2.000	2.250
23	0.920505	3.41290	8.04029	21°34′26.3″	2.000	2.250
24	0.913545	3.43890	7.72389	21°43′22.9″	2.000	2.250
25	0.906308	3.46636	7.43364	21°52′58.7″	2.000	2.250
26	0.898794	3.49534	7.16651	22° 2′44.2″	2.000	2.250
27	0.891007	3.52589	6.91994	22°13′10.6″	2.000	2.250
28	0.882948	3.55807	7.69175	22°24′ 9.0″	2.000	2.250
29	0.874620	3.59195	6.48004	22°35′40.0″	2.000	2.250
30	0.866025	3.62760	6.28318	22°47′45.1″	2.000	2.250

* Darle W. Dudley, "Practical Gear Design," pp. 95–97, McGraw-Hill Book Company, New York, 1954.

Table 9-2 Helical-gear-tooth proportions for $\phi = 20°$*
Diametral pitch $P = 1$; circular pitch $p = 3.14159$ in.

Helix angle ψ, *deg*	*Normal diameter pitch* P_n	*Normal circular pitch* p_n	*Axial pitch* p_x	*Normal pressure angle* ϕ_n	*Working depth, in.*	*Whole depth, in.*
15	1.03528	3.03454	11.72456	19°22′12.2″	2.000	2.350
23	1.0836	2.89185	7.40113	18°31′21.6″	1.840	2.200
30	1.15470	2.72070	5.44140	17°29′42.7″	1.740	2.050
45	1.41421	2.22144	3.14159	14°25′57.9″	1.420	1.700

* Darle W. Dudley, "Practical Gear Design," pp. 95–97, McGraw-Hill Book Company, New York, 1954.

pitch of 1 and a pressure angle ϕ of 20°. Values are given for helix angles of 15 to 45°. More than 45° is not recommended. The 30 and 45° angles are usually used for double-helical gears. These proportions are shown in Table 9-2 and are especially recommended when the noise level must be kept low. They require different tools for each helix angle, even when the diametral pitch is the same.

The face width of helical gears should be at least two times the axial pitch in order to obtain helical-gear action. For face widths which are less than this the action will approach that of spur gears. For very high speeds the face width may be made four or more times the axial pitch, but the accuracy must be increased correspondingly.

It is noted that in a parallel helical gearset, the two gears must have the same helix angle and pitch and must be of opposite hand. The velocity ratio is determined in the same manner as for spur gears.

9-4 CONTACT OF HELICAL-GEAR TEETH

Mating spur-gear teeth contact each other in a line which is parallel to their axes of rotation. As shown in Fig. 9-4, the contact between helical-gear teeth is a diagonal line.

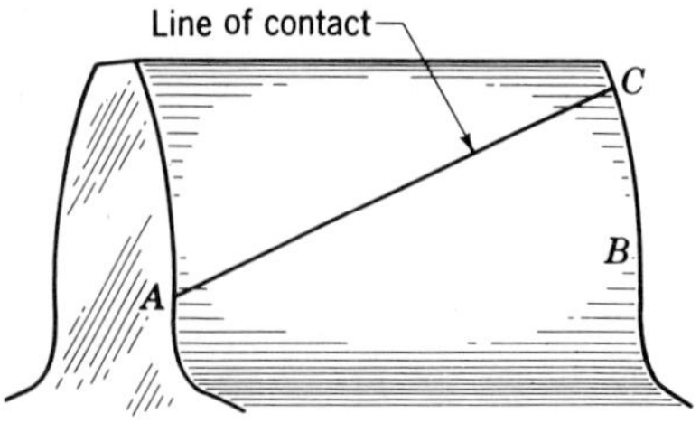

Fig. 9-4 While contact at A is just beginning, contact at the other end of the tooth has already progressed from B to C.

There are several kinds of contact ratios used in evaluating the performance of helical gears. The *transverse contact ratio* is designated by m and is the contact ratio in the transverse plane. It is obtained in exactly the same manner as for spur gears.

The *normal contact ratio* m_n is the contact ratio in the normal section. It is also found in exactly the same manner as for spur gears, but the *equivalent spur gears must be used in the determination.* The *base helix angle* ψ_b and the *pitch helix angle* ψ, for helical gears, are related by the equation

$$\tan \psi_b = \tan \psi \cos \phi \tag{9-7}$$

Then the transverse and normal contact ratios are related by the equation

$$m_n = \frac{m}{\cos^2 \psi_b} \tag{9-8}$$

The *axial contact ratio,* also called the *face contact ratio,* is the ratio of the face width of the gear to the axial pitch. It is given by the equation

$$m_x = \frac{F}{p_x} = \frac{F \tan \psi}{p} \tag{9-9}$$

where F is the face width. Note that the face contact ratio depends only on the geometry of a single gear, while the transverse and normal contact ratios depend upon the geometry of a pair of mating gears.

The *total contact ratio* m_t is the sum of the face and transverse contact ratios. In a sense it gives the average total number of teeth in contact.

9-5 CROSSED-HELICAL GEARS

Crossed-helical, or spiral, gears are used sometimes when the shaft centerlines are neither parallel nor intersecting. They are essentially nonenveloping worm gears, because the gear blanks have a cylindrical form.

The teeth of crossed-helical gears have *point contact* with each other, which changes to *line contact* as the gears wear in. For this reason they will carry only very small loads. Because of the point contact, however, they need not be mounted accurately; either the center distance or the shaft angle may be varied slightly without affecting the amount of contact.

There is no difference between a crossed-helical gear and a helical gear until they are mounted in mesh with each other. They are manufactured in the same way. A pair of meshed crossed-helical gears usually

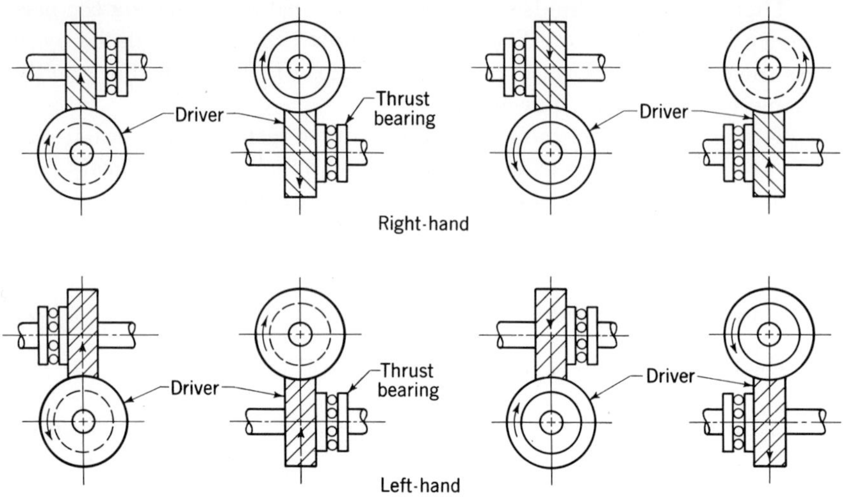

Fig. 9-5 Thrust, rotation, and hand relations in crossed-helical gears. *(Boston Gear Works, Inc., North Quincy, Mass.)*

have the same hand; that is, a right-hand driver goes with a right-hand driven. The relation between thrust, hand, and rotation for crossed-helical gears is shown in Fig. 9-5.

When tooth sizes are specified, the normal pitch should always be used. The reason for this is that when different helix angles are used for the driver and driven, the transverse pitches are not the same. The relation between the shaft and helix angles is as follows:

$$\Sigma = \psi_2 \pm \psi_3 \tag{9-10}$$

where Σ is the shaft angle. The plus sign is used when both helix angles are of the same hand, and the minus sign when they are of opposite hand. Opposite-hand crossed-helical gears are used when the shaft angle is small.

The pitch diameter is obtained from the equation

$$d = \frac{N}{P_n \cos \psi} \tag{9-11}$$

where N = number of teeth
P_n = normal diametral pitch
ψ = helix angle

Since the pitch diameters are not directly related to the tooth numbers, they cannot be used to obtain the angular-velocity ratio. This ratio must be obtained from the ratio of the tooth numbers.

Table 9-3 Tooth proportions for crossed-helical gears
Normal diametral pitch $P_n = 1$;
working depth = 2.400 in.;
whole depth = 2.650 in.;
addendum = 1.200 in.

Driver			
Helix angle ψ_2, *deg*	*Minimum tooth number* N_2	*Driven helix angle* ψ_3, *deg*	*Normal pressure angle* ϕ_n
45	20	45	14°30′
60	9	30	17°30′
75	4	15	19°30′
86	1	4	20°

Crossed-helical gears will have the lowest sliding velocity at contact when the helix angles of the two gears are equal. If the helix angles are not equal, then the gear with the largest helix angle should be used as the driver if both gears have the same hand.

There is no standard for crossed-helical gear-tooth proportions. Many different proportions give good tooth action. Since the teeth are in point contact, an effort should be made to obtain a contact ratio of 2 or more. For this reason, crossed-helical gears are usually cut with a low pressure angle and a deep tooth. Dudley[1] lists the tooth proportions shown in Table 9-3 as representative of good design. The driver tooth numbers shown are the minimum required to avoid undercut. The driven gear should have 20 or more teeth if a contact ratio of 2 is to be obtained.

9-6 WORM GEARS

In Fig. 9-6 is shown a worm and worm-gear application. These gears are used with nonintersecting shafts which are usually at a shaft angle of 90°; however, there is no reason why shaft angles other than 90° cannot be used if the design demands it.

The worm is the member having the screwlike thread, and worm teeth are frequently spoken of as threads. Worms in common use from 1 to 8 teeth, and, as we shall see, there is no definite relation between the number of teeth and the pitch diameter of a worm. Worms may be

[1] Darle W. Dudley, "Practical Gear Design," p. 111, McGraw-Hill Book Company, New York, 1954.

Fig. 9-6 A single-enveloping worm and worm gear. (*Horsburgh and Scott Company, Cleveland, Ohio.*)

designed with a cylindrical pitch surface, as shown in Fig. 9-7, or they may have an hourglass shape, such that the worm wraps around or partially encloses the worm gear.

The worm gear is normally the driven member of the pair and is made to envelop, or wrap around, the worm. If the worm gear is mated with a cylindrical worm, then the gearset is said to be *single-enveloping*. When the worm is hourglass-shaped, the worm gearset is said to be *double-enveloping*, because each member then wraps around the other.

A worm and worm-gear combination is similar to a pair of mating crossed-helical gears except for the fact that the worm gear partially encloses the worm. For this reason they have line contact, instead of the point contact found in crossed-helical gears, and are thus able to transmit more power. When a double-enveloping gearset is used, even more power may be transmitted, theoretically at least, because contact occurs over an area of the tooth surfaces.

In the single-enveloping worm gearset it does not make any difference whether the worm rotates on its own axis and drives the gear by a screw action or whether the worm is translated along its axis and drives the gear through rack action. The resulting motion and contact are the same. For this reason the worm need not be accurately mounted upon its shaft. However, the worm gear should be correctly mounted along its axis of rotation; otherwise the two pitch surfaces would not be concentric about the worm axis.

In a double-enveloping worm gearset both members are throated and must therefore be accurately mounted in every direction in order to obtain correct action.

The nomenclature of a single-enveloping gearset is shown in Fig. 9-7. A mating worm and worm gear with a 90° shaft angle have the same hand of helix, but the helix angles are usually quite different. On the worm, the helix angle is quite large (at least, for one or two teeth) and very small on the gear. Because of this, it is customary to specify the *lead angle* for the worm and the helix angle for the gear. This is convenient because, for a 90° shaft angle, the two angles are equal. The worm lead angle is the complement of the worm helix angle, as shown in Fig. 9-7.

In specifying the pitch of worm gearsets, specify the axial pitch of the worm and the circular pitch of the gear. These are equal if the

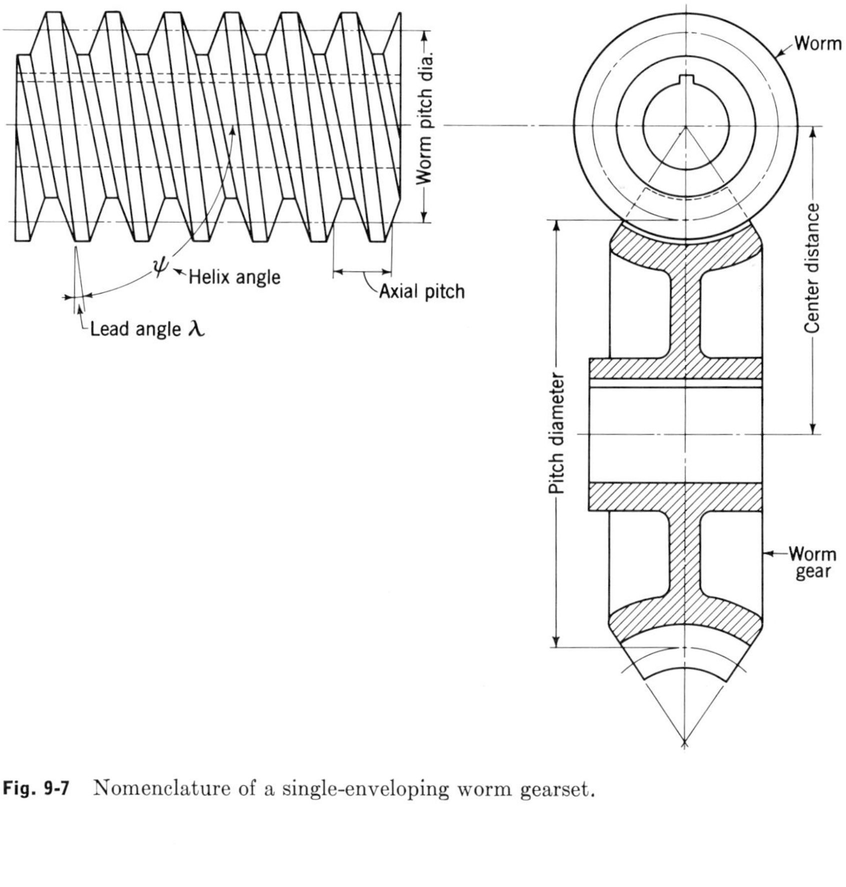

Fig. 9-7 Nomenclature of a single-enveloping worm gearset.

shaft angle is 90°. It is quite common to employ even fractions for the circular pitch, like $\frac{1}{4}$, $\frac{3}{8}$, $\frac{1}{2}$, $\frac{3}{4}$, 1, $1\frac{1}{4}$ in., etc. There is no reason, however, why standard diametral pitches, such as those used for spur gears, cannot be used. The pitch diameter of the gear is the same as that for spur gears; that is,

$$d_3 = \frac{N_3 p}{\pi} \tag{9-12}$$

where d_3 = pitch diameter
N_3 = number of teeth
p = circular pitch
all taken with reference to the worm gear.

The pitch diameter of the worm may be any value, but it should be the same as the hob used to cut the worm-gear teeth. AGMA recommends the following relation between the pitch diameter of the worm and the center distance:

$$d_2 = \frac{(r_2 + r_3)^{0.875}}{2.2} \tag{9-13}$$

where the quantity $r_2 + r_3$ is the center distance. This equation gives a set of proportions which will result in a good power capacity. Equation (9-13) does not have to be used; other proportions will work well too, and in fact power capacity may not always be the primary consideration. However, there are a lot of variables in worm-gear design, and the equation is helpful in obtaining trial dimensions. The AGMA Standard[1] also states that the denominator of Eq. (9-13) may vary from 1.7 to 3 without appreciably affecting the capacity.

The *lead* of a worm has the same meaning as for a screw thread and is the distance that a point on the helix will move when the worm is turned through one revolution. Thus, for a one-tooth worm, the lead is equal to the axial pitch. In equation form,

$$l = p_x N_2 \tag{9-14}$$

where l is the lead in inches, and N_2 is the number of teeth on the worm. The lead and the lead angle are related as follows:

$$\lambda = \tan^{-1} \frac{l}{\pi d_2} \tag{9-15}$$

where λ is the lead angle, as shown in Fig. 9-7.

The teeth on worms are usually cut on a milling machine or on a lathe. Worm-gear teeth are most often produced by hobbing. Except for clearance at the tip of the hob teeth, the worm should be an exact

[1] AGMA Standard 213.02, 1952.

Table 9-4 Pressure angles recommended for worm gearsets

Lead angle λ, *deg*	*Pressure angle* ϕ, *deg*
0–16	$14\frac{1}{2}$
16–25	20
25–35	25
35–45	30

duplicate of the hob in order to obtain conjugate action. This means, too, that where possible the worm should be designed using the dimensions of existing hobs.

The pressure angles used on worm gearsets vary widely and should depend approximately on the value of the lead angle. Good tooth action will be obtained if the pressure angle is made large enough to eliminate undercutting of the worm-gear tooth on the side at which contact ends. Buckingham recommends the values shown in Table 9-4.

A satisfactory tooth depth, which remains in about the right proportion to the lead angle, may be obtained by making the depth a proportion of the normal circular pitch. Based on an addendum of $1/P$ for full-depth spur gears, we obtain the following proportions for worm and worm gears:

$$\text{Addendum} = 0.3183p_n$$

$$\text{Whole depth} = 0.6366p_n$$

$$\text{Clearance} = 0.050p_n$$

The face width of the worm gear should be obtained as shown in Fig. 9-8. This makes the face of the worm gear equal to the length of

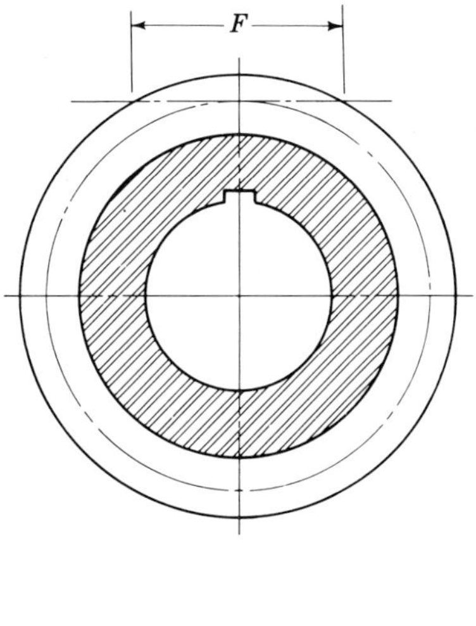

Fig. 9-8

a tangent to the worm pitch circle between its points of intersection with the addendum circle.

9-7 STRAIGHT BEVEL GEARS

When motion is to be transmitted between shafts whose axes intersect, some form of bevel gear is required. Although bevel gears are often made for a shaft angle of 90°, they may be produced for almost any angle. The most accurate teeth are obtained by generation.

Bevel gears have pitch surfaces which are cones; these cones roll together without slipping, as shown in Fig. 9-9. The gears must be mounted so that the apexes of both pitch cones are coincident because the pitch of the teeth depends upon the radial distance from the apex.

The true shape of a bevel-gear tooth is obtained by taking a spherical section through the tooth, where the center of the sphere is at the common apex, as shown in Fig. 9-10. Thus, as the radius of the sphere increases, the same number of teeth must exist on a larger surface; so the size of the teeth increases as larger and larger spherical sections are taken. We have seen that the action and contact conditions of spur-gear teeth may be viewed on a plane surface taken at right angles to the axes of the spur gears. For bevel-gear teeth, the action and contact conditions should be viewed on a spherical surface (instead of a plane surface). We can even think of spur gears as a special case of bevel gears in which the spherical radius is infinite, thus producing a plane surface on which the tooth action is viewed.

It is standard practice to specify the pitch diameter of bevel gears at the large end of the teeth. In Fig. 9-11 the pitch cones of a pair of bevel gears are drawn and the pitch radii given as r_2 and r_3, respectively, for the pinion and gear. The angles γ_2 and γ_3 are defined as the *pitch angles*, and their sum is equal to the shaft angle Σ. The velocity ratio is

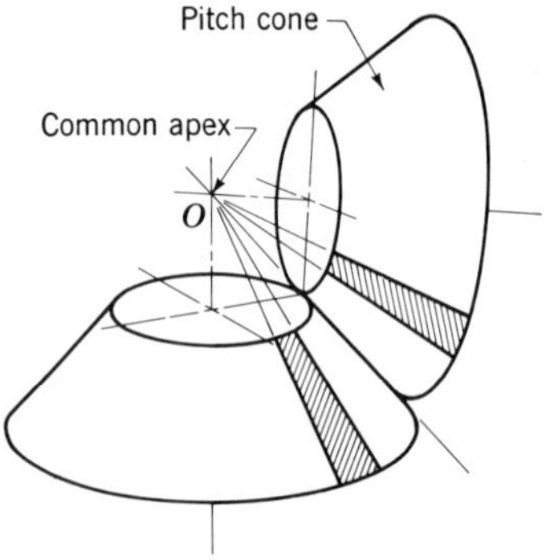

Fig. 9-9 The pitch surfaces of bevel gears are cones which have pure rolling contact.

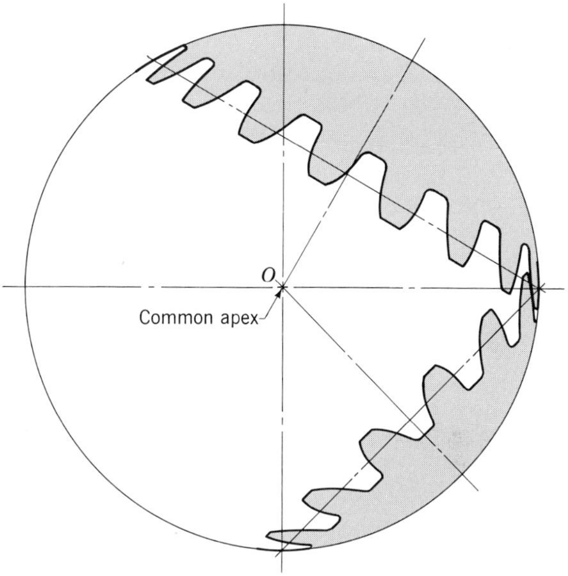

Fig. 9-10 A spherical section of bevel-gear teeth.

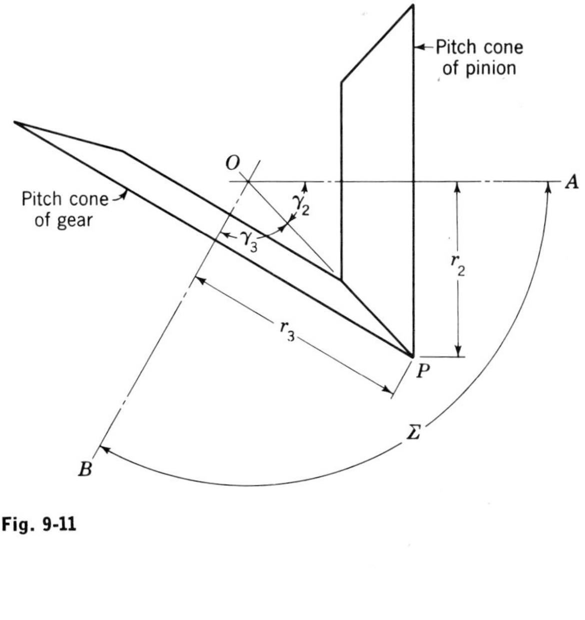

Fig. 9-11

obtained in the same manner as for spur gears and is

$$\frac{\omega_2}{\omega_3} = \frac{r_3}{r_2} = \frac{N_3}{N_2} \tag{9-16}$$

In the kinematic design of bevel gears the tooth numbers of each gear and the shaft angle are usually given, and the corresponding pitch angles are to be determined. Although they may easily be found by a graphical method, the analytical approach gives exact values. From Fig. 9-11 the distance OP may be written

$$OP = \frac{r_2}{\sin \gamma_2} \qquad \text{or} \qquad OP = \frac{r_3}{\sin \gamma_3}$$

so that

$$\sin \gamma_2 = \frac{r_2}{r_3} \sin \gamma_3 = \frac{r_2}{r_3} \sin (\Sigma - \gamma_2) \tag{a}$$

or

$$\sin \gamma_2 = \frac{r_2}{r_3} (\sin \Sigma \cos \gamma_2 - \sin \gamma_2 \cos \Sigma) \tag{b}$$

Dividing both sides of Eq. (*b*) by $\cos \gamma_2$ and rearranging gives

$$\tan \gamma_2 = \frac{\sin \Sigma}{(r_3/r_2) + \cos \Sigma} = \frac{\sin \Sigma}{(N_3/N_2) + \cos \Sigma} \tag{9-17}$$

Similarly,

$$\tan \gamma_3 = \frac{\sin \Sigma}{(N_2/N_3) + \cos \Sigma} \tag{9-18}$$

For a shaft angle of 90° the above expressions reduce to

$$\tan \gamma_2 = \frac{N_2}{N_3} \tag{9-19}$$

$$\tan \gamma_3 = \frac{N_3}{N_2} \tag{9-20}$$

The projection of bevel-gear teeth on the surface of a sphere would indeed be a difficult and time-consuming problem. Fortunately, an approximation is available which reduces the problem to that of ordinary spur gears. This method is called Tredgold's approximation, and as long as the gear has eight or more teeth, it is accurate enough for practical purposes. It is in almost universal use, and the terminology of bevel-gear teeth has evolved around it.

In using Tredgold's method, a *back cone* is formed of elements which are perpendicular to the elements of the pitch cone at the large end of the teeth. This is shown in Fig. 9-12. The length of a back-cone ele-

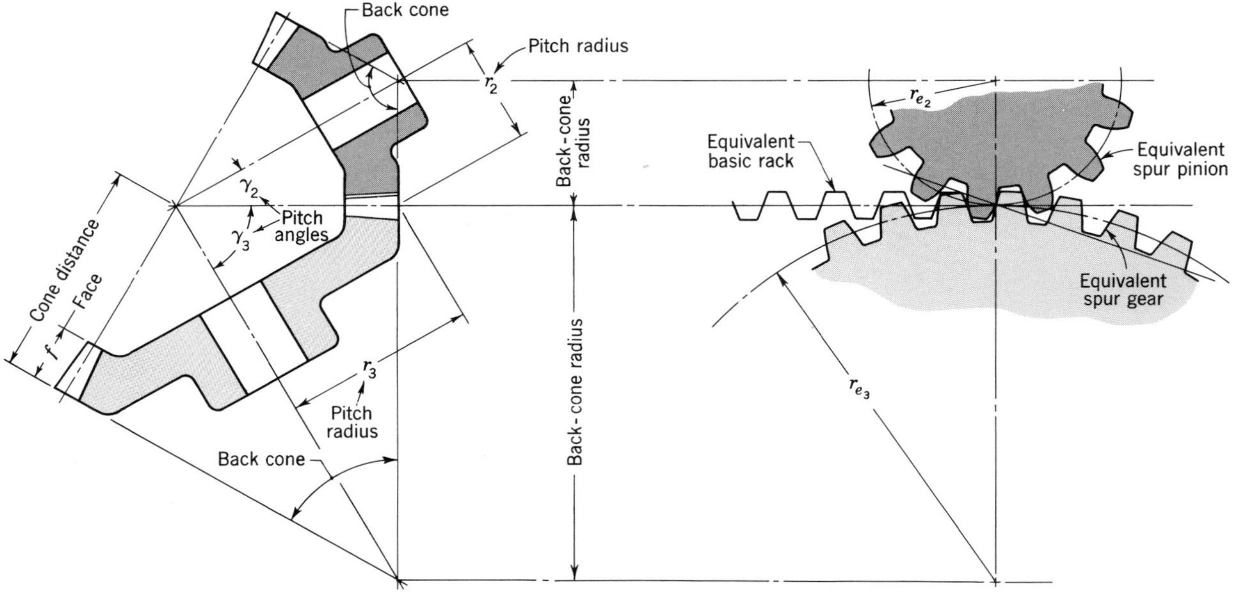

Fig. 9-12 Tredgold's approximation.

ment is called the *back-cone radius*. Now an equivalent spur gear is constructed whose pitch radius r_e is equal to the back-cone radius. Thus from a pair of bevel gears we can obtain, using Tredgold's approximation, a pair of equivalent spur gears, which are then used to define the tooth profiles; they can also be used to determine the tooth action and contact conditions in exactly the same manner as for ordinary spur gears, and the results will correspond closely to those for the bevel gears. From the geometry of Fig. 9-12, the equivalent pitch radii are

$$r_{e_2} = \frac{r_2}{\cos \gamma_2} \qquad r_{e_3} = \frac{r_3}{\cos \gamma_3} \tag{9-21}$$

The number of teeth on the equivalent spur gear is

$$N_e = \frac{2\pi r_e}{p} \tag{9-22}$$

where p is the circular pitch of the bevel gear measured at the large end of the teeth. In the usual case the equivalent spur gears will *not* have an integral number of teeth.

9-8 TOOTH PROPORTIONS FOR BEVEL GEARS

Practically all straight-tooth bevel gears manufactured today use the 20° pressure angle. Since bevel gears are not interchangeable, there is no reason to use the interchangeable tooth form. Thus they are well adapted for using the long-and-short-addendum system. The Gleason company[1] has worked out a set of tooth proportions using the long-and-short-addendum system; they are based on the velocity ratio and have since been adopted as AGMA Standard for bevel gears. These proportions are given in Table 9-5 and are for a shaft angle of 90° and for a pinion of not less than 16 teeth. For other shaft angles and smaller tooth numbers, special gears may be designed using the addendum modifications as demonstrated in Chap. 8 and Tredgold's approximation. For diametral pitches different from unity, divide the values shown in Table 9-5 by the actual diametral pitch.

Bevel gears are usually mounted on the outboard side of the bearings, because the shaft axes intersect, and this means that the effect of shaft deflection on the tooth action is much more pronounced. The effect of shaft deflection is to pull the small end of the teeth away from mesh, causing the large end to take most of the load. Thus the load across the tooth is variable, and for this reason it is desirable to design

[1] Gleason Works, Rochester, N.Y.

Table 9-5 Gear addendums for 1-diametral-pitch straight bevel gears
Velocity ratio = N_3/N_2; pressure angle = 20°; working depth = $2/P$; whole depth = $2.188/P$; shaft angle = 90°

Ratios			*Ratios*		
From	*To*	*Addendum*	*From*	*To*	*Addendum*
1.00	1.00	1.000	1.42	1.45	0.760
1.00	1.02	0.990	1.45	1.48	0.750
1.02	1.03	0.980	1.48	1.52	0.740
1.03	1.04	0.970	1.52	1.56	0.730
1.04	1.05	0.960	1.56	1.60	0.720
1.05	1.06	0.950	1.60	1.65	0.710
1.06	1.08	0.940	1.65	1.70	0.700
1.08	1.09	0.930	1.70	1.76	0.690
1.09	1.11	0.920	1.76	1.82	0.680
1.11	1.12	0.910	1.82	1.89	0.670
1.12	1.14	0.900	1.89	1.97	0.660
1.14	1.15	0.890	1.97	2.06	0.650
1.15	1.17	0.880	2.06	2.16	0.640
1.17	1.19	0.870	2.16	2.27	0.630
1.19	1.21	0.860	2.27	2.41	0.620
1.21	1.23	0.850	2.41	2.58	0.610
1.23	1.25	0.840	2.58	2.78	0.600
1.25	1.27	0.830	2.78	3.05	0.590
1.27	1.29	0.820	3.05	3.41	0.580
1.29	1.31	0.810	3.41	3.94	0.570
1.31	1.33	0.800	3.94	4.82	0.560
1.33	1.36	0.790	4.82	6.81	0.550
1.36	1.39	0.780	6.81	∞	0.540
1.39	1.42	0.770			

a fairly short tooth. Available recommendations limit the face width to about one-fourth or one-third of the cone distance; these are maximum values. We note further that a fairly short face width simplifies the tooling problems in cutting bevel-gear teeth.

Figure 9-13 defines additional terms characteristic of bevel gears. Note particularly that a constant clearance is maintained by making the elements of the face cone parallel to the elements of the root cone of the mating gear. This explains why the face-cone apex is not coincident with the pitch-cone apex in Fig. 9-13. This permits a larger fillet at the small end of the teeth than would otherwise be possible.

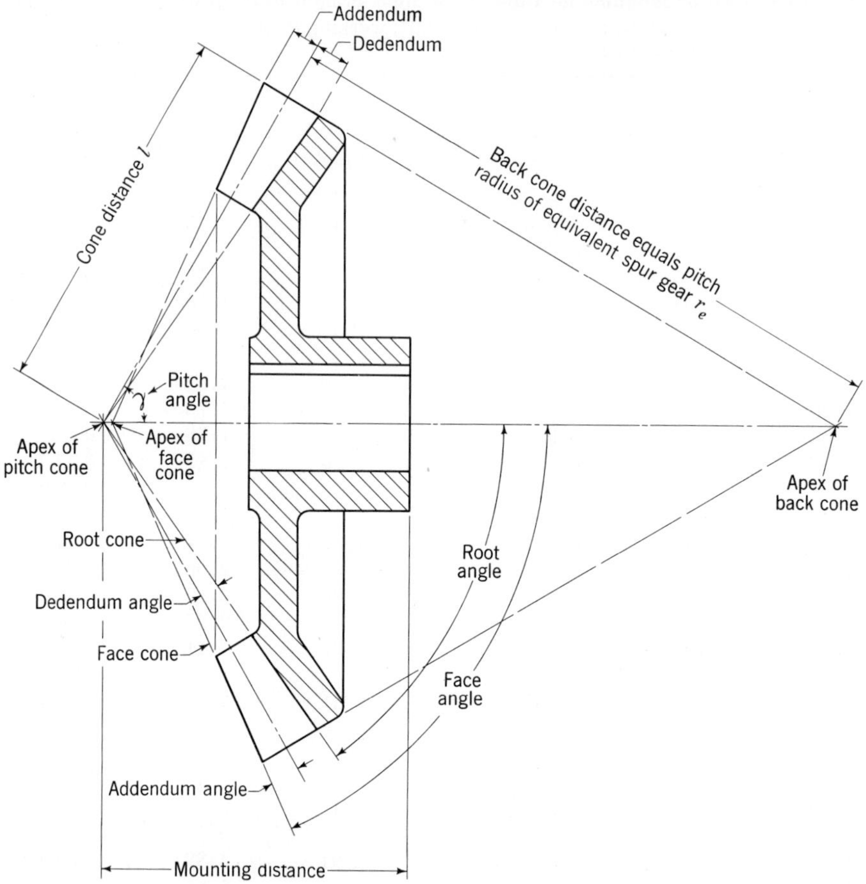

Fig. 9-13

9-9 CROWN AND FACE GEARS

If the pitch angle of one of a pair of bevel gears is made equal to 90°, then the pitch cone becomes a flat surface and the resulting gear is called a *crown gear*. Figure 9-14 shows a crown gear in mesh with a bevel pinion. Notice that a crown gear is the counterpart of a rack in spur gearing. The back cone of a crown gear is a cylinder, and the resulting involute teeth have straight sides, as indicated in Fig. 9-12.

A pseudo-bevel gearset may be obtained by using a face gear in mesh with a spur gear. The shaft angle is at 90°. In order to secure the best tooth action, the spur pinion should be a duplicate of the shaper cutter used to cut the face gear, except, of course, for the additional clear-

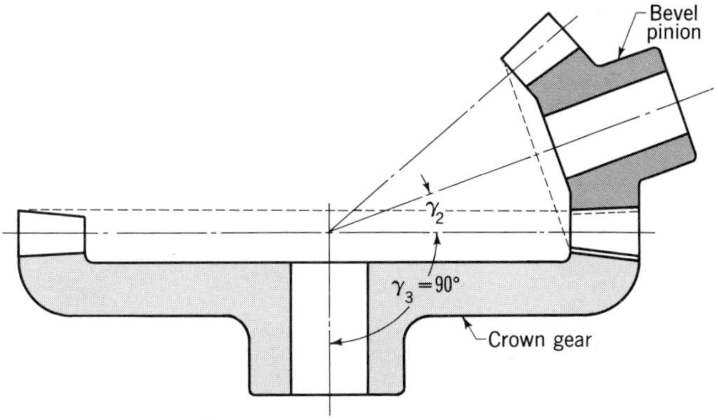

Fig. 9-14 A crown gear and bevel pinion.

ance at the tips of the cutter teeth. The face width of the teeth on the face gear must be held quite short; otherwise the top land will become pointed at the larger diameter.

9-10 SPIRAL BEVEL GEARS

Straight bevel gears are easy to design and simple to manufacture and give very good results in service if they are mounted accurately and positively. As in the case of spur gears, however, they become noisy at the higher values of the pitch-line velocity. In these cases it is often good design practice to go to the spiral bevel gear, which is the bevel counterpart of the helical gear. Figure 9-15 shows a mating pair of spiral bevel gears, and it can be seen that the pitch surfaces and the nature of contact are the same as for straight bevel gears except for the differences brought about by the spiral-shaped teeth.

Spiral-bevel-gear teeth are conjugate to a basic crown rack, which is generated as shown in Fig. 9-16 by using a circular cutter. The spiral angle ψ is measured at the mean radius of the gear. As in the case of helical gears, spiral bevel gears give a much smoother tooth action than straight bevel gears and hence are useful where high speeds are encountered. In order to obtain true spiral-tooth action the face contact ratio should be at least 1.25.

Pressure angles used with spiral bevel gears are generally $14\frac{1}{2}$ to 20°, while the spiral angle is about 30 or 35°. As far as the tooth action is concerned, the hand of the spiral may be either right or left; it makes no difference. However, looseness in the bearings might result in the teeth's jamming or separating, depending upon the direction of rotation and the

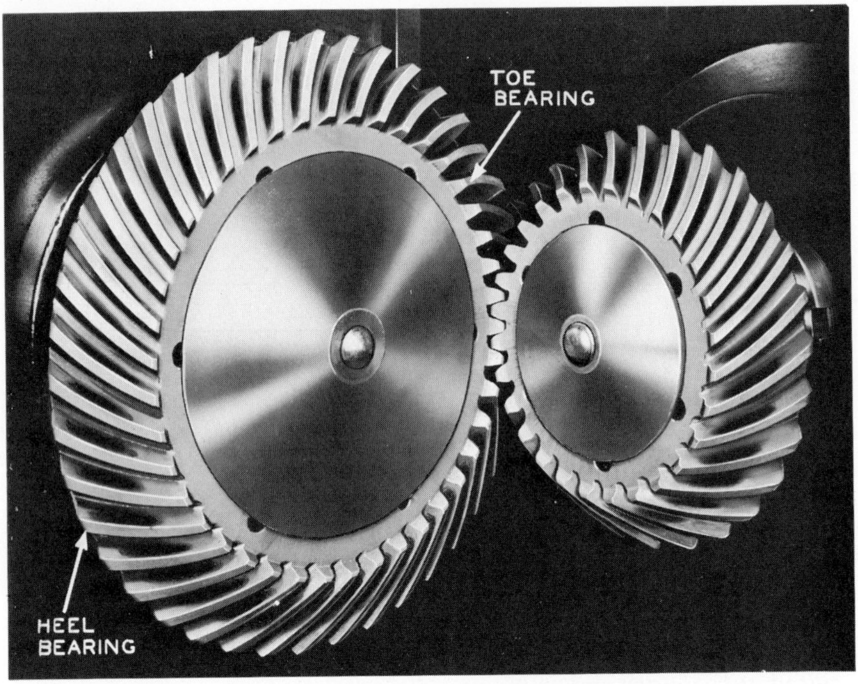

Fig. 9-15 Spiral bevel gears. *(Gleason Works, Rochester, N.Y.)*

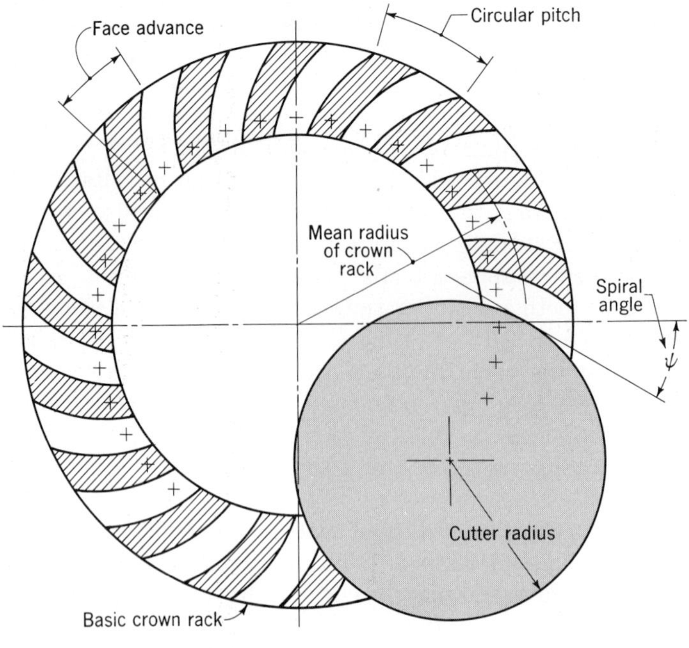

Fig. 9-16 Cutting spiral-gear teeth on the basic crown rack.

hand of the spiral. Since jamming of the teeth would do the most damage, the hand of the spiral should be such that the teeth will tend to separate.

Zerol bevel gears The Zerol bevel gear is a patented gear which has curved teeth but a zero-degree spiral angle. It has no advantage in tooth action over the straight bevel gear and is designed simply to take advantage of the cutting machinery which is used for cutting spiral bevel gears.

9-11 HYPOID GEARS

It is frequently desirable, as in the case of automotive-differential applications, to have a gear similar to a bevel gear but with the shafts offset. Such gears are called hypoid gears because their pitch surfaces are hyperboloids of revolution. The tooth action between such gears is a combination of rolling and sliding along a straight line and has much in common with that of worm gears. Figure 9-17 shows a pair of hypoid gears

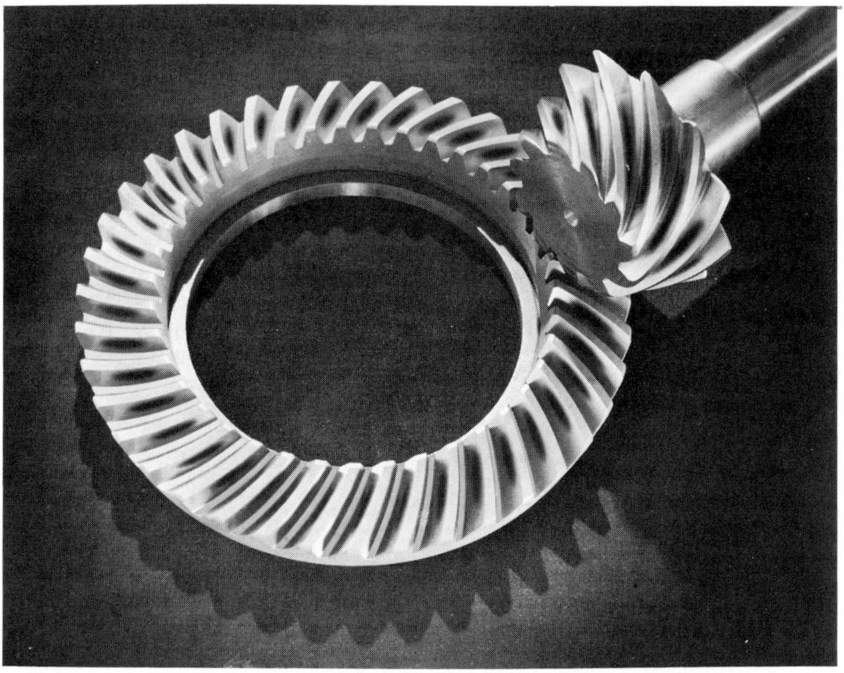

Fig. 9-17 Hypoid gears. (*Gleason Works, Rochester, N.Y.*)

in mesh. For additional details the reader should review some of the references cited earlier in this chapter.

PROBLEMS

9-1. A pair of parallel helical gears are $14\frac{1}{2}°$ normal pressure angle, 6 diametral pitch, and 45° helix angle. The pinion has 15 teeth, and the gear 24 teeth. Calculate the transverse and normal circular pitch, the normal diametral pitch, the pitch diameters, and the equivalent tooth numbers.

9-2. A pair of parallel helical gears are cut with a 20° normal pressure angle and a 30° helix angle. They are 16 diametral pitch and have 16 and 40 teeth, respectively. Find the transverse pressure angle, the normal circular pitch, the axial pitch, and the pitch radii of the equivalent spur gears.

9-3. A parallel helical gearset is made with a 20° transverse pressure angle and a 35° helix angle. The gears are 10 diametral pitch and have 15 and 25 teeth, respectively. If the face width is $\frac{3}{4}$ in., calculate the base helix angle and the axial contact ratio.

9-4. A pair of helical gears are to be cut for parallel shafts whose center distance is to be about $3\frac{1}{2}$ in. to give a velocity ratio of approximately 1.80. The gears are to be cut with a standard 20°-pressure-angle hob whose diametral pitch is 8. Using a helix angle of 30°, determine the transverse values of the diametral and circular pitch and the tooth numbers, pitch diameters, and center distance.

9-5. A 16-tooth helical pinion is to run at 1800 rpm and drive a helical gear on a parallel shaft at 400 rpm. The centers of the shafts are to be spaced 11 in. apart. Using a helix angle of 23° and a pressure angle of 20°, determine values for the tooth numbers, pitch diameters, normal circular and diametral pitch, and face width.

9-6. The catalog description of a pair of helical gears is as follows: $14\frac{1}{2}°$ normal pressure angle, 45° helix angle, 8 diametral pitch, 1-in. face width, 11.31 normal diametral pitch. The pinion has 12 teeth and a 1.500-in. pitch diameter, and the gear 32 teeth and 4.000-in. pitch diameter. Both gears have full-depth teeth, and they may be purchased either right-hand or left-hand. If a right-hand pinion and a left-hand gear are placed in mesh, find the transverse contact ratio, the normal contact ratio, the axial contact ratio, and the total contact ratio.

9-7. In a medium-sized truck transmission a 22-tooth clutch-stem gear meshes continuously with a 41-tooth countershaft gear. The data are 7.6 normal diametral pitch, $18\frac{1}{2}°$ normal pressure angle, $23\frac{1}{2}°$ helix angle, and 1.12-in. face width. The clutch-stem gear is cut with a left-hand helix, and the countershaft gear with a right-hand helix. Determine the normal and total contact ratio if the teeth are cut full depth with respect to the normal diametral pitch.

9-8. A helical pinion is right-hand, has 12 teeth, a 60° helix angle, and is to drive another gear at a velocity ratio of 3. The shafts are at a 90° angle, and the normal diametral pitch of the gears is 8. Find the helix angle and the number of teeth on the mating gear. What is the center distance?

9-9. A right-hand helical pinion is to drive a gear at a shaft angle of 90°. The pinion has 6 teeth and a 75° helix angle and is to drive the gear at a velocity ratio of 6.5. The normal diametral pitch of the gears is 12. Calculate the helix angle and the number of teeth on the mating gear. Also, determine the pitch diameter of each gear.

9-10. Gear 2 in the figure is to rotate clockwise and drive gear 3 counterclockwise at a velocity ratio of 2. Use a normal diametral pitch of 5, a center distance of about

10 in., and the same helix angle for both gears. Find the tooth numbers, the helix angles, and the exact center distance.

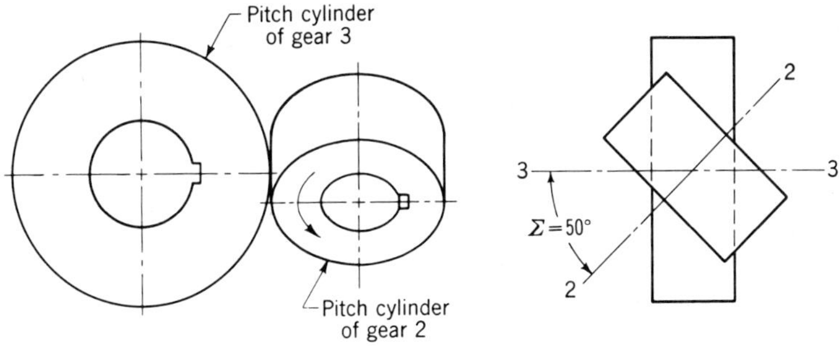

Prob. 9-10

9-11. A worm having 4 teeth and a lead of 1 in. drives a worm gear at a velocity ratio of $7\frac{1}{2}$. Determine the pitch diameters of the worm and worm gear for a center distance of $1\frac{3}{4}$ in.

9-12. Specify a suitable worm and worm-gear combination for a velocity ratio of 60 and a center distance of $6\frac{1}{2}$ in. Use an axial pitch of 0.500 in.

9-13. A 3-tooth worm drives a worm gear having 40 teeth. The axial pitch is $1\frac{1}{4}$ in., and the pitch diameter of the worm is $1\frac{3}{4}$ in. Calculate the lead and lead angle of the worm. Find the helix angle and pitch diameter of the worm gear.

9-14. A pair of straight-tooth bevel gears are to be manufactured for a shaft angle of 90°. If the driver is to have 18 teeth and the velocity ratio is 3, what are the pitch angles?

9-15. A pair of straight-tooth bevel gears have a velocity ratio of 1.5 and a shaft angle of 75°. What are the pitch angles?

9-16. A pair of straight bevel gears is to be mounted at a shaft angle of 120°. The pinion and gear are to have 15 and 33 teeth, respectively. What are the pitch angles?

9-17. A pair of 2-diametral-pitch, straight bevel gears have 19 teeth and 28 teeth, respectively. The shaft angle is 90°. Determine the pitch diameters, pitch angles, addendum, dedendum, face width, and the pitch diameters of the equivalent spur gears.

9-18. A pair of 8-diametral-pitch, straight bevel gears have 17 teeth and 28 teeth, respectively, and a shaft angle of 105°. For each gear calculate the pitch diameter, pitch angle, addendum, dedendum, face width, and the equivalent tooth numbers. Make a sketch of the two gears in mesh. Use the standard tooth proportions as for a 90° shaft angle.

10
Mechanism Trains

Mechanisms arranged in series with each other so that the driven member of one mechanism is the driver for another mechanism are termed *mechanism trains*. The analysis of such trains can proceed in chain fashion by using the methods already developed. However, there are certain shortcuts, applicable to gear trains especially, which we shall explore in this chapter.

10-1 INTRODUCTION

Consider a pinion 2 driving a gear 3. The speed of the driven gear is

$$n_3 = \frac{N_2}{N_3} n_2 = \frac{d_2}{d_3} n_2 \tag{10-1}$$

where n = rpm or number of turns
N = number of teeth
d = pitch diameter

For parallel-shaft gearing, the directions can be kept track of by speci-

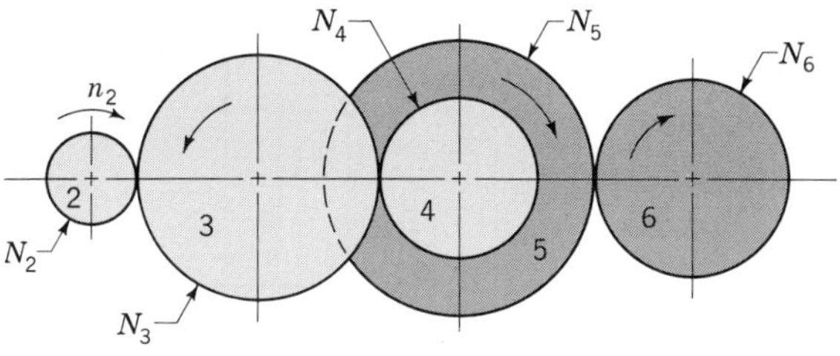

Fig. 10-1

fying that the speed is positive or negative, depending upon whether it is counterclockwise or clockwise. This approach fails with non-parallel-shaft gearing; hence it is often simpler to keep track of the directions by using a sketch of the train.

The gear train shown in Fig. 10-1 is made up of five gears. The speed of gear 6 is

$$n_6 = \frac{N_2}{N_3}\frac{N_3}{N_4}\frac{N_5}{N_6}n_2 \tag{a}$$

Here we notice that gear 3 is an idler, that its tooth numbers cancel in Eq. (*a*), and hence that it affects only the direction of rotation of 6. We notice furthermore that gears 2, 3, and 5 are drivers, while 3, 4, and 6 are driven members. Thus the equation for *train value e* is

$$e = \frac{\text{product of driving tooth numbers}}{\text{product of driven tooth numbers}} \tag{10-2}$$

Note that pitch diameters can be used in Eq. (10-2) as well. When Eq. (10-2) is used for parallel-shaft trains, e is made positive if the last gear rotates in the same sense as the first, and negative if the last rotates in the opposite sense.

Now we can write

$$n_L = en_F \tag{10-3}$$

where n_L is the speed of the last gear in the train and n_F is the speed of the first.

10-2 EXAMPLES OF GEAR TRAINS

In speaking of gear trains it is often convenient to describe a *simple gear train* as one which has only one gear on each axis. A *compound gear train* is one which, like that in Fig. 10-1, has two or more gears on one or more axes.

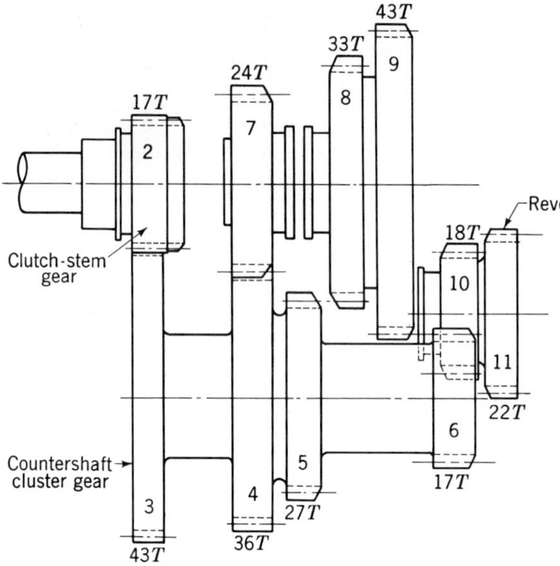

Speed	Drive
1	2-3-6-9
2	2-3-5-8
3	2-3-4-7
4	Straight through
Reverse	2-3-6-10-11-9

Fig. 10-2 A truck transmission. The gears are 7-diametral-pitch spur gears with a $22\frac{1}{2}°$ pressure angle.

Figure 10-2 shows a transmission for small- and medium-sized trucks; it has four speeds forward and one in reverse.

The gear train shown in Fig. 10-3 is composed of bevel, helical, and spur gears. The helical gears are crossed, and so the direction of rotation depends upon the hand of the helical gears.

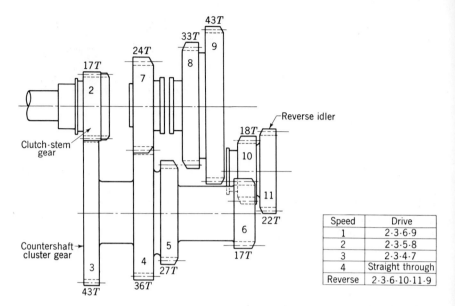

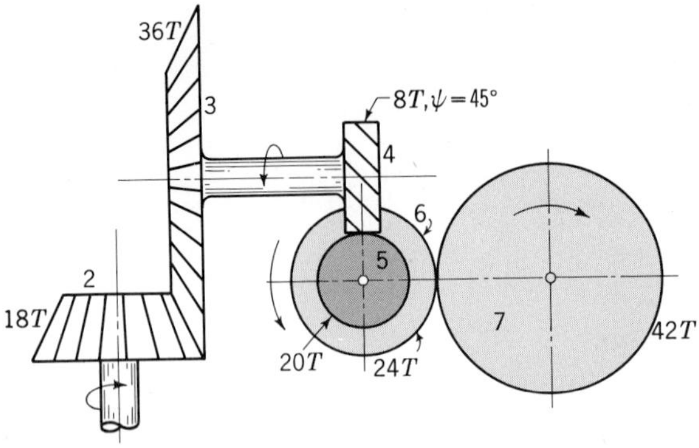

Fig. 10-3 A gear train composed of bevel, crossed-helical, and spur gears.

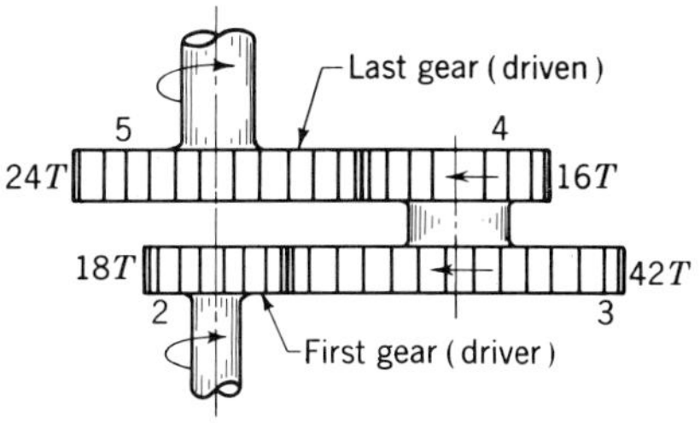

Fig. 10-4 A reverted gear train.

A *reverted gear train* (Fig. 10-4) is one in which the first and last gears are on the same axis. This arrangement produces compactness and is used in such applications as speed reducers, clocks (to connect the hour hand to the minute hand), and machine tools. As an exercise it is suggested that the reader determine a suitable set of diametral pitches for each pair of gears shown in the figure, in order for the first and last gears to have the same axis of rotation.

10-3 DETERMINING TOOTH NUMBERS

If much power is being transmitted through a speed-reduction unit, the pitch of the last pair of mating gears will be larger than that of the first pair because the torque is greater at the output end. In a given amount of space more teeth can be used on smaller-pitch gears; hence a greater speed reduction can be obtained at the high-speed end.

Without going into the problem of tooth strength, suppose we wish to use two pairs of gears in a train to obtain a train value of $\frac{1}{12}$. Let us also impose the restriction that the tooth numbers must not be less than 15 and that the reduction obtained in the first pair of gears should be about twice that obtained in the second pair. This means that

$$e = \frac{N_2}{N_3}\frac{N_4}{N_5} = \frac{1}{12} \tag{a}$$

where N_2/N_3 is the train value of the first pair, and N_4/N_5 that of the second. Since the train value of the first pair must be half that of the second,

$$\frac{N_4}{2N_5}\frac{N_4}{N_5} = \frac{1}{12} \tag{b}$$

or

$$\frac{N_4}{N_5} = \sqrt{\frac{1}{6}} = 0.4082 \tag{c}$$

to four places. The problem is best solved on the slide rule. To 0.408 on D, set the right slide index. Read N_4 on the D scale and N_5 on the C scale. The following tooth numbers are seen to be close:

$$\frac{15}{37} \qquad \frac{16}{39} \qquad \frac{18}{44} \qquad \frac{20}{49} \qquad \frac{22}{54}$$

Of these, $\frac{20}{49}$ is the best approximation; but notice that

$$e = \frac{N_2}{N_3}\frac{N_4}{N_5} = \left(\frac{20}{98}\right)\left(\frac{20}{49}\right) = \frac{200}{2401}$$

which is not quite $\frac{1}{12}$. On the other hand, the combination of $\frac{22}{108}$ for the first reduction and $\frac{18}{44}$ for the second gives a train value of exactly $\frac{1}{12}$. Thus

$$e = \left(\tfrac{22}{108}\right)\left(\tfrac{18}{44}\right) = \tfrac{1}{12}$$

In this case, the reduction in the first pair is not exactly twice that in the second. However, this consideration is usually of only secondary importance.

The problem of specifying the tooth numbers and the number of pairs of gears to give a train value within any specified degree of accuracy has interested many persons. Consider, for instance, the problem of specifying a set of gears to have a train value of $\pi/10$ accurate to eight decimal places. This is an excellent problem for digital computation.

10-4 PLANETARY GEAR TRAINS

Unusual, and perhaps unexpected, effects can be obtained in a gear train by permitting some of the gear axes to rotate about others. Such trains are called *planetary*, or *epicyclic, gear trains*. Such trains always include a *sun gear*, a *planet carrier* or *arm*, and one or more *planet gears*, as shown in Fig. 10-5. If the Grübler criterion is applied to a planetary gearset

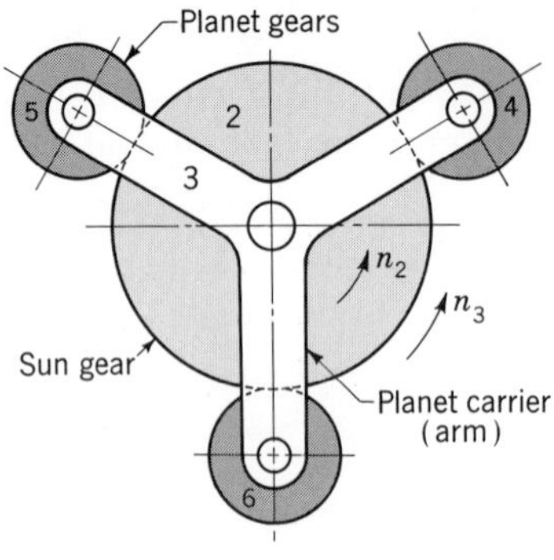

Fig. 10-5 A planetary gearset.

it will be found that there are two freedoms. Thus there must be either one input and two outputs or two inputs and one output.

The usual methods of analyzing these trains are the *formula method* and the *tabulation method*. The formula method is easier to use, but the tabulation method provides more information.

10-5 THE TABULATION METHOD

Figure 10-6 illustrates a planetary gear train composed of a sun gear 2, a planet carrier (arm) 3, a planet gear 4, and an internal gear 5 which is in mesh with the planet. We might reasonably give certain values for the rpm of the sun gear and of the arm and wish to determine the rpm of the internal gear.

The analysis is carried out in the following three steps:

1. Lock all the gears to the arm and rotate the arm one turn. Tabulate the resulting turns of the arm and of each gear.
2. Fix the arm and rotate one or more of the sun gears. Tabulate the resulting turns of each gear.
3. Add the turns of each gear in steps 1 and 2 so that the given conditions are satisfied.

As an example of such a solution, assign tooth numbers as shown in Fig. 10-6, and also let the rpm of the sun gear and arm be 100 and 200 rpm, respectively, both in the positive direction. The solution is shown in Table 10-1. In step 1 the gears are locked to the arm and the arm is given 200 counterclockwise turns. This results in 200 counterclockwise turns for gears 2, 4, and 5, too. In step 2 the arm is fixed. Now determine the turns that gear 2 must make such that when they are added to

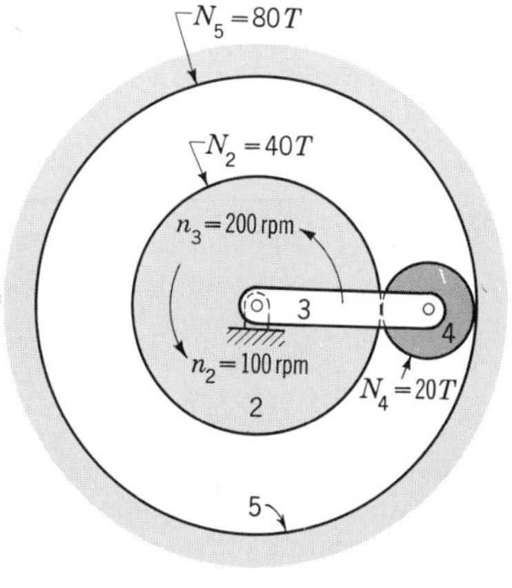

Fig. 10-6

Table 10-1 Solution by tabulation, rpm

Step number	*Arm 3*	*Gear 2*	*Gear 4*	*Gear 5*
1. Gears locked	+200	+200	+200	+200
2. Arm fixed	0	−100	+200	+ 50
3. Results	+200	+100	+400	+250

those in step 1, the result, in this case, will be +100 rpm. This is −100 turns, as shown. To complete step 2, use gear 2 as the driver and find the number of turns of gears 4 and 5. Thus

$$n_4 = (-100)(-\tfrac{40}{20}) = +200 \text{ rpm}$$

and

$$n_5 = (-100)(-\tfrac{40}{20})(\tfrac{20}{80}) = +50 \text{ rpm}$$

These values are entered in the appropriate columns, and steps 1 and 2 are added together to obtain the result.

The following worked-out examples will assist in obtaining a better understanding of this method.

Example 10-1 Find the speed of the external gear of Fig. 10-6 if gear 2 rotates 100 rpm in the clockwise direction instead, and arm 3 rotates 200 rpm counterclockwise.

Solution The results are tabulated below. In step 1 the gears are locked to the arm and the arm is given 200 turns in a counterclockwise direction. This causes gears 2, 4, and 5 to move through 200 counterclockwise turns also.

In step 2 the arm is fixed; so we record zero for the turns of the arm in the first column. In the second column, gear 2 must run in such a way that when its turns are added to those in step 1 the result will be 100 clockwise turns. For this reason, we specify −300 turns for gear 2. Now, treating gear 2 as a driver, the turns of gears 4 and 5 are:

Step number	*Arm 3*	*Gear 2*	*Gear 4*	*Gear 5*
1. Gears locked	+200	+200	+200	+200
2. Arm fixed	0	−300	+600	+150
3. Results	+200	−100	+800	+350

$$n_4 = (-300)(-\tfrac{40}{20}) = +600 \text{ rpm}$$
$$n_5 = (-300)(-\tfrac{40}{20})(\tfrac{20}{80}) = +150 \text{ rpm}$$

When the columns are summed, the result is seen to be

$$n_5 = 350 \text{ rpm ccw}$$

Example 10-2 The planetary gear train shown in Fig. 10-7 is called Ferguson's paradox. Gear 2 is stationary by virtue of being fixed to a frame. The arm 3 and gears 4 and 5 are free to turn upon the shaft. Gears 2, 4, and 5 have tooth

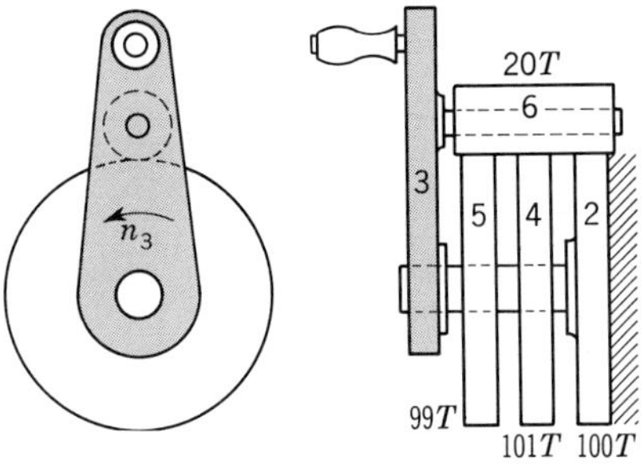

Fig. 10-7 Ferguson's paradox.

numbers of 100, 101, and 99, respectively, all cut from gear blanks of the same diameter so that the planet 6 meshes with all of them. Find the turns of gears 4 and 5 if the arm is given one counterclockwise turn.

Solution The results are shown in the table below.

Step number	*Arm* 3	*Gear* 2	*Gear* 4	*Gear* 5
1. Gears locked	+1	+1	+1	+1
2. Arm fixed	0	−1	$-\frac{100}{101}$	$-\frac{100}{99}$
3. Results	+1	0	$+\frac{1}{101}$	$-\frac{1}{99}$

In order for gear 2 to be fixed, it must be given one clockwise turn in step 2. The results show that as the arm is turned, gear 4 rotates very slowly in the same direction, while gear 5 rotates very slowly in the opposing direction.

Example 10-3 The overdrive unit shown in Fig. 10-8 is used behind a standard transmission to reduce engine speed. Determine the percentage reduction that will be obtained when the overdrive is "in."

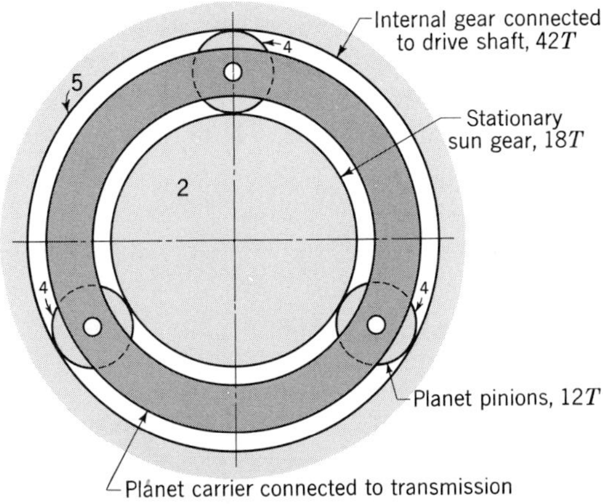

Fig. 10-8 Overdrive unit.

Solution It is convenient to use one turn for the arm. This gives the results shown in the table below. The engine speed corresponds to the speed of the

Step number	*Arm 3*	*Gear 2*	*Gear 4*	*Gear 5*
1. Gears locked	+1	+1	+1	+1
2. Arm fixed	0	−1	+1.5	+0.429
3. Results	+1	0	+2.5	+1.429

arm, and the drive-shaft speed to that of gear 5. Thus

$$\text{Reduction in engine speed} = \frac{1.429 - 1}{1.429}(100) = 30\%$$

10-6 SOLUTION OF PLANETARY TRAINS BY FORMULA

Figure 10-9 shows a planetary gear train which is composed of a sun gear 2, an arm 3, and planet gears 4 and 5. Utilizing Eq. (4-25), we may write that the velocity of gear 2 relative to the arm is

$$n_{23} = n_2 - n_3 \tag{a}$$

Also, the speed of gear 5 relative to the arm is

$$n_{53} = n_5 - n_3 \tag{b}$$

Dividing Eq. (*b*) by (*a*) gives

$$\frac{n_{53}}{n_{23}} = \frac{n_5 - n_3}{n_2 - n_3} \tag{c}$$

Equation (*c*) expresses the ratio of the relative velocity of gear 5 to that of gear 2, and both velocities are taken relative to the arm. Now this ratio is the same and is proportional to the tooth numbers, whether the arm is rotating or not. It is the train value. Therefore we may write

$$e = \frac{n_5 - n_3}{n_2 - n_3} \tag{d}$$

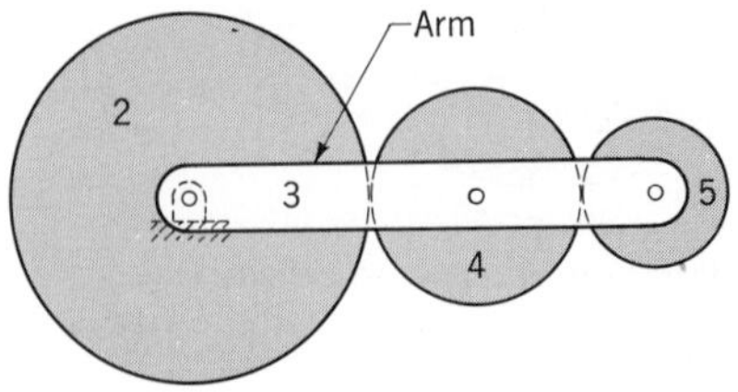

Fig. 10-9

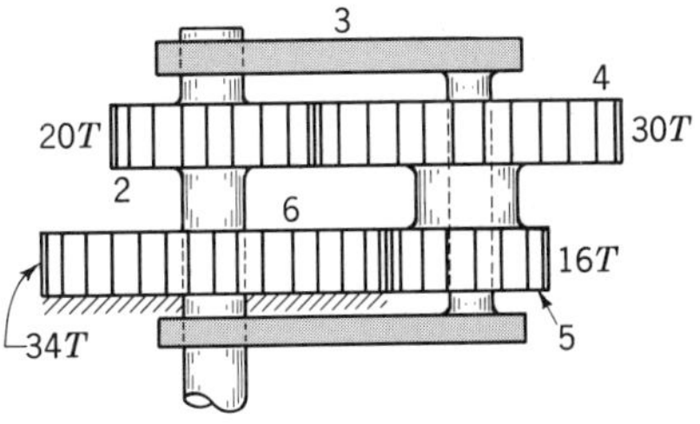

Fig. 10-10

Equation (*d*) is all that we need in order to solve any planetary train. It is convenient to express it in the form

$$e = \frac{n_L - n_A}{n_F - n_A} \tag{10-4}$$

where n_F = rpm of first gear in train
n_L = rpm of last gear in train
n_A = rpm of arm

The following examples will illustrate the use of Eq. (10-4).

Example 10-4 Figure 10-10 shows a reverted planetary train. Gear 2 is fastened to its shaft and is driven at 250 rpm in a clockwise direction. Gears 4 and 5 are planet gears which are joined but are free to turn on the shaft carried by the arm. Gear 6 is stationary. Find the rpm and direction of rotation of the arm.

Solution We must first decide which gears to designate as the first and last members of the train. Since the speeds of gears 2 and 6 are given, either may be used as the first. The choice makes no difference in the results, but once the decision is made, it cannot be changed. We shall choose gear 2 as the first; then gear 6 will be the last. Thus

$$n_F = n_2 = -250 \text{ rpm} \qquad n_L = n_6 = 0 \text{ rpm} \qquad e = \left(\tfrac{20}{30}\right)\left(\tfrac{16}{34}\right) = \tfrac{16}{51}$$

Substituting these values in Eq. (10-4),

$$\frac{16}{51} = \frac{0 - n_A}{-250 - n_A} \qquad n_A = 114 \text{ rpm ccw}$$

Example 10-5 The bevel-gear train shown in Fig. 10-11 is called Humpage's reduction gear. The input is to gear 2, and the output from gear 6, which is connected to the output shaft. The arm 3 turns freely on the output shaft and carries the planets 4 and 5. Gear 7 is fixed to the frame. What is the output speed if gear 2 rotates at 2000 rpm?

Solution The problem is solved in two steps. In the first step we consider the train as made up of gears 2, 4, and 7 and calculate the velocity of the arm. Thus

$$n_F = n_2 = 2000 \text{ rpm} \qquad n_L = n_7 = 0 \text{ rpm} \qquad e = \left(-\tfrac{20}{56}\right)\left(\tfrac{56}{76}\right) = -\tfrac{5}{19}$$

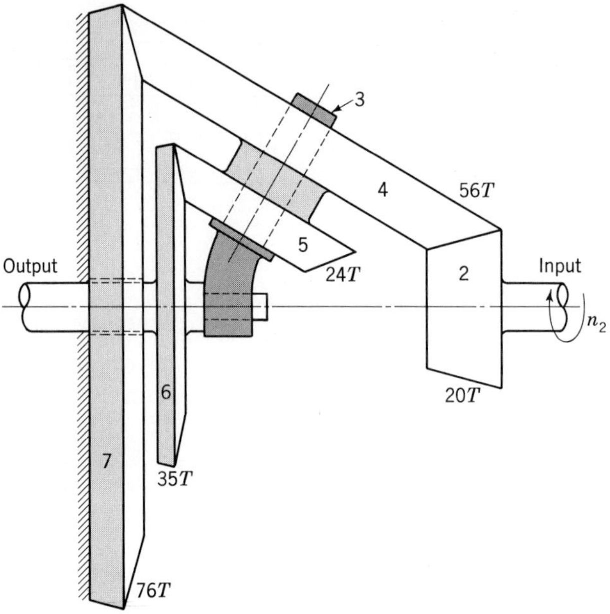

Fig. 10-11

Substituting in Eq. (10-4) and solving for the velocity of the arm gives

$$-\frac{5}{19} = \frac{0 - n_A}{2000 - n_A} \qquad n_A = 416.7 \text{ rpm}$$

Now consider the train as composed of gears 2, 4, 5, and 6. Then $n_F = n_2 = 2000$ rpm, as before, and $n_L = n_6$, which is to be found. The train value is

$$e = (-\tfrac{20}{56})(\tfrac{24}{35}) = -\tfrac{12}{49}$$

Substituting in Eq. (10-4) again and solving for n_L, since n_A is now known,

$$-\frac{12}{49} = \frac{n_L - 416.7}{2000 - 416.7}$$

$$n_L = n_6 = 28.91 \text{ rpm}$$

The output shaft rotates in the same direction as gear 2 with a reduction of 2000:28.91, or 69.2:1.

10-7 DIFFERENTIALS

The class of planetary gear trains known as differentials is used so widely that it deserves special attention. The operation of a differential is illustrated by the schematic drawing of the automotive differential shown in Fig. 10-12. The drive-shaft pinion and the ring gear are normally hypoid gears. The ring gear acts as the planet carrier, and its speed may be calculated as for a simple gear train when the speed of the drive shaft is given. Gears 5 and 6 are connected, respectively, to each rear wheel, and when the car is traveling in a straight line, these two gears rotate in the same direction with exactly the same speed. Thus for straight-line motion of the car, there is no relative motion between the planet gears and gears 5 and 6. The planet gears, in effect, serve only as keys to transmit motion from the planet carrier to both wheels.

When the vehicle is making a turn, the wheel on the inside of the turn makes fewer revolutions than the wheel with a longer turning radius. Unless this difference in speed were accommodated in some manner, one or both of the tires would have to slide in order to make the turn. The differential permits each wheel to rotate at different velocities while, at the same time, delivering power to both. During a turn, the planet gears rotate about their own axes, thus permitting gears 5 and 6 to revolve at different velocities.

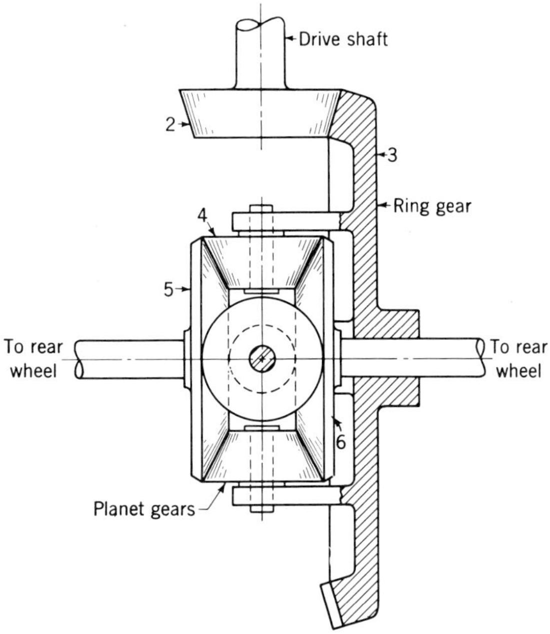

Fig. 10-12 Schematic drawing of a bevel-gear automotive differential.

If the right rear wheel is jacked up so that it is free to rotate and the left wheel left in contact with the road surface, then the effect is to fix gear 5; that is, gear 5 cannot rotate. The reader can easily verify by means of Eq. (10-4) that the right wheel will now turn at twice the speed of the ring gear.

The purpose of a differential is to differentiate between the speeds of the two wheels. In the usual passenger-car differential, the torque is divided equally whether the car is traveling in a straight line or on a curve. Sometimes the road conditions are such that the tractive effort that can be developed by the two wheels is unequal. In this case the total tractive effort available will be only twice that at the wheel having the least traction, because the differential divides the torque equally. If one wheel should happen to be resting on snow or ice, then the total effort available is very small and only a small torque will be required to cause the wheel to spin.

Most automotive manufacturers will provide a *locking differential* as optional equipment to avoid the spinning-wheel problem. The locking differential contains a clutch which is always engaged. This clutch is connected between one of the rear axles and the ring gear. When the car is traveling in a straight line, both rear axles and the ring gear rotate at the same speed and there is no slipping between the plates. During a turn, however, slippage occurs because the rear axles rotate at different speeds than the ring gear. When one or both wheels are on a low-traction surface, such as ice or mud, the friction between the clutch plates transfers some of the torque to the wheel having the most traction, prevents spinning, and assists both wheels in driving.

In automatic-control mechanisms, differentials are often used as error detectors. Suppose, for example, that the negative of a desired motion is fed into the left side of the differential of Fig. 10-12 and the actual motion is fed into the right side. If the two motions are exactly equal and opposite to each other, the ring gear remains stationary. Any rotation of the ring gear is, therefore, a measure of the difference between the two motions or signals.

Differentials are also used in calculators for adding and subtracting. For example, let θ_L be the input to the left shaft, θ_R the input to the right shaft, and θ_T the rotation of the ring gear. Then

$$\theta_T = \frac{\theta_L + \theta_R}{2}$$

Figure 10-13 illustrates a *differential screw*. In this numerical example there are 16 threads per inch on one end and 18 threads per inch on the other. When the crank is turned, the movement of the carriage is con-

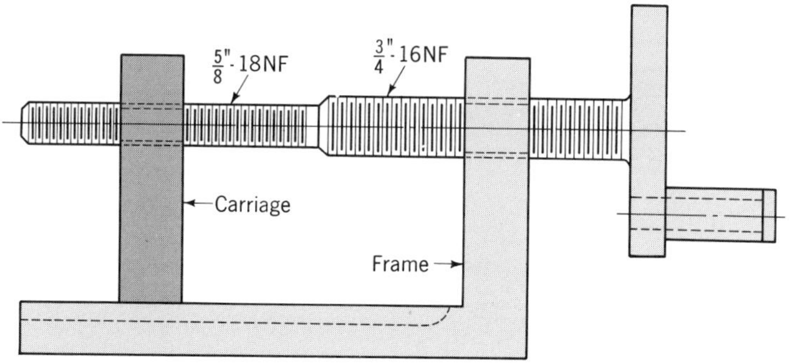

Fig. 10-13 A differential screw.

trolled by the difference in the pitch of the two threads. If the handle is moved one revolution, the screw advances $\frac{1}{16}$ in. into the frame and $\frac{1}{18}$ in. into the carriage. Hence the motion of the carriage is

$$p_1 - p_2 = \tfrac{1}{16} - \tfrac{1}{18} \approx 0.007 \text{ in.}$$

To obtain the same motion with a single screw thread would require a pitch of $\frac{1}{143}$ in.

PROBLEMS

10-1. Find the speed and direction of rotation of gear 8 in the figure. What is the train value?

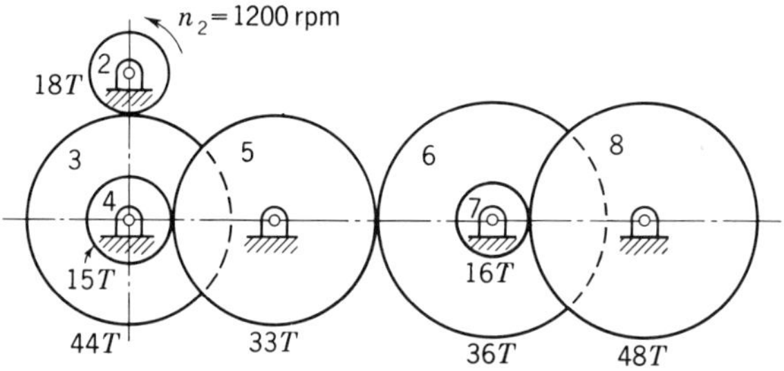

Prob. 10-1

10-2. The figure gives the pitch diameters of a set of spur gears forming a train. Calculate the velocity ratio and the train value. What are the speed and direction of rotation of gear 5? Of gear 7?

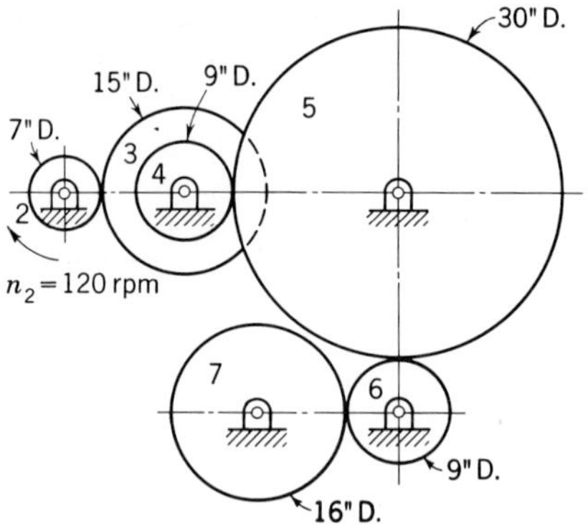

Prob. 10-2

10-3. The figure shows a gear train consisting of bevel gears, spur gears, and a worm and worm gear. The bevel pinion is mounted on a shaft which is driven by a V-belt and pulleys. If pulley 2 rotates at 1200 rpm in the direction shown, find the speed and direction of rotation of gear 9.

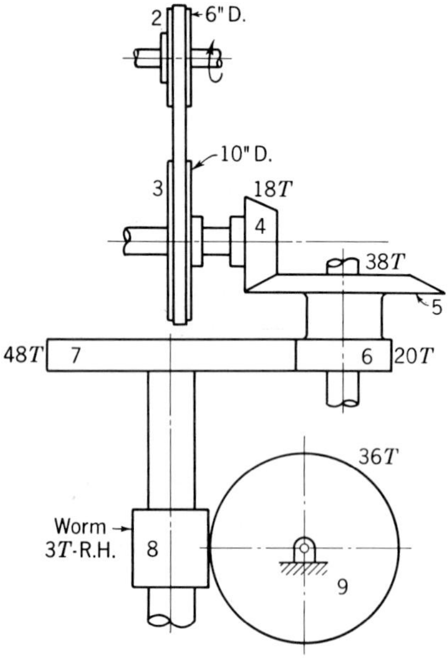

Prob. 10-3

10-4. Use the truck transmission of Fig. 10-3 and an input speed of 3000 rpm. Find the drive-shaft speed for first, second, third, fourth, and reverse gears.

10-5. The figure illustrates the gears in a speed-change gearbox used in machine-tool applications. By sliding the cluster gears on shafts B and C, nine speed changes can be obtained. The problem of the machine-tool designer is to select tooth numbers for the various gears so as to produce a reasonable distribution of speeds for the output shaft. The smallest and largest gears are gears 2 and 9, respectively. Using 20 teeth and 45 teeth for these gears, determine a set of suitable tooth numbers for the remaining gears. What are the corresponding speeds of the output shaft? Notice that the problem has many solutions.

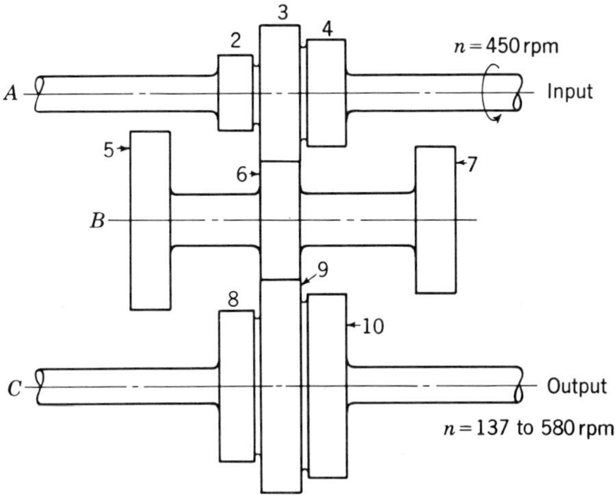

Prob. 10-5

10-6. Determine the speed and direction of rotation of the output shaft shown in the figure. The 104T internal gear is connected to the stationary frame.

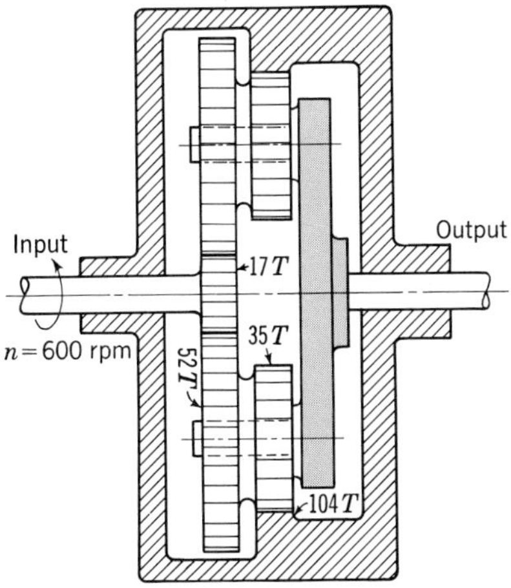

Prob. 10-6

10-7. The internal gear (gear 7) in the figure turns counterclockwise at 60 rpm. What are the speed and direction of rotation of arm 3?

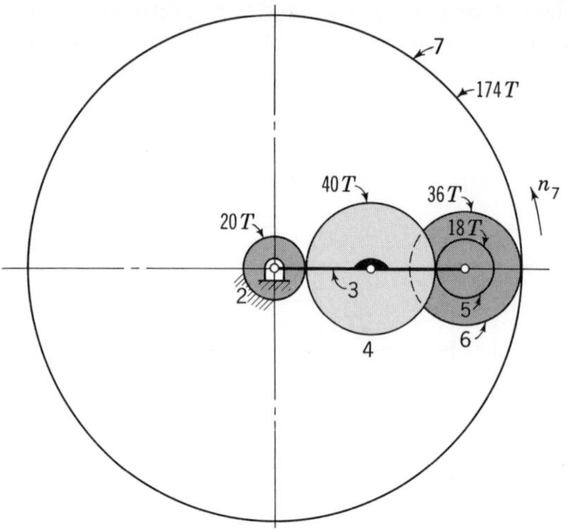

Probs. 10-7 and **10-8**

10-8. If the arm in the figure rotates at 300 rpm in a counterclockwise direction, determine the speed and direction of rotation of the internal gear.

10-9. Shaft C in the figure is stationary. If gear 2 rotates clockwise at 800 rpm, what are the speed and direction of rotation of shaft B?

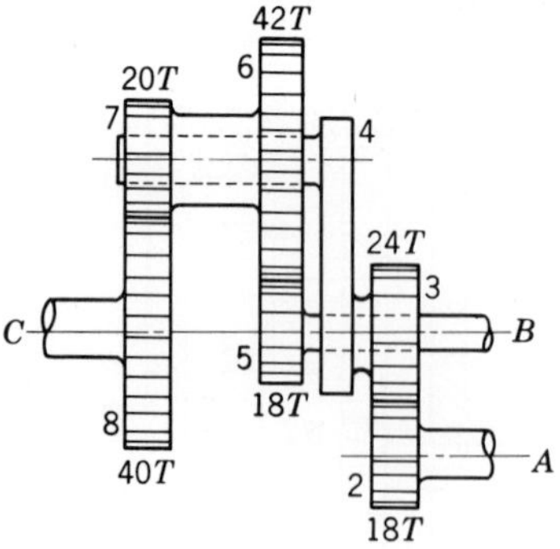

Probs. 10-9 to **10-11**

10-10. If shaft B in the figure is stationary and shaft C is driven 380 rpm in a counterclockwise direction, what are the rpm and direction of rotation of shaft A?

10-11. What are the speed and direction of rotation of shaft C in the figure (a) if shafts A and B rotate at 360 rpm in a counterclockwise direction; (b) if shafts A and B turn at 360 rpm with A turning clockwise and B counterclockwise?

10-12. If gear 2 in the figure is connected to the input shaft and arm 3 to the output shaft, what speed reduction can be obtained? What is the sense of rotation of the output shaft? What changes could be made in the gear train to produce the opposite sense of rotation?

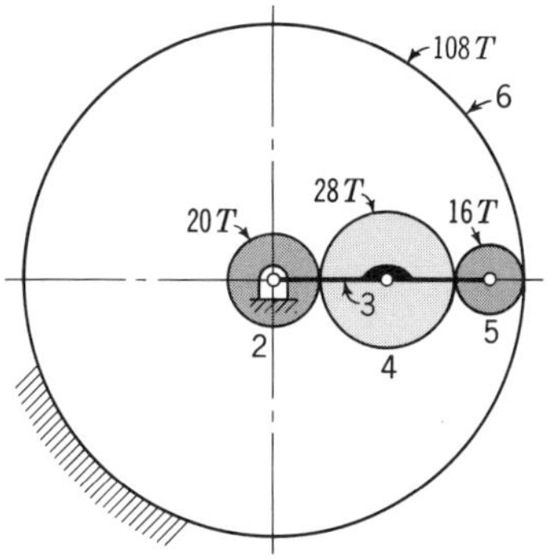

Prob. 10-12

10-13. The tooth numbers for the automotive differential shown in the figure are $N_2 = 11$, $N_3 = 54$, $N_4 = 11$, and $N_5 = N_6 = 16$. The drive shaft turns at 1200 rpm. What is the speed of the right wheel if it is jacked up and the left wheel is resting on the road surface?

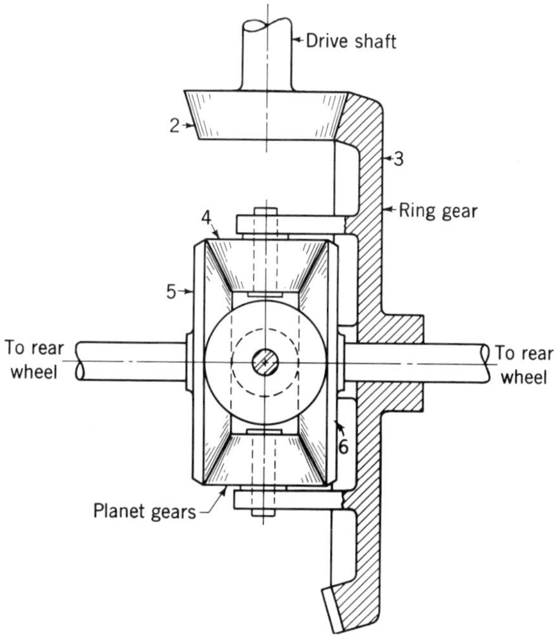

Probs. 10-13 and **10-14**

10-14. A vehicle using the differential shown in the figure turns to the right at a speed of 30 mph on a curve of 80 ft radius. Use the same tooth numbers as in Prob. 10-13. The tires are 15 in. in diameter. Use 60 in. as the center-to-center distance between treads.

(*a*) Calculate the speed of each rear wheel.

(*b*) What is the speed of the ring gear?

11
Synthesis of Linkages

By *kinematic synthesis* we mean the design or the creation of a mechanism to yield a desired set of motion characteristics. For the most part synthesis is accomplished by using the geometric theory of kinematics in Chap. 6 and inversion. Because of the very large number of techniques available, some of which may be quite frustrating, we shall present here some of the more useful methods together with a sampling of the approaches which illustrate the usefulness of the theory.[1,2]

[1] Extensive bibliographies may be found in Kurt Hain (translated by T. P. Goodman, Douglas P. Adams, B. L. Harding, and D. R. Raichel), "Applied Kinematics," 2d ed., pp. 639–727, McGraw-Hill Book Company, 1967, and Ferdinand Freudenstein and George N. Sandor, Kinematics of Mechanisms, in Harold A. Rothbart (ed.), "Mechanical Design and Systems Handbook," pp. 4–56 to 4–68, McGraw-Hill Book Company, New York, 1964.

[2] In the English language, the following are the most useful references on kinematic synthesis:

Rudolf A. Beyer (translated by Herbert Kuenzel), "Kinematic Synthesis of Mechanisms," McGraw-Hill Book Company, New York, 1963.

Alexander Cowie, "Kinematics and Design of Mechanisms," International

11-1 TWO-POSITION SYNTHESIS

If θ_2 is the angular position of link 2 in a four-bar linkage and θ_4 the angular position of link 4, then one of the problems in kinematic synthesis is to find the dimensions of the linkage such that

$$\theta_4 = f(\theta_2) \tag{a}$$

where f is any desired functional relation.

Though this problem has not been solved, it is possible to specify up to five values for θ_2, called *precision points*, and sometimes find a linkage which will satisfy the desired relationship for these points. The process, usually, consists in plotting a graph of the function and then selecting two to five precision points from the graph for use in the synthesis. If the process is successful, the functional relation is satisfied for these points, but deviations will occur elsewhere. For many functions the greatest error can be held to less than 4 percent.

Figure 11-1*a* shows two positions, A_1 and A_2, of point A. Point A can be made to travel from A_1 to A_2 by rotating it about any point on the perpendicular bisector, also called the *midnormal*, of the line A_1A_2. If, for example, a stationary point O is chosen on the midnormal and a link inserted from O to A_1, then OA_2 is another position of the link.

Note that O can be chosen at infinity, and in this case the link is a slider.

In Fig. 11-1*b*, A_1B_1 and A_2B_2 are two positions of the same rigid link AB. The problem here is to use AB as the coupler of a four-bar mechanism designed to enable AB to occupy either of these positions. Draw the perpendicular bisector of A_1A_2 and B_1B_2. These intersect at point R, which is called the *rotopole*. Now choose *any* two points O_2 and O_4 on each of these bisectors. Place a link from O_2 to A and another link from O_4 to B, as shown. Then the linkage O_2ABO_4 is the synthesized four-bar mechanism. Note that there are an infinite number of linkages which satisfy the problem requirements.

In this simple problem we have solved two synthesis problems.

Textbook Company, Scranton, Pa., 1961.

Hain, *op. cit.*

Allen S. Hall, Jr., "Kinematics and Linkage Design," Prentice-Hall, Inc., Englewood Cliffs, N.J., 1961.

R. S. Hartenberg and Jacques Denavit, "Kinematic Synthesis of Linkages," McGraw-Hill Book Company, New York, 1964.

Jeremy Hirschhorn, "Kinematics and Dynamics of Plane Mechanisms," McGraw-Hill Book Company, New York, 1962.

D. C. Tao, "Applied Linkage Synthesis," Addison-Wesley Publishing Company, Inc., Reading, Mass., 1964.

The first problem is that of moving the coupler from one position A_1B_1 to a second position A_2B_2. The second problem that can be solved by this approach is that of getting a specified output or swing angle ϕ_{12} (Fig. 11-1*b*) for a given oscillation ψ_{12} of the driver. This is accomplished by a proper choice of the frame points O_2 and O_4. Note, too, that the choice of O_2 and O_4 dictates whether the result will be a double-rocker, a double-crank, or a crank-and-rocker mechanism.

Inversion is a particularly useful and powerful technique in synthesis. Before illustrating the method for two-position synthesis it will be of value to examine the finished solution to a problem. Figure 11-2*a* shows a four-bar linkage in which motion of the input rocker O_2A through the angle ψ_{12} causes a motion of the output rocker through the angle ϕ_{12}. Two positions of the linkage are shown: the $O_2A_1B_1O_4$ position and the $O_2A_2B_2O_4$ position. By *inverting on the output rocker* we mean that we are going to hold O_4B stationary and permit the remaining links, including the frame, to occupy the positions dictated by the constraints. Thus Fig. 11-2*b* shows the same linkage inverted on the O_4B position. The first position is the same as in Fig. 11-2*a*. Since O_4B is to be fixed, the frame will have to move in order to get the linkage to the second position. In fact, it will have to move *backward* through the angle ϕ_{12}. The second position is, therefore, $O_2'A_2'B_2'O_4$.

Finally, we observe that the coupler AB is rigid and hence A_1B_1 is the same distance as $A_2'B_2'$.

In a synthesis problem we merely reverse this procedure. We have

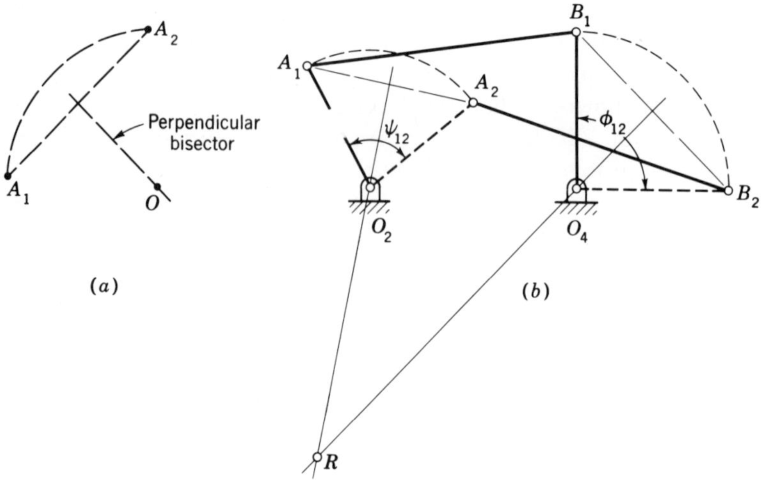

Fig. 11-1

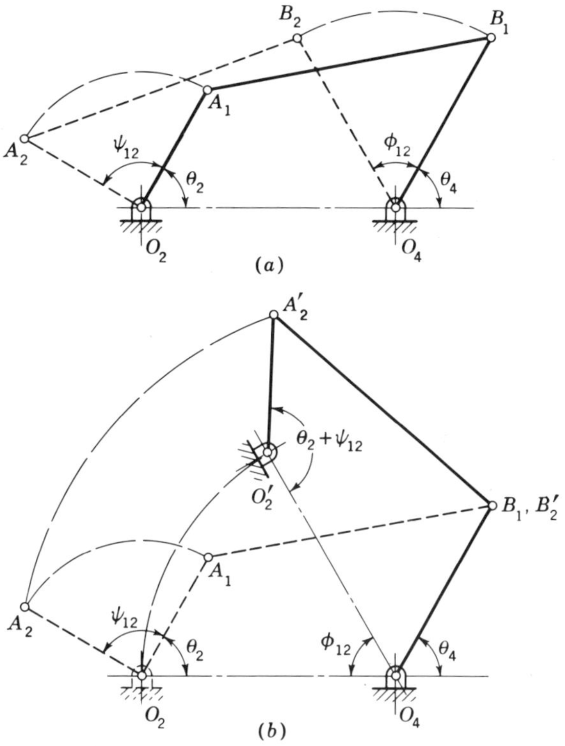

Fig. 11-2 (*a*) Rotation of input rocker O_2A through the angle ψ_{12} causes the output rocker O_4B to rock through the angle ϕ_{12}. (*b*) Linkage inverted on the O_4B position.

given two positions of the input link and the angle ϕ_{12} through which the output link is to move. What, then, are the dimensions of the linkage? In Fig. 11-3 the given quantities are shown in *a* and the result in *b*. Begin by selecting a point O_4 on the frame. Visualize O_4B_1 as fixed and rotate O_2 to O_2' and A_2 to A_2', both backward through the angle ϕ_{12}. Select a convenient coupler length and strike arcs about A_1 and A_2'. These intersect at B_1, defining the distance O_4B_1. The resulting linkage is shown in both positions in the figure.

We might note that an infinite number of solutions are possible; this suggests that additional requirements could have been specified. Also, there is nothing about the process which warrants that B will move directly from B_1 to B_2 while A is moving from A_1 to A_2. In fact, it is possible to obtain impractical designs which require removal of the pin at B in order to get from one position to the other.

11-2 PROPERTIES OF THE ROTOPOLE

In Fig. 11-1*b* we learned that the rotopole R lies at the intersection of the perpendicular bisectors of two finitely separated coupler positions. We might note that the rotopole *is not* the same as the velocity pole except for infinitesimal separations.

The geometry associated with the rotopole is shown in Fig. 11-4. Notice that the midnormals a_{12} and b_{12} to the lines A_1A_2 and B_1B_2, respectively, contain the frame points O_2 and O_4 and subtend the same angle β at the rotopole as does the coupler AB. In addition, the input and output levers O_2A and O_4B each subtend the same angle α at the rotopole though their swing angles ψ and ϕ may differ.

Figure 11-4 illustrates only two positions of coupler AB. Should there also be a third position designated by the subscript 3, then we can define three rotopoles. These are R_{12}, R_{23}, and R_{13}, finitely separated and each associated with a corresponding pair of coupler positions.

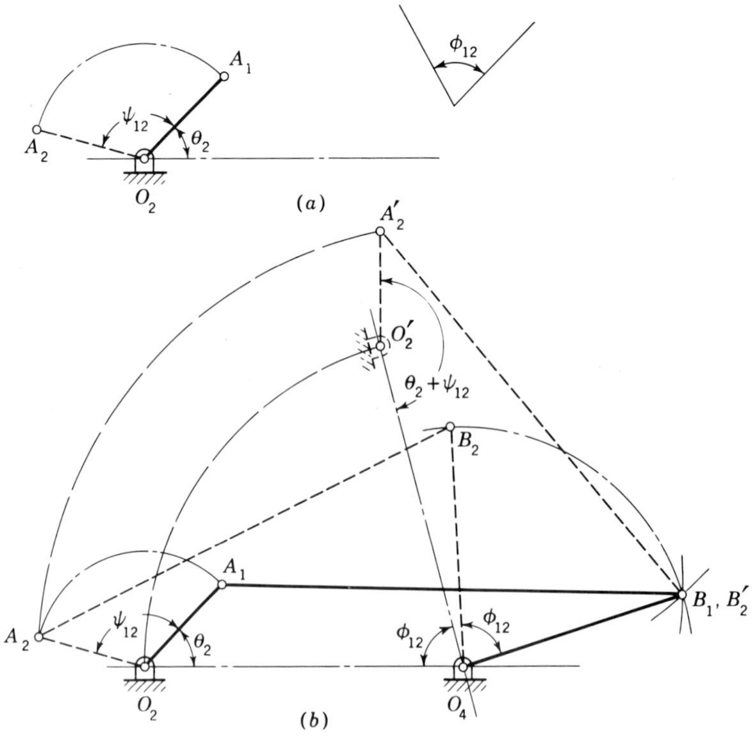

Fig. 11-3 Synthesis using inversion.

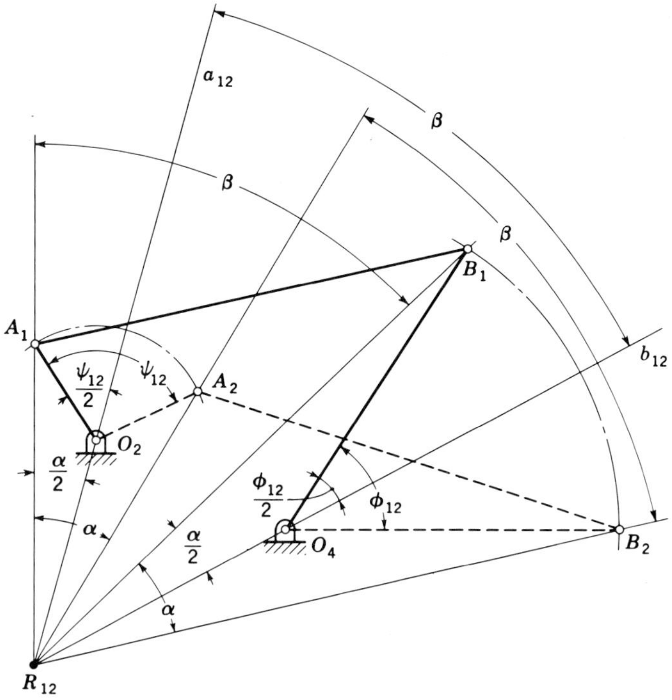

Fig. 11-4 Geometry of the rotopole.

11-3 CHEBYSHEV SPACING

In the general case it is not possible to specify a function $y = f(x)$ and design a linkage which will generate the desired function exactly. Instead, we choose a number of precision points on a graph of the function and design the linkage to pass through these points. Between the points deviations, called *structural errors,* will occur. One of the problems of linkage design is to select a set of precision points for use in the synthesis so as to minimize the structural errors.

For the first trial the best spacing of these points is called *Chebyshev spacing.* For n points in the range $x_0 \leq x \leq x_{n+1}$ the Chebyshev spacing, according to Freudenstein and Sandor,[1] is

$$x_j = \frac{1}{2}(x_0 + x_{n+1}) - \frac{1}{2}(x_{n+1} - x_0)\cos\frac{\pi(2j-1)}{2n} \qquad j = 1, 2, \ldots, n \qquad \text{(11-1)}$$

where x_j are the precision points.

[1] *Op. cit.,* p. 4-27.

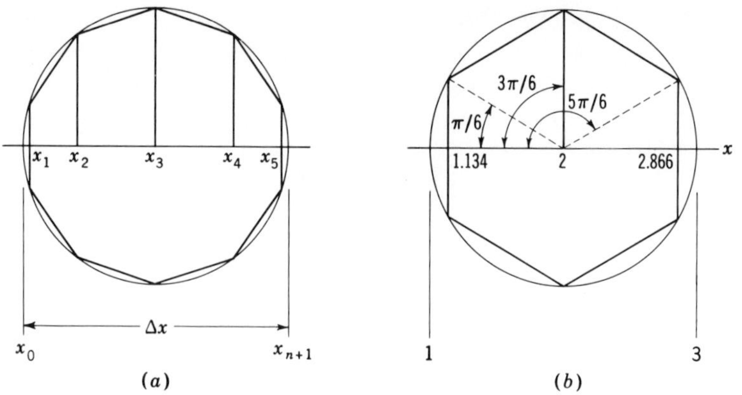

Fig. 11-5 Graphical determination of Chebyshev spacing.

As an example, suppose we wish to devise a linkage to generate the function

$$y = x^{0.8} \tag{a}$$

for the range $1 \leq x \leq 3$ using three precision points. Then, from Eq. (11-1), the three values of x are

$$x_1 = \frac{1}{2}(1 + 3) - \frac{1}{2}(3 - 1)\cos\frac{\pi(2 - 1)}{(2)(3)}$$

$$= 2 - \cos\frac{\pi}{6} = 1.134$$

$$x_2 = 2 - \cos\frac{3\pi}{6} = 2.000$$

$$x_3 = 2 - \cos\frac{5\pi}{6} = 2.866$$

We find the corresponding values of y, from Eq. (a), to be

$$y_1 = 1.106 \qquad y_2 = 1.741 \qquad y_3 = 2.322$$

These accuracy points are easily obtained by using the graphical approach in Fig. 11-5. The method is shown in Fig. 11-5a, where a circle is first constructed whose diameter is the range Δx, given by the equation

$$\Delta x = x_{n+1} - x_0 \tag{b}$$

In this circle inscribe a regular polygon having $2n$ sides. Perpendiculars dropped from each corner will intersect Δx in the precision points. Figure 11-5b illustrates the construction for the numerical example.

In closing this section it should be noted that the Chebyshev spacing is the best first approximation; depending upon the accuracy requirements of the problem, it may be satisfactory. If additional accuracy is required, then, by plotting a curve of the structural error versus x, one can usually determine visually the adjustments to be made in the precision points for the next trial.

11-4 OPTIMIZATION OF THE TRANSMISSION ANGLE

In Sec. 6-4 we learned that the best motion characteristics are obtained when the transmission angle, the angle between the coupler and follower in a four-bar linkage, deviates from 90° as little as possible. In this section we shall illustrate the synthesis of a two-position four-bar mechanism for optimum transmission angle.

First, we shall show that by choosing the frame connection O_4 coincident with the rotopole, equal transmission angles will be obtained at the two design positions.[1] Suppose we wish to design a linkage in which the input lever is to swing through the angle ψ and drive the output member through the angle ϕ. Construct the midnormal a_{12} (Fig. 11-6), and

[1] See Hain, *op. cit.*, p. 362.

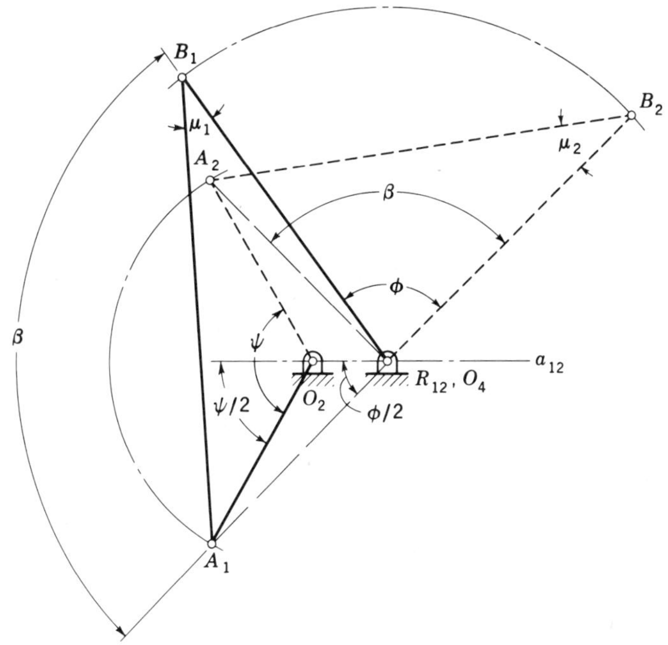

Fig. 11-6

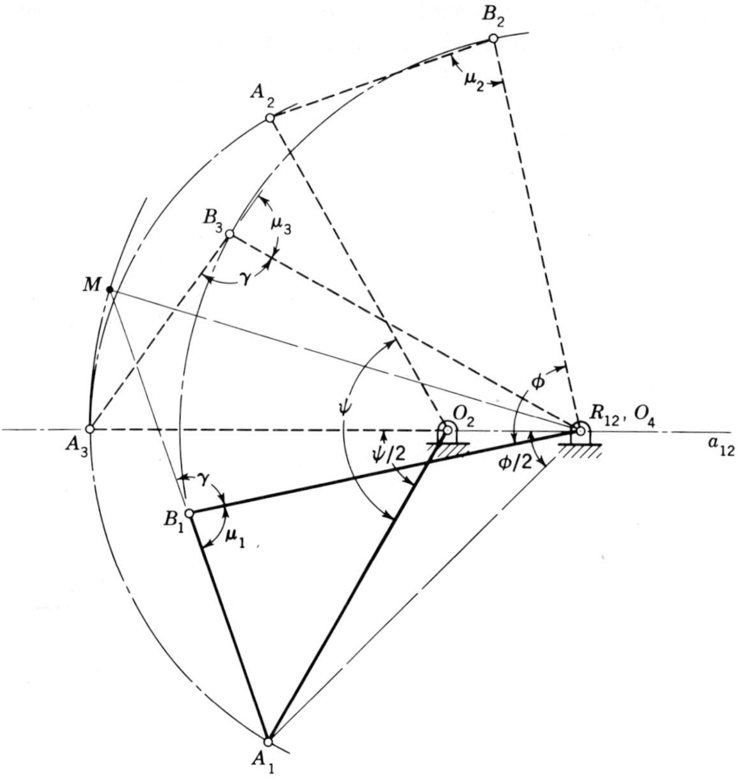

Fig. 11-7

on it locate pivots O_2 and O_4 and designate O_4 also as the rotopole R_{12}. Through O_2 lay off the angle $\psi/2$ and through O_4 the angle $\phi/2$. The legs of these two angles intersect and define point A_1. The next step is to select B_1. Avoid locating it so as to cause the mechanism to lock up or to require the follower to reverse its motion. Except for these restrictions, B_1 can be selected anywhere. When the resulting linkage is turned to the second design position, shown in dashed lines, the transmission angle μ_2 in Fig. 11-6 is seen to be the same as the starting angle μ_1. Inspection of the figure shows that this occurs because the frame connection O_4 was chosen coincident with the rotopole. Triangles $O_4A_1B_1$ and $O_4A_2B_2$ are congruent, and they each subtend the same angle β at the rotopole.

Inspection of Fig. 11-6 shows that the transmission angle μ_1 depends upon the choice of B_1. We suspect that it can be improved by making a better choice for the location of B_1. The optimum situation would be one in which we began with an interior transmission angle μ_1 as close to 90° as possible. During motion of the linkage the angle should improve, pass through 90° and reach an exterior angle $\mu_3 = \mu_1$, and then return to the interior angle $\mu_2 = \mu_1$ at the second design position. Hain[1] shows how to do this.

Figure 11-7 is an optimum solution to the problem solved in Fig. 11-6. We begin in the same manner by locating O_2, A_1, and O_4, with the rotopole coincident with O_4. The difference in the two approaches occurs in the manner of choosing B_1. Notice that we have selected an intermediate position O_2A_3 for the input lever coincident with the mid-normal a_{12}. Through A_3 construct an arc with center at O_4 and choose any point M on this arc. Draw the line A_1M and bisect it, locating B_1. This completes the construction, and the resulting linkage is $O_2A_1B_1O_4$. As before, μ_1 and μ_2 are equal at the two design positions.

We now wish to show that the transmission angle μ_3 in the intermediate position is the same as μ_1. Observe that B_1 is the midpoint of A_1M, that $O_4M = O_4A_3$, and, of course, that $O_4B_1 = O_4B_3$. Therefore, the two triangles O_4B_1M and $O_4B_3A_3$ are congruent. Locating the corresponding angles γ in each triangle, we conclude that μ_3 is indeed the same as μ_1.

The transmission angles μ_1, μ_2, and μ_3 are all *minimum* transmission angles. At intermediate points the actual angle will be closer to 90° and, hence, better.

11-5 THE OVERLAY METHOD

Synthesis of a function generator, say, using the overlay method, is the easiest and quickest of all methods to use. It is not always possible to obtain a solution, and sometimes the accuracy is rather poor. Theoretically, however, one can employ as many points as are desired in the process.

Let us design a function generator to solve the equation

$$y = x^{0.8} \qquad 1 \leq x \leq 3 \tag{a}$$

Suppose we choose six positions of the linkage for this example and use uniform spacing of the output rocker. Table 11-1 shows the values of x and y, to slide-rule accuracy, and the corresponding angles selected for the input and output rockers.

The first step in the synthesis is shown in Fig. 11-8*a*. Use a sheet of tracing paper and construct the input rocker O_2A in all its positions.

[1] *Ibid.*

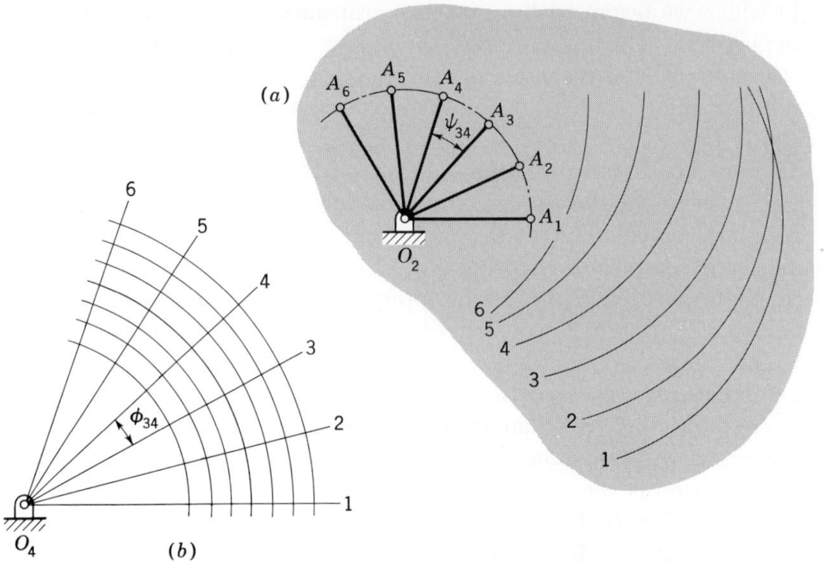

Fig. 11-8

This requires a choice for the length of O_2A. Also, on this sheet, choose a length for the coupler AB and draw arcs numbered 1 through 6 using A_1 to A_6, respectively, as centers.

Now, on another sheet of paper, construct the output rocker, whose length is unknown, in all its positions, as shown in Fig. 11-8*b*. Through O_4 draw a number of equally spaced arcs intersecting the lines O_41, O_42, etc.; these represent possible lengths of the output rocker.

The final step is to lay the tracing over the drawing and manipulate it in an effort to find a "fit." In this case a fit was found, and the result is shown in Fig. 11-9.

Table 11-1

Position	x	ψ, *deg*	y	ϕ, *deg*
1	1	0	1	0
2	1.366	22.0	1.284	14.2
3	1.756	45.4	1.568	28.4
4	2.16	69.5	1.852	42,6
5	2.58	94.8	2.136	56.8
6	3.02	121.0	2.420	71.0

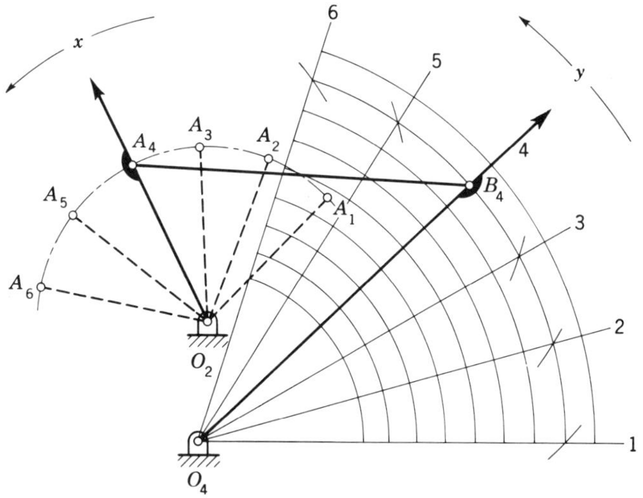

Fig. 11-9

11-6 THREE-POSITION SYNTHESIS

In Sec. 11-1 (Fig. 11-3) we employed two precision points in synthesizing a linkage. Here we shall present a similar procedure using three precision points. For convenience we present the problem and the synthesized linkage in Fig. 11-10. Input lever 2 is to drive the output rocker 4 through three specified positions. The starting angle of the input lever is θ_2; and ψ_{12}, ψ_{23}, and ψ_{13} are the swing angles, respectively, between the design positions 1 and 2, 2 and 3, and 1 and 3. Corresponding angles of

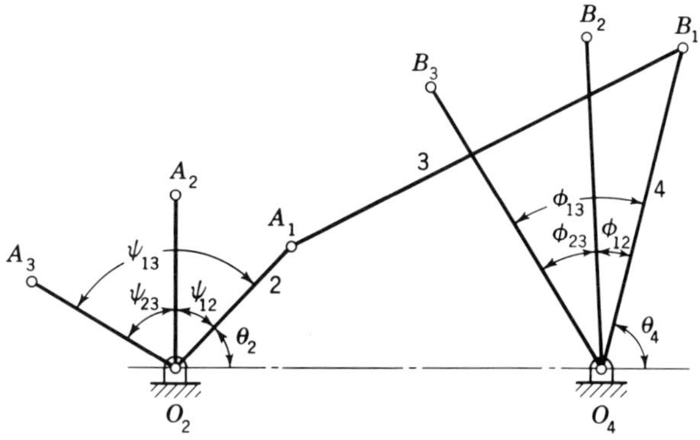

Fig. 11-10

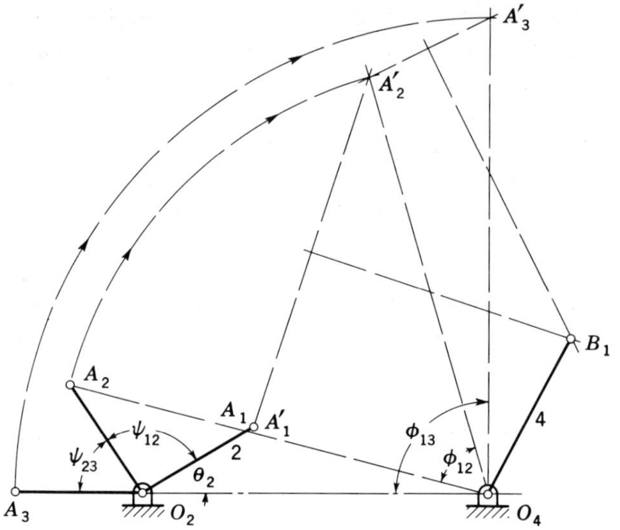

Fig. 11-11

swing ϕ_{12}, ϕ_{23}, and ϕ_{13} are desired for the output lever. The length of link 4 and its starting position θ_4 are to be determined.

The solution to the problem is illustrated in Fig. 11-11, and it is based on inverting the linkage on link 4. Draw the input rocker O_2A in its three specified positions and locate a desired position for O_4. Since we shall invert on link 4 in the first design position, draw a ray from O_4 to A_2 and rotate it backward through the angle ϕ_{12} to locate A_2'. Similarly, draw another ray O_4A_3 and rotate it backward through the angle ϕ_{13} to locate A_3'. Since we are inverting on the first design position, A_1 and A_1' are coincident. Now draw midnormals to the lines $A_1'A_2'$ and $A_2'A_3'$. These intersect at B_1 and define the length of the coupler 3 and the length and starting position of link 4.

11-7 POINT-POSITION REDUCTION—FOUR PRECISION POINTS

In point-position reduction the linkage is made symmetrical about the frame centerline O_2O_4 so as to cause two of the A' points to be coincident. The effect of this is to produce three equivalent A' points through which a circle can be drawn as in three-position synthesis. It is best illustrated by an example.

Let us synthesize a linkage to generate the function $y = \log x$ for $10 \leq x \leq 60$ using an input crank range of 120° and an output range of 90°.

To simplify the presentation we shall not employ Chebyshev spacing. The angle ψ is evaluated for the four design positions from the equation $\psi = ax + b$ and from the boundary conditions $\psi = 0$ when $x = 10$ and $\psi = 120°$ when $x = 60$. This gives $\psi = 2.40x - 24$. The angle ϕ is evaluated in exactly the same manner; thus, we get $\phi = 50y - 115$. The results of this preliminary work are shown in Table 11-2.

For the starting position a choice of four arrangements is shown in Fig. 11-12. In *a* the line O_2O_4 bisects both ψ_{12} and ϕ_{12}; and so, if the output member is turned counterclockwise from the O_4B_2 position, A_1' and A_2' will be coincident and at A_1. The inversion would then be based on the O_4B_1 position. Then A_3 would be rotated through the angle ϕ_{13} about O_4 counterclockwise to A_3', and A_4 through the angle ϕ_{14} to A_4'.

In Fig. 11-12*b* the line O_2O_4 bisects ψ_{23} and ϕ_{23}, while in *d* the angles ψ_{14} and ϕ_{14} are bisected. In obtaining the inversions for each case great

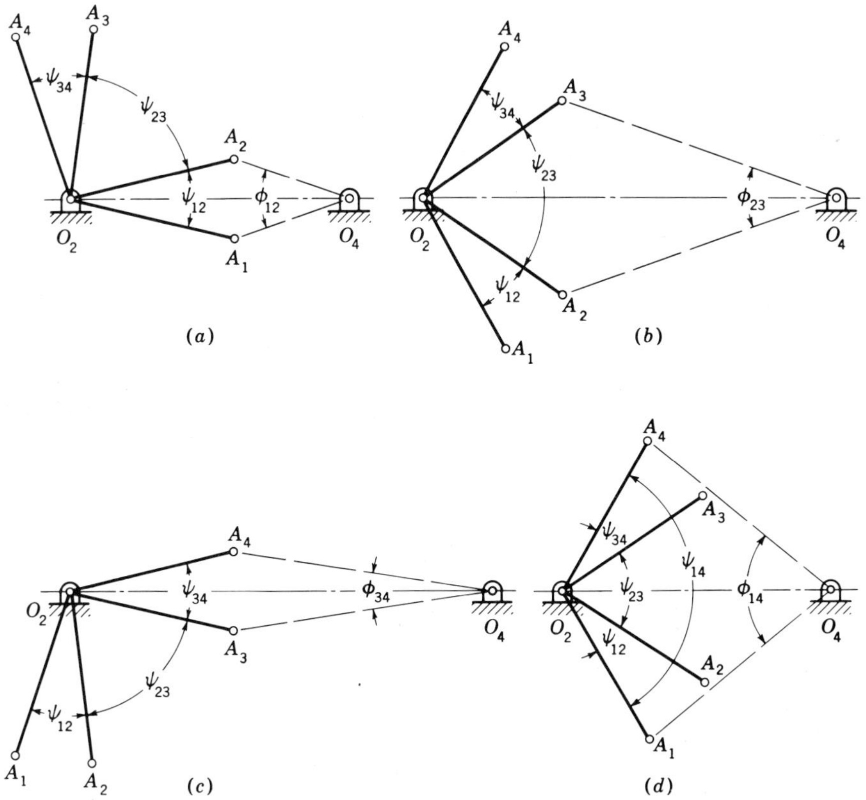

Fig. 11-12

Table 11-2*

Position	x	ψ, *deg*	y	ϕ, *deg*
1	10	0	2.30	0
2	20	24	3.00	35
3	45	94	3.80	75
4	60	120	4.10	90

* $\psi_{12} = 24°$ $\phi_{12} = 35°$
$\psi_{23} = 70°$ $\phi_{23} = 40°$
$\psi_{34} = 26°$ $\phi_{34} = 15°$

care must be taken to ensure that rotation is made in the correct direction and with the correct angles.

When point-position reduction is used, only the length of the input rocker O_2A can be specified in advance. The distance O_2O_4 is dependent upon the values of ψ and ϕ, as indicated in Fig. 11-12. Notice that each

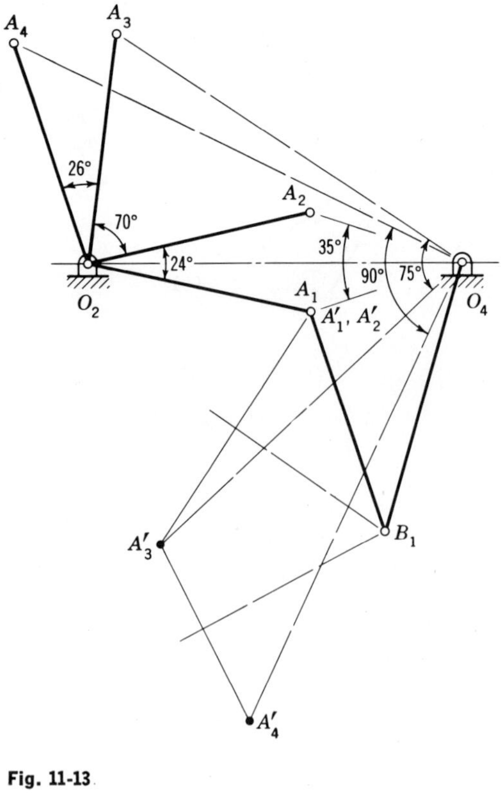

Fig. 11-13

synthesis position gives a different value for this distance. This is really quite convenient because it is not at all unusual to synthesize a linkage which is not workable. When this happens, one of the other arrangements can be tried.

The synthesized linkage is shown in Fig. 11-13. The procedure is exactly the same as that for three precision points, except as previously noted. Point B_1 is obtained at the intersection of the midnormals to $A_1'A_3'$ and $A_3'A_4'$. In this example the greatest error is less than 3 percent.

11-8 COUPLER-CURVE SYNTHESIS—BY POINT-POSITION REDUCTION[1]

In this section we shall employ the method of point-position reduction to synthesize a four-bar linkage so that a tracing point on the coupler will trace any previously specified path when the linkage is moved. Then, in sections to follow we shall discover that paths having certain characteristics are particularly useful in synthesizing linkages having dwells of the output member for certain periods of rotation of the input member.

In synthesizing a linkage to generate a path we can choose up to six precision points on the path. If the synthesis is successful, the tracing point will pass through each precision point. The final result may or may not approximate the desired path.

Two positions of a four-bar linkage are shown in Fig. 11-14. Link 2 is the input member; it is connected at A to coupler 3, containing the tracing point C, and connected to output link 4 at B. Two phases of the linkage are illustrated by the subscripts 1 and 3. Points C_1 and C_3 are two positions of the tracer on the path to be generated. In this example C_1 and C_3 have been especially selected so that the midnormal c_{13} passes through O_4. Note, for the selection of points, that the angle $C_1O_4C_3$ is the same as the angle $A_1O_4A_3$ as indicated on the figure.

The advantage of making these two angles equal is that, when the linkage is finally synthesized, the triangles $C_3A_3O_4$ and $C_1A_1O_4$ are congruent. Thus, if the tracing point is made to pass through C_1 on the path, then it will also pass through C_3.

To synthesize a linkage so that the coupler will pass through four precision points, we locate any four points C_1, C_2, C_3, C_4 on the desired path (Fig. 11-15). Choosing C_1 and C_3, say, we first locate O_4 anywhere on the midnormal c_{13}. Then, with O_4 as a center and any radius R, construct a circle arc. Next, with centers at C_1 and C_3 and any other radius r, strike arcs to intersect the arc of radius R. These two intersections define points A_1 and A_3 on the input link. Construct the mid-

[1] The methods presented here were devised by Kurt Hain and are presented in his book, "Applied Kinematics," *op. cit.*, chap. 17.

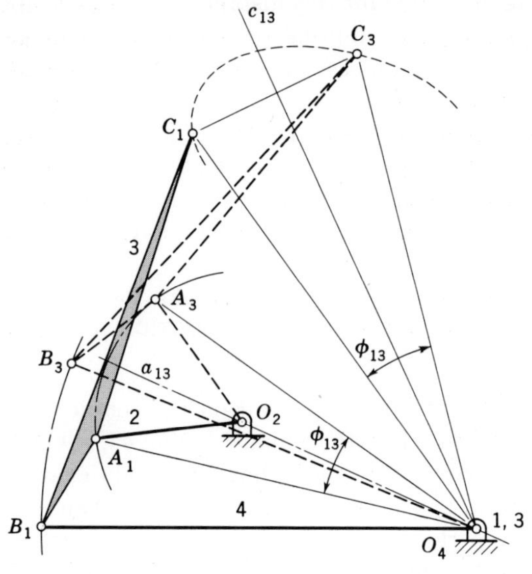

Fig. 11-14

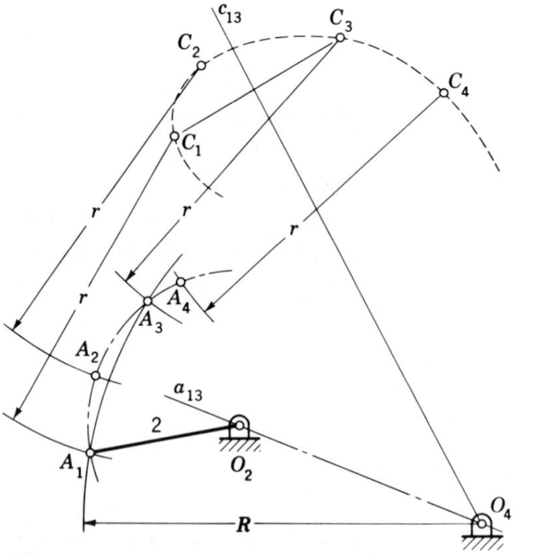

Fig. 11-15

normal a_{13} to A_1A_3 and note that it passes through O_4. Locate O_2 anywhere on a_{13}. This provides an opportunity to choose a convenient length for the input rocker. Now use O_2 as a center and draw the crank circle through A_1 and A_3. Points A_2 and A_4 on this circle are obtained by striking arcs of radius r again about C_2 and C_4. This completes the first phase of the synthesis; we have located O_2 and O_4 relative to the desired path and hence defined the distance O_2O_4. We have also defined the length of the input member and located its positions relative to the four precision points on the path.

Our next task is to locate point B, the point of attachment of the coupler and output member. Any one of the four locations of B may be employed; in this example we shall use the B_1 position.

Before beginning the final step we note that the linkage is now defined. Four decisions were made: the location of O_4, the radii R and r, and the location of O_2. Thus an infinite number of solutions are possible.

Referring to Fig. 11-16, locate point 2 by making triangles $C_2A_2O_4$ and C_1A_12 congruent. Locate point 4 by making $C_4A_4O_4$ and C_1A_14 congruent. Points 4, 2, and O_4 lie on a circle whose center is B_1. So B_1 is found at the intersection of the midnormals of O_42 and O_44. Note that the procedure used causes points 1 and 3 to coincide with O_4. With B_1 located, the links may be drawn in place and the mechanism tested to see how well it traces the prescribed path.

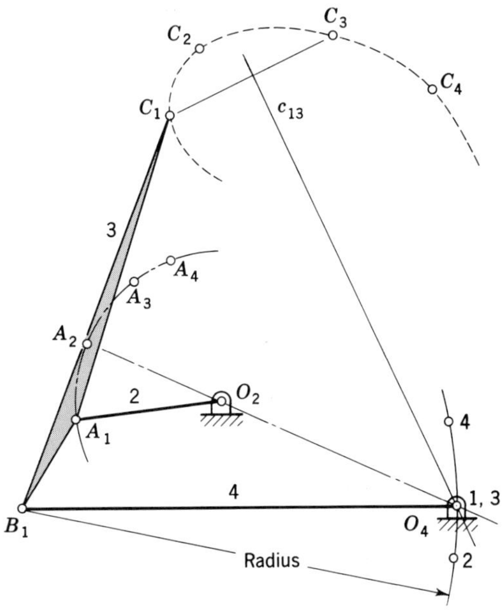

Fig. 11-16

To synthesize a linkage to generate a path through five precision points, it is necessary to make two point reductions. Begin by choosing five points C_1 to C_5 on the path to be traced. Choose two pairs of these for reduction purposes. In Fig. 11-17 we have chosen the pairs C_1C_5 and C_2C_3. Other pairs which could have been used are

$$C_1C_5,\ C_2C_4 \qquad C_1C_5,\ C_3C_4 \qquad C_1C_4,\ C_2C_3 \qquad C_2C_5,\ C_3C_4$$

Construct the perpendicular bisectors c_{23} and c_{15} of the lines connecting each pair. These intersect at point O_4. Note that O_4 may therefore be located conveniently by a judicious choice of the pairs to be used as well as by the choice of the positions of the points C_i on the path.

The next step is best performed by using a scrap of tracing paper as an overlay. Secure the tracing paper to the drawing and mark upon it the center O_4, the midnormal c_{23}, and another line from O_4 to C_2. Such an overlay is shown in Fig. 11-18*a* with the line O_4C_2 designated as O_4C_2'. This defines the angle $\phi_{23}/2$. Now rotate the overlay about O_4 until the midnormal coincides with c_{15} and repeat for point C_1. This defines the angle $\phi_{15}/2$ and the corresponding line O_4C_1'.

Now pin the overlay at O_4, using a thumbtack, and rotate it until a good position is found. It is helpful to set the compass for some convenient radius r and draw circles about each point C_i. The intersection of these circles with the lines O_4C_1' and O_4C_2' on the overlay, and with each other, will reveal which areas will be profitable to investigate. See Fig. 11-18*b*.

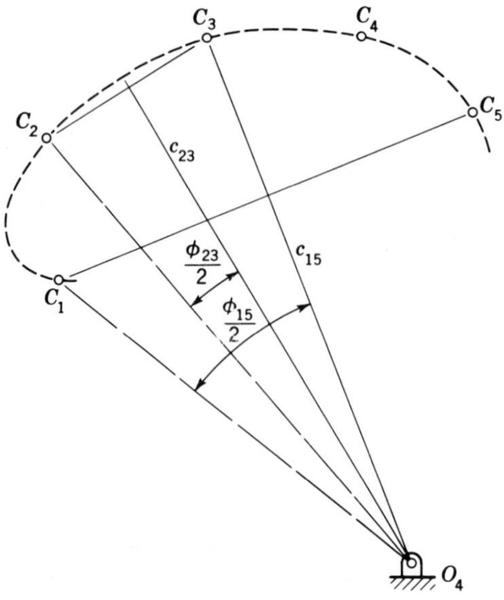

Fig. 11-17

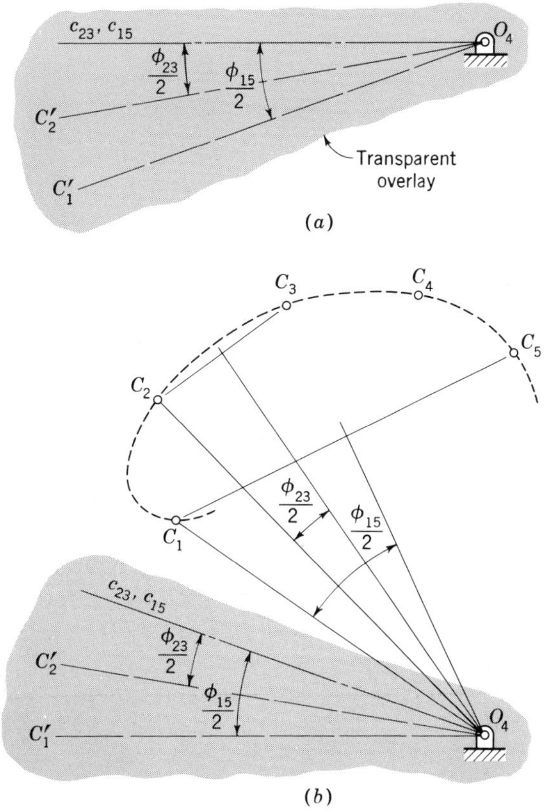

Fig. 11-18

The final steps in the solution are shown in Fig. 11-19. Having located a good position for the overlay, transfer the three lines to the drawing and remove the overlay. Now draw a circle arc of radius r to intersect O_4C_1' and locate A_1. Another arc of the same radius r from C_2 intersects O_4C_2' at A_2. With A_1 and A_2 located, draw the midnormal a_{12}; it intersects a_{23} at O_2, giving the length of the input rocker. A circle through A_1 about O_2 will contain all the design positions of A; use the same radius r and locate A_3, A_4, and A_5 on arcs about C_3, C_4, and C_5.

We have now located everything except point B_1, and this is found in the same manner as before. A double point 2, 3 exists because of the choice of O_4 on the midnormal c_{23}. To locate this point, strike an arc from C_1 of radius C_2O_4. Then strike another arc from A_1 of radius A_2O_4. These intersect at point 2, 3. To locate point 4, strike an arc from C_1 of radius C_4O_4, and another from A_1 of radius A_4O_4. Note that points

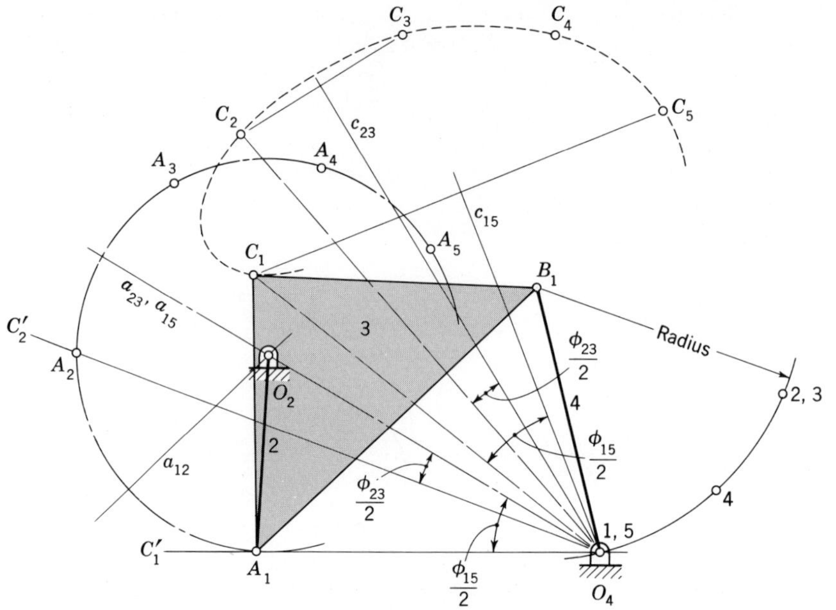

Fig. 11-19

O_4 and the double points 1, 5 are coincident because the synthesis is based on inversion on the O_4B_1 position. Points O_4, 4 and double points 2, 3 lie on a circle whose center is B_1, as shown in Fig. 11-19. The linkage is completed by drawing in the coupler and the follower in the first design position.

11-9 COUPLER-CURVE SYNTHESIS—GENERAL

In the previous section we learned that four-bar linkages can sometimes be synthesized to generate a desired curve. We also learned that an infinite number of solutions are possible. The unfortunate possibility therefore exists that one may spend many days or even weeks in an attempt to generate a particular solution and, at the end of that period, be no closer to a solution than at the beginning. And yet an infinite number of solutions still remain to be investigated! Since the dollar spent for engineering time must be conserved, our purpose in this section is to suggest alternative approaches.

The references cited at the beginning of this chapter contain additional methods of coupler-curve synthesis; if the alternative approaches suggested in this section do not yield solutions, these references should be consulted.

The Hrones and Nelson atlas, noted in Chap. 6, contains 7000 coupler curves and should always be consulted first.[1] This atlas can be used to obtain coupler curves having one and two cusps or having crossovers or figure eights, for curves having segments which approximate circle arcs, and for curves having straight-line segments.

Hunt, Fink, and Nayar[2] give the dimensions of a class of four-bar linkages which generate a symmetrical triangular-shaped path with two of the sides approximating straight lines.

Hartenberg and Denavit[3] and also Hall[4] illustrate most of the classical straight-line generators. Tesar, Vidosic, and Watts[5] have investigated approximate straight-line mechanisms in great detail and have developed a very considerable amount of information for this class of mechanisms.

It was observed in Chap. 6 that Ball's point, the intersection of the cubic of stationary curvature with the inflection circle, is a point of inflection on the coupler curve with stationary curvature. Thus there is a good possibility of getting a rather accurate straight line by locating a coupler point coincident with Ball's point.

In a similar manner, one can usually get an accurate circle arc by choosing a coupler point on the cubic of stationary curvature. Then the inflection circle is used to obtain the exact value of the radius of curvature.

Freudenstein[6] reveals linkage dimensions capable of generating segments of parabolas and ellipses. Timm[7] presents a general method of kinematic analysis employing the electronic analog computer, which has great promise in synthesis. With his approach the coupler paths are displayed on the cathode-ray oscilloscope. With modern high-speed

[1] John A. Hrones and George L. Nelson, "Analysis of the Four-bar Linkage," published jointly by The Technology Press of the Massachusetts Institute of Technology, Cambridge, Mass., and John Wiley & Sons, Inc., New York, 1951.

[2] K. H. Hunt, N. Fink, and J. Nayar, Linkage Geneva Mechanisms: A Design Study in Mechanism Geometry, *Proc. Inst. Mech. Engrs.*, vol. 174, no. 21, pp. 643–668, 1960. See also Hirschhorn, *op. cit.*, pp. 349–353.

[3] *Op. cit.*

[4] *Op. cit.*

[5] D. Tesar and J. P. Vidosic, Analysis of Approximate Four-bar Straight-line Mechanisms, *J. Eng. Ind.*, ser. B, vol. 87, no. 3, August, 1965; Delbert Tesar and E. H. Watts, The Analytical Design of an Adjustable Four-bar Linkage for Variable Straight-line Motions, *ASME Paper* 66-Mech-30, October, 1966; J. P. Vidosic and D. Tesar, The Selection of Watt's Four-bar Straight-line Mechanism, *ASME Paper* 66-Mech-15, October, 1966.

[6] Ferdinand Freudenstein, Higher Path-curvature Analysis in Plane Kinematics, *J. Eng. Ind.*, ser. B, vol. 87, no. 2, May, 1965.

[7] R. F. Timm, Analog Simulation of Rigid Link Mechanisms, *J. Eng. Ind.*, ser. B, vol. 89, no. 2, May, 1967. See also the discussion by Joseph E. Shigley.

analog-hybrid computers it should be possible to display a series of changing coupler paths much like a motion picture. The logic section of the computer can be used to increment the linkage dimensions. Since such computers operate at speeds of over 1000 solutions per second, one should be able to display, say, the entire Hrones and Nelson atlas in just a few minutes.

11-10 SYNTHESIS OF DWELL MECHANISMS[1]

One of the more interesting uses of coupler curves having straight-line or circle-arc segments is in the synthesis of mechanisms having a substantial dwell during a portion of their operating period. By using segments of coupler curves it is not difficult to synthesize linkages having a dwell at either or both of the extremes of its motion or at an intermediate point.

In Fig. 11-20 a coupler curve having approximately an elliptical shape is selected from the Hrones and Nelson atlas so that a substantial portion of the curve approximates a circle arc. Connecting link 5 is then given a length equal to the radius of this arc. Thus, in the figure, points D_1, D_2, and D_3 are stationary while coupler point C moves through positions C_1, C_2, and C_3. The length of output link 6 and the location of the frame point O_6 depend upon the desired angle of oscillation of this link. The frame point should also be positioned for optimum transmission angle.

When segments of circle arcs are desired for the coupler curve, an organized method of searching the Hrones and Nelson atlas should be employed. The overlay, shown in Fig. 11-21, is made on a sheet of tracing paper and can be fitted over the paths in the atlas very quickly. It

[1] See Charles E. Rice and Lee Harrisberger, Precision Six-link Dwell Mechanisms, *ASME Paper* 66-Mech-32, October, 1966.

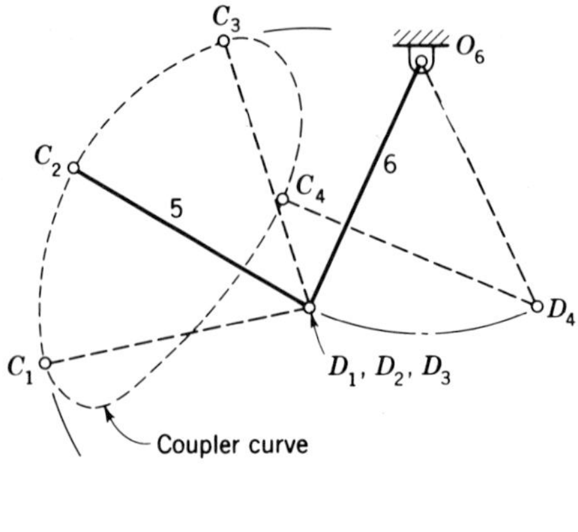

Fig. 11-20 Synthesis of a dwell mechanism; the four-bar linkage which generates the coupler curve is not shown.

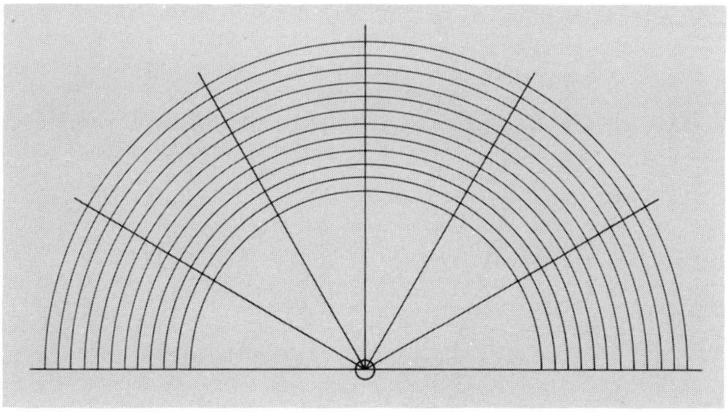

Fig. 11-21 Overlay for use with the Hrones and Nelson atlas.

reveals immediately the radius of curvature of the segment, the location of pivot point D, and the swing angle of the connecting link.

Figure 11-22 shows a dwell mechanism employing a slider. A coupler curve having a straight-line segment is used, and the pivot point O_6 placed on an extension of this line.

The arrangement shown in Fig. 11-23 has a dwell at both extremes of the motion. A practical arrangement of this mechanism is rather difficult to achieve, however, because link 6 has such a high velocity when the slider is near the pivot O_6.

The slider mechanism of Fig. 11-24 uses a figure-eight coupler curve having a straight-line segment to produce an intermediate dwell linkage. Pivot O_6 must be located on an extension of the straight-line segment, as shown.

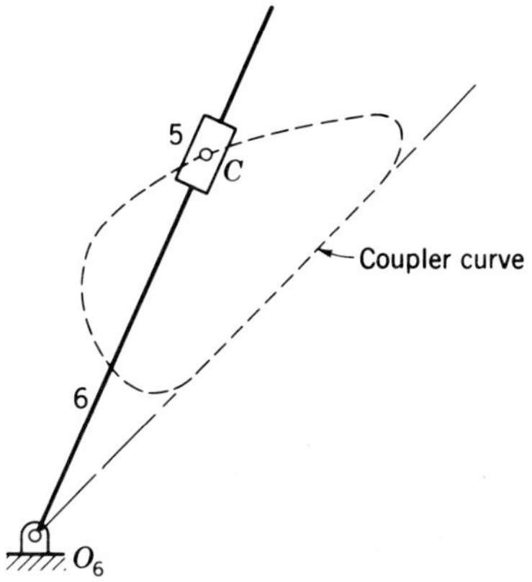

Fig. 11-22

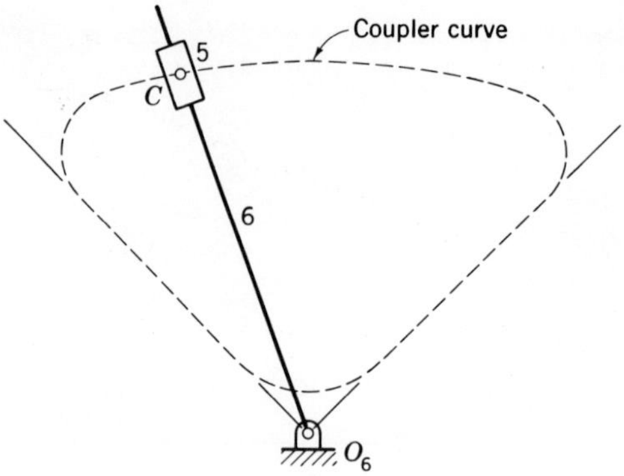

Fig. 11-23

11-11 BLOCH'S SYNTHESIS

Methods of synthesis using complex numbers probably originated with Bloch,[1] and since then they have been used quite extensively.[2–5]

In Fig. 11-25 replace the links of a four-bar linkage by position vectors and write the vector equation

$$\mathbf{r}_1 + \mathbf{r}_2 + \mathbf{r}_3 + \mathbf{r}_4 = 0 \tag{a}$$

[1] S. Sch. Bloch, On the Synthesis of Four-bar Linkages (in Russian), *Bull. Acad. Sci. USSR*, pp. 47–54, 1940.

[2] Rudolf Beyer, "Kinematische Getriebesynthese," pp. 189–192, Springer-Verlag OHB, Berlin, Vienna, 1953.

[3] N. Rosenauer, Complex Variable Method for Synthesis of Four-bar Linkages, *Australian J. Appl. Sci.*, vol. 5, no. 4, 1954.

[4] G. H. Martin and M. F. Spotts, An Application of Complex Geometry to Relative Velocities and Accelerations in Mechanisms, *Trans. ASME*, vol. 79, pp. 687–693, 1957. See also George H. Martin, Four-bar Linkages, *Machine Design*, vol. 30, no. 8, pp. 146–149, April, 1958.

[5] R. S. Hartenberg, Complex Numbers and Four-bar Linkages, *Machine Design*, vol. 30, no. 6, pp. 156–163, March, 1958.

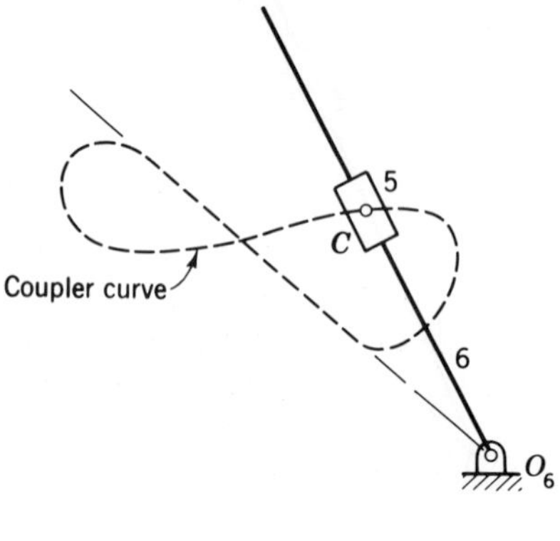

Fig. 11-24

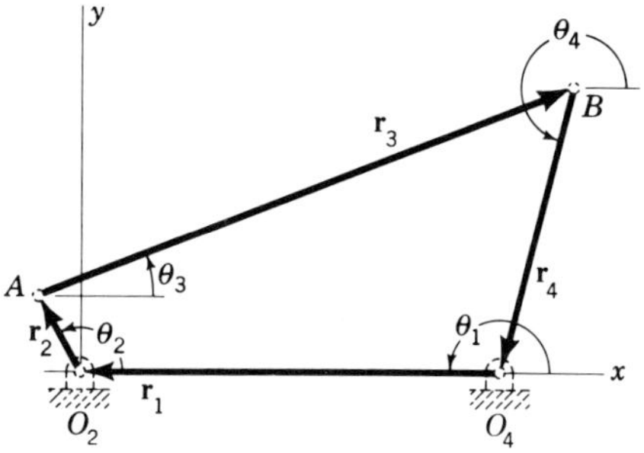

Fig. 11-25

In complex polar notation Eq. (*a*) is written

$$r_1e^{j\theta_1} + r_2e^{j\theta_2} + r_3e^{j\theta_3} + r_4e^{j\theta_4} = 0 \tag{b}$$

The first and second derivatives of this equation are

$$r_2\omega_2e^{j\theta_2} + r_3\omega_3e^{j\theta_3} + r_4\omega_4e^{j\theta_4} = 0 \tag{c}$$

$$r_2(\dot{\omega}_2 + j\omega_2{}^2)e^{j\theta_2} + r_3(\dot{\omega}_3 + j\omega_3{}^2)e^{j\theta_3} + r_4(\dot{\omega}_4 + j\omega_4{}^2)e^{j\theta_4} = 0 \tag{d}$$

If we now transform Eqs. (*b*), (*c*), and (*d*) back to vector notation, we obtain the simultaneous equations

$$\begin{aligned} \mathbf{r}_1 \quad + \mathbf{r}_2 \quad + \mathbf{r}_3 \quad + \mathbf{r}_4 &= 0 \\ \omega_2\mathbf{r}_2 \quad + \omega_3\mathbf{r}_3 \quad + \omega_4\mathbf{r}_4 &= 0 \\ (\dot{\omega}_2 + j\omega_2{}^2)\mathbf{r}_2 + (\dot{\omega}_3 + j\omega_3{}^2)\mathbf{r}_3 + (\dot{\omega}_4 + j\omega_4{}^2)\mathbf{r}_4 &= 0 \end{aligned} \tag{e}$$

This is a set of homogeneous vector equations having complex numbers as coefficients. Bloch specified desired values of all the angular velocities and angular accelerations and then solved the equations for the relative linkage dimensions.

Solving Eqs. (*e*) for $\mathbf{r}_2$ gives

$$\mathbf{r}_2 = \frac{\begin{vmatrix} -1 & 1 & 1 \\ 0 & \omega_3 & \omega_4 \\ 0 & \dot{\omega}_3 + j\omega_3{}^2 & \dot{\omega}_4 + j\omega_4{}^2 \end{vmatrix}}{\begin{vmatrix} 1 & 1 & 1 \\ \omega_2 & \omega_3 & \omega_4 \\ \dot{\omega}_2 + j\omega_2{}^2 & \dot{\omega}_3 + j\omega_3{}^2 & \dot{\omega}_4 + j\omega_4{}^2 \end{vmatrix}} \tag{f}$$

Similar expressions will be obtained for $\mathbf{r}_3$ and $\mathbf{r}_4$. It turns out that the denominators for all three expressions, that is, for $\mathbf{r}_2$, $\mathbf{r}_3$, and $\mathbf{r}_4$, are complex numbers and are equal. In Sec. 2-10 we learned to divide and multiply complex numbers. In division, we divide the magnitudes and subtract the angles. Since these denominators are all alike, the effect of the division would be to change the magnitudes of $\mathbf{r}_2$, $\mathbf{r}_3$, and $\mathbf{r}_4$ by the same factor, and to shift all the directions by the same angle. For this reason, we make all the denominators unity; the solutions then give dimensionless vectors for the links. When the determinants are evaluated, we find

$$\begin{aligned}
\mathbf{r}_2 &= \omega_4(\dot{\omega}_3 + j\omega_3^2) - \omega_3(\dot{\omega}_4 + j\omega_4^2) \\
\mathbf{r}_3 &= \omega_2(\dot{\omega}_4 + j\omega_4^2) - \omega_4(\dot{\omega}_2 + j\omega_2^2) \\
\mathbf{r}_4 &= \omega_3(\dot{\omega}_2 + j\omega_2^2) - \omega_2(\dot{\omega}_3 + j\omega_3^2) \\
\mathbf{r}_1 &= -\mathbf{r}_2 - \mathbf{r}_3 - \mathbf{r}_4
\end{aligned} \tag{11-2}$$

Example 11-1 Synthesize a four-bar linkage to give the following values for the angular velocities and accelerations:

$$\omega_2 = 200 \text{ rad/sec} \qquad \omega_3 = 85 \text{ rad/sec} \qquad \omega_4 = 130 \text{ rad/sec}$$

$$\dot{\omega}_2 = 0 \text{ rad/sec}^2 \qquad \dot{\omega}_3 = -1000 \text{ rad/sec}^2 \qquad \dot{\omega}_4 = -16{,}000 \text{ rad/sec}^2$$

Solution Substituting the given values into Eqs. (11-2) gives

$$\begin{aligned}
\mathbf{r}_2 &= 130[-1000 + j(85^2)] - 85[-16{,}000 + j(130)^2] \\
&= 1{,}230{,}000 - j497{,}000 = 1{,}330{,}000\underline{/-22^\circ} \text{ units} \\
\mathbf{r}_3 &= 200[-16{,}000 + j(130)^2] - 130[0 + j(200)^2] \\
&= -3{,}200{,}000 - j1{,}820{,}000 = 3{,}690{,}000\underline{/-150.4^\circ} \text{ units} \\
\mathbf{r}_4 &= 85[0 + j(200)^2] - 200[-1000 + j(85)^2] \\
&= 200{,}000 + j1{,}955{,}000 = 1{,}965{,}000\underline{/84.15^\circ} \text{ units} \\
\mathbf{r}_1 &= -(1{,}230{,}000 - j497{,}000) - (-3{,}200{,}000 - j1{,}820{,}000) \\
&\quad - (200{,}000 + j1{,}955{,}000) \\
&= 1{,}770{,}000 + j362{,}000 = 1{,}810{,}000\underline{/11.6^\circ} \text{ units}
\end{aligned}$$

In Fig. 11-26*a* these four vectors are plotted to a scale of 10^6 units per inch. In order to make $\mathbf{r}_1$ horizontal and in the $-x$ direction, the entire vector system must be rotated counterclockwise $180 - 11.6 = 168.4°$. The resulting linkage can then be constructed by using $\mathbf{r}_1$ for link 1, $\mathbf{r}_2$ for link 2, etc., as shown in Fig. 11-26*b*. This mechanism has been dimensioned in inches and, if analyzed, will show that the conditions of the example have been fulfilled.

11-12 FREUDENSTEIN'S EQUATION[1]

If Eq. (*b*) of the preceding section is transformed into complex rectangular form and if the real and the imaginary components are separated, then we obtain the two algebraic equations

$$r_1 \cos \theta_1 + r_2 \cos \theta_2 + r_3 \cos \theta_3 + r_4 \cos \theta_4 = 0 \tag{a}$$

$$r_1 \sin \theta_1 + r_2 \sin \theta_2 + r_3 \sin \theta_3 + r_4 \sin \theta_4 = 0 \tag{b}$$

[1] Ferdinand Freudenstein, Approximate Synthesis of Four-bar Linkages, *Trans. ASME*, vol. 77, no. 6, pp. 853–861, 1955.

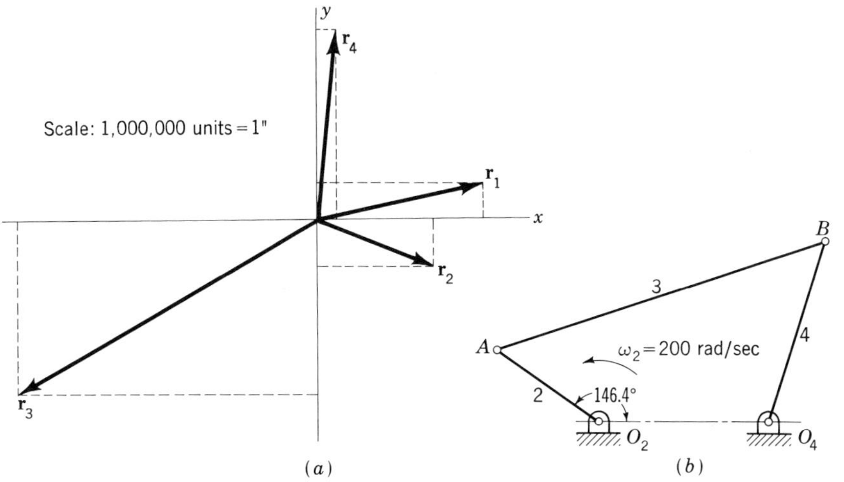

Fig. 11-26 $O_2A = 1.33$ in., $AB = 3.69$ in., $O_4B = 1.965$ in., $O_2O_4 = 1.81$ in.

From Fig. 11-25, $\sin \theta_1 = 0$ and $\cos \theta_1 = -1$; therefore

$$-r_1 + r_2 \cos \theta_2 + r_3 \cos \theta_3 + r_4 \cos \theta_4 = 0 \tag{c}$$

$$r_2 \sin \theta_2 + r_3 \sin \theta_3 + r_4 \sin \theta_4 = 0 \tag{d}$$

In order to eliminate the coupler angle θ_3 from the equations, move all terms except those involving r_3 to the right-hand side and square both sides. This gives

$$r_3^2 \cos^2 \theta_3 = (r_1 - r_2 \cos \theta_2 - r_4 \cos \theta_4)^2 \tag{e}$$

$$r_3^2 \sin^2 \theta_3 = (-r_2 \sin \theta_2 - r_4 \sin \theta_4)^2 \tag{f}$$

Now expand the right-hand sides of both equations and add them together. The result is

$$r_3^2 = r_1^2 + r_2^2 + r_4^2 - 2r_1r_2 \cos \theta_2 - 2r_1r_4 \cos \theta_4 + 2r_2r_4(\cos \theta_2 \cos \theta_4 + \sin \theta_2 \sin \theta_4) \tag{g}$$

Now, note that $\cos \theta_2 \cos \theta_4 + \sin \theta_2 \sin \theta_4 = \cos (\theta_2 - \theta_4)$. If we make this replacement, divide by the term $2r_2r_4$, and rearrange again, we obtain

$$\frac{r_3^2 - r_1^2 - r_2^2 - r_4^2}{2r_2r_4} + \frac{r_1}{r_4} \cos \theta_2 + \frac{r_1}{r_2} \cos \theta_4 = \cos (\theta_2 - \theta_4) \tag{h}$$

Freudenstein writes Eq. (*h*) in the form

$$K_1 \cos \theta_2 + K_2 \cos \theta_4 + K_3 = \cos (\theta_2 - \theta_4) \tag{11-3}$$

with

$$K_1 = \frac{r_1}{r_4} \tag{11-4}$$

$$K_2 = \frac{r_1}{r_2} \tag{11-5}$$

$$K_3 = \frac{r_3^2 - r_1^2 - r_2^2 - r_4^2}{2r_2r_4} \tag{11-6}$$

We have already learned graphical methods of synthesizing a linkage so that the motion of the output member is coordinated with that of the input member. Freudenstein's equation enables us to perform this same task by analytical means. Thus, suppose we wish the output lever of a four-bar linkage to occupy the positions ϕ_1, ϕ_2, and ϕ_3 corresponding to the angular positions ψ_1, ψ_2, and ψ_3 of the input lever. In Eq. (11-3) we simply replace θ_2 with ψ, θ_4 with ϕ, and write the equation three times, once for each position. This gives

$$\begin{aligned} K_1 \cos \psi_1 + K_2 \cos \phi_1 + K_3 &= \cos (\psi_1 - \phi_1) \\ K_1 \cos \psi_2 + K_2 \cos \phi_2 + K_3 &= \cos (\psi_2 - \phi_2) \\ K_1 \cos \psi_3 + K_2 \cos \phi_3 + K_3 &= \cos (\psi_3 - \phi_3) \end{aligned} \tag{i}$$

Equations (*i*) are solved simultaneously for the three unknowns K_1, K_2, and K_3. Then a length, say r_1, is selected for one of the links and Eqs. (11-4) to (11-6) solved for the dimensions of the other three. The method is best illustrated by an example.

Example 11-2 Synthesize a function generator to solve the equation

$$y = \frac{1}{x} \qquad 1 \leq x \leq 2$$

using three precision points.

Solution Choosing Chebyshev spacing, we find, from Eq. (11-1), the values of x and corresponding values of y to be

$$\begin{aligned} x_1 &= 1.067 & y_1 &= 0.937 \\ x_2 &= 1.500 & y_2 &= 0.667 \\ x_3 &= 1.933 & y_3 &= 0.517 \end{aligned}$$

We must now choose starting angles for the input and output levers as well as total swing angles for each. These are arbitrary decisions and may not result in a good linkage in the sense that the structural errors between the precision points may be large or the transmission angles may be poor. Sometimes, in such a synthesis, it is even found that one of the pivots must be removed in order to get from one precision point to another. Generally, some trial-

Table 11-3

x	ψ, deg	y	ϕ, deg
1.000	30.00	1.000	240.00
1.067	36.03	0.937	251.34
1.500	75.00	0.667	300.00
1.933	113.97	0.517	326.94
2.000	120.00	0.500	330.00

and-error work is necessary to discover the best starting positions and swing angles.

For the input lever we choose a 30° starting position and a 90° total swing angle. For the output lever, choose the starting position at 240° and a range of 90° total travel too. With these decisions made, the first and last rows of Table 11-3 can be completed.

Next, to obtain the values of ψ and ϕ corresponding to the precision points, write

$$\psi = ax + b \qquad \phi = cy + d \tag{1}$$

and use the data in the first and last rows of Table 11-3 to evaluate the constants a, b, c, and d. When this is done, we find Eqs. (1) are

$$\psi = 90x - 60 \qquad \phi = -180y + 420 \tag{2}$$

These equations can now be used to compute the data for the remaining rows in Table 11-3 and to determine the scales of the input and output levers of the synthesized linkage.

Now take the values of ψ and ϕ from the second line of Table 11-3 and substitute them for θ_2 and θ_4 in Eq. (11-3). Repeat this for the third and fourth lines. We then have the three equations

$$\begin{aligned} K_1 \cos 36.03 + K_2 \cos 251.34 + K_3 &= \cos (36.03 - 251.34) \\ K_1 \cos 75.00 + K_2 \cos 300.00 + K_3 &= \cos (75.00 - 300.00) \\ K_1 \cos 113.97 + K_2 \cos 326.94 + K_3 &= \cos (113.97 - 326.94) \end{aligned} \tag{3}$$

When the indicated operations are carried out, we have

$$\begin{aligned} 0.8087K_1 - 0.3200K_2 + K_3 &= -0.8160 \\ 0.2588K_1 + 0.5000K_2 + K_3 &= -0.7071 \\ -0.4062K_1 + 0.8381K_2 + K_3 &= -0.8389 \end{aligned} \tag{4}$$

Upon solving Eqs. (4) we obtain

$$K_1 = 0.4032 \qquad K_2 = 0.4032 \qquad K_3 = -1.0130$$

Now, if we choose $r_1 = 1$ in., then, from Eq. (11-4),

$$r_4 = \frac{r_1}{K_1} = \frac{1}{0.4032} = 2.4800 \text{ in.}$$

Similarly, from Eqs. (11-5) and (11-6), we learn

$r_2 = 2.4800$ in. $\quad r_3 = 0.9165$ in.

The result is the crossed linkage shown in the starting position in Fig. 11-27. A graphical analysis of the linkage using 5° increments of the input member will reveal only minute errors in between the precision points. This was no accident! The selection of the starting positions as well as the choice of the swing angles of the input and output levers were *not* arbitrary decisions as the reader has been led to suspect. Instead, they were selected from the published results of a digital-computer program[1] employing five precision points in the synthesis. The program optimizes the ranges of the two swing angles and finds the best starting position for each lever. It then generates the link proportions and computes the maximum output error.

Freudenstein offers the following suggestions which will be helpful in synthesizing such generators:

1. The total swing angles of the input and output members should be less than 120°.
2. Avoid the generation of symmetric functions such as $y = x^2$ in the range $-1 \leq x \leq 1$.
3. Avoid the generation of functions having abrupt changes in slope.

In this example it turns out that the input and output links have the same length. This appears to be pure coincidence. In Freudenstein's synthesis a difference in the length of these two members does not appear until the fourth decimal place.

[1] Ferdinand Freudenstein, Four-bar Function Generators, *Machine Design*, vol. 30, no. 24, pp. 119–123, Nov. 27, 1958.

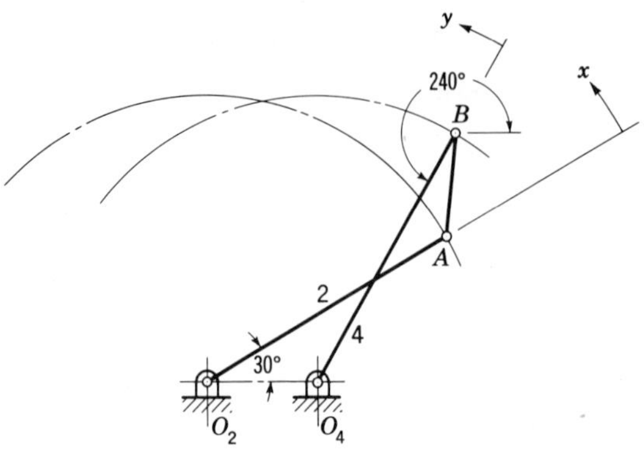

Fig. 11-27

PROBLEMS

11-1. Synthesize a double-crank linkage so that rotation of the driver through 180° ccw results in rotation of the output member through 90° ccw.

11-2. The same as Prob. 11-1 except that the linkage is to be a crank-and-rocker mechanism.

11-3. The same as Prob. 11-2 except that the output member is to move clockwise.

11-4. A function varies from 0 to 1. Find the Chebyshev spacing for two, three, four, five, and six precision points.

11-5. Using two-position synthesis, find a linkage which will generate the function $y = x^{\frac{1}{2}}$ for $0.5 \leq x \leq 1$. Use Chebyshev spacing and an input crank oscillation of 90°. Plot the error curve.

11-6. Synthesize a linkage to generate the function $\theta = \sin^{-1} x$, where $0.25 \leq x \leq 1$. Use two-position synthesis, Chebyshev spacing, and an input crank angle of 150°. Synthesize for optimum transmission angle and plot the error curve.

11-7. Using the overlay method, synthesize a linkage to generate the function $y = \sin x$ for $-\pi/2 \leq x \leq \pi/2$.

11-8. Use the overlay method, Chebyshev spacing, and six positions, and design a function generator to solve the equation $y = x^{1.4}$ for $0.5 \leq x \leq 1.5$.

11-9. Solve Prob. 11-8, using point-position reduction with four precision points.

11-10. The figure shows two positions of a folding seat used in the aisles of buses to accommodate more than the usual number of seated passengers. Design a four-bar linkage to support the seat so that it will lock in the open position and fold to a stable closing position along the side of the aisle.

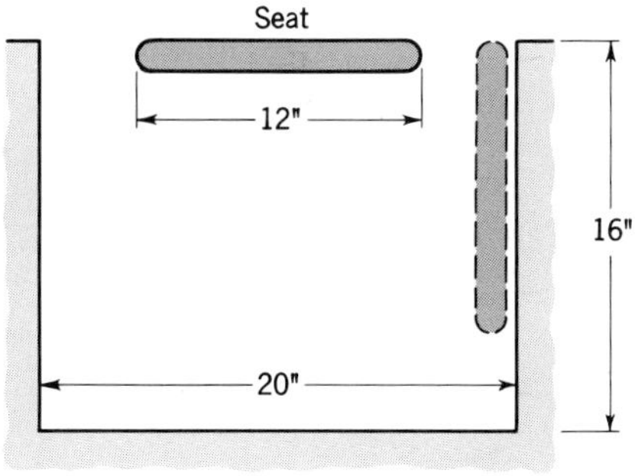

Prob. 11-10

11-11. Design a spring-operated four-bar linkage to support a heavy lid such as the trunk lid of an automobile. The lid is to swing through an angle of 80° from the closed to the open position. The springs are to be mounted so that the lid will be held closed against a stop. And the springs should also hold the lid in a stable open position without the use of a stop.

11-12. Design a crank-and-rocker linkage with a rocking angle of 120° such that the forward motion occurs in 240° of crank rotation. This is termed a quick-return motion because the return will then occur in the remaining 120° of crank angle.

11-13. Design a slider-crank mechanism with a slider displacement of 6 in. in which the forward motion occurs in 210° of crank rotation.

11-14. Synthesize a linkage to move AB from position 1 to position 2 and return (see figure).

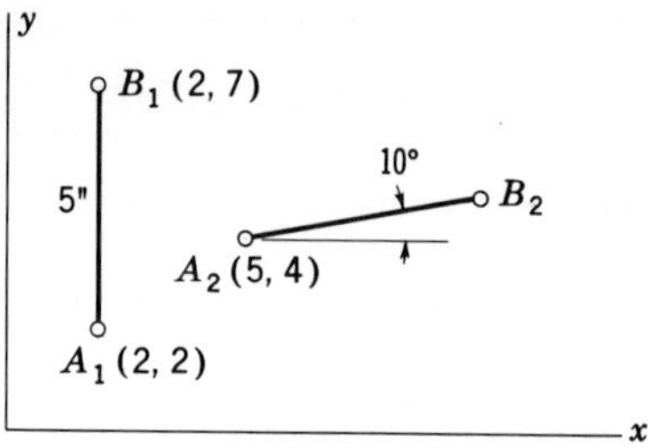

Prob. 11-14

11-15. Synthesize a mechanism to move AB (see figure) successively through positions 1, 2, and 3.

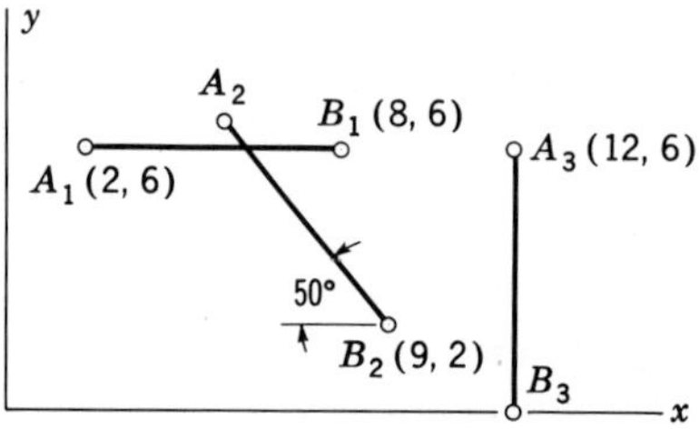

Prob. 11-15

11-16 to **11-25.**[1] The figure shows a function-generator linkage in which the motion of rocker 2 corresponds to x and the motion of rocker 4 to the function $y = f(x)$. Use four precision points and Chebyshev spacing and synthesize a linkage to generate the functions shown in the accompanying table. Plot a curve of the desired function and a curve of the actual function which the linkage generates. Compute the maximum error between them in percent.

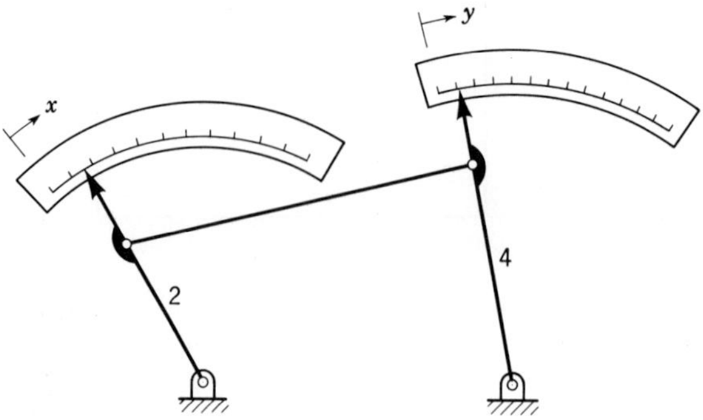

Probs. 11-16 to **11-25**

[1] Digital-computer solutions for these problems were obtained by Dr. Freudenstein of Columbia University. See *ibid.*

Problem number	*Function y*	*Range of x*
11-16, 11-26	$\log_{10} x$	$1 \leq x \leq 2$
11-17, 11-27	$\sin x$	$0 \leq x \leq \pi/2$
11-18, 11-28	$\tan x$	$0 \leq x \leq \pi/4$
11-19, 11-29	e^x	$0 \leq x \leq 1$
11-20, 11-30	$1/x$	$1 \leq x \leq 2$
11-21, 11-31	$x^{1.5}$	$0 \leq x \leq 1$
11-22, 11-32	x^2	$0 \leq x \leq 1$
11-23, 11-33	$x^{2.5}$	$0 \leq x \leq 1$
11-24, 11-34	x^3	$0 \leq x \leq 1$
11-25, 11-35	x^2	$-1 \leq x \leq 1$

11-26 to **11-35.** The same as Probs. 11-16 to 11-25 except use the overlay method.

11-36. The figure illustrates a coupler curve which can be generated by a four-bar linkage (not shown). Link 5 is to be attached to the coupler point, and link 6 is to be a rotating member with O_6 as the frame connection. In this problem we wish to find a coupler curve from the Hrones and Nelson atlas or, by point-position reduction, such that, for an appreciable distance, point C moves through an arc of a circle. Link 5 is then proportioned so that D lies at the center of curvature of this arc. The result is then called a *hesitation motion* because link 6 will hesitate in its rotation for the period in which point C traverses the approximate circle arc. Make a drawing of the complete linkage and plot the velocity-displacement diagram for 360° of displacement of the input link.

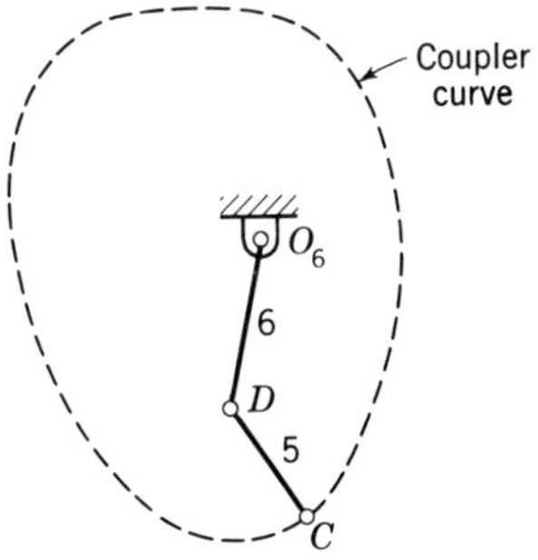

Prob. 11-36

11-37. The same as Prob. 11-36 except use the cubic of stationary curvature to get a more accurate circle arc.

11-38. Synthesize a four-bar linkage so as to obtain a coupler curve having an approximate straight-line segment. Then, using the suggestion included in Fig. 11-22 or 11-24, synthesize a dwell motion. Using an input crank angular velocity of unity, plot the velocity of rocker 6 versus the input crank displacement.

11-39. Synthesize a dwell mechanism using the idea suggested in Fig. 11-20 and the Hrones and Nelson atlas or point-position reduction. Rocker 6 is to have a total angular displacement of 60°. Using this displacement as the abscissa, plot a velocity diagram of the motion of the rocker to illustrate the dwell motion.

11-40. The same as Prob. 11-39 except use the cubic of stationary curvature to obtain a better circle arc for the coupler curve.

11-41. One method of designing a walking vehicle is to use a four-bar linkage for each leg, letting a point on the coupler be the foot. Ideally the coupler path relative to the vehicle should be a symmetric curve having a flat segment, as shown in the figure. Using a method of your choice, see whether you can synthesize a suitable walking mechanism according to these requirements.

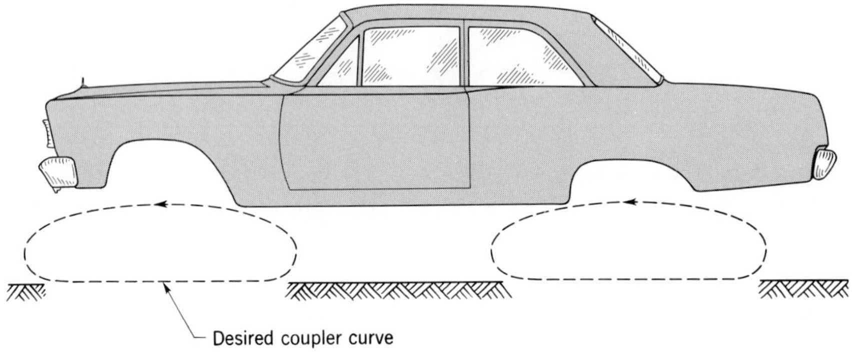

Prob. 11-41

12
Space Mechanisms

Bevel, worm, and spiral gears are examples of space mechanisms which we have already studied. For this reason our attention in this chapter will be directed toward an introduction to space linkages. Though such mechanisms have been in use for many years, it is only recently that kinematicians have become interested in them. We shall find that the addition of the third dimension makes their study infinitely more interesting.

12-1 THE MOBILITY EQUATION

In Sec. 6-3 we learned that the number of freedoms of a kinematic chain is obtained by multiplying the number of links by the freedoms of each link and then subtracting those which are destroyed by each kinematic pair. A space link has six degrees of freedom, three translations and three rotations. The fixed link has none. If we designate the pairs by p_i, where i is the number of freedoms of the pair, then the *general mobility equation* is

$$f = 6(n - 1) - 5p_1 - 4p_2 - 3p_3 - 2p_4 - p_5 \qquad (12\text{-}1)$$

Here f is the number of freedoms of the mechanism and n the number of links. Therefore, for a mechanism, $f = 1$.

Equation (12-1) is an extension of the Grübler criterion and is frequently called the *Kutzbach-Grübler criterion.*[1] It has a long history, however, dating back in its early form to P. O. Somov, and in its present form to I. I. Artobolevskii and V. V. Dobrovolskii.

One of the solutions to Eq. (12-1) is $n = 7$, $p_1 = 7$, $p_2 = p_3 = p_4 = p_5 = 0$. Harrisberger calls this a *mechanism type;* in particular, the $7p_1$ type. Other combinations of the p_i's produce other types of mechanisms. For example, the $3p_1 + 2p_2$ type has five links, while the $1p_1 + 2p_3$ has only three links.

Each mechanism type contains a finite number of *kinds* of mechanisms; there are as many kinds of mechanisms in each type as there are ways of combining the different kinds of pairs. In Table 6-1 we see that there are three different kinematic pairs having one degree of freedom. These are the revolute R, the prismatic P, and the screw S_L. Thus, by using any seven of these pairs, it turns out that there are 36 kinds of type $7p_1$ mechanisms. All together, Harrisberger lists 13 types of mechanisms containing 435 kinds which satisfy the Kutzbach criterion. Not all of these types or kinds are likely to have any practical value. Consider, for example, the $7p_1$ type with all turning pairs. This defines a seven-link mechanism with seven revolute pairs!

For those mechanisms defined by the mobility criterion as having one freedom, Harrisberger has selected nine kinds from two types which appear to be useful. These are illustrated in Fig. 12-1; they are all four-bar linkages having four pairs with either rotating or sliding input and output members. The designation $RGCS_L$ in the legend for Fig. 12-1*f*, for example, identifies the kinematic pairs (see Table 6-1) beginning with the input link and proceeding through the coupler and output member back to the frame. Thus the input crank is pivoted to the frame by revolute R of one freedom and to the coupler by the globular pair G (spherical). The coupler is paired to the output member by the cylinder C. The motion of the output member is then determined by the screw pair S_L. The freedoms of these pairs, from Table 6-1, is $R = 1$, $G = 3$, $C = 2$, and $S_L = 1$.

The mechanisms in Fig. 12-1*a* to *c* are described by Harrisberger as type 1 mechanisms. Each is composed of one single-freedom pair and three double-freedom pairs and hence is a $1p_1 + 3p_2$ type of mechanism. The remaining mechanisms in Fig. 12-1 are type 2 linkages hav-

[1] See L. Harrisberger, A Number Synthesis Survey of Three-dimensional Mechanisms, *J. Eng. Ind.*, ser. B, vol. 87, no. 2, May, 1965. This important paper summarizes many of the known space linkages and defines the present gaps in space-mechanism technology.

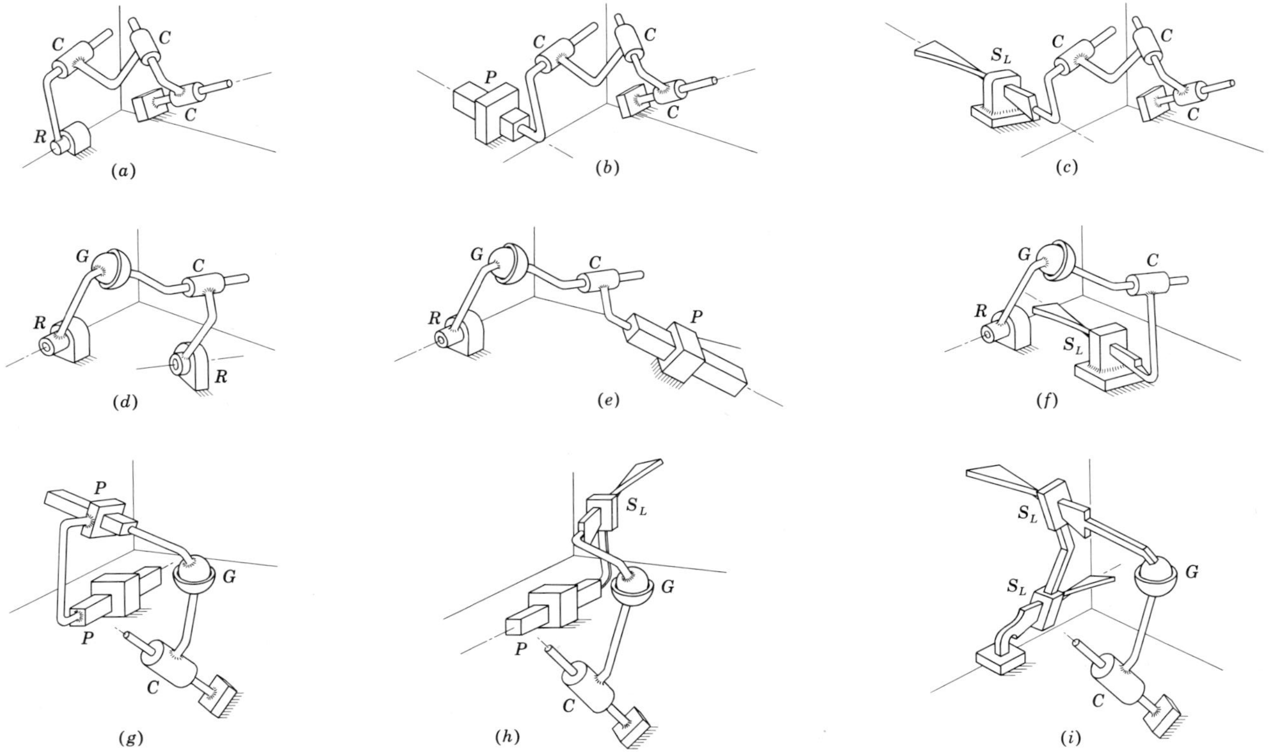

Fig. 12-1 Four-bar linkages having one freedom as defined by the mobility criterion. (*a*) *RCCC*; (*b*) *PCCC*; (*c*) S_LCCC; (*d*) *RGCR*; (*e*) *RGCP*; (*f*) $RGCS_L$; (*g*) *PPGC*; (*h*) PS_LGC; (*i*) S_LS_LGC. *(These figures are from the article, L. Harrisberger, A Number Synthesis Survey of Three-dimensional Mechanisms, J. Eng. Ind., ser. B, vol.* 87, *no.* 2, *May,* 1965, *and have been published with the permission of the ASME and the author of the paper.)*

ing two single-freedom pairs, one double-freedom pair, and one triple-freedom pair. Thus, they are $2p_1 + 1p_2 + 1p_3$ types of mechanisms.

12-2 SPECIAL MECHANISMS

Curiously enough, the most useful space mechanisms which have been found are those which *are not* defined as mechanisms by the mobility criterion. Because of certain geometric conditions, such as a particular ratio of link lengths or the orientation of single-freedom-pair axes, idle freedoms or idle constraints may be introduced.

At least two of the known space linkages which violate the Kutzbach criterion are four-link *RRRR* mechanisms. Thus $n = 4$, $p_1 = 4$, and Eq. (12-1) gives $f = -2$; so we conclude that there are three idle constraints. One of these mechanisms is the spherical four-bar space linkage shown in Fig. 12-2. The axes of all four revolutes intersect at the center of a sphere and the links may be regarded as great-circle arcs existing on the surface of the sphere. Their lengths are then designated as spherical angles. By properly proportioning these angles it is possible to design all the spherical counterparts of the plane four-bar mechanism such as the spherical crank-and-rocker linkage and the spherical drag linkage. The spherical four-bar linkage is easy to design and to manufacture and hence one of the most useful of all space mechanisms. The well-known Hooke, or Cardan, joint, which is the basis of the universal joint, is a special case of the spherical mechanism having input and output cranks which subtend the same angle at the center of the sphere. The *wobble-plate mechanism*, shown in Fig. 12-3, is also a special case.

Bennett's *RRRR* mechanism, shown in Fig. 12-4, is probably one of the most useless of the known space linkages. In this mechanism the

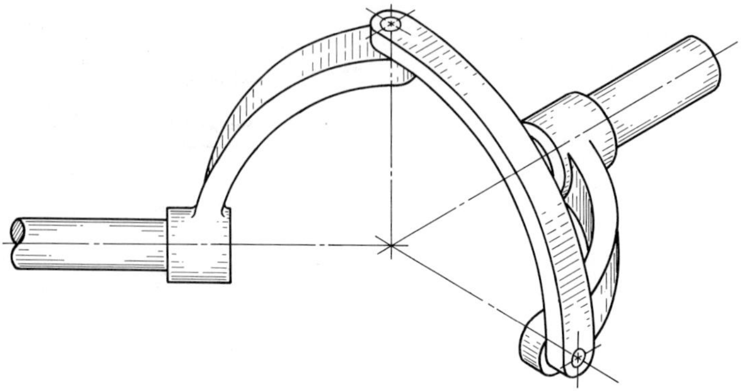

Fig. 12-2 The spherical four-bar linkage.

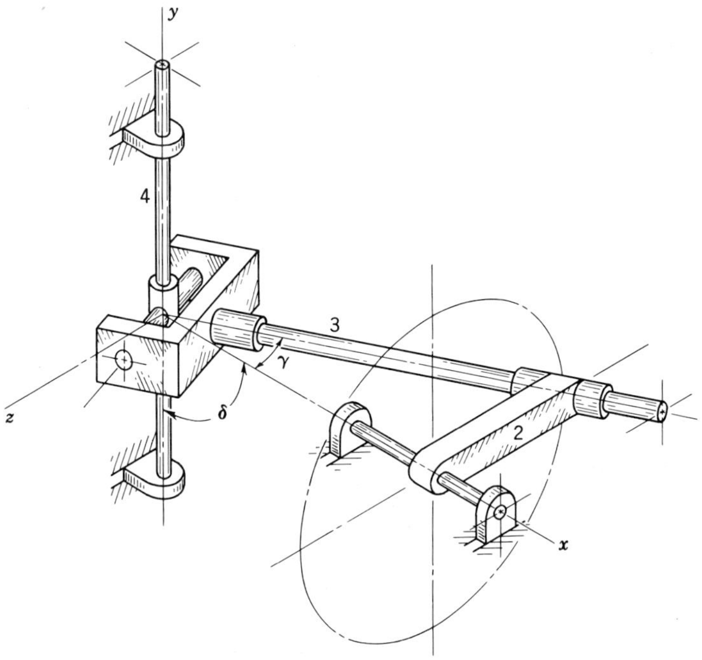

Fig. 12-3 The wobble-plate mechanism; input crank 2 rotates and output shaft 4 oscillates. When $\delta \neq 90°$, the mechanism is called the spherical slide oscillator. If $\gamma > \delta$, the output shaft rotates.

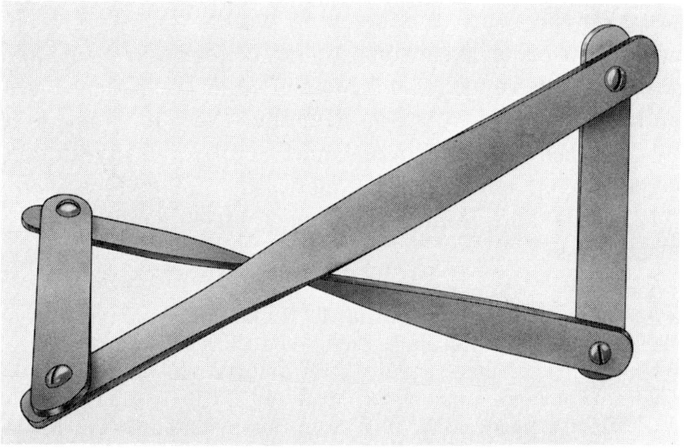

Fig. 12-4 The Bennett four-link mechanism.

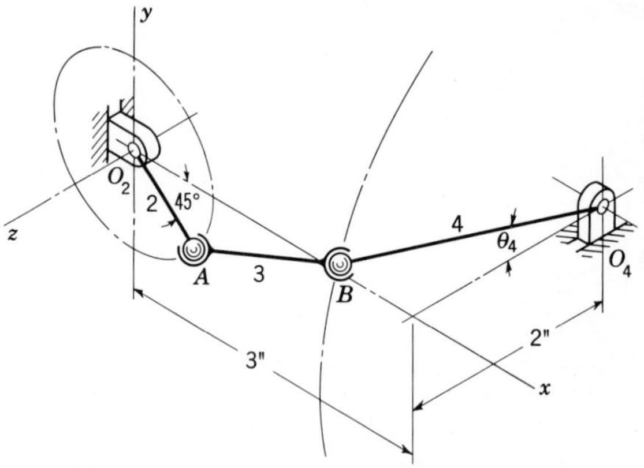

Fig. 12-5 $O_2A = 1$ in., $AB = 3\frac{1}{2}$ in., $O_4B = 4$ in.

opposite links are twisted the same amount and also have equal lengths. The twist angles and the lengths must be related mathematically to each other.[1]

The four-link *RGGR* mechanism of Fig. 12-5 is another important and useful exception to the mobility criterion. Substitution of $n = 4$, $p_1 = 2$, $p_3 = 2$ in Eq. (12-1) gives $f = 2$, indicating the existence of an idle freedom. This extra degree of freedom exists in the coupler; it may rotate about its own axis without affecting the kinematic relationships between the input and output links. Thus the extra freedom does no harm; in fact, it may be an advantage because it is easy to manufacture and the rotation of the coupler about its own axis should equalize the wear on the two ball-and-socket joints.

Still other exceptions to the criterion are the Goldberg (not Rube) five-bar five-revolute mechanism and the Bricard six-bar six-revolute linkage.[2] Again, it is doubtful that these have any practical value.

Recently Harrisberger and Soni[3] have investigated all space linkages having one general constraint. They identified eight types and 212 kinds and found seven new mechanisms which may have some usefulness.

[1] For more details, see Rolland T. Hinkle, "Kinematics of Machines," 2d ed., pp. 343–345, Prentice-Hall, Inc., Englewood Cliffs, N.J., 1960.

[2] For pictures of these, see R. S. Hartenberg and Jacques Denavit, "Kinematic Synthesis of Linkages," pp. 85, 86, McGraw-Hill Book Company, New York, 1964.

[3] Lee Harrisberger and A. H. Soni, A Survey of Three-dimensional Mechanisms with One General Constraint, *ASME Paper* 66-Mech-44, October, 1966. This paper contains 45 references on space mechanisms.

12-3 THE POSITION PROBLEM

The position of a space linkage is defined by the vector equation

$$\mathbf{r} + \mathbf{s} + \mathbf{t} + \mathbf{C} = 0 \tag{12-2}$$

which is called the *vector tetrahedron equation* because the equation defines four of the six edges of a tetrahedron. The equation is three-dimensional and hence can be solved for three unknowns. These may exist in any combination in vectors **r**, **s**, and **t**. As in Sec. 2-17, the vector **C** is the sum of all the known vectors. By using spherical coordinates each of the vectors **r**, **s**, and **t** may be expressed as a magnitude and two angles. Vector **r**, for example, is defined when the magnitude r and the two angles θ_r and ϕ_r are known. Thus, in Eq. (12-2), any three of the nine quantities r, θ_r, ϕ_r, s, θ_s, ϕ_s, t, θ_t, ϕ_t can be unknown. When these are solved it turns out that there are just nine combinations of unknowns which result in different solutions. Chace[1] has solved all these by first reducing each one to a polynomial. He classifies the solutions depending upon whether the unknowns occur in one, two, or three vectors, and tabulates the solution to be obtained as shown in Table 12-1. In this table, the unit vectors $\hat{\boldsymbol{\omega}}_r$, $\hat{\boldsymbol{\omega}}_s$, and $\hat{\boldsymbol{\omega}}_t$ are *known* directions from which the *known* angles ϕ_r, ϕ_s, and ϕ_t are measured. In case 1 vectors **s** and **t** are known and are summed into the vector **C**. In case 2 vector **t** is completely known and is also summed into **C**. Only case 3*d*, involving the solution of an eighth-order polynomial, must be solved by iteration techniques; the remaining cases have explicit solutions.[2]

[1] Milton A. Chace, Vector Analysis of Linkages, *J. Eng. Ind.*, ser. B, vol. 85, pp. 289–297, August, 1963.

[2] Dual numbers and quaternions as well as matrix methods can also be used; see A. T. Yang and F. Freudenstein, Application of Dual-number and Quaternion

Table 12-1 Classification of the solutions to the vector tetrahedron equation

Case number	Unknowns	Known quantities: Vectors	Known quantities: Scalars	Degree of polynomial
1	r, θ_r, ϕ_r	$\mathbf{C}$		1
2*a*	r, θ_r, s	$\mathbf{C}, \hat{\mathbf{s}}, \hat{\boldsymbol{\omega}}_r$	ϕ_r	2
2*b*	r, θ_r, θ_s	$\mathbf{C}, \hat{\boldsymbol{\omega}}_r, \hat{\boldsymbol{\omega}}_s$	ϕ_r, s, ϕ_s	4
2*c*	θ_r, ϕ_r, s	$\mathbf{C}, \hat{\mathbf{s}}$	r	2
2*d*	$\theta_r, \phi_r, \theta_s$	$\mathbf{C}, \hat{\boldsymbol{\omega}}_s$	r, s, ϕ_s	2
3*a*	r, s, t	$\mathbf{C}, \hat{\mathbf{r}}, \hat{\mathbf{s}}, \hat{\mathbf{t}}$		1
3*b*	r, s, θ_t	$\mathbf{C}, \hat{\mathbf{r}}, \hat{\mathbf{s}}, \hat{\boldsymbol{\omega}}_t$	t, ϕ_t	2
3*c*	r, θ_s, θ_t	$\mathbf{C}, \hat{\mathbf{r}}, \hat{\boldsymbol{\omega}}_s, \hat{\boldsymbol{\omega}}_t$	s, ϕ_s, t, ϕ_t	4
3*d*	$\theta_r, \theta_s, \theta_t$	$\mathbf{C}, \hat{\boldsymbol{\omega}}_r, \hat{\boldsymbol{\omega}}_s, \hat{\boldsymbol{\omega}}_t$	$r, \phi_r, s, \phi_s, t, \phi_t$	8

The solution of these polynomials turns out to be equivalent to finding the intersections of straight lines or circles with various surfaces of revolution. Such problems can be easily and quickly solved by using the methods of descriptive geometry. This graphic approach has the additional advantage that the geometry is not concealed by a multiplicity of mathematical operations.

Let us use a four-link *RGGR* crank-and-rocker mechanism in which the knowns are the position and plane of rotation of the input link, the plane of rotation of the output link, and the dimensions of all four links. Such a mechanism is shown in Fig. 12-5. The position problem consists in finding the position of the coupler and rocker, links 3 and 4. If we treat link 4 as a vector, then the only unknown is one angle, because the magnitude and the plane of oscillation are given. Similarly, if link 3 is a vector, then its magnitude is known but there exist two unknowns which are the two angular directions in spherical coordinates. We identify this as case 2*d* in Table 12-1, requiring the solution of a second-degree polynomial and, hence, yielding two solutions.

This problem is solved by using only two orthographic views, the front and profile. If we imagine, in Fig. 12-5, that the coupler is disconnected from B and permitted to occupy all positions relative to A, then B must lie on the surface of a sphere whose center is at A. With the coupler still disconnected, the motion of B on link 4 is a circle about O_4 in a plane parallel to the yz plane. Therefore, to solve this problem we need only find the two points of intersection of a circle with a sphere.

The solution is shown in Fig. 12-6. The subscripts F and P denote projections on the frontal and profile planes, respectively. First locate O_2, A, and O_4 on both views. On the profile view draw a circle of radius $O_4B = 4$ in. about O_{4P}; this is the path of point B. This circle appears as the vertical line $M_FO_{4F}N_F$ on the front view. Next, on the front view construct the outline of a sphere with A_F as a center and the coupler length $AB = 3\frac{1}{2}$ in. as the radius. If $M_FO_{4F}N_F$ is regarded as the trace of a plane normal to the frontal plane, then the intersection of this plane with the sphere appears as the shaded circle on the profile view of diameter M_PN_P. The arc of radius O_4B intersects the circle in two points, yielding two solutions. One of these points is chosen for B_P and projected back to the front view to locate B_F. Links 3 and 4 are now drawn in, as dashed lines in this case, in the front and profile views.

By simply measuring the x, y, and z projections from the graphic

---

Algebra to the Analysis of Spatial Mechanisms, *J. Appl. Mech.*, ser. E, vol. 86, pp. 300–308, 1964, and J. J. Uicker, Jr., J. Denavit, and R. S. Hartenberg, An Iterative Method for the Displacement Analysis of Spatial Mechanisms, *J. Appl. Mech.*, ser. E, vol. 87, pp. 309–314, 1965.

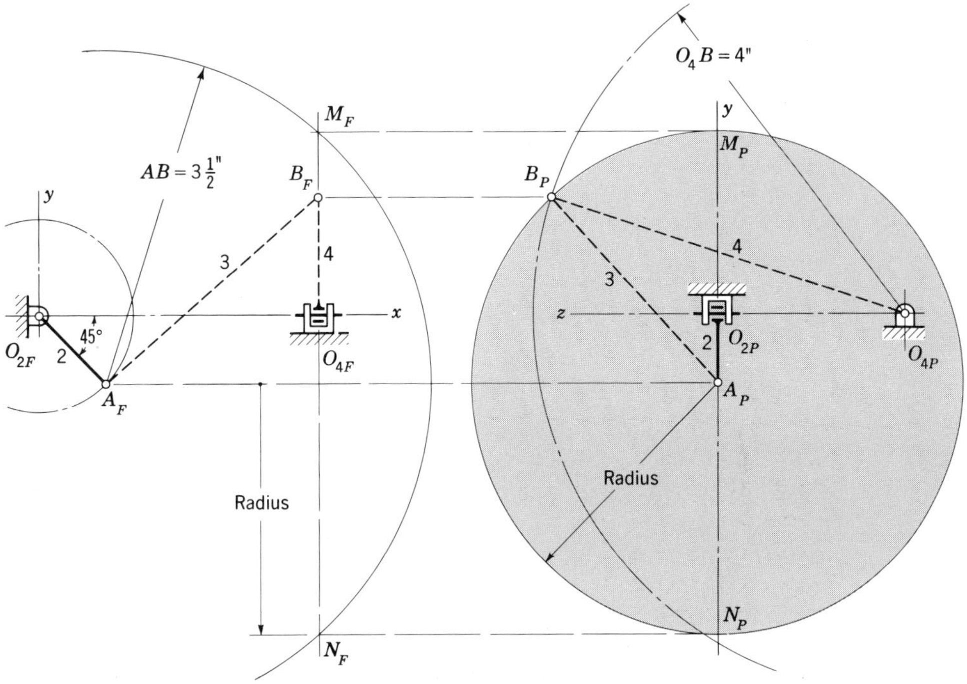

Fig. 12-6

solution, one can write the vector expressions for each link:

$$\begin{aligned}
\mathbf{r}_1 &= 3\hat{\imath} - 2\hat{\mathbf{k}} \\
\mathbf{r}_2 &= 0.707\hat{\imath} - 0.707\hat{\jmath} \\
\mathbf{r}_3 &= 2.30\hat{\imath} + 1.95\hat{\jmath} + 1.77\hat{\mathbf{k}} \\
\mathbf{r}_4 &= 1.22\hat{\jmath} + 3.81\hat{\mathbf{k}}
\end{aligned} \tag{12-3}$$

where $\mathbf{r}_1$, $\mathbf{r}_2$, $\mathbf{r}_3$, and $\mathbf{r}_4$ are directed from O_2 to O_4, O_2 to A, A to B, and O_4 to B, respectively. The components shown above were obtained from a full-size solution; better accuracy would result, of course, by making the drawings two or four times actual size.

The four-revolute spherical four-link mechanism is also case $2d$ and is solved in the same manner when the position of the input link is given.

12-4 VECTOR ANALYSIS OF VELOCITY AND ACCELERATION

Once the positions of all the members of a space mechanism have been found, the velocities and accelerations can be determined by using the

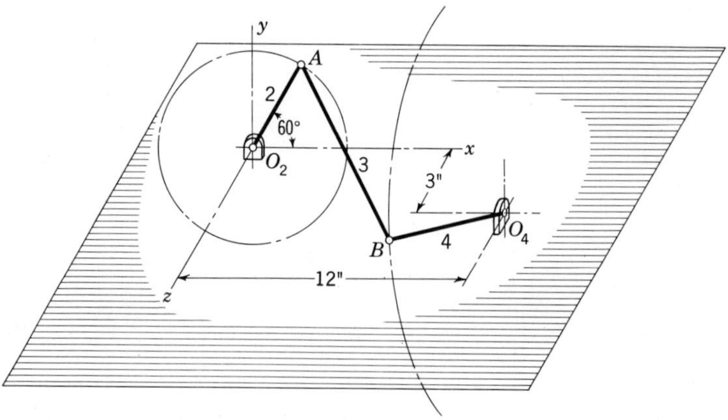

Fig. 12-7 $O_2A = 4$ in., $AB = 15$ in., $O_4B = 10$ in.

methods of Chaps. 4 and 5. In analyzing planar mechanisms the angular velocities and accelerations have only one direction and hence one component when expressed in vector form. In the analysis of space linkages these terms have three components, one for each axis. Otherwise, the methods of analysis are the same. The following example will serve to illustrate these differences.

Example 12-1 The angular velocity of link 2 of the four-bar *RGGR* linkage of Fig. 12-7 is $\boldsymbol{\omega}_2 = 40\hat{\mathbf{k}}$ rad/sec. Find the angular velocity and acceleration of links 3 and 4 and the velocity and acceleration of point *B*.

Solution Using descriptive geometry to solve the position problem as explained in Sec. 12-3 results in the three-view drawing of the linkage shown in Fig. 12-8. We now replace O_2A, AB, and O_4B with vectors $\mathbf{r}_2$, $\mathbf{r}_3$, and $\mathbf{r}_4$, respectively. The components may be read directly from Fig. 12-8:

$$\mathbf{r}_2 = 2\hat{\imath} + 3.46\hat{\jmath}$$

$$\mathbf{r}_3 = 10\hat{\imath} + 2.71\hat{\jmath} + 10.89\hat{\mathbf{k}}$$

$$\mathbf{r}_4 = 6.17\hat{\jmath} + 7.89\hat{\mathbf{k}}$$

From the constraints we see that the angular velocities and accelerations can be written as

$$\boldsymbol{\omega}_2 = 40\hat{\mathbf{k}} \qquad \dot{\boldsymbol{\omega}}_2 = 0$$

$$\boldsymbol{\omega}_3 = \omega_3^x\hat{\imath} + \omega_3^y\hat{\jmath} + \omega_3^z\hat{\mathbf{k}} \qquad \dot{\boldsymbol{\omega}}_3 = \dot{\omega}_3^x\hat{\imath} + \dot{\omega}_3^y\hat{\jmath} + \dot{\omega}_3^z\hat{\mathbf{k}}$$

$$\boldsymbol{\omega}_4 = \omega_4\hat{\imath} \qquad \dot{\boldsymbol{\omega}}_4 = \dot{\omega}_4\hat{\imath}$$

First, we find $\mathbf{V}_A$. Thus

$$\mathbf{V}_A = \boldsymbol{\omega}_2 \times \mathbf{r}_2 = \frac{1}{12}\begin{vmatrix} \hat{\imath} & \hat{\jmath} & \hat{\mathbf{k}} \\ 0 & 0 & 40 \\ 2 & 3.46 & 0 \end{vmatrix} = -11.53\hat{\imath} + 6.67\hat{\jmath} \qquad \text{fps} \tag{1}$$

Similarly,

$$\mathbf{V}_{BA} = \boldsymbol{\omega}_3 \times \mathbf{r}_3 = \frac{1}{12}\begin{vmatrix} \hat{\mathbf{i}} & \hat{\mathbf{j}} & \hat{\mathbf{k}} \\ \omega_3^x & \omega_3^y & \omega_3^z \\ 10 & 2.71 & 10.89 \end{vmatrix}$$

$$= (0.906\omega_3^y - 0.226\omega_3^z)\hat{\mathbf{i}} + (0.833\omega_3^z - 0.906\omega_3^x)\hat{\mathbf{j}} + (0.226\omega_3^x - 0.833\omega_3^y)\hat{\mathbf{k}} \quad (2)$$

And, finally,

$$\mathbf{V}_B = \boldsymbol{\omega}_4 \times \mathbf{r}_4 = \frac{1}{12}\begin{vmatrix} \hat{\mathbf{i}} & \hat{\mathbf{j}} & \hat{\mathbf{k}} \\ \omega_4 & 0 & 0 \\ 0 & 6.17 & 7.89 \end{vmatrix} = -0.657\omega_4\hat{\mathbf{j}} + 0.515\omega_4\hat{\mathbf{k}} \quad (3)$$

The next step is to substitute Eqs. (1) to (3) into the relative-velocity equation

$$\mathbf{V}_B = \mathbf{V}_A + \mathbf{V}_{BA} \quad (4)$$

When this is done, the $\hat{\mathbf{i}}$, $\hat{\mathbf{j}}$, and $\hat{\mathbf{k}}$ components can be separated so as to obtain three algebraic equations. These equations can be solved for only three unknowns. Notice that ω_3^x, ω_3^y, and ω_3^z are scalar components of $\boldsymbol{\omega}_3$; hence they are *not* independent of each other. For this reason we solve the relative-

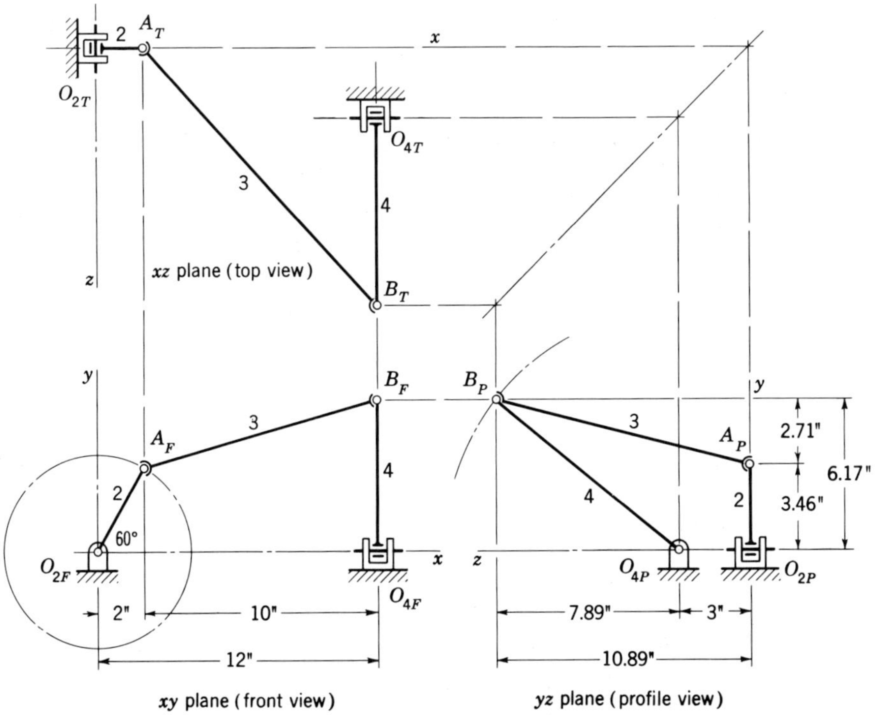

Fig. 12-8

velocity equation for ω_4 and get

$$\omega_4 = -25.5\hat{\imath} \qquad \text{rad/sec}$$

Substituting into Eq. (3) gives $\mathbf{V}_B$ as

$$\mathbf{V}_B = 16.8\hat{\jmath} - 13.2\hat{\mathbf{k}} \qquad \text{fps}$$

and a magnitude $V_B = 21.3$ fps.

Since both $\mathbf{V}_B$ and $\mathbf{V}_A$ are now known, we can rearrange Eq. (4) and solve for $\mathbf{V}_{BA}$ as follows:

$$\begin{aligned}\mathbf{V}_{BA} = \mathbf{V}_B - \mathbf{V}_A &= (16.8\hat{\jmath} - 13.2\hat{\mathbf{k}}) - (-11.53\hat{\imath} + 6.67\hat{\jmath}) \\ &= 11.53\hat{\imath} + 10.13\hat{\jmath} - 13.2\hat{\mathbf{k}}\end{aligned}$$

And the magnitude of this velocity is $V_{BA} = 20.2$ fps. Consequently,

$$\omega_3 = \frac{V_{BA}}{r_3} = \frac{20.2}{\frac{15}{12}} = 16.2 \text{ rad/sec}$$

We can now write the scalar equation

$$\omega_3^2 = (\omega_3^x)^2 + (\omega_3^y)^2 + (\omega_3^z)^2 \tag{5}$$

The next step consists in substituting the value of $\mathbf{V}_{BA}$ into Eq. (2) and separating the components as follows:

$$0.906\omega_3^y - 0.226\omega_3^z = 11.53 \tag{6}$$

$$-0.906\omega_3^x + 0.833\omega_3^z = 10.13 \tag{7}$$

$$0.226\omega_3^x - 0.833\omega_3^y = -13.2 \tag{8}$$

Now solve Eq. (7) for ω_3^z and Eq. (8) for ω_3^y. [Note that Eq. (6) is not really needed.] We get

$$\omega_3^z = 12.20 + 1.089\omega_3^x \tag{9}$$

$$\omega_3^y = 15.85 + 0.272\omega_3^x \tag{10}$$

Equations (9) and (10) are substituted, next, into Eq. (5) together with the value of ω_3. We then have

$$(\omega_3^x)^2 + (15.85 + 0.272\omega_3^x)^2 + (12.20 + 1.089\omega_3^x)^2 = 262$$

or

$$(\omega_3^x)^2 + 15.6\omega_3^x + 61 = 0$$

The roots of this equation are real and equal. The result is

$$\omega_3^x = -7.8 \text{ rad/sec}$$

Substitution of ω_3^x in Eqs. (9) and (10) yields

$$\omega_3^y = 13.73 \text{ rad/sec} \qquad \omega_3^z = 3.72 \text{ rad/sec}$$

Therefore

$$\omega_3 = -7.8\hat{\imath} + 13.73\hat{\jmath} + 3.72\hat{\mathbf{k}}$$

Turning next to the acceleration analysis, we compute the following components:

$$\mathbf{A}_A{}^r = \boldsymbol{\omega}_2 \times \mathbf{V}_A = \begin{vmatrix} \hat{\imath} & \hat{\jmath} & \hat{\mathbf{k}} \\ 0 & 0 & 40 \\ -11.53 & 6.67 & 0 \end{vmatrix} = -267\hat{\imath} - 461\hat{\jmath} \tag{11}$$

$$\mathbf{A}_{BA}{}^t = \dot{\boldsymbol{\omega}}_3 \times \mathbf{r}_3 = \frac{1}{12}\begin{vmatrix} \hat{\imath} & \hat{\jmath} & \hat{\mathbf{k}} \\ \dot{\omega}_3{}^x & \dot{\omega}_3{}^y & \dot{\omega}_3{}^z \\ 10 & 2.71 & 10.89 \end{vmatrix}$$

$$= (0.906\dot{\omega}_3{}^y - 0.226\dot{\omega}_3{}^z)\hat{\imath} + (0.833\dot{\omega}_3{}^z - 0.906\dot{\omega}_3{}^x)\hat{\jmath} + (0.226\dot{\omega}_3{}^x - 0.833\dot{\omega}_3{}^y)\hat{\mathbf{k}} \tag{12}$$

$$\mathbf{A}_{BA}{}^r = \boldsymbol{\omega}_3 \times \mathbf{V}_{BA} = \begin{vmatrix} \hat{\imath} & \hat{\jmath} & \hat{\mathbf{k}} \\ -7.8 & 13.73 & 3.72 \\ 11.53 & 10.13 & -13.2 \end{vmatrix} = -219\hat{\imath} - 60\hat{\jmath} - 237\hat{\mathbf{k}} \tag{13}$$

$$\mathbf{A}_B{}^t = \dot{\boldsymbol{\omega}}_4 \times \mathbf{r}_4 = \frac{1}{12}\begin{vmatrix} \hat{\imath} & \hat{\jmath} & \hat{\mathbf{k}} \\ \dot{\omega}_4 & 0 & 0 \\ 0 & 6.17 & 7.89 \end{vmatrix} = -0.657\dot{\omega}_4\hat{\jmath} + 0.515\dot{\omega}_4\hat{\mathbf{k}} \tag{14}$$

$$\mathbf{A}_B{}^r = \boldsymbol{\omega}_4 \times \mathbf{V}_B = \begin{vmatrix} \hat{\imath} & \hat{\jmath} & \hat{\mathbf{k}} \\ -25.5 & 0 & 0 \\ 0 & 16.8 & -13.2 \end{vmatrix} = -337\hat{\jmath} - 429\hat{\mathbf{k}} \tag{15}$$

These quantities are substituted into

$$\mathbf{A}_B{}^t + \mathbf{A}_B{}^r = \mathbf{A}_A{}^r + \mathbf{A}_{BA}{}^t + \mathbf{A}_{BA}{}^r$$

and the results obtained exactly as before:

$\dot{\boldsymbol{\omega}}_3 = -520\hat{\imath} + 619\hat{\jmath} + 333\hat{\mathbf{k}} \qquad \dot{\omega}_3 = 870 \text{ rad/sec}^2$

$\dot{\boldsymbol{\omega}}_4 = -858\hat{\imath} \qquad \dot{\omega}_4 = 858 \text{ rad/sec}^2$

$A_A{}^r = 524 \text{ fps}^2 \qquad A_{BA}{}^r = 328 \text{ fps}^2 \qquad A_B{}^r = 545 \text{ fps}^2$

$\mathbf{A}_{BA}{}^t = 486\hat{\imath} + 748\hat{\jmath} - 634\hat{\mathbf{k}} \qquad A_{BA}{}^t = 1092 \text{ fps}^2$

$\mathbf{A}_B{}^t = 564\hat{\jmath} - 442\hat{\mathbf{k}} \qquad A_B{}^t = 715 \text{ fps}^2$

$\mathbf{A}_B = 227\hat{\jmath} - 871\hat{\mathbf{k}} \qquad A_B = 900 \text{ fps}^2$

It is interesting to note, in the preceding solution, that the extra freedom of link 3 to rotate about its own axis never entered into the problem.

12-5 VELOCITY AND ACCELERATION ANALYSIS USING DESCRIPTIVE GEOMETRY

The determination of the velocities and accelerations of a space mechanism by graphical means is conducted in the same manner as for a plane-motion mechanism. However, the velocity and acceleration vectors which appear on the standard front, top, and profile views are not usually seen in their true length; that is, they are foreshortened. This means that the last step in constructing the vector polygon must be completed in an auxiliary view in which the unknown vector does appear in its true length.

The directions of the vectors depend upon the directions of the elements of the mechanism; for this reason, it is necessary to project one of the links of the mechanism into the auxiliary view or views too. Also, for this reason, we choose to connect the poles of the vector polygons to a point on one of the links so that the relationship between the vectors and one of the links is evident in all the views.

Once the unknown vector or vectors are obtained in the auxiliary views, they may be projected back into the standard three-view orthogonal system and the lengths of the x, y, and z projections measured directly. The procedure is best illustrated by an example.

Example 12-2 Using descriptive geometry, find the velocity and acceleration polygons for the mechanism of Example 12-1.

Solution The velocity solution is shown in Fig. 12-9, and the notation corresponds with that used in many works on descriptive geometry. The letters F, T, and P designate the front, top, and profile planes, and the numbers 1 and 2 the first and second auxiliary planes of projection. Points projected upon these planes bear the subscripts F, T, P, etc. The steps in obtaining the velocity solution are as follows:

1. Construct the front, profile, and top views of the linkage, and designate each point.

2. Calculate $\mathbf{V}_A$, and place this vector in position with the origin at A on the three views. The velocity of A shows in its true length on the front view. Designate the terminus of $\mathbf{V}_A$ as a_F, and project to the top and profile views.

3. The velocity of B is unknown, but not its direction. The direction is perpendicular to link 4 and in the direction that 4 rotates. When the problem is solved, $\mathbf{V}_B$ will show in its true length in the profile view. Construct a line in the profile view to correspond with the known direction of $\mathbf{V}_B$. Locate any point d_P on this line, and project to the front and top views.

4. The equation to be solved is

$$\mathbf{V}_B = \mathbf{V}_A + \mathbf{B}_{BA} \tag{1}$$

where $\mathbf{V}_A$ and the directions of $\mathbf{V}_B$ and $\mathbf{V}_{BA}$ are known. Note that $\mathbf{V}_{BA}$ is perpendicular to link 3, but its magnitude is unknown. In space the lines perpendicular to link 3 are like the spokes of a wheel, link 3 being the axis of rotation of that wheel. There are, therefore, an infinite number of lines perpendicular to link 3, but we are interested in only one of them. The line we require must originate at the terminus of $\mathbf{V}_A$ and terminate by intersecting the line Ad or its extension. In order to choose this line from the infinite number of available ones we need to examine AB in the direction in which it appears as a point. Therefore in this step we must project AB upon a plane which shows its true length; so, construct the edge view of plane 1 parallel to A_TB_T, and project AB to this plane. In making this projection, note that the distances k and l in the front view are the same in this first auxiliary view. The auxiliary view of AB is A_1B_1, which is its true length. Also project points a and d to this view, but the remaining links need not be projected.

5. In this step choose a second auxiliary plane 2 such that the projection of AB upon it is a point. Then all lines drawn parallel to the plane will be

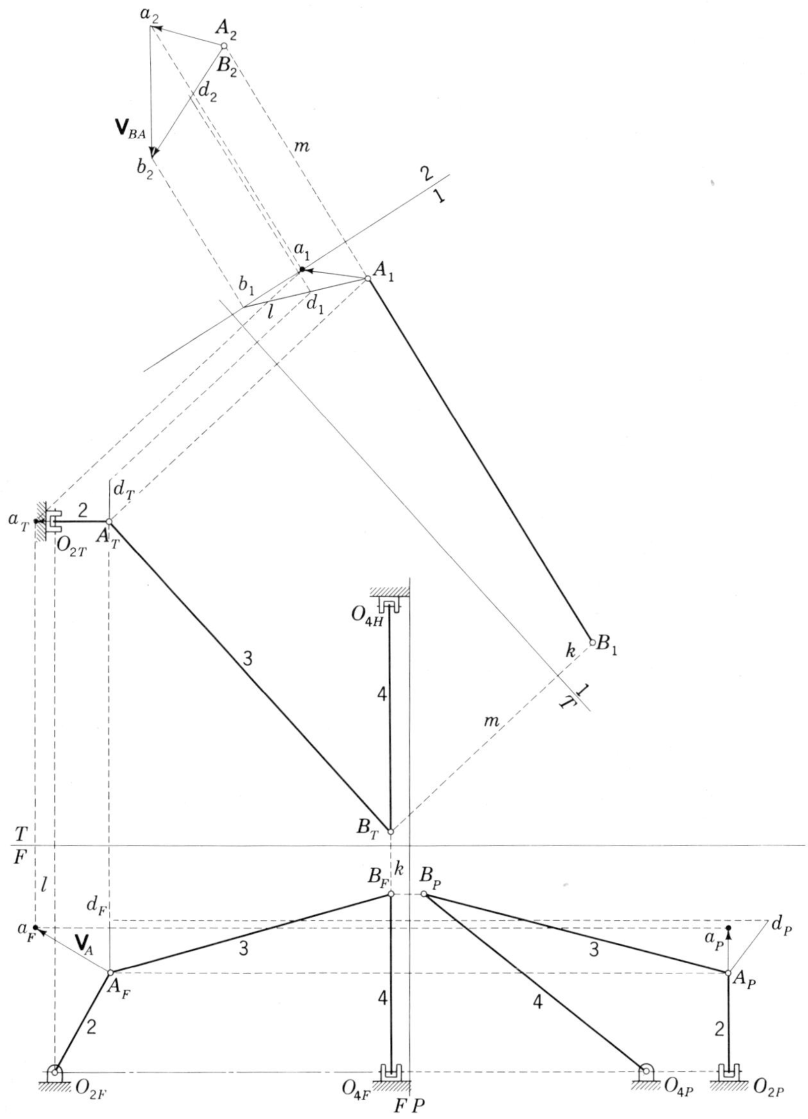

a₂
A₂
B₂
d₂
V_BA
b₂
m
2
1
a₁
A₁
b₁
l
d₁
d_T
2
a_T
O_2T
A_T
O_4H
3
4
k
B₁
1
T
m
B_T
T
F
l
B_F
k
B_P
d_F
a_F
V_A
3
3
d_P
a_P
A_F
A_P
2
4
4
2
O_2F
O_4F
F P
O_4P
O_2P

Fig. 12-9

perpendicular to link 3. The edge view of such a plane is perpendicular to A_1B_1 extended. In this example it is convenient to choose this plane so that it contains point a; so construct the edge view of plane 2 through point a_1 perpendicular to A_1B_1 extended. Now project points A, B, a, and d upon this plane. Note that the distances, m, for example, of points from plane 1 must be the same with respect to plane 2.

6. Extend line A_1d_1 until it intersects the edge view of plane 2 at b_1, and find the projection b_2 of this point in plane 2. Now both a and b lie in plane 2; any line drawn in plane 2 is perpendicular to link 3. Therefore the line ab is $\mathbf{V}_{BA}$, and the view of it in the second auxiliary is its true length. The line A_2b is the projection of $\mathbf{V}_B$ on the second auxiliary plane, but not in its true length because A is not in plane 2.

7. (In order to simplify reading of the drawing, step 7 is not shown; if the reader will carefully follow the first six steps he will have no difficulty with the seventh.) Project the three vectors back to the front, top, and profile views. $\mathbf{V}_B$ may then be measured from its profile view because it appears in its true length in this view. When all the vectors have been projected back to these three views, the x, y, and z projections may be measured directly.

The solution to the acceleration problem is obtained in an identical manner by using the same two auxiliary planes. The equation to be solved is

$$\mathbf{A}_B{}^t + \mathbf{A}_B{}^r = \mathbf{A}_A + \mathbf{A}_{BA}{}^t + \mathbf{A}_{BA}{}^r \tag{2}$$

Since $\mathbf{A}_B{}^r$, $\mathbf{A}_A$, and $\mathbf{A}_{BA}{}^r$ are known or can be found, it is possible to define a vector $\mathbf{C}$ as

$$\mathbf{C} = \mathbf{A}_A + \mathbf{A}_{BA}{}^r - \mathbf{A}_B{}^r \tag{3}$$

so that Eq. (2) becomes

$$\mathbf{A}_B{}^t = \mathbf{C} + \mathbf{A}_{BA}{}^t \tag{4}$$

Upon comparing Eq. (4) with Eq. (1) we see that $\mathbf{A}_B{}^t$ and $\mathbf{A}_{BA}{}^t$ have the same directions, respectively, as do $\mathbf{V}_B$ and $\mathbf{V}_{BA}$. The only difference in the approach, therefore, is that we begin with $\mathbf{V}_A$ when analyzing velocities and with $\mathbf{C}$ when analyzing accelerations.

12-6 A THEOREM ON ANGULAR VELOCITIES AND ACCELERATIONS

Figure 12-10 is a schematic drawing of the general seven-revolute seven-link space mechanism. The orientation of the revolute axes is schematically represented by the unit relative-velocity vectors $\hat{\boldsymbol{\omega}}_{ij}$ on the drawing. We assume that the directions of the revolute axes are such that the linkage has a single freedom.

To develop the theorem on angular velocities we note that

$$\boldsymbol{\omega}_{31} = \boldsymbol{\omega}_{21} + \boldsymbol{\omega}_{32} \tag{a}$$

which is the same as Eq. (4-26) for plane motion.[1] It is convenient to

[1] This important equation is often stated without proof. The proof consists in treat-

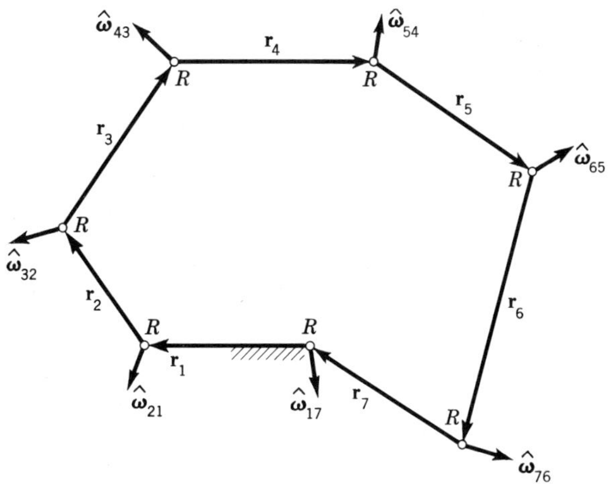

Fig. 12-10

rewrite Eq. (a) as

$$\omega_{21} - \omega_{31} + \omega_{32} = 0 \tag{b}$$

and then, proceeding in a similar manner around the loop, we have

$$\omega_{31} - \omega_{41} + \omega_{43} = 0 \tag{c}$$

$$\omega_{41} - \omega_{51} + \omega_{54} = 0 \tag{d}$$

$$\omega_{51} - \omega_{61} + \omega_{65} = 0 \tag{e}$$

$$\omega_{61} - \omega_{71} + \omega_{76} = 0 \tag{f}$$

$$\omega_{71} - \omega_{11} + \omega_{17} = 0 \tag{g}$$

Noting that $\omega_{11} = 0$ by definition and adding together Eqs. (b) to (g), we obtain

$$\omega_{21} + \omega_{32} + \omega_{43} + \omega_{54} + \omega_{65} + \omega_{76} + \omega_{17} = 0 \tag{h}$$

which states that *the sum of the relative angular velocities around a closed loop in a single-freedom system is zero.* Expressed mathematically this theorem reads

$$\sum_{i=1}^{n} \omega_{i+1,i} = 0 \qquad n + 1 = 1 \tag{12-4}$$

---

ing infinitesimal angular displacements as vectors, in summing these vectors, and in taking the time derivative of the result in a manner quite similar to that used for linear velocities. For a rigorous proof, see L. A. Pars, "A Treatise on Analytical Dynamics," p. 102, William Heinemann, Ltd., London, 1965.

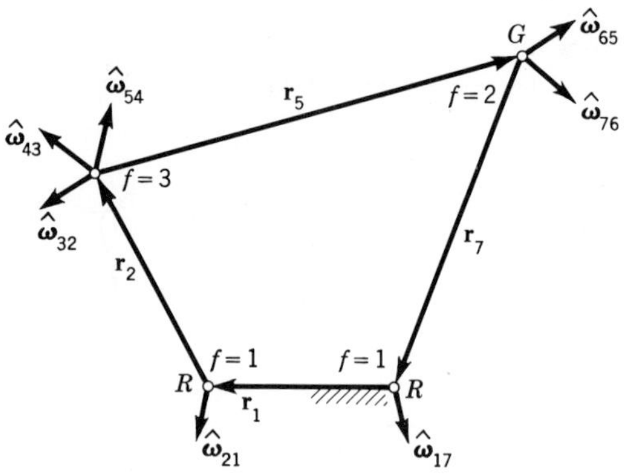

Fig. 12-11

Note that the relative-angular-velocity theorem for space linkages is analogous to the linear-velocity polygon for plane-motion mechanisms.

The relative-angular-velocity theorem is particularly useful for space linkages having two- and three-freedom pairs; see Prob. 12-16, for example. Especial care must be taken, however, to eliminate any idle freedoms before Eq. (12-4) is applied.

The approach for the *RGGR* linkage is illustrated in Fig. 12-11. Notice that the diagram gives the number of freedoms for each pair and that the idle freedom corresponding to one of the globular pairs has been eliminated. The directions $\hat{\boldsymbol{\omega}}_{32}$, $\hat{\boldsymbol{\omega}}_{43}$, and $\hat{\boldsymbol{\omega}}_{54}$ corresponding to the first globular pair need not be orthogonal; in fact, any convenient directions can be assigned as long as they are different from one another.

The *relative-angular-acceleration theorem* can be developed in the same manner. In mathematical form it is written

$$\sum_{i=1}^{n} \dot{\boldsymbol{\omega}}_{i+1,i} = 0 \qquad n + 1 = 1 \tag{12-5}$$

Since

$$\frac{d}{dt}(\omega\hat{\boldsymbol{\omega}}) = \dot{\omega}\hat{\boldsymbol{\omega}} + \omega\dot{\hat{\boldsymbol{\omega}}} \tag{i}$$

the direction of $\dot{\boldsymbol{\omega}}$ is *not* necessarily the same as $\hat{\boldsymbol{\omega}}$ and so care must be used in employing Eq. (12-5).

12-7 THE UNIVERSAL JOINT

Figure 12-12 shows the well-known Hooke, or Cardan, joint. It consists of two yokes which are the driving and driven members and a cross

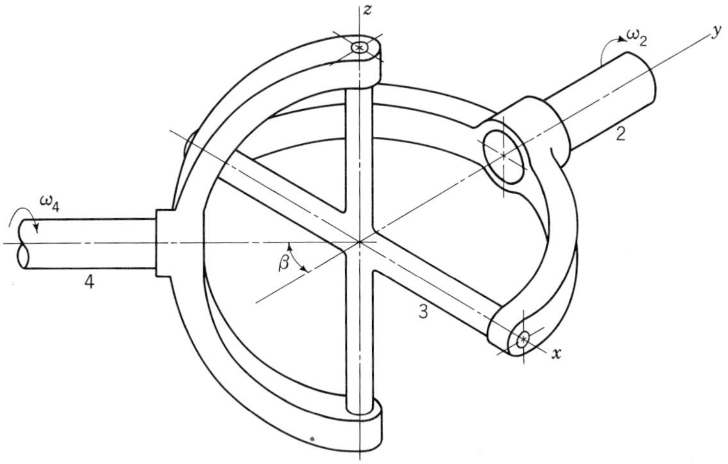

Fig. 12-12

which is the connecting link. One of the disadvantages of this joint is that the velocity ratio is not constant during rotation. Figure 12-13 is a polar angular-velocity diagram which shows the angular velocity of both the driver and the driven for one complete revolution of the joint. Since the driving member is assumed to have a constant angular velocity, its polar diagram is a circle. But the diagram for the driven member is an ellipse which crosses the circle at four places. This means that there

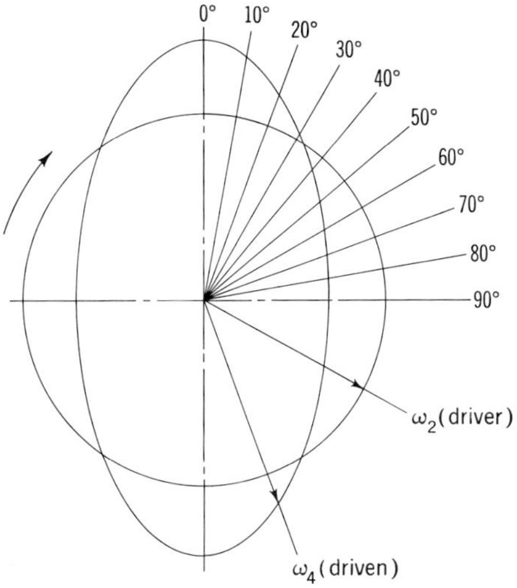

Fig. 12-13

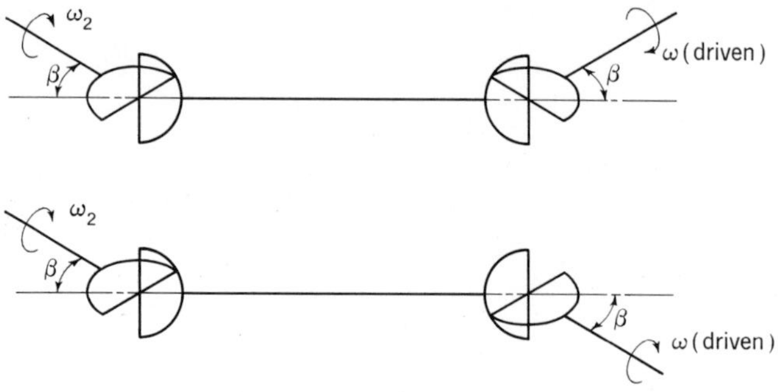

Fig. 12-14

are four instants during a single rotation when the angular velocities of the two shafts are equal. During the remaining time, the driven shaft rotates faster for part of the time and slower for part of the time.

We may think of the drive shaft of an automobile as having an inertia load at each end—the flywheel and engine rotating at constant speed at one end, and the weight of the car running at high speed at the other. If only a single universal joint, working at a finite angle, were used in an automobile, then either the speed of the engine or the speed of the car would have to vary during each revolution of the drive shaft. Both inertias resist this, and so the effect would be for the tires to slip and for the parts composing the line of power transmission to be highly stressed. Figure 12-14 shows two arrangements of universal joints which will provide a uniform velocity ratio between the input and output ends.

Analysis In Fig. 12-15 the driving shaft 2 connects the driven shaft 4 through the connecting cross 3. The shaft centerlines intersect at O, producing the shaft angle β. The ends of the crosspiece connect to the driving yoke at points A and B and to the driven yoke at C and D. During motion the line AB describes a circle in a vertical plane perpendicular to the drawing and the line CD another circle in a plane at an angle β to the vertical plane. These two circles are great circles of the same sphere, the center being at O. Points A and C always remain the same distance apart, that is, 90° of great-circle arc. The maximum deviation in the angular-velocity ratio occurs when either point A or point C lies at the intersection of the great circles.

The two great circles on which A and C travel are illustrated again in Fig. 12-16. The circles intersect at D and are shown separated by the shaft angle β. Let point A travel a distance θ from the point of intersection. Then point C will be located on the great-circle arc AC 90° behind A. Now locate C' 90° ahead of C on the great circle which

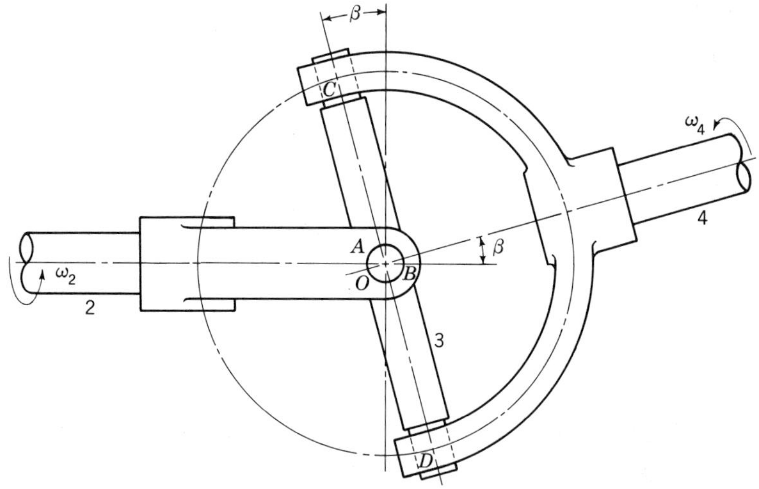

Fig. 12-15

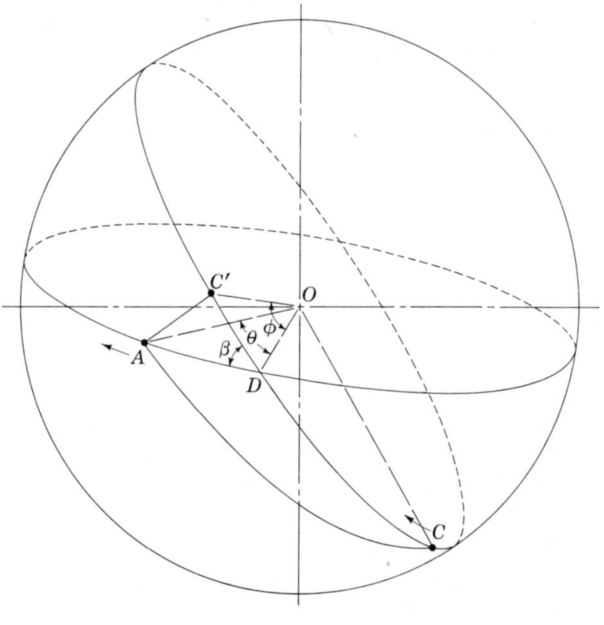

Fig. 12-16

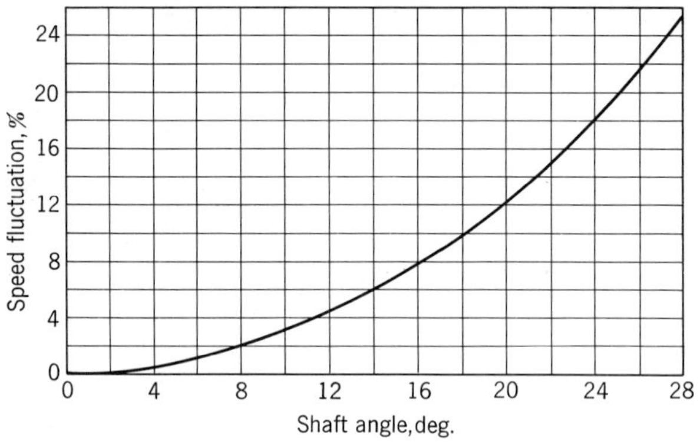

Fig. 12-17 Relationship between the shaft angle and the fluctuation in speed of a universal joint.

C travels. The triangles $AC'D$ and $AC'C$ are spherical triangles. Arcs AC and $C'C$ are both 90°, and therefore angles $C'AC$ and $AC'C$ are both right spherical angles.[1] We then have the right spherical triangle $AC'D$ in which angle $AC'D$ is a right angle, $C'DA$ is the shaft angle β, arc AD is the angle through which the driving shaft turns, and arc $C'D$, designated ϕ, is the arc through which the driven shaft turns. According to a right-triangle formula from spherical trigonometry,

$$\cos \beta = \tan \phi \cot \theta \qquad (12\text{-}6)$$

In order to obtain the relationship between the angular velocities, the equation is rearranged to

$$\tan \phi = \cos \beta \tan \theta \qquad (a)$$

Differentiating,

$$\dot{\phi} \sec^2 \phi = \dot{\theta} \cos \beta \sec^2 \theta \qquad (b)$$

Since $\dot{\phi} = \omega_4$, the angular velocity of the driven, and $\dot{\theta} = \omega_2$, the angular velocity of the driver, the ratio of these velocities is

$$\frac{\omega_4}{\omega_2} = \frac{\cos \beta \sec^2 \theta}{\sec^2 \phi} = \frac{\cos \beta \sec^2 \theta}{1 + \tan^2 \phi} \qquad (c)$$

[1] The sides and angles of a spherical triangle may have any values from 0 to 360°. If one or more of the parts is greater than 180°, the triangle is called a *general spherical triangle*. A triangle having each part less than 180° is called a *spherical triangle*. A spherical right triangle is a spherical triangle one of whose angles is a right angle. The other parts may have any values from 0 to 180°.

It is convenient to eliminate ϕ; substituting Eq. (*a*) in (*c*) gives

$$\frac{\omega_4}{\omega_2} = \frac{\cos\beta}{1 - \sin^2\theta\,\sin^2\beta} \tag{12-7}$$

If we assume the shaft angle β a constant, the maximum value of Eq. (12-7) occurs when $\sin\theta = 1$, that is, when $\theta = 90°, 270°$, etc. The denominator is greatest when $\sin\theta = 0$, and this condition gives the minimum ratio of the velocities.

If the difference between the maximum and minimum ratios of Eq. (12-7) is expressed in percent and plotted against the shaft angle, a curve useful in the evaluation of universal joints is obtained. Figure 12-17 was obtained in this manner for shaft angles up to 28°.

PROBLEMS

12-1. Determine the number of freedoms of the *GGC* chain shown in the figure. Identify any idle constraints or idle freedoms and state how these may be removed. What is the nature of the path described by point *B*?

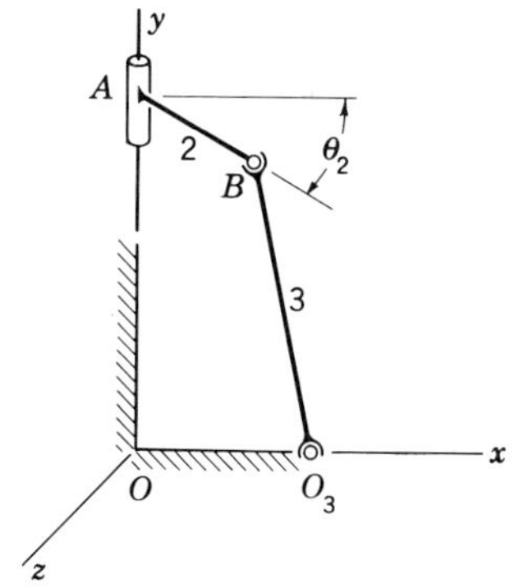

Prob. 12-1

12-2. Using the linkage of Prob. 12-1 with $AB = OO_3 = 3$ in., $BO_3 = 6$ in., $\theta_2 = 30°$, express the position of each link in vector form.

12-3. Using $\mathbf{V}_A = -2\hat{\jmath}$ in./sec, find the angular velocity of links 2 and 3 and the linear velocity of *B* of the mechanism of Prob. 12-2. Use vector analysis.

12-4. Solve Probs. 12-2 and 12-3 by using descriptive geometry.

12-5. The spherical or conical four-revolute linkage shown in the figure has the following dimensions: $O_2A = 3$ in., $OO_2 = 7$ in., $OO_4 = 2$ in., and $O_4B = 9$ in. Link 2 is shown in the *xz* plane and link 4 in the *xy* plane. For better presentation, the linkage has not been drawn to scale. Find the length of link 3 in vector notation. With $\boldsymbol{\omega}_2 = -60\hat{\mathbf{k}}$ rad/sec, make a complete velocity and acceleration analysis of the linkage for the position shown, using vector analysis.

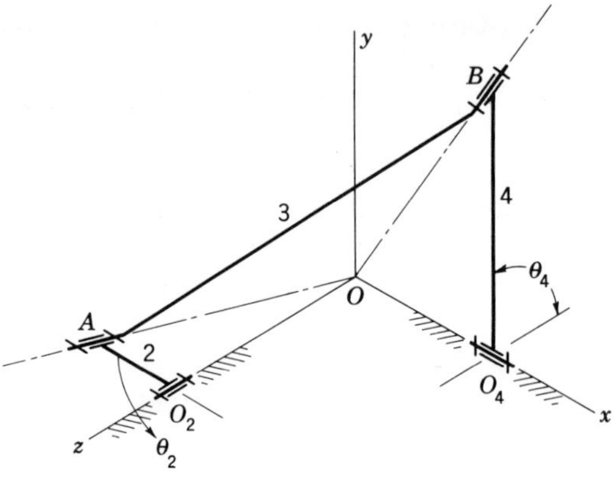

Prob. 12-5

12-6. The same as Prob. 12-5 except use descriptive geometry to solve the problem.

12-7. In Prob. 12-5, what is the total angle of oscillation of link 4? Find the ratio of the time for the forward stroke of link 4 to the return stroke by assuming that link 2 rotates at a constant angular velocity and then finding the ratio of the respective angles through which link 2 must rotate.

12-8. The same as Prob. 12-5 except $\theta_2 = 90°$.

12-9. Investigate the motion of the spherical four-bar linkage shown in the figure and determine whether the crank is free to rotate through a complete turn. If so, find

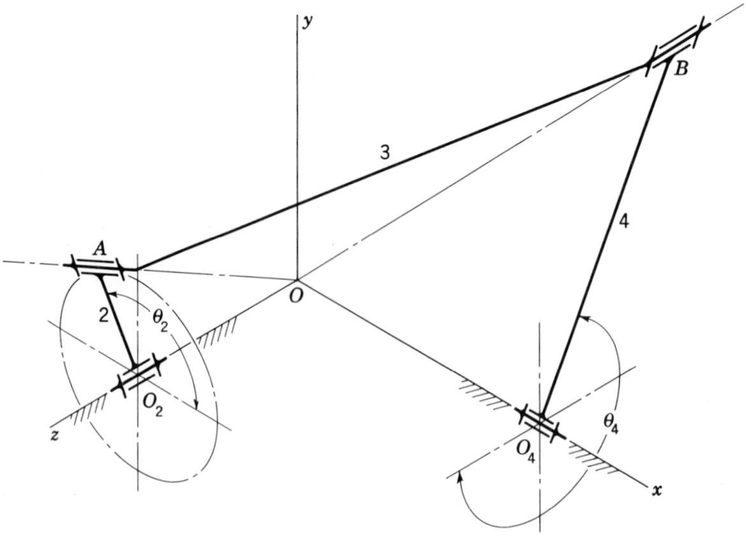

Prob. 12-9 $\theta_2 = 120°$, $O_2A = 3$ in., $OO_2 = 6$ in., $OO_4 = 9$ in., $AB = 16\frac{1}{2}$ in., $O_4B = 10\frac{1}{2}$ in.

the angle of oscillation of link 4 and the ratio of the time of the forward stroke to the time of the return stroke.

12-10. Using $\omega_2 = 36\hat{\mathbf{k}}$ rad/sec, make a complete velocity and acceleration analysis of the mechanism of Prob. 12-9, using vector analysis.

12-11. The same as Prob. 12-10 except you are to use descriptive geometry as the method of analysis.

12-12. The figure shows the front, top, and first auxiliary views of a space slider-crank linkage with two ball joints. In the construction of many such mechanisms the angle β can be varied. Thus the stroke of slider 4 is adjustable from zero, when $\beta = 0$, to twice the crank radius when $\beta = 90°$. In this example $\beta = 30°$, $\theta_2 = 240°$, and $\omega_2 = 24$ rad/sec. Express the links in vector form and make a complete velocity analysis of the mechanism, using vector analysis.

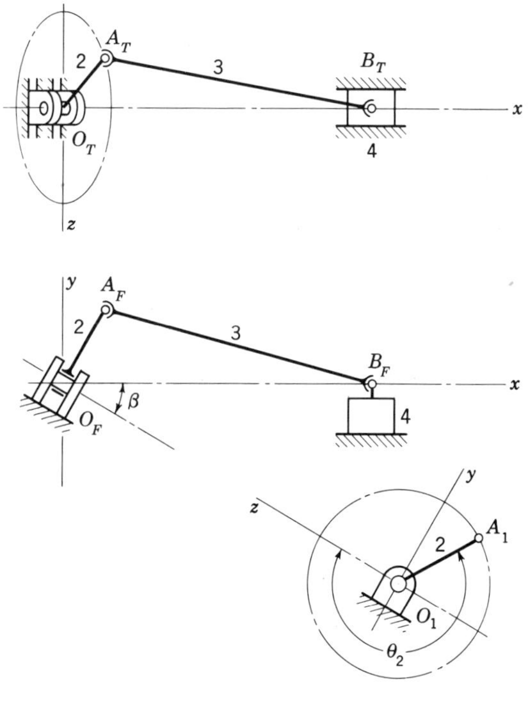

Prob. 12-12 $OA = 2$ in., $AB = 6$ in.

12-13. The same as Prob. 12-12 except use descriptive geometry in the analysis.

12-14. The same as Prob. 12-12 except $\beta = 60°$.

12-15. The figure shows the front, top, and profile views of an *RGRC* crank-and-oscillating-slider mechanism. In the profile view, link 4 is the oscillating slider. It is physically attached to a round rod which has a motion of rotation and sliding in the two bearings.

(*a*) Using the Kutzbach criterion, find the degrees of freedom of the linkage.

(*b*) With crank 2 as the driver, find the total angular and linear travel of link 4.

(*c*) Using $\theta_2 = 40°$, write the vector position equation for the mechanism and solve for all unknown scalars.

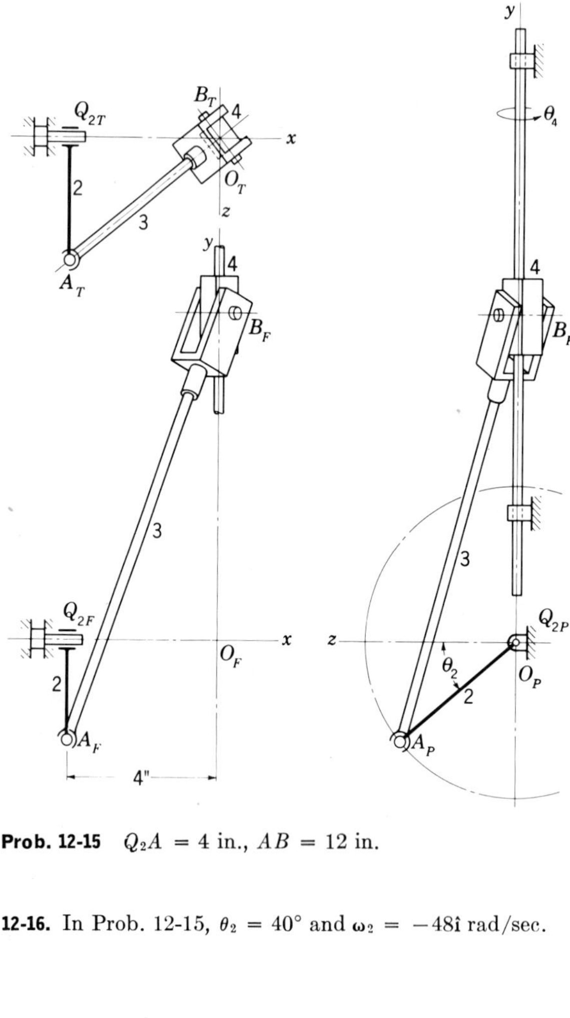

Prob. 12-15 $Q_2A = 4$ in., $AB = 12$ in.

12-16. In Prob. 12-15, $\theta_2 = 40°$ and $\boldsymbol{\omega}_2 = -48\hat{\imath}$ rad/sec. Find $\mathbf{V}_B$, $\boldsymbol{\omega}_3$, and $\boldsymbol{\omega}_4$.

Appendix Involute Functions

Deg	*Inv φ*	*Deg*	*Inv φ*	*Deg*	*Inv φ*	*Deg*	*Inv φ*
00.0	.000000						
00.1	.000000	03.1	.000053	06.1	.000404	09.1	.001349
00.2	.000000	03.2	.000058	06.2	.000424	09.2	.001394
00.3	.000000	03.3	.000064	06.3	.000445	09.3	.001440
00.4	.000000	03.4	.000070	06.4	.000467	09.4	.001488
00.5	.000000	03.5	.000076	06.5	.000489	09.5	.001536
00.6	.000000	03.6	.000083	06.6	.000512	09.6	.001586
00.7	.000000	03.7	.000090	06.7	.000536	09.7	.001636
00.8	.000000	03.8	.000097	06.8	.000560	09.8	.001688
00.9	.000001	03.9	.000105	06.9	.000586	09.9	.001740
01.0	.000002	04.0	.000114	07.0	.000612	10.0	.001794
01.1	.000002	04.1	.000122	07.1	.000638	10.1	.001849
01.2	.000003	04.2	.000132	07.2	.000666	10.2	.001905
01.3	.000004	04.3	.000141	07.3	.000694	10.3	.001962
01.4	.000005	04.4	.000151	07.4	.000723	10.4	.002020
01.5	.000006	04.5	.000162	07.5	.000753	10.5	.002079
01.6	.000007	04.6	.000173	07.6	.000783	10.6	.002140
01.7	.000009	04.7	.000184	07.7	.000815	10.7	.002202
01.8	.000010	04.8	.000197	07.8	.000847	10.8	.002265
01.9	.000012	04.9	.000209	07.9	.000880	10.9	.002329
02.0	.000014	05.0	.000222	08.0	.000914	11.0	.002394
02.1	.000016	05.1	.000236	08.1	.000949	11.1	.002461
02.2	.000019	05.2	.000250	08.2	.000985	11.2	.002528
02.3	.000022	05.3	.000265	08.3	.001022	11.3	.002598
02.4	.000025	05.4	.000280	08.4	.001059	11.4	.002668
02.5	.000028	05.5	.000296	08.5	.001098	11.5	.002739
02.6	.000031	05.6	.000312	08.6	.001137	11.6	.002812
02.7	.000035	05.7	.000329	08.7	.001178	11.7	.002894
02.8	.000039	05.8	.000347	08.8	.001219	11.8	.002962
02.9	.000043	05.9	.000366	08.9	.001262	11.9	.003039
03.0	.000048	06.0	.000384	09.0	.001305	12.0	.003117

Deg	*Inv* ϕ	*Deg*	*Inv* ϕ	*Deg*	*Inv* ϕ	*Deg*	*Inv* ϕ
12.1	.003197	16.3	.007932	20.6	.016337	24.8	.029223
12.2	.003277	16.4	.008082	20.7	.016585	24.9	.029598
12.3	.003360	16.5	.008234	20.8	.016836	25.0	.029975
12.4	.003443	16.6	.008388	20.9	.017089	25.1	.030357
12.5	.003529	16.7	.008544	21.0	.017345	25.2	.030741
12.6	.003615	16.8	.008702	21.1	.017603	25.3	.031130
12.7	.003712	16.9	.008863	21.2	.017865	25.4	.031521
12.8	.003792	17.0	.009025	21.3	.018129	25.5	.031917
12.9	.003883	17.1	.009189	21.4	.018395	25.6	.032315
13.0	.003975	17.2	.009355	21.5	.018665	25.7	.032718
13.1	.004069	17.3	.009523	21.6	.018937	25.8	.033124
13.2	.004164	17.4	.009694	21.7	.019212	25.9	.033534
13.3	.004261	17.5	.009866	21.8	.019490	26.0	.033947
13.4	.004359	17.6	.010041	21.9	.019770	26.1	.034364
13.5	.004459	17.7	.010217	22.0	.020054	26.2	.034785
13.6	.004561	17.8	.010396	22.1	.020340	26.3	.035209
13.7	.004664	17.9	.010577	22.2	.020630	26.4	.035637
13.8	.004768	18.0	.010760	22.3	.020921	26.5	.036069
13.9	.004874	18.1	.010946	22.4	.021216	26.6	.036505
14.0	.004982	18.2	.011133	22.5	.021514	26.7	.036945
14.1	.005091	18.3	.011323	22.6	.021815	26.8	.037388
14.2	.005202	18.4	.011515	22.7	.022119	26.9	.037835
14.3	.005315	18.5	.011709	22.8	.022426	27.0	.038287
14.4	.005429	18.6	.011906	22.9	.022736	27.1	.038696
14.5	.005545	18.7	.012105	23.0	.023049	27.2	.039201
14.6	.005662	18.8	.012306	23.1	.023365	27.3	.039664
14.7	.005782	18.9	.012509	23.2	.023684	27.4	.040131
14.8	.005903	19.0	.012715	23.3	.024006	27.5	.040602
14.9	.006025	19.1	.012923	23.4	.024332	27.6	.041076
15.0	.006150	19.2	.013134	23.5	.024660	27.7	.041556
15.1	.006276	19.3	.013346	23.6	.024992	27.8	.042039
15.2	.006404	19.4	.013562	23.7	.025326	27.9	.042526
15.3	.006534	19.5	.013779	23.8	.025664	28.0	.043017
15.4	.006665	19.6	.013999	23.9	.026005	28.1	.043513
15.5	.006799	19.7	.014222	24.0	.026350	28.2	.044012
15.6	.006934	19.8	.014447	24.1	.026697	28.3	.044516
15.7	.007071	19.9	.014674	24.2	.027048	28.4	.045024
15.8	.007209	20.0	.014904	24.3	.027402	28.5	.045537
15.9	.007350	20.1	.015137	24.4	.027760	28.6	.046054
16.0	.007493	20.2	.015372	24.5	.028121	28.7	.046575
16.1	.007637	20.3	.015609	24.6	.028485	28.8	.047100
16.2	.007784	20.4	.015850	24.7	.028852	28.9	.047630
		20.5	.016092			29.0	.048164

Deg	*Inv* ϕ	*Deg*	*Inv* ϕ	*Deg*	*Inv* ϕ	*Deg*	*Inv* ϕ
29.1	.048702	33.1	.074188	37.1	.108777	41.1	.155025
29.2	.049245	33.2	.074932	37.2	.109779	41.2	.156358
29.3	.049792	33.3	.075683	37.3	.110788	41.3	.157700
29.4	.050344	33.4	.076439	37.4	.111805	41.4	.159052
29.5	.050901	33.5	.077200	73.5	.112828	41.5	.160414
29.6	.051462	33.6	.077968	37.6	.113860	41.6	.161785
29.7	.052027	33.7	.078741	37.7	.114899	41.7	.163165
29.8	.052597	33.8	.079520	37.8	.115945	41.8	.164556
29.9	.053172	33.9	.080305	37.9	.116999	41.9	.165956
30.0	.053751	34.0	.081097	38.0	.118060	42.0	.167366
30.1	.054336	34.1	.081974	38.1	.119130	42.1	.168786
30.2	.054924	34.2	.082697	38.2	.120207	42.2	.170216
30.3	.055519	34.3	.083506	38.3	.121291	42.3	.171656
30.4	.056116	34.4	.084321	38.4	.122384	42.4	.173106
30.5	.056720	34.5	.085142	38.5	.123484	42.5	.174566
30.6	.057267	34.6	.085970	38.6	.124592	42.6	.176037
30.7	.057940	34.7	.086804	38.7	.125709	42.7	.177518
30.8	.058558	34.8	.087644	38.8	.126833	42.8	.179009
30.9	.059181	34.9	.088490	38.9	.127965	42.9	.180511
31.0	.059809	35.0	.089342	39.0	.129106	43.0	.182023
31.1	.060441	35.1	.090201	39.1	.130254	43.1	.183546
31.2	.061079	35.2	.091066	39.2	.131411	43.2	.185080
31.3	.061721	35.3	.091938	39.3	.132576	43.3	.186625
31.4	.062369	35.4	.092816	39.4	.133749	43.4	.188180
31.5	.063022	35.5	.093701	39.5	.134931	43.5	.189746
31.6	.063680	35.6	.094592	39.6	.136122	43.6	.191324
31.7	.064343	35.7	.095490	39.7	.137320	43.7	.192912
31.8	.065012	35.8	.096395	39.8	.138528	43.8	.194511
31.9	.065685	35.9	.097306	39.9	.139743	43.9	.196122
32.0	.066364	36.0	.098224	40.0	.140968	44.0	.197744
32.1	.067048	36.1	.099149	40.1	.142201	44.1	.199377
32.2	.067738	36.2	.100080	40.2	.143443	44.2	.201022
32.3	.068432	36.3	.101019	40.3	.144694	44.3	.202678
32.4	.069133	36.4	.101964	40.4	.145954	44.4	.204346
32.5	.069838	36.5	.102916	40.5	.147222	44.5	.206026
32.6	.070549	36.6	.103875	40.6	.148500	44.6	.207717
32.7	.071266	36.7	.104841	40.7	.149787	44.7	.209420
32.8	.071988	36.8	.105814	40.8	.151082	44.8	.211135
32.9	.072716	36.9	.106795	40.9	.152387	44.9	.212863
33.0	.073449	37.0	.107782	41.0	.153702	45.0	.214602

Answers to Selected Problems

CHAPTER 2

2-1. $11.2\underline{/14^\circ}$.

2-4. $8.1\underline{/-120^\circ}$.

2-7. 7.9, 4.8 on 1.

2-10. $-11.7\hat{\imath} + 2.9\hat{\jmath}$.

2-13. 95 in second quadrant; 97 in fourth.

2-16. 20 in second quadrant; 34.6 in third.

2-19. 10.5 in third quadrant.

2-22. $\mathbf{A} = -250\hat{\imath} + 433\hat{\jmath}$.

2-25. $64\underline{/116.6^\circ}$, $-28.6\hat{\imath} + 57.2\hat{\jmath}$.

2-28. $14.3\underline{/2.6^\circ}$.

2-31. $\mathbf{R} = -0.57\hat{\imath} - 1.71\hat{\jmath}$, $\mathbf{A} = -5\hat{\imath} + 8.66\hat{\jmath}$.

2-35. $r = 3$.

CHAPTER 3

3-1. A counterclockwise spiral passing through $x = 0,\ -a/2,\ a,\ -3a/2,\ 2a,\ \ldots$.

3-4. $\Delta\mathbf{S} = 9.48\underline{/-71.6^\circ}$.

3-7. Clockwise; $x = 4$, $y = 0$; $t = 20$; $\Delta\mathbf{S} = 400\underline{/0^\circ}$.

3-10. $\Delta\mathbf{r} = 6.32\underline{/108.4^\circ}$.

3-12. $\Delta\mathbf{P} = 7.69\underline{/58.5^\circ}$.

CHAPTER 4

4-1. $\dot{\mathbf{R}} = 314\underline{/72^\circ}$ in./sec, $\omega = \pi$ rad/sec ccw.

4-3. $V_{BA} = 84$ mph, N25°E.

4-6. (*a*) 51.7 miles. (*b*) 6:40 $\frac{1}{2}$ PM.

4-10. $\mathbf{V}_C = 13\underline{/59.5^\circ}$ fps, $\omega_3 = 43.3$ rad/sec cw, $\omega_4 = 38$ rad/sec cw.

4-14. $V_C = 1.21$ fps, $V_D = 1.24$ fps.

4-17. $V_B = 0.98$ fps, $V_D = 2.00$ fps, $\omega_6 = 4.00$ rad/sec ccw.

4-20. $V_B = 17$ fps, $\omega_3 = 3.45$ rad/sec ccw.

4-23. $V_B = 5.5$ fps, $V_C = 5.7$ fps, $V_D = 1.8$ fps.

4-26. $\mathbf{V}_B = 1.06\underline{/66^\circ}$ fps.

4-28. $V_B = 6$ fps, $V_C = 6.4$ fps, $V_E = 6.5$ fps, $\omega_3 = 14.3$ rad/sec ccw, $\omega_4 = 14.4$ rad/sec ccw, $\omega_5 = 9.75$ rad/sec cw.

CHAPTER 5

5-1. $-4.01\hat{\imath}$ in./sec².

5-3. $\dot{\mathbf{r}} = 17.6\underline{/117^\circ}$ in./sec, $\ddot{\mathbf{r}} = 1968\underline{/5.5^\circ}$ in./sec².

5-6. $\dot{\mathbf{r}} = -2.4\hat{\jmath} + 3\hat{\mathbf{k}}$ fps, $\ddot{\mathbf{r}} = -8\hat{\jmath}$ fps², $\dot{r} = 3.84$ fps, $\hat{\lambda} = -0.625\hat{\jmath} + 0.782\hat{\mathbf{k}}$, $\rho = 2.36$ ft.

5-9. $3790\underline{/161.6^\circ}$ fps².

5-12. $\mathbf{V}_B = 12\underline{/-90^\circ}$ fps, $\mathbf{V}_C = 8.30\underline{/11.8^\circ}$ fps, $\mathbf{A}_B = 399\underline{/163.8^\circ}$ fps², $\mathbf{A}_C = 207\underline{/-121.1^\circ}$ fps².

5-15. $\mathbf{V}_B = 11.6\underline{/-90^\circ}$ fps, $\mathbf{A}_B = 142\underline{/-90^\circ}$ fps².

5-18. $\mathbf{A}_C = 4650\underline{/-113.6^\circ}$ fps², $\omega_3 = 27.2$ rad/sec ccw, $\omega_4 = 104$ rad/sec ccw, $\dot{\omega}_3 = 6220$ rad/sec² ccw, $\dot{\omega}_4 = 15{,}200$ rad/sec² ccw.

5-21. $\dot{\omega}_3 = 563$ rad/sec² ccw, $\dot{\omega}_4 = 124$ rad/sec² ccw.

5-24. $\theta_3 = 105.3^\circ$, $\theta_4 = 159.6^\circ$, $\omega_3 = +0.809$ rad/sec, $\omega_4 = +0.525$ rad/sec, $\dot{\omega}_3 = +0.230$ rad/sec², $\dot{\omega}_4 = +0.008$ rad/sec².

5-27. $\theta_3 = 28.3^\circ$, $\theta_4 = 55.9^\circ$, $\omega_3 = -0.632$ rad/sec, $\omega_4 = -2.16$ rad/sec, $\dot{\omega}_3 = +7.82$ rad/sec², $\dot{\omega}_4 = +6.70$ rad/sec².

5-30. $\theta_3 = 72.9^\circ$, $\theta_4 = 134.2^\circ$, $\omega_3 = +1.60$ rad/sec, $\omega_4 = +8.13$ rad/sec, $\dot{\omega}_3 = -225.1$ rad/sec², $\dot{\omega}_4 = -80.4$ rad/sec².

5-33. $V_B = 19.2$ fps, $\omega_4 = 0.975$ rad/sec cw, $A_{B_3} = 77$ fps², $\dot{\omega}_4 = 39.6$ rad/sec² cw.

5-36. $\dot{y} = 12.55$ in./sec, $\ddot{y} = 0$.

5-39. $A_B = 171$ fps², $A_C = 270$ fps², $A_D = 163$ fps², $\dot{\omega}_3 = 305$ rad/sec² cw, $\dot{\omega}_4 = 304$ rad/sec² ccw.

CHAPTER 7

7-16. $\dot{y} = \pi d\omega/2\beta$, $\dddot{y} = -(d/2)(\pi\omega/\beta)^3$, $\ddot{y} = (d/2)(\pi\omega/\beta)^2$.

7-19. $d_y = 8.17$ in., $\dot{y}_{\max} = 5$ fps, $\ddot{y}_{\max} = 157$ fps².

7-22. $r_B(\min) = 6.2$ in.

CHAPTER 8

8-2. 54, 0.349 in.

8-5. 9.19 in.

8-8. 4.833 in.

8-11. $q_a = 1.07$ in., $q_r = 0.99$ in., $q_t = 2.06$ in., $m_c = 1.64$.
8-13. (*a*) $q_a = 1.54$ in., $q_r = 1.52$ in., $q_t = 3.06$ in., $m_c = 1.95$. (*b*) $q_a = 1.12$ in., $q_r = 1.31$ in., $q_t = 2.44$ in., $m_c = 1.55$.
8-20. $t_b = 1.124$ in., $t_a = 0.4394$ in., $\varphi_a = 33°58'45''$.
8-24. $a = 0.1810$ in.
8-27. $m_c = 1.345$.
8-30. $a_2 = 0.2359$ in., $a_3 = 0.0413$ in.

CHAPTER 9

9-1. $p_t = 0.523$ in., $p_n = 0.370$ in., $P_n = 8.48$, $d_2 = 2.5$ in., $d_3 = 4$ in., 42.4, 67.8.
9-4. $P_t = 6.93$, $p_t = 0.453$ in., $N_2 = 17$, $N_3 = 31$, $d_2 = 2.45$ in., $d_3 = 4.48$ in.
9-7. $m_n = 1.79$, $m_t = 2.87$.
9-10. $N_2 = 30$, $N_3 = 60$, $\psi_2 = \psi_3 = 25°$ left-hand, $(d_2 + d_3)/2 = 9.93$ in.
9-13. $l = 3.75$ in., $\lambda = 34°22'$, $\psi = 34°22'$, $d_3 = 15.90$ in.
9-16. 27°, 93°.
9-18. $d_2 = 2.125$ in., $d_3 = 3.500$ in., $\gamma_2 = 34.8°$, $\gamma_3 = 70.2°$, $a_2 = 0.1612$ in., $a_3 = 0.0888$ in., $F = 0.559$ in.

CHAPTER 10

10-1. $n_8 = 68.2$ rpm cw, $e = -\frac{5}{88}$.
10-3. $n_9 = 11.82$ rpm cw.
10-7. $n_A = 78$ rpm ccw.
10-9. 2200 rpm cw.
10-11. (*a*) $n_8 = 135$ rpm cw; (*b*) $n_8 = 289$ rpm ccw.
10-13. 488 rpm.

CHAPTER 12

12-2. $\mathbf{r}_A = 5.796\hat{\jmath}$ in. (O to A), $\mathbf{r}_2 = 2.6\hat{\imath} + 1.5\hat{\mathbf{k}}$ in. (A to B), $\mathbf{r}_3 = -0.4\hat{\imath} + 5.796\hat{\jmath} + 1.5\hat{\mathbf{k}}$ in. (O_3 to B), $\mathbf{r}_1 = 3\hat{\imath}$ in. (O to O_3).
12-5. $\mathbf{r}_1 = -2\hat{\imath} + 7\hat{\mathbf{k}}$ in. (O_4 to O_2), $\mathbf{r}_2 = -3\hat{\imath}$ in. (O_2 to A), $\mathbf{r}_3 = 5\hat{\imath} + 9\hat{\jmath} - 7\hat{\mathbf{k}}$ in. (A to B), $\mathbf{r}_4 = 9\hat{\jmath}$ in. (O_4 to B), $r_3 = 12.45$ in., $\mathbf{V}_A = 180\hat{\jmath}$ in./sec, $\mathbf{V}_B = -231\hat{\mathbf{k}}$ in./sec, $\mathbf{V}_{BA} = -180\hat{\jmath} - 231\hat{\mathbf{k}}$ in./sec, $\boldsymbol{\omega}_3 = -21.56\hat{\imath} + 7.45\hat{\jmath} - 5.8\hat{\mathbf{k}}$ rad/sec, $\omega_3 = 23.5$ rad/sec, $\boldsymbol{\omega}_4 = -25.7\hat{\imath}$ rad/sec, $\mathbf{A}_A{}^r = 10{,}800\hat{\imath}$ in./sec², $\mathbf{A}_B{}^r = -5950\hat{\jmath}$ in./sec², $\mathbf{A}_{BA}{}^r = -2784\hat{\imath} - 4990\hat{\jmath} + 3870\hat{\mathbf{k}}$ in./sec², $\mathbf{A}_B{}^t = -3087\hat{\mathbf{k}}$ in./sec², $\mathbf{A}_{BA}{}^t = -8020\hat{\imath} - 950\hat{\jmath} - 6960\hat{\mathbf{k}}$ in./sec², $\dot{\boldsymbol{\omega}}_3 = -447\hat{\imath} + 588\hat{\jmath} + 436\hat{\mathbf{k}}$ rad/sec², $\dot{\omega}_3 = 852$ rad/sec², $\dot{\boldsymbol{\omega}}_4 = -343\hat{\imath}$ rad/sec².
12-12. $\mathbf{r}_2 = 0.866\hat{\imath} + 1.5\hat{\jmath} - \hat{\mathbf{k}}$ in., $\mathbf{r}_3 = 5.73\hat{\imath} - 1.5\hat{\jmath} + \hat{\mathbf{k}}$ in., $\mathbf{r}_4 = 6.59\hat{\imath}$ in., $\boldsymbol{\omega}_2 = 20.8\hat{\imath} - 12\hat{\jmath}$ rad/sec, $\boldsymbol{\omega}_3 = 2.31\hat{\imath} + 6.66\hat{\jmath} - 3.23\hat{\mathbf{k}}$ rad/sec, $\omega_3 = 7.76$ rad/sec, $\mathbf{V}_B = 13.83\hat{\imath}$ in./sec.

Name Index

Subject Index